Energy Efficiency
Towards the End of Demand Growth

Energy Efficiency
Towards the End of Demand Growth

Edited by

Fereidoon P. Sioshansi
Menlo Energy Economics

AMSTERDAM • BOSTON • HEIDELBERG • LONDON
NEW YORK • OXFORD • PARIS • SAN DIEGO
SAN FRANCISCO • SINGAPORE • SYDNEY • TOKYO
Academic Press is an imprint of Elsevier

Academic Press is an imprint of Elsevier
The Boulevard, Langford Lane, Kidlington, Oxford, OX5 1GB, UK
225 Wyman Street, Waltham, MA 02451, USA

First edition 2013

Notice
No responsibility is assumed by the publisher for any injury and/or damage
to persons or property as a matter of products liability, negligence or otherwise,
or from any use or operation of any methods, products, instructions or ideas
contained in the material herein. Because of rapid advances in the medical
sciences, in particular, independent verification of diagnoses and drug dosages
should be made.

British Library Cataloguing-in-Publication Data
A catalogue record for this book is available from the British Library

Library of Congress Cataloging-in-Publication Data
A catalog record for this book is available from the Library of Congress

ISBN: 978-0-12-397879-0

For information on all Academic Press publications
visit our website at http://store.elsevier.com

Printed and bound in United States of America

13 14 15   10 9 8 7 6 5 4 3 2 1

**Working together to grow
libraries in developing countries**

www.elsevier.com | www.bookaid.org | www.sabre.org

ELSEVIER   BOOK AID
International   Sabre Foundation

Contents

Hunt Allcott is an Assistant Professor of Economics at New York University and a Faculty Research Fellow at the National Bureau of Economic Research. He is also a Scientific Director of ideas42, a think tank that applies insights from psychology and economics to problems in international development, health care, consumer finance, and the energy industry. He has worked as a consultant with Cambridge Energy Research Associates, Arthur D. Little, and the World Bank.

Dr. Allcott's interests include environmental economics, industrial organization, behavioral economics, and international development. His most recent research centers on consumer behavior, business strategy, and regulatory policy in energy markets including how firms set prices in electricity markets, how consumers value energy, and insights from behavioral economics and psychology to conserve energy.

He holds a Ph.D. from Harvard University and a B.S. and M.S. from Stanford University.

Ren Anderson is the Residential Research Group Manager at the National Renewable Energy Laboratory (NREL) as well as the Department of Energy Building America National Technical Lead. He leads the development of advanced residential energy efficiency systems targeting energy savings of 30 to 50 percent.

His research focus is on the development of innovative emerging system solutions that improve the durability, reliability, and energy efficiency of new and existing buildings. This is reflected in his previous positions at NREL including Solar Buildings Systems Integration Technology Leader and Buildings Energy Technology Manager. He also supervised graduate thesis work for Mechnical Engineering at Clemson University, University of Colorado, and Colorado State University. He is a member of ASME, ASHRAE, NAHB, NCHI, and has been an ACEEE session chair.

He received a B.A. in Engineering Physics and an M.S. and Ph.D. in Mechanical Engineering from the University of Colorado.

Frits Bliek is a principal consultant at DNV KEMA Energy & Sustainability in the Netherlands and the program director of the Smart Energy Collective, a joint industry initiative of over 30 companies to develop smart energy solutions in an open innovation environment.

He is the founding father of PowerMatching City, the first full concept smart energy demonstration and is focused on innovation and business development in smart energy systems with over 15 years of experience along the entire energy value chain.

Dr. Bliek graduated from the FOM Institute for Plasma Physics at University of Utrecht in Energy Physics and received his Ph.D. at the Kernfysisch Versneller Instituut, Rijskuniversiteit Groningen on Atomic Physics and Thermonuclear Fusion.

F. Stuart Bresler is Vice President, Market Operations and Demand at PJM, LLC, where he is responsible for Forward and Real-Time Market Operations, the Financial Transmission Rights Markets, Market Settlements, the development of the technical systems utilized in markets and operations, and Demand Response integration.

He has been involved in the development and implementation of PJM markets for ancillary services and the integration of demand response resources into PJM markets.He has also led systems development enhancements required to expand PJM's markets and the coordination of those markets with neighboring regional transmission organizations.

Mr. Bresler graduated from Pennsylvania State University with a B.S. degree in Electrical Engineering.

Ralph Cavanagh is co-director of NRDC's energy program, which he joined in 1979. In this capacity, he has worked extensively with utilities across North America on strategies for integrating energy efficiency and renewable resources in their procurement planning, capital budgeting, and business models. He has been a visiting professor of law at Stanford and UC Berkeley, a faculty member for the Utility Executive Course at the University of Idaho, and a lecturer on law at the Harvard Law School. From 1993 to 2003 he served on the U.S. Secretary of Energy's Advisory Board.

His board memberships include the Bipartisan Policy Center, the Bonneville Environmental Foundation, the California Clean Energy Fund, the Center for Energy Efficiency and Renewable Technologies, the Northwest Energy Coalition, and the Renewable Northwest Project. His public service awards include the Heinz Award for Public Policy, the National Association of Regulatory Utility Commissioners' Mary Kilmarx Award, and the Bonneville Power Administration's Award for Exceptional Public Service.

He is a graduate of Yale College and the Yale Law School.

Paul Centolella is a Vice President in the Energy and the Environment and Natural Resources practices of Analysis Group, an economic and strategy consulting firm. He is a former commissioner on the Public Utilities Commission of Ohio. Before his appointment to the commission, he was a senior economist at Science Applications International Corporation, and earlier in his career was the Senior Energy Policy Advisor and a Senior Utility Attorney for the Office of the Ohio Consumers' Counsel.

He is a member of the Secretary of Energy's Electricity Advisory Committee. He has served on the boards of the Organization of PJM States and of the Smart Grid Interoperability Panel, a public-private partnership of over

700 organizations created by the National Institute of Standards and Technology to accelerate the development of standards for the smart grid. He represented the National Association of Regulatory Utility Commissioners (NARUC) on the Electric Power Research Institute's Advisory Council and was a member of the NARUC Energy Resources and Environment Committee, and the FERC/NARUC Smart Response Collaborative.

Mr. Centolella earned a B.A. degree from Oberlin College and a Juris Doctor from the University of Michigan Law School.

Youngho Chang is an assistant professor of economics at the Division of Economics and an adjunct senior fellow at the S. Rajaratnam School of International Studies (RSIS), Nanyang Technological University, Singapore. He is a member of Technical Committee for Clean Development Mechanism (CDM) Designated National Authority (DNA), National Environment Agency, Singapore.

He specializes in the economics of climate change, energy and security, oil and macroeconomy, and the economics of electricity market deregulation. His current research interests include the impact of oil price fluctuations in macroeconomic performance, energy security, energy use, and climate change. He has numerous publishing credits and co-edited two books – *Energy Conservation in East Asia*: *Towards Greater Energy Security* (World Scientific, 2011) and *Energy Security*: *Asia Pacific Perspectives* (Manas Publications, 2010).

Dr. Chang received a B.A.Sc. in Landscape Architecture from the Seoul National University, a M.A. in Economics from the Yonsei University, and a Ph.D. in Economics from the University of Hawaii at Manoa.

Noelle C. Cole is the founder and principal of Resource Refocus LLC, a consulting company focused on energy efficiency and zero net energy research, policy, and program implementation. In this capaciy, she serves as a consultant to the Pacific Gas and Electric Company (PG&E), engaged in the Zero Net Energy (ZNE) Pilot Program, which conducts a range of activities in support of the California Public Utilities Commission's ZNE goals.

Previously, she worked with the City of Emeryville Economic Development & Housing Department, and with StopWaste.Org, a public agency combining Alameda County's Waste Management Authority and Source Reduction Board. Her area of interest is in the flow of resources – water, energy, and waste – through communities and the design and policy strategies that shape them.

She holds master's degrees in City Planning and Landscape Architecture from University of California, Berkeley, and a B.F.A. from Scripps College in Claremont, California.

Susan Covino is a senior consultant, Emerging Markets at PJM. She previously served as Manager for Demand Side Response.

She has been actively involved in the development of demand response in the PJM region and nationally. She has served as PJM's liaison to the Mid-Atlantic Distributed Resources Initiative (MADRI) and has led the development of multiple iterations of the Demand Response Road Map for the PJM Region.Ms. Covino has also led PJM stakeholder outreach in demand response, organizing three demand response and smart grid symposia over the last 5 years.She currently serves as a board member of the Association for Demand Response and Smart Grid.

Ms. Covino earned a B.A. with a double major in Economics and History from the University of Connecticut and a Juris Doctor degree from Dickinson School of Law.

Ahmad Faruqui is a principal at The Brattle Group where he focuses on assessing the economics of dynamic pricing, demand response, advanced metering infrastructure, and energy efficiency in the context of the smart grid. He pioneered the use of experimentation in understanding customer behavior, and his early work on time-of-use pricing experiments is frequently cited in the literature.

He has assisted the Federal Energy Regulatory Commission (FERC) in the development of the National Action Plan on Demand Response and led the effort that resulted in A National Assessment of Demand Response Potential. He co-authored the Electric Power Research Institute's (EPRI) national assessment of the potential for energy efficiency and the Edison Electric Institute's (EEI) report on quantifying the benefits of dynamic pricing. He has assessed the benefits of dynamic pricing for the New York Independent System Operator, worked on fostering economic demand response for the Midwest ISO and ISO New England, reviewed demand forecasts for the PJM Interconnection, and assisted the California Energy Commission in developing load management standards.

Dr. Faruqui holds a Ph.D. in economics from the University of California, Davis.

Frank A. Felder is Director of the Center for Energy, Economic and Environmental Policy and Associate Research Professor at the Bloustein School of Planning and Public Policy at Rutgers University.

He directs applied energy and environmental research. Ongoing and recent projects include energy efficiency evaluation studies, economic impact of renewable portfolio standards, and power system and economic modeling of state energy plans. He is also an expert on restructured electricity markets. He has published widely in professional and academic journals on market power and mitigation, wholesale market design, reliability, transmission planning, market power, and rate design issues. He was a nuclear engineer and submarine officer in the U.S. Navy.

Dr. Felder holds undergraduate degrees from Columbia College and the School of Engineering and Applied Sciences and a master's and doctorate from M.I.T. in Technology, Management, and Policy.

David Fridely is a senior staff scientist at the China Energy Group of the Environmental Energy Technologies Division of Lawrence Berkeley National Laboratory, where his research involves extensive collaboration with the Chinese on end-use energy efficiency, industrial energy use, government energy management programs, data compilation and analysis, and medium and long-term energy policy research.

He led the China Energy Project at the East-West Center in Hawaii primarily in the area of petroleum supply and demand, refinery analysis and modeling, international oil trade and energy policy, and concurrently was a consultant with Fesharaki Associates. From 1993 to 1995, he worked as Business Development Manager in refining and marketing for Caltex China. He has 25 years of experience workingand living in China.

He has a master's degree in East Asian Studies from Stanford University.

Clark W. Gellings is a fellow at the Electric Power Research Institute (EPRI), responsible for technology strategy in areas concerning energy efficiency, demand response, renewable energy resources, and other clean technologies.

He joined EPRI in 1982 progressing through a series of technical management and executive positions including seven vice president positions. Prior to joining EPRI, he spent 14 years with Public Service Electric & Gas Company in NJ. He is the recipient of numerous awards, has served on numerous boards and advisory committees, both in the United States and internationally, and is a frequent speaker at industry conferences.

Mr. Gellings has a B.S. in Electrical Engineering from Newark College of Engineering in New Jersey, a M.S. degree in Mechanical Engineering from New Jersey Institute of Technology, and a M.M.S. from the Wesley J. Howe School of Technology Management at Stevens Institute of Technology.

Michael Greenstone is the 3 M Professor of Environmental Economics in the Department of Economics at M.I.T. and serves on the M.I.T. Energy and Environmental Research Councils. In addition, he is the Director of The Hamilton Project and a research associate at the National Bureau of Economic Research. He served as the Chief Economist for President Obama's Council of Economic Advisors in the first year of his Administration and was an editor of *The Review of Economics and Statistics*.

His interests include environmental, public, development, and labor economics. His research is focused on estimating the costs and benefits of environmental quality. He is currently engaged in a project to estimate the economic costs of climate change, the impact of air pollution on life expectancy, and the costs and benefits of emissions trading schemes.

Dr. Greenstone received a Ph.D. in Economics from Princeton University and a B.A. in economics with High Honors from Swarthmore College.

Robert C.G. Haffner is a principal manager in PwC's Strategy and Economics practice specializing in regulatory, market, and competition analyses using both quantitative and qualitative tools. He has more than 20 years of experience in this area as he has worked for two ministries, a university and a regulator. In his last assignment prior to joining PwC, he was the chief economist of the Dutch energy regulator. As a result, the energy sector has been a key focus areawith projects in the field of energy regulation, renewable energy, energy finance, and carbon caputure and storage. He has also co-authored a book on the economic impact of privatization and deregulation.

Mr. Haffnergraduated *cum laude* in economcs at Erasmus University Rotterdam, the Netherlands.

Steven G. Hauser is the Chief Technology Officer at New West Technologies, LLC and serves as President Emeritus of The GridWise Alliance. He is responsible for new corporate strategies to create a smarter and cleaner electricity grid and is a recognized expert on transforming the power sector to meet future economic, environmental, and energy security mandates. He has been active in various innovative energy technology developments and contributed a chapter to an edited volume by the same editor on smart grids.

Mr. Hauser was formerly a VP at National Renewable Energy Laboratory and the driving force behind the creation of the GridWise® Alliance and in bringing together a broad coalition of companies globally to establish the Global Smart Grid Federation. Previously, he held senior management positions at GridPoint, Battelle SAIC, and CH2M Hill. He serves as an advisor to several smart energy organizations.

He received a B.S. in Engineering Physics from Oregon State University and an M.S. in Chemical Engineering from the University of Washington.

Stephen Healy is a senior lecturer at the School of Humanities involved in the Environmental Humanities Program of UNSW's Faculty of Arts and Social Science. He is also Research Coordinator of the Centre for Energy and Environmental Markets.

Dr. Healy has worked for Greenpeace International, the NSW EPA, and Middlesex University, UK, where he led the Science, Technology and Society Program. His research interests include climate change, energy, risk and uncertainty, and public participation. He in interested in historical development of contemporary systems of energy provision, energy institutions, governance and politics, and re-envisioning energy consumption. He has published across a number of fields by CCH Australia and co-authored *Guide to Environmental Risk Management*, published in 2006.

He has a B.S.c. (Hons) in Physics and an Electrical Engineering Ph.D. in Photovoltaics from UNSW.

Theodore Hesser was formerly with Bloomberg New Energy Finance's Energy Smart Technologies research practice in North America. His work focuses on market trends, technology, and policy analysis within the emerging smart grid, energy efficiency, battery storage, and advanced transportation sectors.

Prior to joining Bloomberg he worked with the Natural Resources Defense Council in their Center for Market Innovation focusing on the impact of slated EPA regulations on coal retirements and authored a chapter on "Renewable Integration through Direct Load Control and Demand Response" in *Smart Grid: Integrating Renewable, Distributed, & Efficient Energy* published by Elsevier in 2012. Mr. Hesser received a B.A. in physics from The Colorado College and a M.S. in Civil and Environmental Engineering from Stanford University.

Heidi M. Keller, currently at Johns Hopkins University, was a program analyst with the National Oceanic & Atmospheric Administration (NOAA) where she managed NOAA's scientific performance metrics and communicates the results to the White House and Congress.

Previously, she worked in NOAA's National Ocean Service as a physical scientist studying tidal and current data in relation to coastal restoration, sea-level trends, and the location and extent of harmful algal bloom movement in the Gulf of Mexico. As a part of her graduate studies, she collaborated with Skip Laitner on re-inventing the electric utility business model.

Ms. Keller has a B.S. degree in Environmental Science from the University of Vermont and a M.S. degree in Energy Policy and Climate from Johns Hopkins University.

Stefanie Kesting is the Manager for Strategy & Business Development in the area of Sustainable Use at DNV KEMA Energy & Sustainability. In her career, she has advised clients in different segments of the energy chain as well as policymakers and regulators.

In the last years, her focus has shifted from traditional fuels to solving clients' interests in the transition towards renewable energy, decentralized generation, meeting carbon reduction goals, energy efficiency, and smart energy solutions. In her current role, her focus is on sustainable end use, energy efficiency, and related policies.

Before joining DNV KEMA Energy & Sustainability, she worked, *inter alia*, at the energy research institutes of the Universities in Cologne (EWI), Bremen (BEI), and E.ON. Dr. Kesting holds a master's degree in Business Administration from University of Cologne and a Ph.D. in Economics from International University Bremen. She has published on different energy market topics and was awarded the GEE Prize of the Berlin Energy Forum for her Ph.D. work.

Chris King is President, eMeter Strategic Consulting and Chief Regulatory Officer of eMeter, a Siemens Business. He leads Siemens's global smart grid policy efforts and eMeter's marketing and business strategy.

He has over 30 years experience with smart meters and Smart Grids, including as director of rate design at Pacific Gas and Electric (PG&E), executive positions at CellNet Data Systems (now owned by Landis + Gyr), and founder and CEO of electricity retailer utility.com Inc. His interests include smart grid regulatory policy, technology, consumer benefits, and business cases. He chairs the Brussels-based Smart Energy Demand Coalition, and is a director of the Demand Response and Smart Grid

Alliance, Association for Demand Response and Smart Grids, and Demand Response National Action Plan Coalition. He has testified before the U.S. Congress and domestic and international regulatory bodies.

Mr. King holds advanced degrees in Science, Business, and Law from Stanford University and Concord Law School.

John A. "Skip" Laitner is currently a Principal at Economic and Human Dimensions Research Associates. Previously he was the Director of Economic and Social Analysis for the American Council for an Energy-Efficient Economy (ACEEE) where his research has focused on developing a more robust technology and behavioral characterization of energy efficiency resources for use in energy and climate policy analyses and within economic policy models. Prior to ACEEE he served as a Senior Economist for Technology Policy for the U.S. Environmental Protection Agency where he was awarded EPA's Gold Medal in 1998 for his work on evaluating the impact of different strategies in the implementation of greenhouse gas emissions reduction policies.

His 2004 paper "How Far Energy Efficiency?" catalyzed new research in the proper characterization of efficiency as a long-term resource. Author of nearly 300 reports, journal articles, and book chapters, he has more than 40 years of involvement in the environmental, energy, and economic policy arenas.

Mr. Laitner has a master's degree in Resource Economics from Antioch University in Yellow Springs, OH.

Alex Laskey is President and co-founder of Opower, an energy efficiency technology company, responsible for engaging utility and government partners with the company's purpose and products. He is a frequent speaker at energy-related conferences including an invitation to the White House to meet with President Obama and discuss innovation and job creation in the green economy.

Prior to founding Opower, he was engaged in politics and policy, serving as a campaign manager, strategist, and public-opinion analyst for several candidates. He provided strategic consulting on several statewide ballot measures for The Nature Conservancy, The Trust for Public Land, and The League of Conservation Voters, and he worked for the White House and on a presidential campaign. Before that, he produced the award-winning feature film, *Assisted Living*, and worked as the director of new business and strategy for The Romann Group, a New York-based advertising agency.

He received his B.A. in History of Science from Harvard.

Anna M. LaRue is the founder and principal of Resource Refocus LLC, a consulting company focused on energy efficiency and zero net energy research, policy, and program implementation. She is also the editor of the blog Zero Resource.

Previously, she was a senior regulatory analyst with Pacific Gas and Electric Company (PG&E) in charge of the Zero Net Energy (ZNE) Pilot

Program in California. Prior to that, she was an architectural program coordinator with the PG&E Pacific Energy Center. She has been an instructor for Boston Architectural College, a researcher on demand response technology through the UC Berkeley Center for the Built Environment, and has worked for Building Science Corporation and the Cleveland Green Building Coalition.

Ms. LaRue earned a M.S. in Architecture with a concentration in Building Science from the University of California, Berkeley, and a B.A. in Physics from Smith College.

Mark Levine is Senior Staff Scientist and founder and group leader of the China Energy Group of Lawrence Berkeley National Laboratory. He has authored more than 100 publications dealing with energy efficiency policy, building energy efficiency, and energy demand and energy efficieincy policies in China, and he has been instrumentalin the creation of the China Sustainable Energy Program of the Energy Foundation and the Beijing Energy Efficiency Center (BECon).

Dr. Levine has led a series of high-profile energy analysis activities: he had overall responsibility for the IPCC chapters on mitigating carbon emissions in buildings (second assessment report) and shared responsibility (fourth assessment report); he was co-leader of the report "Scenarios for a Clean Energy Future" for the United States; and co-leader of a major study of energy and carbon futures of China.

He graduated *summa cum laude* from Princeton University, earned a Ph. D. from the University of California, and has been the recipient of a Fulbright scholarship, a Woodrow Wilson Fellowship, and National Institutes of Health doctoral awards.

Yanfei Li is a research fellow at the Energy Research Institute of Nanyang Technological University, Singapore where he conducts research in energy economics and economics of technological change, serving both academic and consulting constituents. He has collaborated with government agencies of Singapore such as the Energy Market Authority, Land Transport Authority, National Climate Change Secretariat, and Economic Research Institute for ASEAN and East Asia.

Dr. Yanfei Li's current research covers oil prices, regional gas trade modeling, regional power generation planning and trade, economic impacts of information and communication technology, and roadmapping for electrical vehicle technology adoption in Singapore. He has numerous publications in these areas.

He received a B.A. in Economics from Peking University and a Ph.D. in Economics from Nanyang Technological University.

Iain MacGill is an associate professor at the School of Electrical Engineering and Telecommunications at the University of NSW and Joint Director (Engineering) for the University's Centre for Energy and Environmental Markets (CEEM). His teaching and research interests include electricity industry restructuring and the Australian National Electricity Market, sustainable energy technologies, and energy and climate policy.

CEEM undertakes interdisciplinary research in the monitoring, analysis, and design of energy and environmental markets and their associated policy frameworks. It brings together UNSW researchers from the faculties of Engineering, Business, Science, Law and Arts, and Social Sciences. Dr. MacGill leads work in two of CEEM's three research areas: Sustainable Energy Transformation, including energy technology assessment and renewable energy integration and Distributed Energy Systems including "smart grids" and "smart" homes, distributed generation, and demand-side participation. He has published widely in these and related areas.

Dr. MacGill has a bachelor's degree in Engineering, a master's in Engineering Science from the University of Melbourne, and a Ph.D. in Electricity Market Modeling from UNSW.

Matthew T. McDonnell is a consultant with Economic Research and Human Dimensions Research Associates where he specializes in regulatory policies as they impact both environmental costs and the development of renewable energy technologies.Prior to law school, McDonnell worked for several years in the financial community.As part of his law school experience he completed an internship with Arizona Corporation Commission. He has collaborated on the development of a benefit-cost framework for future regulatory assessments and served as the executive editor for the *Arizona Journal of Environmental Law & Policy* (AJELP).

Mr. McDonnell received a B.A. in Finance from Michigan State University and earned his Juris Doctor degree from the University of Arizona's College of Law.

William C. Miller works for Lawrence Berkeley National Laboratory where he is assigned to the Energy Efficiency and Renewable Energy Program at the U.S. Department of Energy (DOE). Prior to his current position, he managed the Strategic Regulatory Issues group for the Customer Energy Efficiency department at Pacific Gas and Electric Company (PG&E) where he oversaw the company's Zero Net Energy Pilot Program. Before joining PG&E, he spent 10 years in academia, engaged in teaching and research.

He has worked for over 25 years in the areas of energy forecasting, strategic market analysis, energy efficiency planning, measurement, policy, and litigation. From 1990 to early 2010, he managed the evaluation, policy, and regulatory activities for the energy efficiency department at PG&E.

Dr. Miller received a B.S. in Economics from Stanford University and a Ph.D. in Economics from the University of Minnesota.

David Mooney is Director of the National Renewable Energy Laboratory's Electricity, Resources, and Building Systems Integration Center. He currently leads efforts to identify and address technical issues associated with the large-scale integration of renewable and efficiency technologies into the existing energy infrastructure.He leads 180 researchers conducting $100 M in R&D annually and has testified on these issues before the U.S. House of Representatives' Subcommittee on Energy and Environment.

Dr. Mooney has held a variety of positions at NREL from a PV materials and devices researcher, to program manager, to assistant to the director, to an assignment to technically support the DOE's Office of the Assistant Secretary for Efficiency and Renewables. He also worked as the director of business development for a U.S. PV manufacturing equipment supplier.

He has a Ph.D. in Physics from the University of Arkansas and a B.S. in Engineering Physics from the University of Tennessee.

Steven Nadel is the Executive Director of the American Council for an Energy-Efficient Economy (ACEEE), a non-profit research organization that works on programs and policies to advance energy-efficient technologies and services. Prior to ACEEE he planned and evaluated energy efficiency programs for New England Electric, directed energy programs for the Massachusetts Audubon Society and ran energy programs for a community organization in New Haven, CT.

He has worked in the energy efficiency field for 30 years and has over 100 publications.He has worked with utilities, regulators, and legislators on utility efficiency programs, building codes, and equipment efficiency standards. He has testified before Congress on energy efficiency subjects and has contributed to national energy legislation passed by Congress and remains engaged in pending energy legislation.

Mr. Nadel has an M.S. in Energy Management from the New York Institute of Technology, an M.A. in Environmental Studies, and a B.A. in Government from Wesleyan University in CT.

Paul H. L. Nillesen is a partner with PwC based in Amsterdam. He is part of the firm's global Energy Utilities & Mining Practice and the co-leader of PwC's Global Renewables Practice. He is also a member of the International Energy Agency's Renewable Energy Working Group.

He specalizes in regulatory economics, strategic analyses, advanced market and demand analyses, and business simulation and modeling primarily focused on energy and utilities. He has advised companies, governments, regulators, national and international institutions, and has published numerous articles in the field of energy and regulatory economics.

He holds master's degrees in Economics from Edinburgh and Oxford and a Ph.D. in Economics from Tilburg University.

Fatih Cemil Ozbugday is a Ph.D. researcher at Tilburg University. He also works as a member of Tilburg Law and Economics Center and an external researcher at the Office of the Chief Economist of the Netherlands Competition Authority.

His main specialization field is competition and regulatory economics. He has been involved in various projects concerning competition and regulatory issues and his papers have appeared in internationally renowned journals.

Mr. Cemil holds a M.A. degree in Economics from Bilkent University, M.S. degree in International Economics and Finance, and a research master's degree in Economics from Tilburg University.

Robert Passey is a senior research associate with the Centre for Energy and Environmental Markets (CEEM) at the University of New South Wales and a project manager atIT Power (Australia). In both positions he is responsible for research and analysis in the areas of renewable energy, distributed generation, electric vehicles, energy efficiency, and carbon markets.

Dr. Passey's specific interests include policy mechanisms to reduce greenhouse emissions, increase uptake of low emission technologies and drive energy efficiency; distributed generation and its integration into electricity networks, including smart grids; electric vehicles and their impact on distribution networks; technical assessments of low emissions technologies; and the characteristics of end-user decision-making relevant to uptake of distributed generation and energy efficiency.

He holds a B.S.c. and a Ph.D. from the University of New South Wales and a B.S. and M.S. from Murdoch University.

Michael R. Peevey was appointed to the California Public Utilities Commission (CPUC) by Governor Davis in March 2002 and designated President of the Commission in December 31, 2002. In December 2008 Governor Schwarzenegger re-appointed Mr. Peevey for another six-year term.

As President of the CPUC, Mr. Peevey is committed to environmental protection and reducing greenhouse gas emissions by maximizing energy efficiency, demand response, and renewable energy. From 1995 until 2000, Mr. Peevey was President of NewEnergy Inc. Prior to that, Mr. Peevey was President of Edison International and Southern California Edison Company. Mr. Peevey has served on the boards of numerous corporations and non-profit organizations.He has also received many awards for his leadership in energy policy and promoting diversity in the energy business, including a "Distinguished Citizen Award" from the Commonwealth Club of California for achievements in sustainable energy.

Mr. Peevey holds B.A. and M.A. degrees in Economics from the University of California, Berkeley.

Glenn Platt leads the Local Energy Systems theme within Commonwealth Scientific & Industrial Research Organisation (CSIRO) Energy Transformed Flagship, developing technologies for dramatically reducing the carbon emissions and increasing the uptake of renewable energy around the world. The theme's work ranges from solar cooling, electric vehicles, smart grids, and the integration of large-scale solar systems, through to understanding people's response and uptake of particular low-carbon energy options.

Prior to CSIRO, Glenn worked in Denmark with Nokia Mobile Phones on the standardization and application of cutting-edge mobile communications technology and prior to that he was employed in an engineering capacity for various Australian engineering consultancies, working on industrial automation and control projects.

He holds Ph.D., M.B.A., and Electrical Engineering degrees from the University of Newcastle Australia, and is an adjunct professor at the University of Technology, Sydney.

Lynn Price is a staff scientist and deputy leader of the China Energy Group of the Environmental Energy Technologies Division, of Lawrence Berkeley National Laboratory. Her research focuses on industrial energy efficiency, policy analysis, and evaluation, and she works collaboratively with a variety of Chinese research organizations.

She has been a member of the Intergovernmental Panel on Climate Change, which won the Nobel Peace Prize in 2007, and was an author on the industrial sector chapter of IPCC's Fourth Assessment Report on Mitigation of Climate Change. She led an effort of introducing the international energy efficiency target-setting to China through development of a pilot program with two steel mills in China, which set the foundation for national scale-up in China. She is currently providing technical assistance for China through the Energy Foundation's China Sustainable Energy Program as well as for a number of projects focused on improvement of energy efficiency and emissions reductions for the U.S. Department of Energy, U.S. Environmental Protection Agency, U.S. State Department, the World Bank, United Nations Development Program, and U.S. Agency for International Development.

She has a M.S. in Environmental Science from the University of Wisconsin-Madison.

Nicholas B. Rajkovich is a Ph.D. candidate in the Urban and Regional Planning Program at the University of Michigan. His current research focuses on the coupling of climate mitigation and adaptation strategies to reduce the vulnerability of low-income populations.

Prior to entering the Ph.D. program, he was responsible for coordinating a Zero Net Energy Pilot Program for Pacific Gas and Electric Company (PG&E) in San Francisco, CA. He also taught courses on energy efficiency and daylighting for the PG&E Pacific Energy Center. Before working for PG&E, he taught courses on building mechanical and electrical systems in the Department of Architecture at Cornell University, and served as an advisor to their 2005 and 2007 Solar Decathlon teams.

He received a bachelor's degree in Architecture from Cornell University and a master's degree in Architecture from the University of Oregon. He is a licensed architect and a LEED Accredited Professional.

Roland J. Risser is the Director for the U.S. Department of Energy (DOE) Building Technologies Program. By working with the private sector, state and local governments, national laboratories, and universities, the

Building Technologies Program improves the efficiency of buildings and the equipment, components, and systems within them.

Before joining DOE, he was Director of Customer Energy Efficiency for Pacific Gas and Electric Company (PG&E). In this role, he was responsible for developing and implementing energy efficiency and demand response programs. He also managed a building and appliance codes and standards program, as well as the Pacific Energy Center, Energy Training Center, and Food Service Technology Center.

Mr. Risser received a B.S. degree from the University of California, Irvine and a M.S. degree from the California Polytechnic State University in San Luis Obispo. He also graduated from the Haas School of Business, Executive Program, at the University of California, Berkeley.

Daniel Rowe is an engineering analyst and project manager within Commonwealth Scientific & Industrial Research Organisation (CSIRO) Demand Side Energy Systems (DSES) group. His work involves energy for buildings, energy efficiency, renewable energy, and grid integration and he leads CSIRO's Virtual Power Station trial, Solar Desiccant Cooling Demonstration, and ASI Intermittency Characterization projects.

His research interests include energy management technologies, solar cooling technology, solar forecasting, and grid integration techniques. He recently led the publication of CSIRO's landmark report *Solar Intermittency: Australia's Clean Energy Challenge*.

Mr. Rowe has a B.A. in Electrical Engineering from The University of Newcastle and currently serves as Secretariat for the Australian Solar Cooling Interest Group and National Representative on the board of Young Engineers Australia.

Bruce Sayler is Manager of Regulatory and Governmental Affairs at Connexus Energy. He has been in the energy field for over 23 years managing areas from their largest commercial accounts to recently launching several residential pilot programs focusing on behavioral modifications through the use of home energy reports and in-home displays. He is also a commissioner on the City of Elk River Energy City Commission, Vice President of the Anoka Technical College Foundation Board, and a member of the Minnesota Smart Grid Coalition.

Mr. Sayler holds a bachelor's degree from Bethel University and an M.B.A. from the University of Saint Thomas.

Allan Schurr is Vice President, Strategy & Development in IBM's Global Energy & Utilities Industry. In this capacity he is responsible for IBM's market strategy and regulatory policy for the global electric, gas, and water industries. IBM's offerings in these markets include customer management, grid operations, work and asset management, and power generation. He also leads IBM's utility initiatives in the role of energy in a Smarter Planet including development of smart grid and the integration of renewable energy sources and distributed energy assets like plug-in vehicles.

He testified before the U.S. Congress on the benefits of smart grid technology and impediments to its development. Prior to joining IBM he held management and executive positions at Pacific Gas and Electric (PG&E) Energy Services, Silicon Energy, and Itron.

Mr. Schurr received a bachelor's degree in Mechanical Engineering from the University of California Davis and a master's degree in Business Administration from St. Mary's College in California. He is a registered engineer in California.

Robert B. Segar is Assistant Vice Chancellor for Campus Planning at the University of California, Davis, and since 1989 has directed planning for the 5,300-acre campus. Before coming to UC Davis, he worked in the campus planning office at Stanford University.

Trained as a landscape architect, Mr. Segar is responsible for long-range land use planning, master planning for campus physical growth and development, plans for neighborhood districts, and site planning for open space, circulation, and buildings. He has led efforts to create a performing arts district on campus, to celebrate the university's Centennial in 2006, and to develop the new West Village neighborhood.

Mr. Segaris a graduate of Stanford University and holds a master's degree from the University of Michigan.

Fereidoon P. Sioshansi is President of Menlo Energy Economics, a consulting firm and the editor and publisher of *EEnergy Informer*, a newsletter with international circulation. His professional experience includes working at Southern California Edison Company, the Electric Power Research Institute, National Economic Research Associates, and Global Energy Decisions (GED), acquired by ABB.

His interests include climate change and sustainability, energy efficiency, renewable energy technologies, smart grid, dynamic pricing, regulatory policy, and integrated resource planning. He has edited five other books: *Electricity Market Reform: An International Perspective*, with W. Pfaffenberger (2006), *Competitive Electricity Markets: Design, Implementation, Performance* (2008), *Generating Electricity in a Carbon Constrained World* (2009), *Energy, Sustainability and the Environment: Technology, Incentives, Behavior* (2011) and *Smart Grid: Integrating Renewable, Distributed & Efficient Energy* (2011), all published by Elsevier.

Mr. Sioshansi has degrees in Engineering and Economics, including an M.S. and a Ph.D. in Economics from Purdue University.

Robert Smith is the Manager, Economics & Strategy at Ausgrid, the largest electricity distribution company in Australia based in Sydney. He has over 25 years experience as an economist including a dozen years working in electricity market design, regulation, energy efficiency, and demand management. While with Ausgrid he has been involved in creating two Energy Efficiency Centres, web-based energy efficiency and e-commerce tools, Australia's first mass market CFL giveaways, implementing full retail competition for electricity, electric vehicles, and the Smart Home Family project.

His interests include cost-benefit analysis and understanding how economics, technology, incentives, regulation, and customers' behavior to create change. He is a regular speaker and author on demand management, energy efficiency, electric vehicles, and economic issues.

Mr. Smith has a graduate degree in Econometrics, a master's in Economics and postgraduate qualifications in Finance from the University of New South Wales.

Paul M. Sotkiewicz is the Chief Economist in the Market Services Division at the PJM Interconnection where he provides analysis on PJM's market design and performance including the implications of various federal and state policies. Prior to his present post he served as the Director of Energy Studies at the Public Utility Research Center, University of Florida, and as an economist at the Federal Energy Regulatory Commission.

He has led initiatives to reform scarcity pricing as mandated by FERC Order 719, in examining transmission cost allocation and the potential effects of climate change policy on PJM's energy market, and in developing proposals for compensating demand response resources under FERC Order 745.

Dr. Sotkiewicz received a B.A. degree in History and Economics from the University of Florida andan M.A. degree and a Ph.D. in Economics from the University of Minnesota.

Jessica Stromback is the founder and Executive Director of the Smart Energy Demand Coalition (SEDC) responsible for liaising with the European Commission (EC) and regulators to enable demand-side participation throughout European energy markets. She is also the Chairman of VaasaETT, a consultancy based in Finland.

She specializes in smart metering, demand-side programs, market structure issues, and policy issues. In her various capacities, she has participated in or managed multiple projects for the EC and companies including EDF, Enel, Microsoft, Nuon, EDF, Capgemini, RWE, Landis & Gyr, Panasonic, CREIPI, British Gas, Bord Gais, Union Fenosa, and BC Hydro. These have included projects measuring the potential of the smart grid to lower greenhouse gas emissions, a comparison of 100 residential pilot projects involving over 450,000 households, and a review of the development of residential smart metering and demand response in 23 national markets.

Ms. Stomback has a degree in English from Vaasa University in Finland.

Peter W. Turnbull is Principal Program Manager in charge of Pacific Gas and Electric Company's (PG&E) Zero Net Energy Pilot Program. He has 30 years of experience promoting energy efficiency.

He has extensive experience in developing and managing energy efficiency rebate programs in commercial markets, in emerging technologies, codes and standards, analytic tool development, and market research, especially in the areas of cool roofing, daylighting, and commercial food service.

He is a current Board member of the Cool Roof Rating Council. More recently, Peter spearheaded PG&E's efforts in the area of automatic benchmarking using the ENERGY STAR *Portfolio Manager* Benchmarking tool.

Mr. Turnbull holds B.A. and M.A. degrees in English and a B.S. in Civil Engineering, the latter from Montana State University.

Josh Wall leads the Building Controls & Wireless Sensors research project as part of a wider Energy for Buildings research program within CSIRO's Energy Transformed Flagship, developing innovative ways of applying advanced information, communications, and control technology to improving the way we distribute and utilize energy in buildings. He has experience with the development of state-of-the-art energy efficiency and demand management technologies with a particular focus on advanced HVAC&R controls for the built environment.

Dr. Wall is a member of the Australian Institute of Refrigeration, Air-conditioning & Heating (AIRAH), and contributed to Best Practice Guideline for HVAC Controls (DA28) in Australia and a corresponding member of the American Society of Heating, Refrigeration & Air-conditioning Engineers (ASHRAE) technical committee TC7.5 on Smart Building Systems, looking at the next generation of control strategies and techniques for low-energy buildings.

He holds Ph.D., Electrical Engineering, and Computer Science degrees from the University of Newcastle Australia.

Stephen M. Wheeler is an associate professor in the Landscape Architecture Program at the University of California, Davis (UCD). He has taught urban and regional planning, urban design, and sustainable development at UCD, the University of New Mexico, and UC Berkeley since 1996.

His books include *Climate Change and Social Ecology* (Routledge, 2012), *Planning for Sustainability* (Routledge, 2004), and *The Sustainable Urban Development Reader* (Second Edition: Routledge, 2009). He has served as an urban planning consultant, as editor of *The Urban Ecologist* journal, and as a lobbyist for environmental organizations in Washington, D.C. His awards include the 2009 William R. and June Dale Prize for Excellence in Urban and Regional Planning. He is a member of the American Institute of Certified Planners.

Dr. Wheeler holds Ph.D. and Master of City Planning degrees from the University of California, Berkeley and an undergraduate degree from Dartmouth College.

Gregory Wikler is Director of Regulatory Affairs for EnerNOC, a world-leading energy management company. He supports electric and gas regulatory policy related to energy efficiency and demand response programs at various levels including state commissions, electric utilities, and government agencies, both nationally and internationally.

In his 25 years of experience in this field, he has been involved in strategic planning assessments and program implementation support on various energy efficiency and demand response program initiatives for utilities and grid operators throughout North America and Asia. He serves on the Executive Board of the California Energy Efficiency Industry Council and is a Board member for the Association of Energy Service Professionals.

Mr. Wikler holds M.S. and Master of Urban Planning degrees from the University of Oregon and an undergraduate degree in Energy Economics from the University of California, Davis.

Nina Zheng is a Senior Research Associate in the China Energy Group of Lawrence Berkeley National Laboratory. She specializes in residential end-use efficiency research, end-use energy modeling and scenario analysis.

Her research interests include China's appliance efficiency standards and labeling programs, building energy efficiency, bottom-up energy end-use modeling, energy supply and energy policy analysis. Her recent work focused on evaluating low carbon city development and alternative energy development in China and quantifying the potential energy and emissions impacts of efficiency and low carbon policies in China.

She holds a M.S. in Energy and Resources from the University of California at Berkeley and a B.S. in Science, Technology and International Affairs from Georgetown University.

Nan Zhou is a scientist in the China Energy Group of Lawrence Berkeley National Laboratory and the deputy director of the U.S.-China Clean Energy Center-Building Energy Efficiency. Dr. Zhou' work has focused on modeling and evaluating China's low-carbon development strategies. Additional work includes evaluation and development of China's appliance standards and labeling programs, energy efficiency in industry and buildings, and assessments of energy efficiency policies.

Dr. Zhou's recent work involves developing a guidebook and tools for local governments in China to create low-carbon development plans, as well as providing trainings to policymakers in China on energy efficiency and low carbon, and forecasting low-carbon future outlook for the critical period extending to 2050.

She has an Architecture degree from Xi'an University of Architecture and Technology, a master's in Architecture from Kyushu University in Japan, and a Ph.D. in Engineering from Kyushu Sangyo University, Japan.

Is Zero Energy Growth in Our Future?

In the year 2010, 50 energy experts were polled about the likely impact of energy efficiency on future electricity use in the year 2020.[1] They said that it would lower energy consumption by 5 to 15 percent and that peak demand would be lowered by 7.5 to 15 percent.[2]

So what do the results portend? If these reductions are applied to the 1 percent annual growth rate in energy consumption and peak demand that seems to be the consensus projection these days, the result is zero growth.

Some of us recall the book, *The Limits to Growth*, which was published in 1972 by the Club of Rome. Partly for its messianic tone and partly for its Malthusian conclusions, the book was immediately panned by academics. With some critics, the "systems dynamics" model at the heart of *Limits* did not cut the mustard. William Nordhaus' entitled his critique in *The Economic Journal* "Measurement without data" (http://nordhaus.econ.yale .edu/worlddynamics.pdf).

Acceptance of the book's policy conclusions was slow. In 1976, Amory Lovins lamented that no one was taking the so-called "soft path" to solving the world's constrained resource problems.[3] Concurrently, at the California Energy Commission, a band of graduate students from UC Davis were assembling the first genre of econometrically structured end-use models under the tutelage of professors Leon Wegge and Dan McFadden. Using these models, they prognosticated the end of the Seven Percent Growth Era. The "culprits" would be many and would include price elasticity (which some insisted back then was zero since electricity was a necessity), appliance codes, and building standards (which some insisted equated to a form of communism that Lenin would have cherished).

---

1. Chapter 1 includes further details of this survey.
2. http://papers.ssrn.com/sol3/papers.cfm?abstract_id = 2029150
3. "Energy Strategy: The Road Not Taken," *Foreign Affairs*, 1976; http://www.foreignaffairs. com/articles/26604/amory-b-lovins/energy-strategy-the-road-not-taken.

Nobody believed us then. We were called neo-Malthusians by at least one academic (who would became famous for propounding the Rebound Effect Theory, which averred that energy efficiency programs would actually cause consumption to bounce back to prior levels).

However, four decades on, the *Limits to Growth* book has proved prescient. In 2008 Graham Turner examined the past thirty years of data with the forecasts made by the *Limits* and found that changes in industrial production, food production, and pollution were all in line with the book's predictions. That is remarkable given that *Limits* was written when green was just a color and not the name of a movement and when climate change was not a household word, let alone the focus of a film.

The future remains impossible to predict. Some continue to believe that new end-uses, not yet invented in Silicon Valley, will overwhelm improvements in energy efficiency brought on by smarter appliances and fuel another boom in electricity growth. But more and more analysts are anticipating a slow-down in growth, maybe even a crawl to zero growth.

The debate about zero growth will linger. Participants on both sides of the debate, passionate in their pursuit of the truth, would do well to heed the wisdom of management guru Peter Drucker, "The best way to predict the future is to create it." Energy efficiency and demand response provide arguably the best way of creating it. Even with the likelihood of a zero growth world, concerns about the limitations of our natural resources will always necessitate having energy efficiency and demand response policies and programs in place. Furthermore, there is a mountain of evidence indicating that these policies and programs are effective, economically attractive, and appealing to customers.

Ahmad Faruqui
The Brattle Group

Preface

In the 20th century, mankind accomplished amazing feats of invention and engineering. The automobile, the airplane, the telephone, radio, television, satellites, nuclear energy, cell phones and the Internet, all were invented and made commercially available in just 100 short years. In that time, the average person went from living a life of hard labor and few amenities to one of comfort, free time, easy travel and luxury. These are accomplishments to be proud of. Technology has freed millions of people from lives of toil and allowed them to enjoy a life only the wealthiest individuals could live a century ago.

Of course, those higher standards of living have come at a cost. Increased income and new devices, along with a booming population, have led to a massive increase in energy consumption. And as the 21st century dawned a little over a decade ago, mankind awoke to the reality that its own energy consumption would be at best unsustainable, and at worst threaten its own survival. We awoke to the realization that global climate change was not some far-off distant threat, but a real problem that is already changing life on the planet even as we seek to study and better understand it.

Some will try to deny that reality. Others will ignore it and hope it goes away, and some will wait for someone else to take action. At some point, however, mankind must confront the problems that are created by our burgeoning energy use. And if the consequences of current energy consumption levels aren't enough of a problem, consider the impact of billions of people in the developing world who are trying to achieve those standards of living we enjoy in Europe and the United States — people that, most would agree, deserve to achieve those high standards of living.

It is truly a monumental challenge. It is a problem the scale of which dwarfs any challenge mankind has previously faced. The cold war, the moon landing ... these almost seem like simple tests compared with the challenge of meeting the energy demands of a growing world in a way that avoids an environmental calamity. If the major accomplishment of the 20th century was creating a high standard of living for the average person, the greatest accomplishment of the 21st century will be maintaining those same standards of living at a much lower cost in terms of energy use and environmental impact.

Truly, the challenge is daunting. But we Californians are not ones to sit around and wring our hands when faced with a seemingly impossible challenge. We are not the ones who wait for someone else to take action and then follow meekly behind. We get up. We act. And when we make mistakes in the course of those early efforts, we learn from those mistakes, correct course and keep going. And sooner or later, others will follow. That is really the story of how California has come to be the national leader in energy efficiency.

California began on this journey following the energy crisis of the 1970s. While others were content to be at the mercy of dirty energy producers in unstable parts of the world, California embarked on a different path. We began to implement new policies to reduce our energy use that were ground-breaking at the time but are now quite commonplace. In 1974, California became the first state to establish appliance energy efficiency standards, introducing standards for refrigerators, freezers, and air conditioners. The federal government followed suit years later, first adopting mandatory appliance standards in 1987.

California again led the way in 1982, when it became the first state in the union to adopt the policy of decoupling, which broke the link between electricity sales and utility profits. This policy was a giant leap forward in the promotion of energy efficiency, since it removed the utility's incentive to sell more power. Around the same time, utilities began to think about demand side management, offering financial incentives for customer purchase of efficient appliances. By the 1990s, we were already pursuing market transformation by which utilities sought to spur the production of highly efficient appliances and literally "transform" the market for these products.

Along the way, California has made mistakes. For instance, some of the market transformation programs of the 1990s were met with less than expected demand, and some of the early appliance rebate programs had problems with "free-ridership." But where we made mistakes, we have learned from them, and we have constantly sought to make our programs better and more effective.

And as a result of taking these actions – of not simply sitting on our hands and waiting for someone else to act on the problem – California has already achieved the milestone that is the subject of this book. That is to say, we have achieved zero demand growth, at least on a per capita basis.

California today has roughly the same per capita energy use that we had in the 1970s, even as per capita electricity use increased by 50 percent in the rest of the country. The fact that California has managed to stabilize per capita electricity use over the past 40 years despite the strong economic growth and a vast increase in the availability in consumer electronics is a substantial achievement that we are quite proud of in my state.

Unfortunately that achievement is not enough to solve the problems we are facing. We need to do much more. Stabilizing per capita demand is a first step, but it is only a first step. In order to stabilize total demand, per capita demand actually has to decrease to account for population growth. And to really make a dent in our energy-related problems, we need to not only stabilize total demand growth – we need to actually reduce the amount of energy every household uses without reducing their standard of living.

That's a tall order. But it's more than just an order — it's the law. California's Global Warming Solutions Act of 2006 directs us to reduce emissions 80 percent below 1990 levels by 2050. Such drastic emissions reductions cannot possibly be accomplished without creating an economy where each person enjoys the same or greater standard of living while using significantly less energy than they do today.

And that's why in the past 10 years California has sought to look at energy efficiency as more than just a demand management program. We've begun to view it as a resource we can tap to meet our future energy needs. In 2005, California established a "loading order" for procurement of new resources that makes energy efficiency the first place utilities look to meet new energy demand.

At the same time, we implemented policies that turn efficiency into a resource in the eyes of the utilities. Under the Risk Reward Incentive Mechanism, California utilities have the opportunity to earn a profit for deploying energy efficiency, similar to how they would earn a return for building a new power plant. As a result, California utilities are now investing $1 billion per year in cost-effective energy efficiency measures. That means for every dollar spent on energy efficient appliances, the utility ratepayers will receive at least as much benefit in the form of cost savings.

The impact of this investment is impressive. In just the current three-year funding cycle, the California investor-owned utilities are making enough energy efficiency investments to offset 1,982 MW of peak capacity, alleviating the need for four large (500 MW) power plants. This savings will allow us to avoid 4.9 million tons of CO_2 emissions.

But of course, we're not stopping there. Recently, the CPUC approved a vision for the next generation of energy efficiency programs, which moves aggressively in the area of financing. The recent decision directs the utilities to create a $200 million fund using private capital to finance energy efficiency improvements that could be repaid with the savings from the efficiency improvements. The decision also focuses on implementing "deeper" efficiency retrofits and greatly expands a successful program that promotes "whole house" energy upgrades.

And that's just for retrofits. In the area of new construction, California is paving the way toward a future where homes and businesses use no more energy than they produce themselves. It sounds like futuristic science fiction, but California is a place that dreams big and where futuristic fantasy becomes reality. In fact, our state policy is to have all new residential construction be zero net energy by 2020 and all new commercial construction zero net energy by 2030.[1]

---

1. Several chapters in the book further describe California's ZNE requirements.

Imagine the economic potential that could be unleashed when spending is freed up by the energy savings of zero net energy buildings. Of course, at this stage, zero net energy for all new construction is still a dream, but dreaming big is the first step to accomplishing big things. And we're actually closer to that goal that you might think. In fact, just last year the nation's largest zero net energy community opened at the University of California in Davis.

The UC Davis West Village will house 3,000 students, faculty and staff, provide 42,000 square feet of commercial space, a recreation center and village square – all without consuming more energy than it produces. This is truly an amazing accomplishment, and it shows what is possible with bold thinking and boundless ambition.[2]

For most people, the idea of a future with no growth in electricity demand is likely a far-out notion that shares an imaginary future realm with jet packs and flying cars. And indeed, an end to total growth in electricity demand would be a milestone never before seen the day Edison illuminated the first light bulb. But here in California, we've already caught a glimpse of the future, and we see that milestone is fast approaching.

Michael Peevey
President, California Public Utilities Commission

2. Wheeler et al further describe this project.

Fereidoon P. Sioshansi

Menlo Energy Economics

1 ELECTRICITY DEMAND GROWTH

Electricity demand growth has been steadily falling within the OECD block for 30+ years. In many developed economies including the United States, the EU and Japan, future demand is projected to grow at a tepid rate of 1 percent or less. In the case of United States, the Energy Information Administration (EIA) projects 0.7 percent growth per annum under a business-as-usual scenario for the period to 2035. On a per-capita basis, electricity consumption has been flat – as in California – or falling in some cases. Moreover, there is a growing recognition that it is *feasible*, and many experts believe, *desirable* to eliminate electricity demand growth cost-effectively and with minimal intervention.

It is, of course, a rather different story for the rapidly growing developing economies. It can, however, be argued that the imperative to influence electricity demand growth in these economies is even *more* compelling. Otherwise, they will end up with inefficient and wasteful infrastructure, appliances and buildings – for which they will have to pay dearly over a very long time.

This book's main focus is to describe why and how we can *influence* the future of electricity[1] demand growth globally, based on what is feasible and often cost effective.

2 THEME OF THE BOOK

For quite some time, a number of influential thinkers have been saying that our traditional approach of investing ever-increasing amounts on the "supply-side" to meet ever-growing demand may be unsustainable – economically and ecologically. A number of these same thinkers have been vouching for improvements in *energy efficiency* – not just in more efficient cars, appliances, and buildings – but also more clever and focused ways of using energy to deliver the *energy services* that we desire while using as little *energy* as possible – and wasting less in the process.

---

1. The book's focus is on electricity demand, although many of the approaches would apply to energy – broadly speaking.

Despite some progress, many studies suggest that energy efficiency remains largely untapped, under-utilized and under-appreciated. There are many reasons for this, some institutional, others behavioral or cultural. Many forms of energy, in general, and electricity in particular, are *underpriced* – for example, because environmental externalities are not fully accounted for – *inappropriately priced* – for example, when one class of customers are charged more than others or when electricity tariffs remain flat even though costs vary – and/or are artificially *subsidized* – as is prevalent in many countries.

Among the topics examined in this book is why energy efficiency, including active customer demand-side participation (DSP), has not made as much of an impact as its proponents would like. What can and should be done to eliminate the *energy efficiency gap* – defined as how much is actually captured vs. how much is available yet untapped – to the extent that it exists.[2]

The overall intent of the book is to examine what can be done in both developed and developing economies to increase the energy efficiency of electricity generation and consumption with the ultimate aim of reducing and eventually eliminating demand growth – when and where it makes sense to do so, is feasible, and economically cost-justified.

2.1 Against the Grain

Perhaps a logical starting point for a book on energy efficiency is to ask why a book on this topic is needed in the first place? If energy efficiency is cost effective as is often claimed, why is it underappreciated by consumers and firms? It is as if there were $20 bills sitting on the sidewalk, and no one is bothering to pick them up, as Amory Lovins,[3] an energy efficiency guru, rhetorically asks. Lovins goes even further, frequently claiming that energy efficiency is not only a *free* lunch, but in fact a lunch we are *paid to eat*. So why aren't we eating the free lunch?

The answer turns out to be more complicated than it seems.[4] A number of reasons are frequently provided.

One explanation may be that historically, the industry's focus has been on production and delivery of ever-larger volumes of electricity through an ever-expanding infrastructure that typically spans from primary fuels to the

---

2. Ironically, Chapter 6 concludes that the energy efficiency gap may not be as large as some prior studies have suggested.

3. Amory Lovins, the founder of Rocky Mountain Institute (RMI), is often cited among the pioneers of energy efficiency. He famously coined the word negawatts, as opposed to megawatts, in promoting what is now universally accepted, namely that the cheapest and cleanest kWhr is the one we do not use. His latest book, *Reinventing Fire*, presents a scenario where the United States can cost-effectively phase out it reliance on oil and coal by 2050.

4. Jared Diamond wrote a book trying to answer a simple question asked by an illiterate nomad from Papua New Guiney.

ubiquitous outlets on customers' walls. The industry's massive expansion in the 1900s, accompanied by major technological advances in generation, transmission, and distribution, allowed customers to benefit from enormous economies of scale and scope.

During much of this period, per unit costs of supplied electricity were falling, while incomes were broadly rising. The mentality behind the famous saying that electricity would be "too cheap to meter" became ingrained[5] in the culture of the stakeholders and institutionalized through rate-of-return regulations. Why bother with conserving energy when prices were constantly falling? Many utilities offered — and some still do — declining block tariffs: the more you use, the lower the per unit costs.[6]

There are also many institutional and regulatory reasons. For example, in many parts of the world, utilities have mild to strong financial incentives to sell more — rather than less — all else being equal.[7]

Partially as a result of these institutional and regulatory policies, utilities in nearly all parts of the world, with a few exceptions, are entirely or mostly focused on selling energy or kWhrs, instead of services that are ultimately desired and valued by consumers.[8] The distinction between *energy* and *energy services*, of course, is critical to any discussion of energy efficiency.

Even though Lovins was not the first to focus on the virtues of energy efficiency or the importance of *energy services* — the fact that what consumers really want and need is *the cold beer and the hot shower*, using his colorful metaphor — he must be credited with a bold vision at a time in the late 1970s and early 1980s when the power industry was overwhelmingly supply-side focused and the demand-side and energy efficiency were rarely mentioned or seriously considered in resource planning process. In those days, the power industry's mission was to build as much capacity as was remotely necessary to meet the demand — which was often treated as a "given."

The challenges facing the early promoters of energy efficiency continues to this day partly because energy efficiency or negawatts run against the electric utility grain — long-held beliefs and established regulatory and

---

5. The famous quote is often attributed to Lewis Strauss, the Chairman of the U.S. Atomic Energy Commission, who predicted that "our children will enjoy in their homes electrical energy too cheap to meter." He was, of course, talking about the promise of cheap nuclear energy touted as the ultimate answer to man's insatiable demand for electricity in the context of President Dwight Eisenhower's Atoms for Peace initiative, which he announced at the United Nations General Assembly in December 1953.

6. The reverse is now the case in California, for example, where rates rise significantly at higher consumption levels, intended to discourage heavy use.

7. This is further explained in Chapter 6.

8. Thomas Edison, the inventor of the fabulously inefficient incandescent light bulb and among the industry's pioneers, famously wanted to sell energy services, not kWhrs. His vision was that consumers would pay for light or light bulbs not kWhrs consumed.

institutional protocols that have gone mostly unchallenged since the formative days of the industry.

The challenge was, and still remains, to convince a hostile industry and occasionally skeptical regulators that the customers, the society and the environment *can* be better off if the industry's prime focus were to change from selling volumetric energy, be it kWhrs or therms – or, for that matter, gallons of petrol – to serving customers' energy service needs at the least cost, which often also means least amount of environmental costs.

Another fundamental issue putting energy efficiency at a distinct disadvantage under current regulatory regimes and rate-of-return paradigm has to do with how energy efficiency is currently measured and rewarded. When a kWhr is consumed, it is easy to measure and record. Moreover, in this case, someone is able to invoice for it and get paid. The bigger the number of kWhrs, the bigger the invoice, and the happier the utility – not necessarily the consumer.

In contrast, it is complicated to measure and record when a kWhr is conserved or *not* consumed. All manner of convoluted questions arise as to why the kWhr was *not* used, why was it conserved, who caused, encouraged, assisted, or prompted the conservation to take place, was it the result of a behavior modification, an investment in more efficient technology or appliance, or did the consumer actually settle for a lower level of service, suffer some discomfort, or sacrifice convenience to achieve it?

As a number of chapters in this book explain, the measurement and reward system for megawatts and negawatts are vastly asymmetric – and all else being equal – selling a kWhr is far more rewarding than not. Which leads to the point that, with the exception of car mileage standards, there are few programs encouraging oil companies to sell nega-gallons. In the case of gasoline, retail prices are supposed to provide adequate incentives to consumers to conserve, to buy smaller and more efficient cars, to car pool, use mass transit, to walk or bicycle, or avoid driving.[9]

The energy efficiency battle, which is still being waged in different parts of the world with various degrees of intensity, involves changing the market rules and mechanisms that have historically encouraged the energy industry to invest in infrastructure to meet customers' growing demand. This is still the norm in many parts of the world.

Until and unless the paradigm is changed to remove the incentives for stakeholders to sell more kWhrs, energy efficiency will face an uphill battle. Similarly, until a way or ways can be found to cause the industry to focus on consumers' energy service needs – rather than selling kWhrs – megawatts will be favored over negawatts.

---

9. Gasoline prices, as is broadly recognized, do not fully capture all externalities associated with driving, including pollution and congestion. Moreover, in many cases, gasoline prices are subsidized, further eroding the strength of the price signal.

3 ORGANIZATION OF THE BOOK AND CHAPTER SUMMARIES

This book, which consists of a collection of contributions from a number of scholars, experts, and practitioners from around the world, is mostly[10] focused on the demand or consumption side of the electricity equation. The book's overarching objective is to examine approaches that can lead to more efficient utilization of electricity, resulting in lower future demand growth, ultimately bringing an end to demand growth when and where it makes sense, is feasible, and economically justified. The book is organized into four parts:

Part I: End of Demand Growth is within Reach sets the book's context by presenting the potential scale and scope of energy efficiency opportunities and the constructive role that it can play in modifying and defining our energy future.

In Chapter 1, **Will Energy Efficiency make a Difference?**, **Fereidoon Sioshansi**, with contributions by **Ahmad Faruqui** and **Gregory Wikler**, sets the context for the book. The chapter makes references to material in the book while suggesting that the answer to the question posed by the chapter's title is a definite yes. The how much and how will depend on choices we make as individuals and societies and by decisions, policies, standards, enabling technologies, and prices selected by regulators and policymakers.

In Chapter 2, **Utility Energy Efficiency Programs: Lessons from the Past, Opportunities for the Future**, **Steven Nadel** describes how utilities have been offering energy efficiency programs for more than 30 years. He summarizes these developments, focusing in particular on recent and projected trends.

The author's main contribution is to highlight that in recent years programs have expanded in terms of the number of utilities offering programs and many utilities are achieving growing savings each year. In fact, some utilities are already saving more each year than underlying load growth, resulting in load decline, not growth. Furthermore, the number of utilities targeting and hitting this milestone is increasing.

The chapter concludes that these levels of savings can likely be sustained in the future, but in order to do so strong regulatory support is needed and program approaches need to continue evolving and improving.

In Chapter 3, **A Global Perspective on the Long-term Impact of Increased Energy Efficiency**, **Paul Nillesen**, **Robert Haffner**, and **Fatih Ozbugday** provide an overview of the potential contribution of increased energy efficiency to a more sustainable energy system using empirical evidence from a number of countries around the world, and alternative future scenarios.

---

10. The only exception is Chapter 15.

The authors provide an overview of the potential scope of energy efficiency, the targets that have been set, and an economic analysis of the costs and benefits.

The chapter's main conclusion is that energy efficiency offers substantial potential from a welfare perspective and a significant contribution to the sustainability targets. However, there remain implementation challenges in making savings stick and locking in the benefits over the longer term.

In Chapter 4, **Carpe Diem – Why Retail Electricity Pricing Must Change Now**, **Allan Schurr** and **Steven Hauser** envision a future where the industry's time-honored sales growth paradigm can no longer be assumed in many markets due to expected energy efficiency effects, self-generation, and structural changes in the composition of the economy.

The authors examine the ramifications of such a scenario for a mostly fixed-cost industry with a rising cost structure for labor, materials, and environmental compliance. Any erosion in sales growth will place new pressure on prices, and on managers and regulators to mitigate costs, which must be spread over a shrinking volume of sales.

The chapter's main conclusion is to point to potential reductions in sales growth, which will vary geographically, but nonetheless suggest a consistent pattern in most of the OECD countries with significant implications for the price structure and viability of this important industry.

In Chapter 5, **Is There an Energy Efficiency Gap?**, **Hunt Allcott** and **Michael Greenstone** examine the extensive literature on the scope of *energy efficiency gap*, namely the difference between cost-effective energy efficiency *potential* and what is actually captured in market. The sheer existence of this gap, reported to be significant in numerous prior studies, and its persistence, appears paradoxical, to say the least.

The authors analyze the empirical evidence on the magnitude of profitable yet unexploited energy efficiency investments and the reasons that cause consumers and firms *not* to fully exploit them. The analysis leads them to conclude that the claims of a massive energy efficiency gap may be overrated considering the net present value of savings and other unmeasured costs and benefits.

The chapter's main conclusion is that the magnitude of the energy efficiency gap may be much smaller than many engineering-accounting studies suggest. These problems notwithstanding, there are ample opportunities for policy-relevant research to estimate the returns to energy efficiency investments and their welfare effects.

Part II: The – Frustratingly Slow – Evolution of Energy Efficiency, provides historical perspective on why it has taken so long to get traction on energy efficiency, particularly highlighting why selling energy efficiency has been an uphill battle despite well-articulated and documented understanding of the causes.

In Chapter 6, **Making Cost-Effective Energy Efficiency Fit Utility Business Models: Why has it Taken So Long?, Ralph Cavanagh** explains what is needed to align utilities' and customers' financial interests in securing cost-effective energy savings, and why the necessary reforms have been slow to emerge.

The author's main contribution is to highlight the multiple conflicts of interest that traditional price regulation unintentionally creates for utilities as energy efficiency partners and promoters, and to identify ways of overcoming obstacles to proven reforms. Utilities want and need both to break long-standing linkages between retail sales volumes and financial health, and to see energy efficiency success as a potential earnings driver. Regulators have been accumulating much useful experience with solutions, but significant resistance remains among some stakeholder groups.

The chapter's conclusions are that utility business models are already changing to accommodate expanded energy efficiency investment and results, that stakeholder concerns are far from insuperable, and that the ultimate outcomes will be well worth the wait.

In Chapter 7, **The Evolution of Demand-Side Management in the United States, Frank Felder** examines how DSM arouse from the oil price shocks of the 1970s combined with stricter environmental regulations, overcapacity, and cost over-runs of nuclear power plants.

By exploring the history and evolution of DSM, the author highlights that even with DSM's history of 40 years, there are substantial open questions regarding the need for DSM programs, how they should be structured, and how they coexist with electricity markets.

The author concludes that DSM proponent must directly engage these questions and issues if DSM is to achieve its potential as a tool to address critical energy, environmental and economic needs.

In Chapter 8, **China: Energy Efficiency Where it *Really* Matters, Mark Levine, Nan Zhou, David Fridley, Lynn Price** and **Nina Zhen** examine to what extent China's rapid economic growth is linked to electricity demand growth, specifically focusing on credible policies that can delink the two.

The authors examine two scenarios of future growth — an *expected* future and another that assumes *strengthened energy efficiency policies* — using an end-use energy model developed at the Lawrence Berkeley National Laboratory (LBNL). The model considers sales and saturation of energy-using equipment, efficiency standards, usage, turnover rates, size, and other factors, which are updated every decade. The inputs are based on historical experiences of developed countries and assumptions about technological change and human preferences.

The chapter's main contribution is to suggest that a *plateauing* of energy use somewhere in the 2025—2030 period combined with a reduction of growth in electricity demand is possible, countering the *prevailing wisdom*. This is supported by sensitivity analyses, which demonstrate the significant role that aggressive energy efficiency policies can play in lowering the growth rate of energy and electricity demand and related carbon emissions.

In Chapter 9, **Rapid Growth at What Cost? Impact of Energy Efficiency Policies in Developing Economies, Youngho Chang** and **Yanfei Li** give an overview of electricity demand growth under different scenarios in several developing Southeast Asian countries and examine how alternative energy efficiency policies can make a difference in the outcome.

The authors contrast demand projections under a business-as-usual scenario vs. alternative scenarios where demand growth is managed and reduced due to policies such as appliance efficiency standards, building codes, pricing, and mandatory quotas.

The chapter's main conclusion is to show that without aggressive policies, electricity demand in the developing countries will grow for the foreseeable future. With aggressive policies, however, significant reductions can be achieved often accompanied by cost savings. The key question is which policies would be more effective in managing electricity demand growth and how they can best be implemented.

Part III: Case Studies of Low-Energy Communities and Projects, presents examples of planned and/or existing efforts to achieve low energy or zero net energy status with empirical evidence that it is technically feasible and economically viable.

In Chapter 10, **The Prospect of Zero Net Energy Buildings in the United States, Nicholas Rajkovich, William Miller,** and **Roland Risser** debate whether efforts to achieve zero net energy represent a sustainable response to climate change or are a diversion from other necessary efforts to reduce greenhouse gas emissions.

The authors' main contribution is to show how zero net energy buildings, as currently conceived, may or may not help to propel the market forward. As one vision for achieving carbon neutrality, zero net energy buildings are in the process of being tested and may ultimately be rejected in favor of alternative approaches to a low-carbon future.

The chapter's conclusion is that existing regulatory frameworks may impede progress to a lower carbon future and that the utility rate case process needs to be altered to reflect a quickly moving and uncertain shift to zero electrical demand growth.

In Chapter 11, **What If this *Actually* Works? Implementing California's Zero Net Energy Goals, Anna LaRue, Noelle Cole,** and **Peter Turnbull** examine the bold zero net energy (ZNE) regulatory goals of California and its implications for the relationship between buildings and the electric grid.

The authors explain the current status of ZNE in California, including the effort to drive California's Title 24 building energy codes towards ZNE. The authors discuss how, as government agencies and utilities push towards those goals, stakeholders are also wrestling with what the ZNE definition should ultimately be for California and at what scale the ZNE goals should be implemented.

The chapter concludes by examining how the relationship between utilities and customers will change in a future world of high performance buildings, distributed generation, and ZNE performance, where it is currently unclear how customers will pay for use of the grid and how decisions about generation infrastructure will be made.

In Chapter 12, **Zero Net Energy At A Community Scale: UC Davis West Village, Stephen Wheeler** and **Robert Segar** show how the ZNE concept has been implemented on a large scale at the U.C. Davis West Village, a new ecological neighborhood for 4,200 students, faculty, and staff of the University of California, Davis.

The authors describe the first phase of the project, inaugurated in 2011, which aims for ZNE status using highly energy efficient construction, passive solar design, and photovoltaic arrays on roofs and parking canopies. The technology required for the project was readily available in the late 2000s, although it took creative partnerships to develop and implement the concept. Several additional features are still in the development stage.

This chapter's main conclusion is to show that ZNE status can be achieved through collaborative public/private partnership with no increased costs to the residents. This advanced eco-district bears similarities to some European examples such as Hammarby in Sweden and Vauban in Germany, and points the way toward a new generation of ZNE neighborhoods in North America.

In Chapter 13, **Crouching Demand, Hidden Peaks: What's Driving Electricity Consumption in Sydney?, Robert Smith** explores how electricity consumption and peak demand for residential customers in Sydney and across Australia has plateaued since 2007. He describes this departure from the long-term post-war trend of energy and peak demand growth and unpacks the possible causes, focusing on the role of energy efficiency versus price.

Looking at a period of volatile weather, unstable economic conditions, soaring electricity prices, erratic peak demand growth, and flat or falling energy consumption, the author sifts out the key underlying causes and outlines the implication for the future. The review highlights how low-profile regulations and minimum energy performance standards (MEPS) have delivered lasting savings compared to impacts of higher profile, short-term, and expensive one-off programs.

The chapter concludes that the forthcoming end of large price increases will test the resilience of recent energy savings and how, without corresponding falls in peak demand, it will be difficult to translate the benefits of energy savings into long-term savings in customer bills.

In Chapter 14, **From Consumer to *Prosumer*: Netherland's PowerMatchingCity Shows the Way, Stefanie Kesting** and **Frits Bliek** describe a smart energy pilot project in the Netherlands where dynamic pricing is employed to optimize energy use at customer level.

The authors explain an experiment where the end users in a multi-goal optimization "game" can produce and/or consume energy, hence the term "prosumers." The game allows prosumers to decide between using energy or selling it to their neighbors or to a local market.

The chapter's main contribution is to show how a decentralized energy future can look like in reality. It also confirms that people act in a smarter, and more energy efficient and sustainable way when they get the instruments to do so.

In Chapter 15, **Back to Basics: Enhancing Efficiency in the Generation and Delivery of Electricity, Clark Gellings** points out that approximately 11 percent of electricity is consumed in the production and delivery process before it reaches consumers in the United States. The figure, which includes auxiliary loads as well as losses in transmission and distribution network, varies from country to country but is significant given the industry's scale, accounting for nearly 50 percent of primary energy used in many advanced economies.

The author's main contribution is to highlight the potential for improving the electrical efficiency of power production and delivery by examining the electricity portion of the industry's long value chain, from power production to delivered electricity at the customer's premises.

The chapter's conclusions are to examine opportunities for reducing these uses through applications of appropriate technologies.

In Chapter 16, **Smarter Demand Response in RTO Markets: The Evolution Toward Price Responsive Demand in PJM, Stuart Bresler, Paul Centolella, Susan Covino**, and **Paul Sotkiewicz** describe how PJM has succeeded in creating a new option for market participation by load reduction capability in the largest organized U.S. electricity market.

The authors examine how the convergence of advanced metering infrastructure (AMI) deployments and dynamic retail rates combined with the growth of demand response resources and the rapid emergence of smart grid technology has required PJM to create price responsive demand, or PRD. The chapter lays out the mechanics of how PRD will integrate the wholesale and retail markets and describes how customers empowered by timely and detailed usage information will be able to make decisions about when and how to use electricity in response to market prices.

The chapter's main contribution is to demonstrate the impact of PRD in reconnecting the wholesale and retail markets in new and valuable ways.

Part IV: Opportunities and Remaining Obstacles, discusses a number of challenges and obstacles that remain to be addressed.

In Chapter 17, **Shifting Demand: From the Economic Imperative of Energy Efficiency to Business Models that Engage and Empower Consumers, John Laitner, Matthew McDonnell**, and **Heidi Keller** examine the need for developing new utility business models to drive critical gains in energy efficiency.

The authors provide an overview of the economic imperative of energy efficiency and suggest that to achieve the requisite decrease in the cost of energy services while simultaneously increasing the amount of *useful energy* consumption, utilities must shift the business model from that of a static deliverer of electrons to one of a dynamic provider of energy services.

The chapter's main conclusion is that engaging and empowering consumers through the development of new business models is essential to move from anemic levels of *inefficiency* toward levels that enable a more robust economy in the future.

In Chapter 18, **What Comes After the Low-Hanging Fruit? Glenn Platt, Daniel Rowe** and **Josh Wall** point out that whilst at the moment there remains huge opportunity to realize significant savings through existing efficiency technologies such as compact fluorescent lights, insulation, more efficient appliances, and so on, there remains a significant question of what to do after these options have saturated the market.

The authors introduce a number of technologies that so far have received little attention with significant future potential including solar cooling systems, energy recommender technologies, new types of heating, ventilation and air-conditioning control that use fundamentally different set points to traditional approaches.

The chapter's main conclusion is that these promising technologies are being trialed in major projects around the world and will soon be ready for significant uptake including case studies that demonstrate the potential savings and cost benefits while examining technical and regulatory hurdles.

In Chapter 19, **Energy Convergence: Integrating Increased Efficiency with Increased Penetration of Renewable Generation, Ren Anderson, Steven Hauser**, and **Dave Mooney** describe future trends in energy systems integration and the complementary nature of efficiency and renewable energy resources while also accounting for the disruptive impacts of emerging innovations on residential building energy use.

The authors' main contribution is to highlight the key barriers that limit successful capture of the value of distributed energy resources and the development of integrated systems modeling tools that identify least cost pathways to maximize total energy savings.

The chapter's conclusion is that systems interactions that seem to limit achievement of cost effective energy savings can also be used to drive energy transformation provided that market strategies focus on total system-level energy savings rather than overly constrained optimization of individual sub-systems and components.

In Chapter 20, **Energy Efficiency Finance: A Silver Bullet Amid the Buckshot? Theodore Hesser** examines the progress being made on avoided cost financing mechanisms and explores the necessary steps to assimilating energy efficiency finance into the mainstream of the capital markets.

The author surveys the activity of all existing financing models and contrasts this expenditure with the raw potential of each financing mechanism. The chapter points out that despite years of progress, energy efficiency finance has only achieved 1 percent of its technically achievable potential.

The chapter concludes that successful energy efficiency finance mechanisms could dramatically reduce annual energy consumption in the United States and catalyze a feedback loop whereby rate increases further incentivize energy efficiency, which further begets rate increases, and so on until new utility business models emerge.

In Chapter 21, **The Holy Grail: Customer Response to Energy Information, Chris King** and **Jessica Stromback** describe how access to detailed energy usage data, especially when integrated with other data sources, enables utilities and others to deliver actionable information needed for consumers to achieve energy efficiency's true potential. They describe a vision for "Intelligent Efficiency" and the associated information infrastructure supported by the empirical evidence of a wide range of feedback and other studies in delivering quantifiable savings.

The authors' main contribution is to synthesize the range of available data program offerings into a structured framework comprising policy, technology, data services, security, privacy, and standards. Beyond that, the authors include a meta-analysis of the extensive available literature on pilot program results and show the link between the theoretical framework and real-world results.

The chapter concludes that achieving the maximum achievable energy efficiency potential requires empowering consumers with a "triad" of information, pricing options, and automation, while focusing on the data element of that triad. This further requires the right data collection and access infrastructure. Their contribution is timely as policymakers are adopting "intelligent efficiency" policies to help achieve efficiency targets from a combination of user behavior adaptation, investments in efficiency and distributed renewables, and automated usage reduction.

In Chapter 22, **Trading in Energy Efficiency – A Market-Based Solution, or Just Another Market Failure? Iain MacGill, Stephen Healy,** and **Robert Passey** describe the underlying theory, rationale, and mixed performance to date of policy measures that establish a trading market in *energy savings*. Such approaches, going under names including White Certificates, Energy Efficiency Portfolio Standards, Energy Savings, and Energy Efficiency Certificate Trading, are receiving growing attention in Europe, the United States, and Australia.

The authors highlight the many challenges – technical, economic, and social – in implementing such *designer* markets that attempt to commodify energy savings including measurement and additionality as well as difficulties in the complex interaction between technologies and consumer behavior inherent in energy use and in the *financialization* of energy savings.

The chapter concludes that considerable care is required with such approaches to energy efficiency policy lest governments merely add yet another market failure to those already existing. More important, policy makers need to move beyond framing energy efficiency within conventional economic terms of market failure, and address the broader challenge of engaging energy users on their behavior including, for many, an ever-growing desire for energy services.

In Chapter 23, **The Ultimate Challenge: Getting Consumers Engaged in Energy Efficiency**, **Alex Laskey** and **Bruce Sayler** describe how consumers have been left out of the energy efficiency equation and how innovative utilities, like Connexus, are engaging and motivating customers.

The authors' main contribution is to highlight that many energy efficiency advocates have focused on pushing new technologies to engage customers in using energy more judiciously and sparingly. Although such initiatives are an essential piece of the puzzle, there is a critical missing ingredient: how to engage consumers. The authors show how Connexus, working with Opower, has succeeded to engage their customers and is generating real energy savings.

The chapter's conclusions are that to deliver real energy efficiency savings, consumers need to be informed of their energy use and the information on how they can save energy and money.

In the book's epilogue, **How Do We Get There From Here? Fereidoon Sioshansi** sums up what needs to take place for the vision espoused by this book's contributors to become a reality.

Part I

End of Demand Growth is within Reach

Chapter 1

Will Energy Efficiency make a Difference?

Fereidoon P. Sioshansi
Menlo Energy Economics. With contributions from Ahmad Faruqui, The Brattle Group, and, Gregory Wikler, EnerNOC

1 INTRODUCTION

Interest in energy efficiency got a significant boost following the 1973 Arab oil embargo when rich economies collectively got a wake-up call on just how vulnerable they were to the whims of the oil exporting countries or potential disruptions in the flow of oil — whether through the Strait of Hormuz or Malacca — or broader conflicts in the Middle East and beyond. The Iranian Revolution of 1979 provided a second wake-up call.[1] Surprisingly, the vulnerability persists, despite four decades of efforts to address it.[2]

The initial response of the oil-importing countries was entirely supply-side focused. The knee-jerk reaction was to decide how best to ration limited supplies of oil among the needy and who should stockpile how much to cushion a sudden disruption of future supplies.[3] A minority decided to address the demand, rather than the supply-side of the equation.[4] President Jimmy Carter, to his credit, gave a famous TV address from the White House, sitting by the fireplace wearing a sweater and encouraging the American public to turn down their thermostats. He called energy conservation "the moral equivalent to war.[5]"

---

1. Felder provides further details and a historical perspective on the evolution of energy efficiency since 1973.
2. Among the first attempts by OECD countries was to form the International Energy Agency (IEA) whose original mission was to stockpile oil and manage shortages in case of major disruptions — a rather narrow and strictly supply-focused approach.
3. The United States and other major oil-consuming countries established strategic petroleum reserves — typically sufficient to supply 90 days of consumption in case of future emergencies.
4. California established the California Energy Commission (CEC) an agency to develop future energy policy, with a major focus on energy efficiency which endures to this day.
5. See Felder in this volume.

Energy Efficiency. DOI: http://dx.doi.org/10.1016/B978-0-12-397879-0.00001-3

Mr. Carter was ridiculed at the time by many who saw the problem as not having sufficient domestic supplies of oil or sufficient stockpiles to withstand through a prolonged embargo. Others identified the need for a bigger Navy to keep the bad guys at bay and keep oil tankers flowing.[6]

Carter was also ridiculed for wearing a sweater and encouraging his fellow citizens to conserve energy. Conservation, after all, has a *negative* connotation — it suggests deprivation, sacrifice, hardship, doing without, lowering one's standard of comfort and so on. Decades later, during another debate about U.S. energy policy, then Vice President Dick Cheney said, "Conservation may be considered a personal virtue, but should not be part of a country's energy policy."

This book, of course, is not about oil or broad energy policy issues, but the preceding discussion is important because:

- First, most observers agree, that the 1973 Arab oil embargo was a game changer because, for the first time, it signified the importance of energy supplies and energy security on a global scale;
- Second, the ensuing debate about how best to respond to energy insecurity, supply vulnerabilities and rising prices — also for the first time — raised the significance of energy efficiency and/or conservation;
- Third, it led to new awareness of the importance of the demand-side, not just *wasting* less, but using energy more efficiently, more frugally, and more wisely — topics further explored in this book.

The electric power sector, a prominent consumer of energy,[7] has been debating the virtues of energy efficiency ever since.[8] As further explained in Box 1.1, the issues have evolved and the debate about energy efficiency has matured over the years. Likewise a lot has been learned from what works and what does not, yet many perplexing questions still remain.

Box 1.1 The Evolution of Energy Efficiency

Efficiency pessimists contend that there is little potential for further improvements in energy efficiency, since all the low-hanging fruit has been harvested. Ergo, the solution to meeting the nation's future energy needs in a carbon-constrained future, is to build more power plants

---

6. According to some estimates, as much as a quarter of Pentagon's roughly $700 billion annual budget is directly aimed at such measures, which does not directly show up at the price of oil at the gas pump.

7. The chapter by Gellings examines opportunities for improving the efficiency with which electricity is generated from primary fuels and delivered to consumers.

8. This book uses the term energy efficiency — rather than energy conservation — since the focus is not on deprivation or lower living standards but maintaining or even increasing comfort and services through more judicial and efficient energy use.

(preferably those that don't burn coal), transmission lines, and distribution systems.

Efficiency optimists, on the other hand, contend that energy efficiency is essentially an inexhaustible well and we have a long way to go before the bottom is reached. Their viewpoint suggests that enhancements in energy efficiency may eliminate the need to make investments in the power supply system, except for routine maintenance and upgrades.

The truth is probably somewhere in between. The question of how much energy efficiency is available continues to come up, since we in the United States have been encouraging energy efficiency in one form or another ever since the first oil shock of 1973. The first wave of programs involved moral exhortations — as in the famous call to put on a sweater by President Jimmy Carter[1] — information dissemination, and energy audits. The actors were government agencies and community organizations and the slogan was "energy conservation." Federal legislation was passed in 1978 to give an impetus to conservation. National efforts at cutting back use were redoubled when the second oil shock hit in 1979.

The second wave was led by the utilities and gifted the somewhat clunky term demand-side management (DSM) to future generations. The focus of DSM was on improving energy efficiency and not on asking consumers to make do with less (i.e., energy conservation). Conservation was frowned upon because it meant that consumers would have to change their behavior, which might be perceived as an unwelcome intrusion into their lives and even considered un-American by some. Incentives in the form of rebates and low-interest financing were used to encourage consumers to buy more efficient equipment and buildings. Utility spending on DSM programs peaked in 1993 as the industry prepared for restructuring, which arrived in the mid-to-late 1990s.

The energy crisis that plagued California's energy markets in the years 2000 and 2001 set in motion a third wave of programs that revolved around the concept of demand response. Customers would be provided incentives either through dynamic pricing or cash payments to curtail their usage during times when the power system was stressed, typically because of a shortage of capacity caused by natural conditions such as prolonged heat storms and droughts. Some of these programs could be instituted with existing meters while others required the deployment of smart meters. As of this writing, some 22 million smart meters have been deployed in the United States, and there is an evolving consensus that the number will rise three-fold in the next five years. Demand response programs represented a major change in the industry's conception of customers as a resource.

Changing consumer behavior, along with behavioral economics, is now in vogue and appealing to an increasingly widening circle of folks who are active in the energy domain. A fourth wave of programs, sometimes called integrated DSM (iDSM), is upon us.

This new wave of iDSM programs include those that inform consumers about how their energy spending compares with a group of peers and helps them to establish targets for optimizing use. These types of consumer-focused programs are finding widespread acceptance, saving one or two percent of energy consumption simply by inducing consumers to change behavior. Consumer-focused programs may save even more by enhancing consumers' awareness about where their energy dollar goes and by directing consumers toward rebates and low-interest financing that may be available from utility DSM programs.

As a result of iDSM programs, Energy Star labels on appliances are almost ubiquitous in "big box" stores, steering consumers toward efficient purchases. Zero-energy homes are being constructed that produce enough energy through renewable sources to meet their own needs.

Aggressive codes and standards are making an impact on the building construction and appliance manufacturing industries, especially in states such as California, which have their own supplemental codes and standards that push the envelope beyond federal requirements.

Nowhere is this transformation in consumer buying habits more visible than in the imminent phase out of incandescent bulbs, called upon by the Energy Independence and Security Act of 2007. Compact fluorescent lamps are expected to get a substantial boost from this legislation, but light-emitting diode (LED) lamps may lead to even greater savings. A recent issue of *Wired* magazine features them on the cover.[2] They are expensive, but prices are expected to come down as scale grows.

And this fourth wave, iDSM, is not confined to the residential sector. It is not uncommon to see full page ads in the mainstream media by companies such as Johnson Controls and Schneider Electric touting projects in which they helped large commercial and industrial facilities reduce their energy bills by up to 30 percent.

Finally, it is important to note that electric rates are now being redesigned to incentivize efficient energy use, with inclining block rates and time-varying rates being two concepts that are receiving increasing interest by utilities and policymakers. So the fourth wave, more comprehensive than the first three, is built around five policy instruments: (a) information, (b) codes and standards, (c) technological change, (d) rebates and low-interest financing, and (e) rate design.

Ahmad Faruqui, The Brattle Group
and **Gregory Wikler,** EnerNOC

1. Felder describes Carter's famous fireside speech and also talks about the 5-waves of DSM in his chapter.
2. http://www.wired.com/magazine/2011/08/ff_lightbulbs/

This chapter is rather narrowly focused on answering the question, "will energy efficiency make a difference?" It is organized into three sections: Section 2 provides a brief historical perspective on prior studies on the scope of energy efficiency potential and the debate on its cost-effectiveness with references to relevant chapters in the book. Section 3 provides a range of estimates offered by several prominent recent studies and surveys that offer alternative scenarios or visions of the future scope of energy efficiency, also making references to relevant chapters. Section 4 is focused on a number of measures and policies, which taken together, can make a difference in the future course of demand growth. The chapter's conclusions are summarized in section 5.

Two important caveats are in order at the outset:

- First, this chapter, and the book, are mostly focused on energy use in the electric power sector — which is a significant energy user in most developed and developing countries;
- Second, the book is mostly but *not* exclusively focused on developing countries.[9] It is rather ironic that the rich countries, where per capita energy and electricity consumption is already at rather high levels are trying, for different reasons and to different degrees, to reduce their energy footprint, mostly by reducing unnecessary and wasteful consumption. Developing countries, where current per capita consumption is low, can leapfrog by skipping the historical trajectory of rapid and wasteful energy consumption patterns followed by more rational energy use. This, however, may prove difficult in practice.

Box 1.2 Eliminating Inefficiency: Markets or Mandates?

Among the main obstacles to improving energy efficiency is the upfront investment costs. It typically costs more to build a more efficient home, office, or factory or buy a more efficient appliance or light bulb. But once the investment is made, there are recurring savings in lower operating costs as well as improved performance and higher comfort levels.

The tradeoff is not different than buying a more efficient car that gets superior mileage — more than recovering the upfront investment over its life. A trucking company or airline would not think twice if given a choice for a truck or plane that gets better mileage. Lower fuel costs over the long life and heavy use of the truck or the plane would be considered wise investments. Why shouldn't the same math apply to a new house, new boiler, new refrigerator, new air conditioner, or a new LED[1] light fixture? That is a perplexing question begging an explanation, and goes to the core of the debate about the existence and scope of the so-called *energy efficiency gap*, further explored by a number of authors in this volume.

---

9. At least two chapters in the book are devoted to developing countries.

LEDs eventually pay for themselves, if you wait long enough

Cost and performance comparison of alterative lighting technologies replacing a 60 W, 800 lumen incandescent bulb

	Avg. Cost $	Life – Hrs	Energy used –Watts	Electricity Cost[*], $
Traditional incandescent bulb	0.25–0.50	750–1,000	60	4.80
Halogen incandescent bulb	1.50–2.00	1000–3,000	43	3.50
Compact fluorescent bulb (CFL)	2–4.00	8,000–10,000	13–14	1.20
Light emitting diode (LED)	20–55	25,000	13–14	1.00

[*]Assume 2 Hrs/day & 11 cents/kWhr avg. electricity cost. Actual cost varies from place to place and depends on prevailing tariffs. Savings will be considerably higher for light bulbs in continuous operation in places such as hotel or commercial building lobbies, parking lots, or street lighting.
Source: The Wall Street Journal (June 1, 2011) based on data from GE, Philips, Osram, Home Depot & DOE

Opinions vary. On the one extreme are those who argue that markets, as imperfect as they may be, are generally functioning and there is no compelling justification for government or regulatory intervention. According to this line of thinking, the energy efficiency gap, if it exists at all, is not significant and should not be artificially reduced or eliminated. Those who adhere to such logic would probably *not* support mandatory car efficiency standards either, based on the same line of reasoning.

At the other extreme are those who argue that markets – and consumers – if left on their own, would make suboptimal decisions with adverse long-term societal consequences. Everyone suffers when individuals make the wrong decisions or choices and that provides sufficient justification in regulatory intervention. Among the reasons often cited for the wrong individual decisions are imperfect information and abnormally high discount rates implied by consumers, builders, and investors. A consumer shopping for appliances may not be able – or sufficiently motivated – to compare upfront costs vs. lower life-long operating costs, or may not be willing or able to make the extra initial investment.[2]

According to this line of thinking, consumers – and the society – would benefit from mandatory appliance labeling, which would address the first issue. Extending the argument, there is justification in market intervention to impose minimum energy efficiency standards for appliances and buildings. Going one step further, some advocates of regulatory intervention would favor outright banning of certain inefficient appliances. Notoriously inefficient incandescent light bulbs, for example, have been banned in Europe, Australia, and the United States based on the argument that everyone is better off when the inefficient choice is simply eliminated from the market.

1. LED refers to highly efficient and long-lasting light emitting diode technology.
2. Chapter by Hesser discusses deficiencies in financing energy efficiency investments.

2 MIND THE GAP, THE ENERGY EFFICIENCY GAP[10]

Energy and electricity, of course, are valued for a range of services, products and comforts they offer, not in and of themselves. This, the notion of *energy services*, is well-understood and is helpful in any discussion of energy efficiency. Ordinary consumers, factories, farms, office and commercial buildings use energy to deliver or derive services – heating, cooling, lighting, motor power, etc. Amory Lovins' famous saying that people simply and mostly want "A hot shower and a cold beer" captures it all.

The question of how much or how little energy is needed to provide the desired services – keeping the beer cold and the shower hot – however, becomes rather circular. For example, if sufficient natural lighting can be provided – at least for the hours when the sun is shining – without resorting to artificial lights, then electricity consumption can be reduced without affecting comfort or productivity. But to make better use of natural lighting, the building's orientation, location and design of windows, shading louvers, and design of working areas must be modified. Some of these modifications may require extra investments and/or adjustments in consumer/occupant behavior.

As long as the lower energy costs, which tend to be recurring, exceed the extra costs, which typically involve a one time, up-front investment, then one can conclude that the measure is economic or cost-justified (Box 1.2). The fundamentals are straightforward at first glance, but can become convoluted in practice.

As further described in the chapter by Allcott and Greenstone, who review an extensive number of studies on the scope of *energy efficiency gap* – which they define as the difference between what may be the socially desirable level of energy efficiency investment to what is actually observed – there are different ways and means of measuring the costs and the benefits,

10. Allcott and Greenstone's chapter titled, *Is There an Energy Efficiency Gap?*, covers similar ground.

which explains why the extent of the energy efficiency gap remains controversial.[11]

In the electric power sector, which is the focus of this volume, a number of complicating factors make achieving efficient delivery of energy services an especially problematic challenge for two critical and a number of collateral reasons:

- The most important of the reasons is that the prevailing regulations, accounting practices, and electricity tariffs motivate the providers of services to *sell more electricity*, rather then less, all else being equal.[12]
- Equally important, the providers of services delivered by electric energy traditionally defined the electric meter not only as their cash register, but as the demarcation line, the end of the industry's long value chain. According to this long-held belief, the business figuratively and literally ends at the customer meter. Once electrons are delivered to the customers' meter – and measured – it is up to the customer to decide *how much* to use, *how* to use it, and in whatever form or fashion they please.[13]

These two issues, as well as a number of other factors described below, are critical obstacles to delivering *efficient energy services* at least cost – an underlying theme in many chapters of this book. The first favors *volumetric consumption* – the more the merrier. The second limits the industry's focus to delivering kWhs to the meter as opposed to providing "hot shower and cold beer" at least cost – which is what consumers ultimately want and value.

That is not all. A number of other factors, often mentioned in the literature when addressing the energy efficiency gap, also get in the way of delivering low-cost energy services as opposed to selling more kWhrs. A few major ones are mentioned, not necessarily in order of significance:

No fuel gauge, no tank, no pump – Motorists are routinely reminded by the fuel gauge on how much gas remains in the tank. A warning light typically comes on when the tank is nearly empty. When they stop at

---

11. Nadel, Felder, Laitner et al, and others also examine these issues.
12. Cavanagh elegantly described this problem.
13. It can be argued that this is how all businesses essentially treat the demarcation point with consumers. Oil companies, who also have a very long value chain, for example, leave it to consumers to decide what to do with the gasoline once it is paid for at the pump. The major difference, however, is that in this case the consumer actually sees the per-unit cost and the volume purchased, whereas typical electricity consumer is clueless about how much is consumed for a given task and at what cost.

FIGURE 1.1   For electricity consumers there is no equivalent to a fuel gauge, fuel tank, and no pump Photo of driver getting refueled at gas pump.

gas station, they are confronted by a highly visible price signal and literally *experience* money flowing from their bank to their tank (Figure 1.1). It is a powerful and often painful signal. By contrast, the typical electricity consumer gets the illusion of being connected to a virtual bottomless tank of electricity that never runs out, no matter how much is used, when, or for how long.[14] For example, the consumer is literally clueless on how many kWhrs are consumed when the air conditioner is turned on or how much it costs to cool the building. The electric bill typically arrives with a long delay[15] and in a form/format that is virtually incomprehensible to the average consumer. A number of chapters in this book address the means and methods to address this critical deficiency.

Lack of a fuel gauge, a tank with limited capacity, and a fuel pump with a highly visible, frequent, timely, and painful, price signal is among the main obstacles to active consumer engagement in the electric power sector.

Lack of visible price signal – The absence of a gauge, a tank, and a pump, referring to the above metaphor – virtually eliminates or significantly reduces motivation on the part of consumers to control or modify electricity usage. It is a business maxim that if you cannot easily measure or monitor something, you cannot effectively control it. This is another major deficiency that remains elusive despite many attempts to rectify.

---

14. This so-called "illusion of plenty," which is equally applicable to other network resources such as water and natural gas, explains why it is difficult to convince consumers to conserve water during a drought – since the faucet never runs dry.

15. Until the recent wide-spread introduction of smart meters, in some countries like England, most residential consumers were billed quarterly, and many on *estimated* consumption, which was only reconciled/adjusted based on a singe annual meter reading.

Box 1.3 Engaging the Disengaged Consumer[1]

Among the most stunning recent examples of the *paradigm change* is the evolution of thinking at the Federal Energy Regulatory Commission (FERC), the closest thing we have in the United States for a federal energy regulator.

In a recent interview with *The New York Times* (November, 29 2010), in an article titled *Making the consumer an active participant in the grid*, FERC chairman Mr. Jon Wellinghoff said, "The energy future of the U.S. looks radically different from its past," partly because consumers will become "… active parts of the grid, providing energy via their own solar panels or wind turbines, a system called **distributed generation**; stabilizing the grid by adjusting demand through intelligent appliances or behavior modification, known as **demand response**; and storing energy for various grid tasks." Mr. Wellinghoff is not only supportive of such schemes but believes that "consumers should get paid to provide these services."

Wellinghoff's quotable quotes – Excerpted from *New York Times* (November, 29, 2010)

"I believe that for **markets to be competitive**, we need to have as many different types of resources in those markets as possible."

"We're doing what we can to the extent that we have jurisdiction to ensure that there are no barriers to **distributed generation** becoming part of wholesale markets."

"To the extent that you can put **demand response** in the system – that is, have consumers control their loads at times when the system is stressed – you can reduce substantially the amount of fossil fuel generators that are needed to relieve that stress."

"If a **battery** or a dishwasher or a water heater or an aluminum pot or a compressor in a Wal-Mart can respond on a microsecond basis, and it takes the generator a minute to respond, that faster response should be rewarded a higher payment because, in fact, it's providing a better service."

"We're reviewing the **economic benefits of storage** and how storage should be compensated for the various services it can provide to the grid."

While such ideas are not necessarily new or novel, coming from the FERC chairman, they get noticed. Moreover, FERC has taken the unusual step of actively promoting these ideas, not just through interviews and public pronouncements, but through published reports, studies, surveys as well as a number of *orders*, which in no uncertain terms obliges the regional transmission organizations (RTOs) and independent system operators (ISOs) to implement the concepts in practice. In the past couple of years alone, FERC has published several seminal studies documenting the substantial potential for demand response (DR) in the United States and has issued a number of orders that are prompting fundamental changes at organized U.S. electricity markets.

As a result of FERC's incessant prodding in the past several years, most organized wholesale market operators have developed various forms of DR programs, have incorporated demand-side bidding into their predominantly supply-focused auctions, and are broadly – although not always effectively or successfully – supportive of the growing significance of the demand side of the market.

It sounds simple and trivial today – but the idea of getting consumers to become active participants in the market is still novel to many in the industry and even more so to the average consumer who has been successfully trained to be a passive consumer.

1. Excerpted from Feb 2011 issue of *EEnergy Informer*.

Misaligned motivations – In many cases, the energy consumer is effectively *disengaged* from critical decisions on energy consuming devices and/or investments in the capital stock including buildings that they occupy. Moreover, in the case of many renters or occupants of commercial or office buildings, the motivations of the stakeholders may be *misaligned*. For example, the landlord may prefer a low-cost, inefficient air conditioner, less insulation, or an old refrigerator since the renter usually bears the higher energy costs.[16]

Disengaged – For over a century, consumers have been effectively trained to act as passive actors rather than being engaged or proactive in any form or fashion (Box 1.3). Rapid technological advances promise to change this.[17]

Pricing not cost reflective – Electricity prices are far from cost-reflective for most consumers most of the time. They tend to be flat for great majority of consumers, which means prices do not reflect significant variations in generation and delivery costs at different times (Box 1.4). They also tend to be set for broad groups or class of consumers, which means some groups get subsidies at the expense of others. They tend to be uniform as in postage-stamp pricing, which means cross-subsidies depending on customer location. They do not adequately capture different components of service costs – fuel, generation, transmission, and local network costs – which mean that literally no consumer actually pays anything resembling the actual costs. Additionally, many externality costs are not fully captured in prices.[18]

16. Consider the difference in attitude and behavior between a truck driver who owns and operates the vehicle vs. one who is merely paid on an hourly basis to drive it.

17. Advances in technology and communications not only offers opportunities for consumers to become active participants in the electricity markets, but also to become producers – hence the term *prosumers*, further described in chapter by Kesting and Bliek.

18. Similar problems, of course, afflict pricing for most products and services, to varying degrees.

Box 1.4 Universal Dynamic Pricing of Electricity

The free market has often enough been condemned as a snare and a delusion, but if indeed prices have failed to perform their function in the context of modern industrial society, it may not be because the free market will not work, but because it has not been effectively tried.[1]

Dynamic pricing incentivizes electricity customers to lower their usage during peak times, especially during the top 100 "critical" hours of the year, which may account for between eight and eighteen percent of annual peak demand. Lowering peak demand in those hours means avoiding capacity and energy costs associated with the installation and running of combustion turbines in the long run and lowering wholesale market prices in the short run.

Dynamic pricing encompasses many different pricing options from nearly instantaneous, hour-ahead pricing designs (often called real-time pricing or RTP) to simple time-of-use (TOU) pricing designs in which the time periods and prices are often fixed at least a year in advance. In between lies critical peak pricing (CPP), in which the prices during the top 60 to 100 hours are known ahead of time, but the time in which they will be called is only known on a day-ahead (and sometimes day-of) basis. A variant on CPP is called critical peak rebates (CPR), in which the standard rate applies but customers can earn a rebate by reducing usage during the critical peak hours. In yet another variant, the price during the critical peak hours is based on real-time conditions, yielding variable peak pricing (VPP).

Each of the dynamic pricing options represents a different combination of risks and rewards for the customer, with RTP rates offering potentially the highest reward compared to a flat rate but also the highest risk. Conversely, a TOU rate offers the least potential reward at the lowest risk. Depending on their risk preferences, customers can self-select into the appropriate rate design, thereby maximizing economic welfare. The set of pricing options can be plotted out in the risk-reward space, yielding the pricing possibilities frontier, as shown in Figure 1.2.

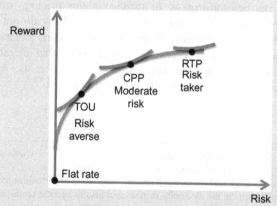

FIGURE 1.2   Pricing possibilities frontiers with indifference curves.

Until fairly recently, the lack of smart meters for residential customers posed a technical barrier to the deployment of these rate designs because almost all dynamic pricing designs require the use of smart meters. As of 2009, less than 9 percent of customers had smart meters.[2] A rapid deployment of smart meters is now underway, pulled by the need to update an aging and increasingly unreliable infrastructure and pushed by the federal stimulus of nearly five billion dollars in smart grid grants. According to the Institute of Electric Efficiency, by 2015 approximately half of the nation's 125 million residential customers will have smart meters and by 2020, nearly all customers will be on smart meters.[3] Thus, a major technical barrier to dynamic pricing should be lifted in the next 5 to 10 years.

While there is wide support for dynamic pricing among academics and consultants, lingering doubts remain about its efficacy among utilities and the state commissions that regulate them. In regulatory hearings, critics routinely contend that residential customers do not respond to dynamic pricing, that dynamic pricing will hurt low-income customers who spend a lot of time at home, and that customers simply do not want to be placed on rates that fluctuate with market conditions.[4]

In the acrimonious atmosphere within which such hearings are often held, a negative mythology has taken root. This negativism has prevented dynamic pricing from germinating. Only 4 of 1,755 respondents to a 2010 survey commissioned by the Federal Energy Regulatory Commission (FERC) indicated they had non-experimental dynamic pricing programs in place for residential customers.[5] Traditional TOU pricing was more widespread, but even that rate design had only garnered a million residential customers, or less than 1 percent of the national population.

It is often argued that dynamic pricing is a novel concept confined to the four corners of the economics classroom and that customers have never encountered in real-day life. That is not true. In his classic book on revenue management, published in the late nineties, Robert Cross highlighted the trend toward setting prices dynamically to maximize profit.[6] During airline deregulation in the 1970s, Cross first used revenue management to dynamically set airline tickets so that his clients, the newly deregulated airlines, could compete in the competitive market. Today, dynamic prices are used consistently by airlines, hotels, rental car companies, and railroads. Customers understand that they will have to pay more when demand is higher; for example, plane tickets cost more on Friday nights, and hotel room rates are higher on Friday and Saturday nights. At the same time, customers also understand the benefit: price-sensitive customers can plan trips around low-priced times and save significant amounts of money.

Dynamic pricing is spreading to a huge number of capital-intensive industries, including broadcasting, manufacturing, and cruise lines. Even professional sports are moving towards dynamic pricing. Since 2009, tickets for San Francisco Giants baseball games have varied according to the value of the game. According to the Giants' website, "market pricing applies to all tickets … rates can fluctuate based on factors affecting supply and demand." While sunny weekend games against big rivals cost more than the average game, fans benefit from cheaper prices during other games. Ticket prices fluctuate according to an algorithm that takes into account a number of factors including the interest in the opponents and weather conditions. After the Giants introduced dynamic pricing in 2009, the Minnesota Twins and St. Louis Cardinals followed suit, and more teams are considering this new option. Concert tickets work the same way: Ticketmaster recently introduced a new technology to allow artists to change the ticket price based on demand observed during the initial sales.[7]

Consumers are used to paying different amounts during different times of day in a variety of settings. In large cities, drivers pay more for parking when there is higher demand, such as during the day or during special events. New parking meters have the technology to adjust to charge different amounts depending on the time of day. Similarly, toll charges on major bridges increase during commuting hours, and drivers who can wait to drive across the bridge during off-peak hours will save money. Customers even acknowledge that they will pay more for using their cell phone minutes during weekdays rather than nights and weekends.

In each of these settings, higher prices during some times are balanced out by lower prices during other times, giving consumers the opportunity to save money by altering their behavior. Customers are used to this, and benefit from it, and for the most part, want it — which leads us to the next myth.

It is also argued that electricity is a commodity and that customers are simply happy with the status quo, which involves flat rates and that they have no desire to switch to dynamic pricing. Naturally, there is some inertia that makes customers reluctant to actively desire to switch pricing plans. However, among customers who have experienced dynamic pricing in pilots, customer satisfaction is strong.

In CL&P's 2009 Plan-It-Wise pilot, carried out in Connecticut, post-pilot surveys and focus groups were carried out to determine how customers felt about their participation in the pilot. Residential customers who participated in the survey had an overall satisfaction rating of 5.1 out of a possible 6, with 92 percent saying they would participate again. Commercial and industrial customers had an average satisfaction rating of 4.1 with 73.5 percent indicating they would participate again. The focus groups revealed that what they liked most about the program was that it saved them money.[8]

Consumers Energy's 2010 Dynamic Pricing Pilot, carried out in Lower Michigan, tested a critical peak pricing rate and a critical peak rebate. The utility surveyed participants to determine satisfaction with the program. The survey found that 78 percent of customers were extremely satisfied or somewhat satisfied with the program and that 92 percent were likely to participate in the same program again.[9]

BGE's surveys among customers in the Smart Energy Pricing pilot found that 92 percent of the customers in 2008 and 93 percent of the customers in both 2009 and 2010 reported that they were satisfied with the program. Furthermore, 98 percent, 99 percent, and 97 percent in the three years, respectively, were overwhelmingly interested in returning to a similar pricing structure the following year.[10]

When the California SPP pilot ended two years after its initiation in 2003, participants were offered the opportunity to continue with some form of dynamic pricing rate or return to the standard tariff. Of the customers who were on the CPP rate, 78 percent chose a time-differentiated rate (either CPP or TOU).[11]

Related to the myth that customers do not want dynamic pricing is the idea that customers will have to resort to extreme measures to save money on dynamic rates, such as getting up at 2 in the morning to run the laundry. Unless a rate were designed such that the peak period was during all waking hours, customers have no need to change their sleeping schedules to save money. In a recent survey of customers who participated in the Hydro One TOU pilot, 72 percent wanted to remain on the TOU rates, and only 4 percent found the changes in their daily activities to be inconvenient.

The bottom line?

At the national level, an assessment carried out for the FERC 2 years ago showed that the universal application of dynamic pricing in the United States had the potential for quintupling the share of U.S. peak demand that could be lowered through demand response, from 4 percent to 20 percent.[12] Another assessment quantified the value of demand response and showed that even a 5 percent reduction in U.S. peak demand could lower energy costs $3 billion a year.[13]

However, progress on dynamic pricing is stalled due to the negative mythology discussed in this article. In the aftermath of the energy crisis in California ten years ago, a group of economists issued a manifesto calling for the institution of dynamic pricing, among other reforms. While California's dynamic pricing experiment concluded in 2004, and meter deployment is rapidly underway, large-scale deployment of dynamic pricing has yet to take place. Hot weather and rapid economic growth can surely precipitate another crisis. It is true that the state has expanded its portfolio of incentive-based reliability-focused programs and rolled out dynamic pricing to large commercial and industrial customers. However, by excluding its residential customers from dynamic pricing, it has left a large share of peak demand exposed to higher costs.

Across the Pacific, Japan lies engulfed in a severe power shortage that has forced people to drastically rotate their work schedules, often switching weekends with weekdays in an effort to lower peak demands by 20 percent.

As noted by the *Wall Street Journal*:

> To prevent blackouts [during the summer], the government is legally mandating that Tokyo Electric Power Co.'s large customers, such as factories, cut their usage by 15 percent from 9 a.m. to 8 p.m. on weekdays. It's asking others, including households, to do the same. Similar steps are being asked of Tohoku Electric Power users. Together, the two utilities supply an area accounting for nearly half of the country's economic output.[14]

An early estimate of the value of lost production due to the power crisis in Japan is a staggering $60 billion.[15] If a regimen of smart metering and smart prices had been in place, the demand-supply balance would have been restored at much less economic cost.

With the national deployment of smart meters, a major barrier to the mass deployment of dynamic prices has been lifted. As Commissioner Rick Morgan of the District of Columbia asked in a widely cited article two years ago, there is no longer any reason for deploying dumb rates with smart meters?[16]

Winston Churchill famously averred, "The future, while imminent, is obscure." While several misperceptions have to be dispelled in the regulatory arena before dynamic pricing will be deployed on a large scale, we wish to note that three recent signs have emerged that create some grounds for optimism.

First, at a recent meeting, the National Association of Regulatory Commissioners passed a resolution on smart grid investments which calls on state commissions to "consider whether to encourage or require the use of tools and innovations that can help consumers understand their energy usage, empower them to make informed choices, and encourage consumers to shift their usage as appropriate. These tools may include dynamic rate structures, energy usage information and comparisons, in-home devices, and web-based portals."[17] Even the inclusion of the words "dynamic rates" would have been unthinkable just a few years ago.

Second, two state commissions, one in the District of Columbia and one in Maryland, have approved in principle the full-scale rollout of peak-time rebates to all residential customers. And, third, a survey of more than 100 senior utility executives carried out in the United States and Canada by the consulting firm Cap-Gemini, in conjunction with Platts, found that dynamic pricing was one of the top five issues on the minds of the respondents as they pondered the future.[18]

Even if there is burgeoning agreement on the end-state, doubts remain about how to make the transition from flat rates to dynamic pricing rates. One possible way is to begin informing the public about the benefits of dynamic pricing and then start rolling out smart prices with smart meters but under the umbrella of full bill protection in the first year. That is, customers would pay the lower of the flat rate bill and the dynamic pricing bill. The bill protection would then be phased out over a three- to five-year period.

Ahmad Faruqui, The Brattle Group
and **Gregory Wikler,** EnerNOC

---

1. William Vickrey, "Responsive Pricing of Public Utility Services," *Bell Journal of Economics and Management Science,* 2, 1971, pp. 337-346.

2. Dean Wight, "Overview of demand response in the United States," *Metering International,* Issue 2, 2011.

3. Institute for Electric Efficiency, "Utility Scale Smart Meter Deployments, Plans, & Proposals," September, 2010. http://www.edisonfoundation.net/IEE

4. Consumer advocates often agree that demand response should be pursued but contend that the best solution is not dynamic pricing, because it is "punitive." Instead, they argue for pursuing traditional programs such as direct load control of central air conditioners which incentivize

customers through monthly rebates. For an exposition, see the viewpoint of Mark Toney, executive director of TURN. http://www.vimeo.com/20206833.

5. Federal Energy Regulatory Commission, *Assessment of Demand Response and Advanced Metering Staff Report,* February 2011. Available at http://www.ferc.gov/industries/electric/indus-act/demand-response/2010/survey.asp

6. Robert Cross, *Revenue Management: Hard-Core Tactics for Market Domination,* Crown Business, 1997.

7. Ben Sisario, Ticketmaster Plans to Use a Variable Pricing Policy, *The New York Times,* April 18, 2011. Available at http://www.nytimes.com/2011/04/19/business/19pricing.html

8. Jessica Brahaney-Cain, "Plan-It Wise Customer Experience," Appendix C, Docket 05-10-03RE01, Compliance Order No. 4, Filing of Connecticut Light & Power Company, Connecticut Department of Public Utility Control, November 30, 2009.

9. Consumers Energy, "Count on Us: 2010 Customer Pilot Results," March 21, 2011.

10. Email communications with BGE personnel.

11. Dean Schulz and David Lineweber, "Real Mass Market Customers React to Real Time-Differentiated Rates: What Choices Do They Make and Why?" 16th National Energy Services Conference, February 2006, San Diego.

12. FERC Staff, "A National Assessment of Demand Response Potential," report submitted to the U.S. Congress, June 2009.

13. Ahmad Faruqui, Ryan Hledik, Sam Newell, Johannes Pfeifenberger, "The Power of Five Percent," *The Electricity Journal,* Volume 20, Issue 8, October 2007.

14. James Simms, "Perverse incentives skew the power of utilities in Japan," *Wall Street Journal,* July 18, 2011 and "Japan needs smart power," *Wall Street Journal,* July 4, 2011.

15. http://www.businessweek.com/magazine/content/11_15/b4223015043715.htm

16. Rick Morgan, "Rethinking 'Dumb' Rates," *Public Utilities Fortnightly,* March, 2009.

17. NARUC, Resolution on Smart Grid Principles, http://summer.narucmeetings.org/2011SummerProposedResolutions.pdf

18. http://www.us.capgemini.com/news-events/press-releases/plattscapgemini-study-north-american-utilities-most-concerned-about-regulation-infrastructure-workforce-pricing/

Volumetric metrics – As Cavanagh points out, if we were to design a pricing scheme from scratch today to allocate costs and recover revenues for the various stakeholders in the electric value chain, we would probably not select a volumetric metrics, or would include other features to better capture cost causation and relate it to revenue collection. This was not as much of an issue in the old days when the industry was predominantly vertically integrated and a cents/kWhr multiplier would suffice, since the proceeds would then be distributed within the same company or state-owned enterprise. The volumetric metrics is particularly inappropriate for stakeholders whose costs are mostly fixed and do not rise with consumption, such as network or distribution companies. The issue becomes especially acute when such companies operate independently of generation or retailing.[19]

19. Schurr et al describe some of the issues rising in this context when network operators lose revenues when customers use less – due to energy efficiency investments – and/or generate on-site by investing in solar PVs and alike.

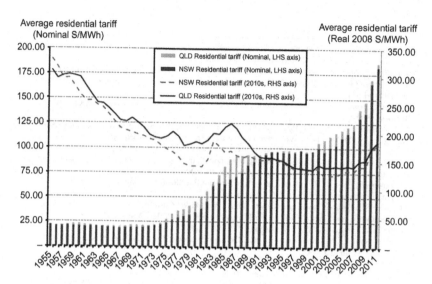

FIGURE 1.3 Per unit electricity costs from Queensland & New South wales, Australia Residential tariffs in NSW and QLD, 1955–2011, in nominal Aus$/MWh (left scale) and inflation-adjusted (right scale). *Source: ESAA, Comparing Australian and international electricity prices, at http://www.esaa.com.au/content/detail/internationalAustralianelectricityprices based on paper by Simshauser and Laochumnanvanit (2011).*

As the preceding discussion illustrates, these issues create special challenges in delivery of energy efficiency in the electric power sector, topics that are further explored in this volume. Returning to the first two items, it is clear that for over a century, virtually all stakeholders in the industry's long value chain – from fuel procurement to generation to transmission, distribution, and retailing – had incentives to invest in bigger and more efficient supply-side *infrastructure*, which resulted in lower per unit costs.[20] But there was little or no motivation or incentives to invest in efficient end-use, on the customer side of the meter. This problem persists to this day, notwithstanding a few proposals to offer incentives for investing in energy efficiency on the customer-side of the meter.[21]

Until recent years, the per-unit cost of electricity delivered to end consumers typically fell when adjusted for inflation (Figure 1.3). The entire industry was built around the notion that bigger is better, and the more consumed,

---

20. Felder and others provide further details.
21. Duke Energy's Save-A-Watt scheme, further described by Nadel, Cavanagh, Felder and Hesser, among others, covers this issue.

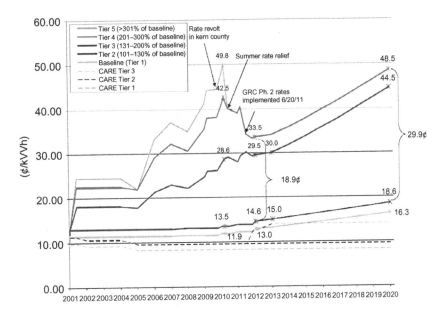

FIGURE 1.4 Rising tiered tariffs for residential consumers in CA. *Source: PG&E.*

the lower the per-unit costs. During the industry's so-called *golden years*, many utilities offered *declining block tariffs*, which awarded heavy consumers with lower per-unit costs. In a declining cost industry, demand growth was considered beneficial and was promoted — conditions that are no longer applicable in nearly all developed economies.

Declining block tariffs are, for the most part, history as more utilities resort to *rising block tiers* to encourage conservation while penalizing heavy consumption, an example of which is shown in Figure 1.4.

The graph shows the rapid rise in top 3 tiers since the bottom two were frozen after the 2000–01 electricity crisis for Pacific Gas & Electric Company (PG&E). The rates have been adjusted downward, but the projected trends suggest returning to high levels by 2020.

Not surprisingly, as Cavanagh and others explain, utilities were historically rewarded for making investments exclusively in supply-side infrastructure.[22] The buck literally stopped at the meter. The more they had in the so-called *rate-base*, the more they were allowed to *earn* through regulated tariffs. With a few minor exceptions, this is still the prevailing mechanism for most utilities in many parts of the world. The only exceptions are in a few jurisdictions that

---

22. The problem, of course, is more pronounced for investor-owned utilities (IOUs), as opposed to government-owned entities, customer-owned cooperatives, or municipal utilities.

have implemented revenue *decoupling*[23] and/or in countries or states where the industry is disaggregated – i.e., is no longer vertically integrated – and where stakeholders no longer operate on regulated tariffs.[24]

Given this context, energy efficiency pioneers started thinking about mechanisms to encourage utilities[25] to focus on delivering *least-cost energy services*, as opposed to maximizing kWh sales and revenues.[26] Making this important transition, however, remains elusive with a few minor exceptions.[27] The simple notion that "utilities" should be rewarded for meeting customers' *energy services* – heating, cooling, lighting, etc. – by offering the best combination of equipment, appliances, investments in energy efficiency, and/or distributed generation at the lowest overall costs is virtually nowhere to be found.[28]

The result is that only sophisticated consumers with the adequate means, resources, and financing have the luxury to examine their overall energy service needs and identify the best combination of supply-side, demand-side, on-site generation, and energy efficiency options that would meet those needs at the least cost. Large energy-intensive industries or businesses where energy costs are a major component of the overall cost of running the business are typically the only ones who may undertake such an analysis or hire an expert to do it for them.

Perhaps as a result of these obstacles – institutional, regulatory, financial, and motivational – most individual consumers are left to their own resources to decide what is best for them based on the limited information and resources that may be at their disposal. Currently, depending on where a customer lives and who provides energy, whether it is a combined electric and gas utility, whether it is privately owned or operated by a not-for-profit municipality or government-owned enterprise, the prevailing regulations, building codes, appliance efficiency standards and applicable tariffs, the consumer must decide how best to get what they need and want at the lowest cost.

Making matters more complicated for the uninformed and/or unmotivated consumer is the growing availability of customer-side distributed generation (DG), also called on-site generation, options, which offer opportunities to

---

23. The chapters by Cavanagh and Hesser address "decoupling" in more detail.

24. Even in so-called deregulated markets, some stakeholders are still paid by the amount of investment in assets and networks, or collect fees based on volumetric electricity consumption – which provides incentives to sell more, rather then less.

25. The term "utilities" is used to refer to stakeholders on the up-stream side of the meter, recognizing that players in some countries and regions remain vertically integrated while they have been partially or totally disaggregated in others.

26. See Felder and others.

27. The chapter by Cavanagh, Hesser, Laitner et al and Nadel, among others, provides further details on remaining barriers.

28. Laitner et al and Hesser describe Save-a-Watt, a scheme proposed by Duke Energy to address this barrier by rewarding utilities for making energy efficiency investments on the customer side of the meter, just as they currently get for investments on the supply-side.

produce a portion of their energy needs — be it hot water or electricity — from devices installed at their premises. These options are subject to many limitations including the prevailing net-metering laws, building and zoning codes, but also supported by tax subsidies, feed-in-tariffs, and other incentives — which are not necessarily easy to decipher.

Because multiple entities and stakeholders are engaged in different parts of the value chain, and put limitations on the customers' time and resources, it is not clear if the society is anywhere near or at the socially optimal level. A consumer living in a highly efficient home, for example, could gain from subsidies to install solar rooftop photovoltaic (PV) panels and rebates to invest in efficient appliances subsidized by other consumers who may not be able to take advantage of similar opportunities. This may lead to a reverse Robin Hood phenomenon, where wealthier consumers may be effectively subsidized by the less fortunate due to the unintended consequences of regulations, incentives, and opportunities.[29]

The empirical evidence, while mixed, shows significant scope for cost-effective energy efficiency as illustrated in Box 1.5.

Box 1.5 If the Empire State Building Can Do It, So Can Everybody Else[1]

In 2009, New York City's iconic Empire State Building embarked on a major remodeling effort to replace its antiquated windows and upgrade its barely functional energy management system with the aid of Johnson Controls Inc. and assisted by a cadre of energy efficiency experts including Amory Lovins' Rocky Mountain Institute (RMI) with some funding from the Clinton Climate Initiative.

Having replaced 6,500 windows and retrofitting lighting and other energy-using devices at a cost of $20 million, the building is now 20 percent more efficient and, according to Johnson Controls, similar improvements can be achieved in most commercial buildings. The building has saved $24.1 million in the first full year, 5 percent better than was initially projected. The results are part of an expected 38 percent reduction in the building's energy use within 3 years, roughly saving $44 million in annual operating costs – a 5-year payback.

According to Dave Myers, who heads Johnson Controls' building efficiency business, the company will now begin working with the building's tenants to manage energy use. Under the terms of its performance contract, the company has guaranteed 20 percent savings for 15 years.

More important, Mr. Myers claims that what has been achieved at Empire State building can be replicated virtually at *any* office or commercial

---

29. A June 4, 2012 article in *New York Times* covering net metering quoted David K. Owens, executive vice president of the Edison Electric Institute saying, "Low-income customers can't put on solar panels — let's be blunt." He added, "So why should a low-income customer have their rates go up for the benefit of someone who puts on a solar panel and wants to be credited the retail rate?"

building *anywhere*. "It's very clear in existing buildings (that) there is a tremendous opportunity to save energy." He has every reason to be bullish. Retrofits of U.S. commercial buildings are projected to become a $16 billion business by 2020.

1. Excerpted from June 2012 issue of *EEnergy Informer*.

As the preceding discussion suggests, the proponents of energy efficiency have a long list of issues to consider. These issues have been examined and, to varying degrees, addressed in the past four decades. But by and large, plenty remains to be done on the way to a new service delivery and pricing paradigm where the interests of consumers are driving all the decisions along the industry's long, complicated, and frequently fragmented value chain.[30] Like achieving world peace or eliminating global poverty or hunger, it is a lofty but elusive goal.

3 WHAT SCOPE FOR ENERGY EFFICIENCY?

An impressive cast of contributors, including researchers, scholars, academics, and practitioners offer their perspectives and insights in this book on ways to move towards a more energy efficient future. But what is the scope for potential energy efficiency gains? More important, will it — or can it — make a difference in the context of the global energy picture? This section is an attempt to address the first question, namely: what is the scope for energy efficiency. The next section addresses the second question: *can* it make a difference?

The existing literature on the scope of energy efficiency is extensive and growing. The range of estimates varies considerably based on assumptions made and the methods employed in implementing, financing, and capturing energy efficiency opportunities. Moreover, the results vary depending on how energy inefficiency is defined (Box 1.6).

Box 1.6 Which Definition?

There are at least three ways to estimate the energy efficiency potential: *Technical, economical,* and *achievable.*

Some studies focus on the *technical* potential, while others focus on what may be *economically justified* — and the definition of what is economic depends on whose costs and benefits are considered. The final measure, usually

30. The problem becomes even more daunting in countries with liberalized or restructured electricity markets where generation, network and supply function have been disaggregated.

closer to what can be expected in practice, is called *achievable* energy efficiency potential.

Technical energy efficiency potential attempts to determine how much is technically possible regardless of costs – which is enormous but mostly irrelevant. In simple terms, technical potential would replace all energy using capital stock with the most efficient technology without considering the cost-effectiveness of the measures. For example, all lighting would be replaced by the most efficient kind, instantly.

Economic potential attempts to replace existing appliances with more efficient varieties so long as the extra costs are justified by the expected savings. Achievable potential attempts to capture delays, behavioral issues and other reasons that generally results in only a fraction of what is economically justified to be realized. For example, it may be difficult to convince a homeowner to replace a perfectly functional washing machine or refrigerator with a more efficient one until and unless it breaks down and has to be replaced.

These three methods produce rather different results, with the last one closer to what may be realistic in practice (Figure 1.5). For these reasons, it is not trivial to unequivocally claim what is the energy efficiency potential. That, however, does not stop people from trying.

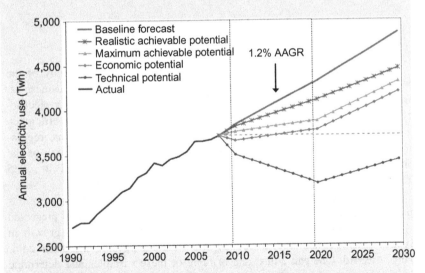

FIGURE 1.5   Achievable energy efficiency potential in the United States, in billion kWhrs under different assumptions about the potential using 4 definitions of technical (bottom), economic (second from bottom) & two versions of achievable, labeled as maximum and realistic achievable (next two) and business-as-usual on top (solid line). More recent projections assume a 0.8 percent growth rate under BAU (Figure 1.6). *Source: Assessment of achievable potential from energy efficiency and demand response programs in the U.S., EPRI, Jan 2009.*

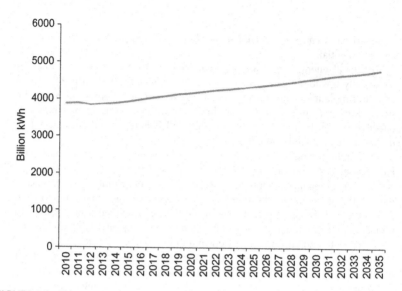

FIGURE 1.6   Forecast of U.S. electricity demand to 2035, in billion kWhrs. *Source: EIA 2012.*

Returning to the first central question of this chapter, any study of the potential scope of energy efficiency must start with a baseline, a business-as-usual, or reference scenario. What would future demand growth for electricity look like in the absence of major changes in policy, appliance efficiency standards, building codes, price changes, behavioral modifications, regulations, etc.? A good place to start may be the long-term predictions of the Energy Information Administration (EIA) covering the period to 2035 (Figure 1.6).

Having a baseline, one can examine the effect of alternative assumptions on key variables affecting demand. A list of the *prime suspects* is presented by Smith in his chapter examining the puzzling end of demand growth in Sydney metropolitan area served by Australia's biggest distribution company, Ausgrid. One can adjust one or more of these variables and determine their impact on the referenced forecast. Levine et al, consider alternative future scenarios in their chapter on China, using a sophisticated model to arrive at different outcomes.

The Institute of Electric Efficiency (IEE), for example, examined the effect of rigorously applying building codes and appliance energy efficiency standards and concluded that the projected demand growth can be virtually eliminated or even reduced over time (Figure 1.7), as described in Nadel.

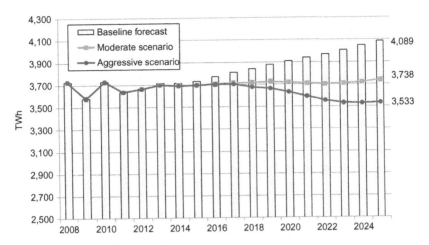

FIGURE 1.7   Potential impact of codes and standards on U.S. electricity consumption.
Source: Institute for Electric Efficiency, Edison Electric Foundation, 2011.

Likewise, in a 2009 study the Electric Power Research Institute examined a range of scenarios and options suggesting significant scope for reducing electricity consumption (Figure 1.5).

As described in the chapter by Nillsen et al, there are many similar studies of energy efficiency potential covering electricity or energy for the United States or on regional or global level using similar techniques, including the International Energy Agency, which summarizes its findings in annual World Energy Outlook. The U.S. Academy of Sciences published a major study of the energy efficiency potential in 2009. Not to be outdone, major oil companies, notably BP and ExxonMobil, publish long-term projections of global energy supply and demand annually.

If there is anything these studies agree on is that demand growth for energy in the rich countries is approaching saturation or has already plateaued. Developing countries, while on a rapid growth path, also shows signs of reaching such a plateau within a couple of decades. The chapter by Levine et al, for example, suggests that such a fate may be awaiting China in a not a too distant future.

Additionally, there are numerous other studies taking different approaches and scenarios − too many to mention. One recent study arriving at a rather remarkable conclusion that the United States can essentially wean itself from use of nearly all fossil fuels, except for a marginal reliance on natural gas, by 2050 mostly through energy efficiency can be found in Amory Lovins' latest book *Reinventing Fire*. Due to its timeliness and direct relevance to the present volume, a review of the book is presented in Box 1.7.

Box 1.7 Book Review: Reinventing Fire[1]

Amory Lovins, the sage of Snowmass, Colorado and the founder of Rocky Mountain Institute (RMI), made a name for himself some four decades ago with what – at the time – was a rather unorthodox and provocative message. Among his arguments was that the cheapest kWh is the one we do *not* use. To drive his point, he referred to energy efficiency as the *free lunch*. As he got better known and bolder, he went further, claiming that energy efficiency was not only a free lunch, but a *lunch we are paid to eat*. Along the way, he wrote numerous seminal articles and books and coined the word *negawatts*, which he prefers to more megawatts.

Initially, the mainstream within the electric power sector found him irritating and a nuisance, to say the least. He was – and still is – an engaging speaker, which many utility CEOs found difficult to challenge in a live debate. His message was deeply unsettling to those who wanted to maintain the status quo. Why was he picking on the electric power sector who, after all, merely wanted to sell more kWhrs and make more money? Why wasn't he, for example, harassing oil companies to sell nega-gallons?

Amory, of course, eventually got tired of dealing with the electric power people, and moved into the transportation sector, challenging car companies, oil companies, and whoever else he could pick on. But his overall message remained unchanged. Why are we using, and more important, wasting so much energy?

His latest book, **Reinventing Fire**, is probably the boldest and most ambitious among many he has written so far, as reflected in its subtitle – Bold business solutions for the new energy era. To start off, the book is printed on paper containing at least 10 percent recycled fiber. It claims that in the process, the printers saved 32 trees, 14,809 gallons of wastewater, 13 million BTUs of energy, 939 pounds of solid waste, and 3,284 pounds of greenhouse gases. You get the gist of it?

The book's aim is stated clearly and concisely upfront; namely, to "answer two questions:

- Could the United States realistically stop using oil and coal by 2050? and
- Could such a vast transition toward efficient use of renewable energy be led by business for durable advantage?"

The answer to both questions – according to Lovins – is a resounding yes. No surprise for anyone remotely familiar with his work.

The book's two questions are particularly timely and germane to the broader debate taking place across the globe on which direction we should be heading, energy-wise, given the environmental, ecological, economic, and geopolitical constraints that will increasingly be facing us in the years ahead. Even though the book is U.S.-centric, the insights resonate in Europe, Asia, Africa, or the Middle East.

Lovins is not particularly fond of nuclear energy, and has never been. So when he talks about phasing out oil and coal, he might as well include nuclear to the list. Oil, of course, is the main fuel in transportation, while coal and nuclear are significant primary fuels for electricity generation. The use of natural gas is barely tolerated as a convenient transition fuel. Hence, the book is a lot more than moving away from fossil fuels. It is mostly about how we can transition to a future predominantly fueled by renewables. At least half of the answer is by using far less energy if we can, and use it far more efficiently if we must.

Among the biggest surprises of the book are that he appears to have the endorsement of a number of prominent executives in the oil and utility business, Pentagon, and former President **Bill Clinton**, who talks generously about Amory's vision, referring to him as "my friend."

Along the way, Lovins does an admirable job of describing the *true cost* of our addiction to oil and our dependence on coal for power generation. As others have pointed out, the price Americans pay at the pump is misleadingly low. "But this seemingly cheap fuel has been a dangerous illusion."

He lists the subsidies to the oil industry and the Pentagon's budget, a good portion of which is to keep the would-be troublemakers away from oil tankers and critical shipping lanes, subsidies to the auto industry, the highway builders and keepers, and so on. He puts the economic costs of U.S. oil dependence at $1.5 trillion a year, roughly 12 percent of the GDP, slightly below the 16 percent spent the United States spends on a bloated and incredibly inefficient healthcare system − another unsustainable scheme.

Oil, energy security, and geopolitics are, of course, closely intertwined. Lovins quotes the former chairman of the Fed, Alan Greenspan, as saying in 2007 that the Iraq war, which was just officially declared as over, "is largely about oil."

The popular press put the price tag at $800 billion. Economist Joseph Stiglitz, in his latest book, puts the price tag closer to $3 trillion by taking a broader definition of the *costs*. Whatever the truth, in a colorful sentence referring to the first Persian Gulf War, Lovins says, "but surely it [the United States] wouldn't have sent a half-million troops to Kuwait in 1991 if Kuwait just grew broccoli." It surely wouldn't have.

The absurdity of our dependence on oil, and the extremes we go to keep the habit, permeates our everyday life and culture. A recent article in *The Wall Street Journal*, for example, reported that U.S. troops stationed in remote parts of Afghanistan can only be supplied with fuel and other necessities by air drops from transport planes at a cost of $400 per gallon. In military circles, the saying is that you do not measure the mileage of heavy armored personnel carriers or tanks in miles per gallon, but rather in gallons per mile. At $400 per gallon, that is no laughing matter.

In another passage, the cost of our dependence on cars, which are in turn currently almost entirely dependent on oil, is estimated at $820 billion a year, equivalent to the entire 9-year cost of the second Iraq war.

Lovins is equally harsh on coal, primarily used to generate electricity. He cites a U.S. National Academy of Sciences study estimating the health-related cost of burning coal − not counting its environmental costs − at $68 billion per annum in 2005. A 2010 study by the Clean Air Task Force estimated 13,000 premature deaths attributed to pollution from coal-fired plants, increasing U.S. healthcare costs by $100 billion per annum.

Kevin Parker, global head of asset management for Deutsche Bank, is quoted as saying, "Coal [-fired generation] is a dead man walking. Banks won't finance them. Insurance companies won't insure them. The EPA is coming after them ... and the economics to make it clean [carbon capture and storage] don't work."

Having established the evils of oil and coal, how can we live without them? Lovins' answer is to use far less energy and to switch to renewables. The United States currently consumes 93 Quadrillion BTUs of energy. Under a business-as-usual scenario, this is projected to grow to 117 by 2050.

Lovins says we can get by with a mere 71 Quads by 2050, less than what we use today, most of which generated from renewables – he allows some natural gas in the mix as a transitional fuel. And not only is this feasible, he claims it will cost $5 trillion *less* than the business-as-usual scenario in 2010 net present value (NPV) terms.

How can this be possible? Lovins starts by a thorough examination of how energy is currently used – and wasted – in the transportation sector, in buildings, in industry, and in the electricity sector. The key metric is how much useful value we get at the end of the value chain for every unit of energy input used. It is classic Lovins at work: how much *primary energy* ends up as useful *energy services* at the point of end use?

No stones are unturned in search for current waste and inefficiency, which can be captured and saved. One is often astonished to find how little useful energy services are delivered at the end of the long and convoluted energy value chain. It is a mixed bag, and the trivial things are thrown in with anecdotal examples as testimonials that it can be done, or has already been demonstrated.

For the electricity sector, Lovins examines four alternative scenarios, labeled *maintain*, *migrate*, *renew*, and *transform*, starting with an extrapolation of current trends to a vastly different future envisioned under the transform scenario. The first three require virtually no new capacity additions to 2050 given the effect of energy efficiency but undergo change in the mix of generation fuels.

The last scenario, the most exotic, however, assumes relatively little centralized and a lot of distributed generation in all imaginable forms and scales. In this case, installed U.S. capacity would more than double, from current 1,000 GW to more than 2,100 including massive amounts of rooftop and utility-scale PVs, considerable amounts of wind, offshore, land-based, and distributed plus storage to manage the intermittency of renewable generation.

Envisioning alternative futures and playing around with the energy mix is fun and easy to do, and these are Lovins' great strengths. Implementing them in practice is not so easy, and this leaves a void, which Lovins attempts to fill in the book's final chapter. He does this by asking a number of rhetorical questions, such as "Is the Reinventing Fire vision economically and technically viable?" for which the answer is always a resounding *yes*. It is not very reassuring if you are an oil or coal company executive or an energy-intensive industry.

Among the rhetorical questions is, "Will Reinventing Fire destroy fossil-fuel companies?" Lovins starts with the categorical statement, "Take oil, which could be obsolete by 2050." He says the writing is already on the wall for those who are looking at the telltale signs. Referring to *peak demand for oil*, he cites a 2009 Deutsche Bank forecast that claims world oil use will *peak* around 2016 and will subsequently *drop* 40 percent by 2030 – ending 8 percent below 2009 level.

"The oil industry," he says, "faces what economist Joseph Schumpeter called *'creative destruction'* – wrenching change that comes when entirely new businesses replace the old ones." Just as traditional booksellers or video rental companies have already experienced the impact of electronic books and movies streamed to TVs, computer screens, or mobile handheld devices, oil companies will have to confront a world where demand for their product falls, perhaps dramatically.

Lovins offers two pieces of advice:

1. Diversify away from oil and into natural gas, renewables, and hydrogen; and
2. Avoid high-risk frontier exploration and production.

Think *hydrogen*, as opposed to *hydrocarbons*, and think twice before investing massive amounts in offshore deepwater platforms, sand oils, tight oil, and so on – not his words but our take on what he may be getting at. In this context, the $7 billion 1,700 mile Keystone XL pipeline bringing carbon-loaded Albertan sand oil to Houston does *not* appear to be a good investment. It locks in Canada to produce and for the United States to consume more of what we should be getting away from. Again, not his words, but ours.

The book's major strength is that it challenges the status quo and the powerful lobbies that wish to prolong the illusion of low-cost, plentiful, climate-changing fossil fuels. Its main weakness is that, despite all the reassurances that it is technically feasible and economically justified, it is not clear how one would go about making the challenging transition to a world without oil and coal – and no nuclear power – by 2050.

Nor is it clear, for example, if the transition will be through a mandatory federal law, implemented and enforced by individual states, or initiated by gradual and voluntary transition by the private sector, who will independently arrive at the same conclusion, namely to get off oil and coal? One is left with the impression that these are trivial and mundane details not worthy of much attention – certainly not Lovins' attention.

Yet, one can argue that much of what Lovins advocates is beginning to take place, albeit not on the scale or timeframe of *Reinventing Fire*. Over 30 states in the United States, for example, already have mandatory renewable portfolio standards (RPS), in some cases calling for a third of retail electricity sales to be met from *new* renewable sources by 2020.

One can easily imagine much higher targets for 2050, approaching what Lovins is advocating. Germany, in fact, has already stated that it wants to get 80 percent of its electricity from renewables by 2050. Denmark is talking about a 100 percent renewable future by 2050.

Similarly, there is more interest in energy efficiency, whether in appliances, building codes, or car mileage standards. California's zero net energy (ZNE) requirement, which applies to all *new* residential buildings by 2020 and new commercial buildings by 2030, is a step in that direction. Europe is on target to do the same, as are a number of other countries.

The clean energy standard (CES) proposal by Senator Jeff Bingaman is yet another way to push for cleaner fuels and away from coal. California's climate bill, which requires state-wide greenhouse gas emissions to be reduced to 1990 levels by 2020, and further reduced to 80 percent of 1990 levels by 2050, provide other examples of the transition to a cleaner and more sustainable future already happening, however sporadically.

One can argue that businesses that ignore these bellwether signals have no one to blame but themselves. Lovins' book offers a fast-forward look at a more sustainable 2050. The global energy mix has changed dramatically in the past in response to external stimuli, prices, technological developments, and demand. *Reinventing Fire* is attempting to accelerate the future course of change and direct it towards lower carbon fuels.

Lovins quotes Buckminster Fuller as saying, "You never change things by fighting existing reality. To change something, build a new model that makes the existing model obsolete." This, more than anything else, conveys the message of the book – replacing what Lovins considers to be an obsolete system dependent on unsustainable climate-changing fuels.

For anyone who may think that this rather lengthy book review is sufficient, nothing can be further from the truth. If you are in the energy mining, extracting, refining, conversion, transport, distribution, or retailing or if you are a major energy user or carbon emitter, this book is a must read even if you do not agree with its premise or conclusions.

1. Book review of *Reinventing Fire* by Amory Lovins, Chelsea Green Publishing Co., 2011 excerpted from Jan 2012 issue of *EEnergy Informer*.

As these examples illustrate, the answer to what is the scope for energy efficiency depends, to a great extent, on who you ask, the context in which the question is asked, and the implicit and explicit assumptions that go into answering the question. A recent survey of "experts" for example, provides one set of answers (Box 1.8).

Box 1.8 A Survey of Expert Opinion

The European Union hopes to reduce energy consumption by 20 percent by the year 2020.[1] While a lack of political consensus prevents the United States as a whole from having a single target, some states such as Arizona, Connecticut,

Maryland, and Pennsylvania have established specific numerical targets, while other states such as California have made energy efficiency and demand response part of their energy action plan, provide regulatory incentives for utilities to undertake such programs, and institute stringent codes and standards to promote efficient energy use.

It is clear that efforts to reduce growth in energy consumption are beginning to take hold. As one commentator noted in the fall of 2011:

A plan for a new coal plant in Michigan is being mothballed. Wind-energy developers in Oregon are cutting back on projects. And in Colorado, Xcel Energy says it may no longer need a proposed high-tension power line to the San Luis Valley. All these projects are being hit by a shift in the utility industry created by increased energy efficiency, small generation projects, such as rooftop solar, and changes in public policy, industry analysts and executives say.

To be sure, the recession and a weak economy have contributed to a slowdown in the demand for electricity, but analysts say there also is a fundamental change afoot. "We are entering a new era," said Douglas Larson, executive director of the Western Interstate Energy Board, adviser to governors of 12 Western states. "I think we'll see fewer transmission lines and power plants built." In its 7-year resource plan filed Oct. 31, Xcel cut the estimated need for additional generation to 292 megawatts by 2020 — down from the 1,000 megawatts forecast in 2010.[2]

What is the likely impact of energy efficiency and demand response going to be on customer usage by the year 2020? The answer depends on who you ask. To get a robust response, we asked some 200 experts in the United States and Canada representing all facets of the industry — from academics to consultants, utilities to regulators, and from consumer activists to environmentalists. A total of 50 experts responded, giving us a wide cross-section of views.

The survey was carried out by economists at *The Brattle Group*, in conjunction with Global Energy Partners, an EnerNOC company.[3] Specifically, we asked the experts to tell us what they expected would happen in terms of energy efficiency and demand response during the next decade relative to each respondent's view of what energy usage would have been in the absence of these new activities.[4] It is important to note that we did not ask them to tell us what *should* happen, a normative question, or what is the economic or technical potential for energy efficiency and demand response. Instead, we asked them to simply tell us what they thought was *likely* to happen, a positive question. We included questions about natural gas as well as electricity.

Key Findings

Overall, the experts expect that U.S. electric consumption will decline by between 5 and 15 percent by the year 2020, compared to what it would have been absent additional energy efficiency measures.

Demand response programs are expected to lower U.S. peak demand for electricity by 7.5 to 15 percent, compared to what it would have been otherwise.

Regional Variation

There is considerable regional variation (the regions are the same as the U.S. census) in the results. For example, the West North Central region of the United States is expected to only see electric energy savings in electricity consumption in the 1.5 to 2.5 percent range, while the Mountain region is expected to see savings in the 5 to 16 percent range (Figure 1.8).

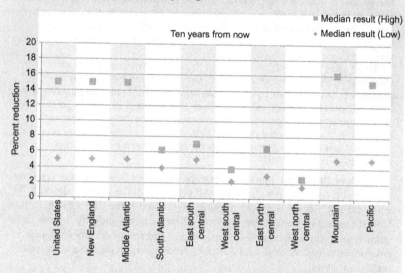

FIGURE 1.8 Percentage reduction projected in survey.

Sectorial Variation

There is also variation across the sectors. For example, the residential sector is expected to see electric savings in the 10 to 12 percent range (Figure 1.9).

It is expected that 40 percent of consumers will buy high efficiency air conditioners and 60 percent will buy high efficiency lighting systems. Some 50 percent of commercial and industrial consumers will buy high efficiency HVAC systems and approximately 70 percent of large commercial and industrial consumers will buy high efficiency electric motors (Figure 1.10).

Demand Response

In the demand response domain, the United States is forecasted to have total system peak demand savings in the range of 7.5 and 15 percent (Figure 1.11).

Within this broad domain, direct load control programs are expected to reach 10 to 15 percent of residential consumers. However, dynamic pricing

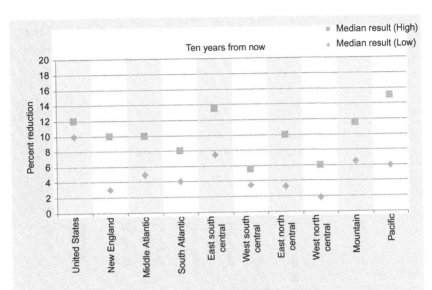

FIGURE 1.9   Variations in residential energy efficiency savings in survey by region.

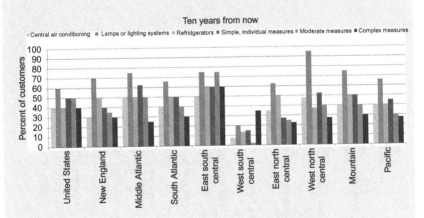

FIGURE 1.10   Percent of residential high-efficiency electric equipment purchases.

programs, which currently reach less than 1 percent of residential consumers, are expected to garner between 7.5 to 20 percent of residential consumers in the United States as a whole and the range could be as high as 12.5. to 45 percent in the East North Central region of the country. Participation rates for commercial and industrial consumers in dynamic pricing programs will be higher than in residential markets, as one would expect, ranging from 10 to 30 percent (Figure 1.12).

While the experts were not explicitly asked what factors would drive improvements in energy efficiency and peak demand, it is quite likely that these

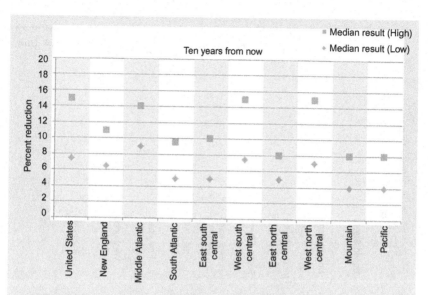

FIGURE 1.11   Total electric demand response peak demand savings.

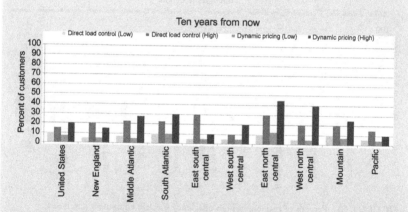

FIGURE 1.12   Percent of residential customers who will be engaged with demand response.

reductions are going to be brought about by factors such as (a) long-standing policy drivers such as rising fuel and capital costs, (b) rapid advances in appliance and building technology, brought on partly by government mandates and partly by competitive economics, and (c) cultural shifts in American values which encourage behavioral change.

The survey results confirm that the age of increasing energy efficiency has not yet come to an end. In fact, they herald a period of acceleration for energy efficiency. Furthermore, the surveyed experts anticipate that demand response will become a valuable resource, with much higher participation rates than today, to manage peak demand.

Ahmad Faruqui, *The Brattle Group*
and **Gregory Wikler,** EnerNOC

1. http://ec.europa.eu/energy/efficiency/action_plan/action_plan_en.htm
2. Mark Jaffe, "Utilities Power Down on Building Plans," *The Denver Post*, November 6, 2011.
3. This survey was conducted as part of an energy efficiency potential study for the New Mexico Energy, Minerals and Natural Resources Department. Copies of this work can be found online at: http://www.emnrd.state.nm.us/ecmd/Multimedia/PublicationsandReports.htm. Global Energy Partners was the prime contractor on the project and contributed to this survey in many ways. They assisted in the development of the survey questions, in carrying out the survey and in the interpretation of the results. Most importantly, they also integrated the results of this study into the LoadMAP model, which assessed the energy efficiency and demand response potential for New Mexico.
4. Each respondent's forecast is assumed to include, in net, growth in overall energy consumption and reductions in energy use due to current and planned future programs.

There are a number of reasons for the broad optimism expressed in the survey results reported above and a number of other indicators that suggests that energy efficiency may be — at last — getting the traction that it deserves. Among the reasons for optimism are the following:

Rising awareness of role of demand — There is a growing recognition that modifying and managing demand — as opposed to relying exclusively on supply-side options to meet customer demand — may be a cost-effective option. Demand, of course, can be modified both in terms of quantity and pattern of use — or energy and peak demand. Depending on the circumstances, sometimes one is preferred to the other — but both ways of managing and modifying are helpful (Box 1.9). Regulators in many parts of the world are gradually recognizing the importance of managing demand and offering utilities — and consumers — the needed incentives to tackle the issues.

Box 1.9 Wellinghoff: A Watt Saved is a Watt Earned — only Better[1]

Jon Wellinghoff, the current chairman of FERC, has a vision for the future of the industry where consumers are paid to save energy. The idea is to cut electricity demand instead of spinning more turbines. Pulling this off will require a smart grid in which customers and utilities communicate real-time information about prices and electricity use.

In an interview with *Technology Review* (April, 17 2012), Wellinghoff said, in part:

"Utilities are going to have to change or die. Traditionally, their business model has been vertically integrated; they generate, distribute, and sell energy. Now, you're seeing opportunities for utility customers – commercial building owners, the Wal-Marts and Safeways of the world – to fully participate in energy markets and go head to head with utilities. Ultimately, you'll have companies helping homeowners install technologies to facilitate their participation. Because of this competition, utilities will have to determine how they are going to continue to make a profit."

Wellinghoff told *Technology Review's* business editor Jessica Leber:

"A number of large utilities are starting to understand that. Still, there are wide swaths of the country where we don't have these markets at all. Customers in those areas are going to have to demand them."

A few excerpts from the interview with *Technology Review* follow:

Q: Does a negawatt have a tangible value?

A: "It absolutely is tangible. We issued an order (Order 745) to say that a negawatt – or reducing a kilowatt of energy demand – is equal to ramping up a kilowatt of energy production. Someone who creates a negawatt should be paid for it. My mission personally has been to integrate negawatts into the wholesale energy market. If we can give the right market signals, entrepreneurs will develop ways to save energy in response to the grid's needs."

Q: Do you have energy apps on your phone?

A: "I have an app on my iPhone, from a company called GreenNet, that allows me to monitor things like my air-conditioning, dishwasher, DVR, and sump pump. I use it all the time. I'm also about to have installed the capability to control them from my phone."

Q: Will more people want to know what their sump pump is up to?

A: "Most people aren't going to be as much of an energy geek as I am. I readily admit that. Some of the most compelling and convenient apps you're seeing now are **Wi-Fi thermostats** you can control from anywhere; you can buy them at Home Depot. Ultimately, to the extent that we can install these types of control devices, residential consumers will be able to volunteer their information to third-party aggregators who can help automatically manage their energy loads."

Q: How far can reducing consumption get us to solving bigger energy problems?

A: "It can get us a long way. Utility commissioners in Massachusetts recently told me they are looking at potentially zero energy-load growth, because they're using smart meters and other devices and have very aggressive energy efficiency programs. I think we're seeing a dramatic shift in the whole energy dynamic in the country. In the next 5 to 10 years, we'll have the ability to manage our energy so that we need very few new traditional resources."

Mr. Wellinghoff is not just pontificating and waiting for the industry to catch up with his vision of the future. He is in the privileged position where he can essentially force his vision on the industry. Under his leadership, in March

2011, FERC issued Order 745, a controversial ruling that obliges wholesale electricity market operators in the United States to implement demand response (DR) programs. Negawatts will increasingly compete with megawatts.

1. Excerpted from June 2012 issue of *EEnergy Informer*.

Rising cost of supply-side options – Cost of meeting customers' energy services are universally rising due to higher fuel costs, higher costs associated with low carbon energy resources such as intermittent technologies and more stringent environmental regulations, increased costs of expanding, updating and maintaining networks, just to name a few. In the words of Amory Lovins, the cheapest kWh is the one we do not use (Figure 1.13).

Enabling technologies allowing customer engagement – A new range of technologies, virtually unheard of and unimaginable until a few years ago, are now allowing consumers to monitor and modify their usage in real time and remotely if they elect to do so[31] (Figure 1.14).

"Prices to devices"[32] revolution – With rapid penetration of smart meters in many countries, it is no longer farfetched to imagine a future

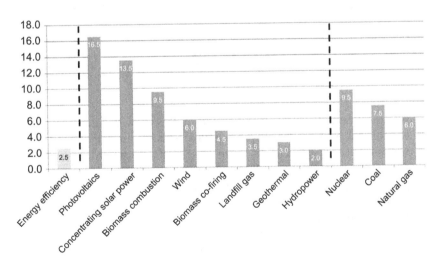

FIGURE 1.13   Relative costs of alternative generation technologies compared to energy efficiency*. *Source: U.S. renewable energy quarterly report, ACORE, Oct 2010.*
*Includes current federal & state level incentives, natural gas price is assumed at $4.50/ MMBTU.

---

31. The chapter by Laskey et al provides further detail.
32. This term was coined at EPRI some years ago.

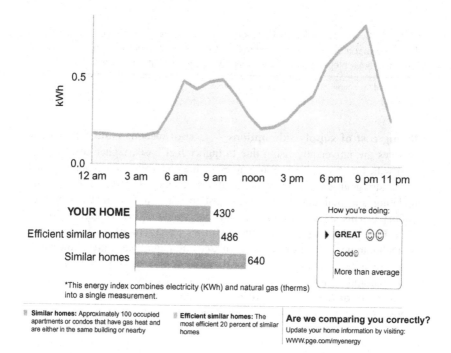

FIGURE 1.14   An increasing number of utilities in the United States and elsewhere are offering customers with timely and easy to understand feedback on their electricity consumption such as the ones illustrated above by Pacific Gas & Electric Company in CA. *Source: PG&E.*

scenario where smart prices communicate directly with smart devices to control and manage energy consumption without direct customer intervention yet consistent with customers' preferences and instructions. Sometimes referred to as "set-and-forget," customers can instruct or program how they wish to manage their overall energy usage and bill without having to actively monitor and adjust devices in real time.

Growing interest in dynamic pricing and cost reflective tariffs – There are hopeful signs that an increasing number of consumers may be willing to engage in dynamic pricing schemes where they are rewarded for shifting some discretionary loads to off-peak periods in response to pricing signals.

Changing consumer attitudes towards energy and the environment – Numerous surveys of customer attitudes from different parts of the world indicate that increasing numbers of customers are concerned about the impact of their lifestyles on the environment.[33] This awareness can be

---

33. Laitner et al, for example, provide recent survey results.

effectively harnessed to reduce energy waste and promote more sustainable habits.

4 WHAT ENERGY FUTURE?

The future – especially the distant future – is not pre-ordained. We can modify it, shape it, or make it as we would like it to be. That said, however, does not mean it will be easy or necessarily cheap. The biggest challenge is deciding who is "we" and who "owns" the problem.

To influence future energy consumption on a global scale is a monumental task, requiring the blessing and participation of countless number of stakeholders – virtually every energy producer and consumer.

The fundamentals are relatively simple. People's needs and desires for energy services tend to multiply with rising income and living standards. Like money, it is not clear if there is a limit, a bliss point, or state of energy Nirvana where once reached satisfies demand. Some anthropologists speculate that human desire for energy – like the desire for money, status, or power – is expandable with virtually no end.[34] According to this, humans are doomed to want more of everything forever, and may suffer the negative consequences of their excesses, sooner or later.

The empirical evidence, while circumstantial, leads to a different conclusion, and a more promising one. While true that people generally prefer more of things they need, want, and like, they typically reach satiation levels, where they virtually cannot consume any more. Lets briefly examine – and speculate – on the evidence for three of the most basic of human needs: food, shelter, and transportation.

Food, or calories, are among the most basic of human needs. For most of human history, life has been mostly about survival, simply getting enough to live, to keep "body and soul together" as the saying goes. But with a few exceptions, humanity has reached a state where an increasing number of people – and not just in rich countries – are obese.[35] The evidence is overwhelming, and the consequences of chronic obesity among growing numbers of people in a growing number of countries is simply alarming.

Can we conclude that humans have reached, or breeched, the limit to how many calories their bodies can reasonably consume? Would this mean that, as more of humanity reaches this state, the focus will shift from simply obtaining more calories to better mix of calories, improved food quality, and a better balance between physical needs and calorie intake? In rich countries,

---

34. Bartiaux et al provide an anthropological perspective on human need for energy in Energy, Sustainability and the Environment: Technology, Incentives, Behavior, Sioshansi (ed), 2011.
35. An interesting observation is that, perhaps for the first time in human history, the rich tend to be thin – or trying – while a growing number of the poor are getting fat, mostly as a result of poor diet rich in cheap and unhealthy calories.

the wealthy and educated appear more concerned about what they eat than how much.

In similar vein, alcohol consumption on a per capita basis is actually declining in a number of developed countries. People are not drinking as much beer, wine, or hard liquor as they used to for a variety of reasons, health-related, social, demographic, and behavioral. Among the wealthy citizens of rich countries, people drink less while moving up − drinking less beer but better wines − for example, with rising income levels.

Shelter, next to food, is among the necessities of life. Throughout history, people of means have always lived in larger and more lavish surroundings than the poor. The lifestyles and norms have changed in the recent past, with smaller family size, aging populations in many OECD countries, and smaller dwelling size for shrinking family size. A growing percentage of people in advanced economies live alone or with a single partner. Larger homes − while still a status symbol and good investment − are not suitable to many old or young people who prefer simpler lifestyles in city centers with short commute to work and desired amenities.

The average size of newly built homes in the United States, which grew steadily since World War II, has apparently reached a plateau. The same phenomenon is happening in Europe and Japan, where typical homes have always been much smaller than in the United States. Can this be a sign of saturation in dwelling size? Will people living in congested cities of the future find bliss in a small apartment in city center close to work, their friends, entertainment, shopping, and other necessities to a larger castle in the countryside?

Transportation, the need to move goods and people around, shows mixed trends. While demand for transportation of passengers, goods via trucks, ships, and air cargo continue to rise, fuel demand for personal transportation in rich countries is dormant or declining. Lee Schipper referred to this phenomenon as "peak transportation," as in "peak oil."

This has a lot to do with demographics, the rising cost of gasoline, and improved efficiency of cars. Older people do not drive as often or as far as the young. High gasoline prices encourage mass transit and shorter commuting. More efficient cars get more miles to the gallon.

But that is not the end of the story. In many congested cities there is simply no place to park a car, making car ownership a mere nuisance. In many developing countries, Beijing, New Delhi, Jakarta, or Bangkok − just to name a few − getting a car is the easy part, getting from one part of the city to another is another matter. In the United States and Japan, there are simply too many cars for the number of drivers (Figure 1.15).

What about energy? Are we near or approaching a saturation point, at least in developed countries? Recent data for the United States shows a practically flat per capita electricity consumption (Box 1.10). Is this an aberration of longer-term trends or the sign of the approach of demand saturation?

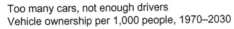

Too many cars, not enough drivers
Vehicle ownership per 1,000 people, 1970–2030

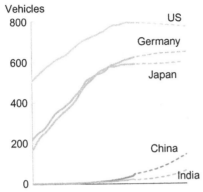

Car ownership per 1,000 population in selected countries

FIGURE 1.15   Car ownership per 1,000 population in selected countries.

Box 1.10 The Utility Growth Paradigm *has* already Shifted[1]

In a blog posted on April 6, 2012, Chris King, an executive at eMeter, wrote:

"An esteemed colleague suggested the other day that we may be approaching zero growth in electricity consumption, mainly as a result of energy efficiency and demand response."

"I'm a data guy, so I figured I'd check it out ..."

"Many of us have seen the famous "Rosenfeld Effect" chart, which shows per capita electricity consumption in California from 1960 to 2008 (Figure 1.16). Energy consumption for Californians looks pretty flat since the mid −1970s − but for people in the rest of the country, it appears to be steadily rising."

"What happens if we look at the past two decades, using Department of Energy and Census Bureau figures?"

"The growth line for the entire U.S. looks pretty flat, especially since 2000. And the statistics agree:

• By the decade spanning 1990 to 2000, the average annual increase in consumption (the compound annual growth rate) had dropped to near zero: 0.34 percent increase per year.

• Then, from 2000 to 2011, the CAGR even went negative: −0.67 percent per year. This means that average consumption fell.

"What about total U.S. electricity consumption − total deliveries to all customers, including the increased population over time? Over the past 10 years, that CAGR was 0.32 percent. This means that utilities sold a little bit more power, but not much (Figure 1.17)."

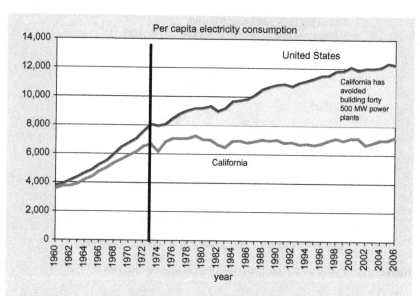

FIGURE 1.16 Per capita electricity consumption in CA (bottom line) and the United States (top line) over time. *Source: Art Rosenfeld, reproduced on C. King blog, April 6, 2012, Blog: http://www.emeter.com/category/blog/*

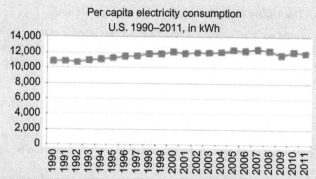

FIGURE 1.17 U.S. per capita electricity Consumption in kWhr/capita, 1990–2011. *Source: April 6, 2012, Blog: http://www.emeter.com/category/blog/*

"Change has happened. The paradigm for utility industry growth has shifted.

"This is good news for the environment and for sustainability – but regulators have some real work to do to ensure that utilities remain financially healthy in this new paradigm."

1. Excerpted from May 2012 issue of *EEnergy Informer*.

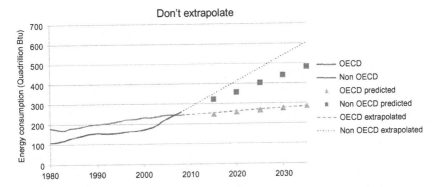

FIGURE 1.18   Alternative growth scenarios for OECD and non-OECD regions, 1980−2035, in Quads. *Source: "How Will Energy Demand Develop in the Developing World?" by Catherine Wolfram, Orie Shelef, and Paul Gertler, Working Paper 226, Haas Energy Institute at University of California, Berkeley, available at http://ei.haas.berkeley.edu/pdf/working_papers/ WP226.pdf.*

To be sure, there is a lot that can be done to influence future demand growth by tinkering with the variables that matter and by adopting policies that can make a difference. Wolfram et al, for example, suggest that the pace of demand growth in developing countries can be adjusted downward with significant impact over time (Figure 1.18).

A promising first step may be to reduce or eliminate energy subsidies, where they exist. There is no rational justification to subsidize energy consumption in this day and age. There are many others, including building codes and appliance efficiency standards. Their potential scope of savings − and their cost-effectiveness − appear overwhelming, even in areas with existing stringent standards such as in California (Box 1.11).

Box 1.11 If Standard is the Way to go, then Which Standard?[1]

Among the favorite tools of ardent proponents of energy efficiency are codes and standards that virtually eliminate the less efficient − and socially undesirable − option from the marketplace. Instead of spending lots of time, money, and effort to inform and encourage consumers to make the *correct* choices, and, say, buy a more efficient compact fluorescent light (CFL) instead of the famously inefficient incandescent type, why not ban the inefficient option from the market? That, in effect, is what Australia, Europe, and the United States have essentially done.

The same logic, it would seem, applies to other energy consuming devices. Why not, for example, ban inefficient fridges, TVs, or air conditioners? It would seem far more efficient than fiddling with labeling and standards − the argument one often hears.

The same type of arguments applies to building codes. Aside for energy used for transportation, the bulk of energy – certainly most of electricity – is used in buildings, homes, shops, offices, factories, and so on. If we are serious about energy efficiency, make buildings more efficient and, *voila*, everything becomes more efficient. Moreover, since buildings and what's in them tend to have long lives, small investments made today to make buildings more efficient will bear fruit for many years to follow. With rising energy prices, it is only a matter of time before the extra initial investment is recouped, and all future savings are mere icing on the cake.

That is among the long-held beliefs of the California Energy Commission (CEC), the government agency that sets long-term energy policy for the most populous state in the union, often years if not decades ahead of other states and the U.S. federal government. California has maintained lower energy intensity relative to the United States average for years (Figure 1.19).

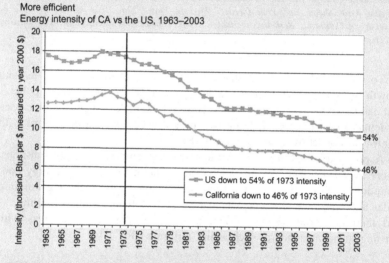

FIGURE 1.19 Energy intensity of California vs. United States, 1963–2003. *Source: 2013 Building Energy Efficiency Standards, CEC, 31 May 2012.*

There may be many factors contributing to this, however, and not all result from energy efficiency policies. For example, California's mild climate, lack of heavy industry – some would say industry, period – and higher retail prices are among the reasons contributing to the state's low energy intensity. Furthermore, not everyone is always happy with the agency's decisions or policies, which sometime appear heavy-handed, and are seen as making the state more expensive for many businesses to operate in (Figure 1.20).

In late May 2012, the CEC adopted a new set of standards, which goes into effect in 2014 with the aim of making *new* residential buildings 25 percent more efficient; 30 percent for *new* commercial buildings. These goals are to be reached mostly through what the CEC calls "common sense improvements" such as more efficient windows, insulated hot water pipes, whole house fans, better air-flow for air conditioning ducts, and increased wall insulation.

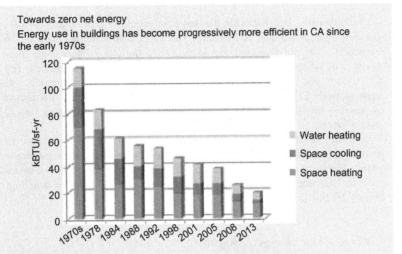

FIGURE 1.20   Energy consumption in California buildings by major use. *Source: 2013 Building Energy Efficiency Standards, CEC, 31 May 2012.*

Recognizing the popularity of solar rooftop PVs, new California roofs will have to be *solar ready* – which means the builders must avoid pipes and other features that will make it more difficult or expensive to install PVs. The new codes are specific to climate zones – which in the case of California vary from mild Mediterranean to harsh dry desert climate (Figure 1.21).

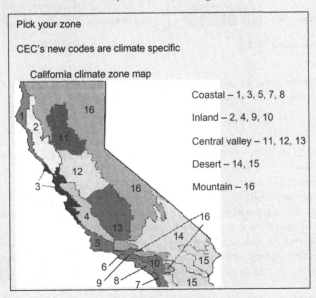

FIGURE 1.21   California's building codes are climate-specific. *Source: 2013 Building Energy Efficiency Standards, CEC, 31 May 2012.*

Building codes and standards, like taxes and regulations, are not universally popular, especially if they add to costs. The CEC's new building codes are no exception. They are projected to add $2,290 to the cost of an average new home, a trivial sum given the relatively high cost of housing in California. The CEC points out that, assuming a 30-year mortgage, the standard will add approximately $11 per month in costs but is expected to save $27 per month on monthly heating, cooling, and lighting bills — a net $6,200 in energy savings over 30 years.

The state of California is projected to save 14 GWhrs of electricity over 30 years as a result of the new standard, eliminating the need for six major power plants, according to CEC estimates. Most of the savings in the residential sector are expected from lower cooling and heating costs in multi-family dwellings (Figure 1.22).

Where are the savings coming from?

Expected savings from various end uses for single and multiple dwelling housing

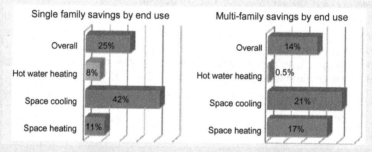

FIGURE 1.22   Expected savings from various uses in single-family and multi-family dwellings. *Source: 2013 Building Energy Efficiency Standards, CEC, 31 May 2012.*

Referring to the new standard in a June 1, 2012 editorial, *Sacramento Bee* said, "All in all, not a bad day for California regulators, for the environment and for consumers." The new standard has received high praise from many sources and relatively few complaints.

Having upgraded the building code, the CEC's eyes are on new standards for electronic gadgets. The reasons are simple. The number of electric gadgets in typical American home has been rising from 9.7 in 1990 to 17.4 by 2000 and 24 today, representing roughly 31 percent of the residential energy consumption. An estimated 13 percent is attributed to consumer electronics — a sub-category, which includes TVs and entertainment devices.

According to an article in the San Jose Mercury News (June 11, 2012), the California Energy Commission (CEC) is focusing on setting energy efficiency standards on some 15 new products, from game consoles and computer monitors to outdoor streetlights and water-using products like toilets and pool pumps. If history is any indication, other states and the federal government will eventually follow. With 38 million consumers, standards adopted by California tend to become *de facto* national standards simply because it is too expensive to make one model of a product for California and another for the rest of the United States.

According to the same article, the CEC is in the early phases of setting new standards that are likely to come for a final vote in 2013. "Energy efficiency is our No. 1 priority," said Ken Rider, an expert in the CEC's appliances and process energy office. "It's the cleanest, cheapest way to take care of increased energy demand in the state."

The reaction to new standards falls along familiar lines. Energy efficiency advocates favor more challenging standards and would like to see them sooner while the manufacturers are opposed. Andrew deLaski, Executive Director of the Appliance Standards Awareness Project, said, "Consumer plug load is on the rise as traditional loads like refrigerators and dishwashers are becoming more efficient. Consumer electronics are making up a bigger piece of the home energy pie."

The Consumer Electronics Association (CEA) who represents manufacturers of devices is opposed. Doug Johnson, VP of technology policy at the CEA, contends that, "The CEC makes poor assumptions about the direction of technology," adding, "Convergence is a good thing for efficiency. Smartphones are now a necessity, and many products are used in ways that save energy."

The CEA points out that the convergence of devices that collapses multiple products into one actually saves energy. The ability to check email and take photographs with a smartphone, for example, means you don't have to turn on the home computer or keep a digital camera. Similarly, many game consoles provide access to the Internet.

No disagreement, but the fact remains that the CEA and its members primarily want to sell more gadgets and make more money – energy efficiency of the devices they market are probably the last item on the list of priorities, if at all. According to the Mercury article, the CEC estimates that the new standards could save $7 billion each year in reduced electricity costs.

1. Reprinted from July 2012 issue of *EEnergy Informer*.

The answer to can – or will – energy efficiency make a difference, is that it indeed can. In developing countries, demand for energy appears to be approaching saturation levels in nearly all sectors of the economy for a variety of reasons (Figure 1.23). How much of a difference energy efficiency can make is probably limited only by our imagination.

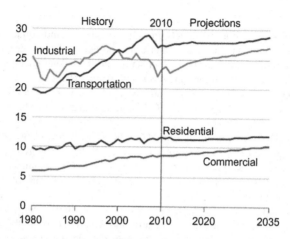

FIGURE 1.23   Mostly flat Energy consumed by major economic sectors, 1980–2035, in quadrillion BTUs/yr. *Source: EIA, Annual Energy Outlook 2012, January, 2012.*

5 CONCLUSIONS

The preceding discussion, repeating and borrowing from what is to follow in the book, offers a mostly upbeat assessment of the significant potential for energy efficiency gains. A recurring theme – and source of frustration – is the fundamental misalignment between the financial interests of those supplying and consuming electricity.

Suppliers – virtually all stakeholders upward of the meter – are generally better off when a larger number of kWhrs are delivered and consumed. They see the meter as the end point, and prefer if it were spinning faster – even in an age when spinning meters are being replaced by the non-spinning variety. By and large, they have not made the mental transition that more kWhrs do not make consumers happier, especially if prices are rising. Nor have they embraced the notion that consumers want their beer cold and their shower hot at the lowest possible cost – they couldn't care less about the kWs, kWhrs, or the therms – which they vaguely understand.

Consumers – and their dumb devices on the downward side of the meter, who have historically been disengaged, uninformed and uninterested, are gradually becoming engaged, informed, and interested in ways that were not imaginable until the arrival of the smart grid, smart meters, and smart prices communicating with smart devices.

With supportive policies, pricing, regulations, standards, and enabling technologies that inform, motivate, and engage customer participation and influence consumer behavior, there *is* considerable scope for optimism.

Chapter 2

Utility Energy Efficiency Programs: Lessons from the Past, Opportunities for the Future

Steven Nadel
Executive Director, American Council for an Energy Efficient Economy

1 INTRODUCTION

Utility energy efficiency programs in the United States began with limited efforts following the 1973 Arab oil embargo. Programs became more common in the 1980s and 1990s and following retrenchment in the second half of the 1990s, have rapidly grown since about 2005. In 2010, approximately $4.6 billion was invested in U.S. electric sector energy efficiency programs – plus an additional $0.9 billion in natural gas programs [1]. Spending on energy efficiency programs over time is summarized in Figure 2.1. In 2010, these programs reduced U.S. electricity use by an estimated 88 billion kWh, which represents 2.3 percent of sales that year [2]. Annual energy savings are summarized in Figure 2.2. However, some states have used energy efficiency programs to reduce electric consumption by more than 10 percent, including a few places where energy efficiency savings are greater than underlying sales growth. As a result consumption is actually declining, hence the subtitle of this volume. In this chapter, it is argued that similar results can be achieved in much of the United States, and in other somewhat similar countries.

This chapter reviews the history of utility sector energy efficiency programs and summarizes some of the key lessons learned. In particular, this chapter focuses on recent efforts by leading utilities to maximize cost-effective energy savings. These results are interpreted in light of underlying sales growth trends, showing how some utilities are saving enough through energy efficiency to actually result in sales decline, not growth. The chapter

Energy Efficiency. DOI: http://dx.doi.org/10.1016/B978-0-12-397879-0.00002-5

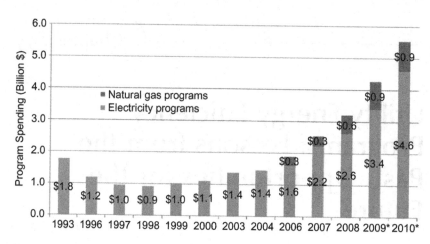

FIGURE 2.1 Annual Spending on U.S. Utility Sector Energy Efficiency Programs. *Source: [1].*

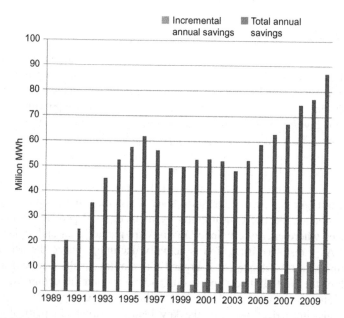

FIGURE 2.2 Annual Energy Savings from U.S. Utility Sector Energy Efficiency Programs. *Notes: "Incremental annual savings" are the savings from energy-saving measures installed that year. "Total annual savings" include savings from all measures in place in a given year, regardless of when the measures were installed. The latter measure accounts for the fact that the typical energy efficiency measure has a service life of multiple years. These figures only include savings from programs operated by electric utilities that report to the Energy Information Administration. Programs operated by non-utility entities (e.g., the New York State Energy Research and Development Authority) are generally not included, even if programs are funded out of utility rates. Source: ACEEE analysis using data from [2] (1999–2010 data) and [3] (1989–1998 data).*

ends with a look to the future, discussing longer-term energy-savings opportunities and strategies to capture these savings.

More specifically, section 2 provides a brief history of utility energy efficiency efforts, discussing four different eras of utility programs. Section 3 discusses the major reasons many utilities invest in energy efficiency. Section 4 discusses recent efforts by leading states, efforts that are flattening, and in some cases eliminating sales growth. Section 5 summarizes some of the key lessons learned as a result of these three decades of efforts. Section 6 summarizes the results of several recent studies that estimate the potential for energy efficiency savings out to 2050. Section 7 discusses a variety of issues that utilities will need to address as they seek to capture these savings. Section 8 discusses a variety of program strategies for the coming decades to address these issues and achieve the available savings. Finally, sections 9 and 10 discuss these results and draw conclusions.

2 A BRIEF HISTORY OF U.S. UTILITY ENERGY EFFICIENCY EFFORTS

U.S. electric utility energy efficiency programs began in the 1970s following the first "energy crisis" triggered by the Arab oil embargo. These initial programs primarily provided information to consumers and businesses, but did not offer financial incentives. In the early 1980s, some utilities recognized that only modest amounts of energy savings resulted from information-type programs and began to include in financial incentives in order to increase participation levels and savings. For example, a review of evaluation results by Collins et al. [4] found that most information programs resulted in energy savings of 0 to 2 percent. On the other hand, a review by Geller [5] of several programs in which different groups received no financial incentives, low incentives, or high incentives, found much higher participation with incentives. For example, Geller describes a study of refrigerator rebates by New York State Electric and Gas (NYSEG) that found the market share for efficient refrigerators was 60 percent in an area receiving information, advertising, and $50 rebates, 49 percent in an area with information, advertising, and $35 rebates, 35 percent in an area with information and advertising, and 15 percent in a control area without any information, advertising, or incentives. These more robust programs began in New Jersey — where replacement power was needed after the Three Mile Island reactors were shut down — the Northwest following the cancelation of multiple nuclear power plants there — and California. Such programs gradually spread to about half the states in the early 1990s, aided by improved understanding of what worked and what didn't (see, e.g., [6]) and initial efforts to remove disincentives to utility investment in energy efficiency and to provide shareholder incentives ([7]; also chapter by Cavanagh).

In the mid-1990s, however, utility investments in energy efficiency contracted (see Figure 2.1), driven by efforts to restructure the electric utility industry in many states. Restructuring began in California in 1994 and was intended to introduce more competition into electricity markets, with the objective of using competition to lower prices. In some states, such as California, utilities sold their power plants, leaving only a regulated distribution utility. Other power providers were free to market directly to customers and many utilities cut their energy efficiency spending in order to cut short-term prices, allowing them to better compete for customers. There was also a belief among some restructuring adherents that improving the operations of the market would unleash increased investment in energy efficiency, without reliance on utility programs (multiple examples of this belief are cited by [8]).

However, over time, it became clear that restructuring did not reduce rates on average nor did it spur increased investments in energy efficiency. For example, a paper by Cain and Lester [9] found that restructured states had lower rate increases in the early years, but by 2008 average rate increases in restructured and non-restructured states were the same. And while these were averages, some well-publicized failures, such as in California, took the wind out of restructuring's sails (see, e.g., [10] and [11] for views from the left and right). Furthermore, data presented by Kushler et al. [8] found that "private market actors each face significant limitations in their interest and ability to deliver energy efficiency and thus far have demonstrated no realistic capability to replace government/regulatory policies and programs to provide energy efficiency."

In the 1990s, the market transformation approach to program design also gained increased attention. The market transformation approach seeks to identify and remove barriers to specific energy-saving technologies and practices, so that the energy-efficient measures can thrive in the market and become business-as-usual. The market transformation approach was found to often have large savings and low costs per kWh saved, although the approach could only be applied in some markets and not all [12].

Starting in 1999, energy efficiency spending by utilities began to increase again, and this increase has picked up speed since 2006 (Figure 2.1). This increase is driven by two factors – the spread of energy efficiency to many more states and increases in savings goals in states that had been offering programs. The spread of programs to more states is illustrated in Figure 2.3, which shows states with significant commitments to energy efficiency before and after 2007, where "significant commitment" means either enactment of mandatory energy-savings targets for utilities, often called energy efficiency resource standards (EERS) and/or incremental annual savings achieved of at least 0.2 percent of electricity sales. Growth in savings in specific states is illustrated by Vermont, where savings have steadily ramped-up to incremental annual savings of about 2 percent of sales each year (Figure 2.4).

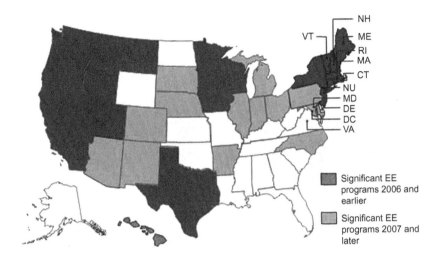

FIGURE 2.3   States With a Significant Commitment to Utility Sector Energy Efficiency Programs. *Source: ACEEE analysis.*

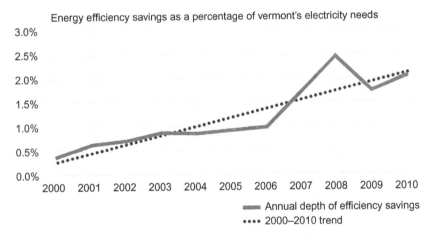

FIGURE 2.4   Energy efficiency saving as a percentage of Vermont's electricity needs. *Source: [13].*

3   WHY DO UTILITIES INVEST IN ENERGY EFFICIENCY?

There are multiple reasons many utilities invest in energy efficiency, with the motivations varying from utility to utility. This section summarizes some of the main reasons, with an emphasis on how these motivations have

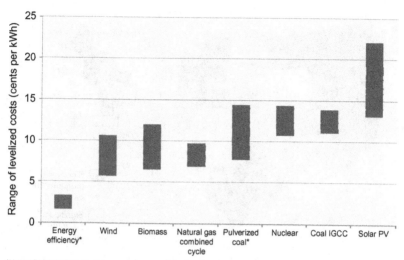

*Notes: Energy efficiency average program portfolio data from [77]: All other data from [81]. High-end range of advanced pulverized coal includes 90% corbon capture and compression.

FIGURE 2.5   Levelized Cost Per kWh for Different Electricity Resources.

increased in recent years. Major rationales for utility investments in energy efficiency include the following:

- Energy efficiency is generally the lowest cost resource, less expensive than new power plants.[1] This is illustrated in Figure 2.5, which compares the range of costs to the utility per kWh for energy efficiency and new power plants. In recent years, the cost of new power plants has increased, but energy efficiency costs have not changed much.[2]
- Energy efficiency is popular with many customers as it is something that helps customers lower their energy bills. For example, Leuthauser and Weaver [16] found that demand-side management programs were a leading contributor to improved customer satisfaction ratings for Mid-American Energy from 2001 to 2005.
- It can be difficult and take many years to get all the necessary approvals for new power plants. Getting approvals for energy efficiency is generally easier.

---

1. In this book Allcott and Greenstone argue that utility demand-side management programs have higher costs, but they are looking at this just from the consumer perspective and not the utility perspective. For example, they include both consumer and utility costs for demand-side management and use discount rates above the utility cost of capital. For a fuller discussion of these issues, including a critique of some other points raised by Allcott and Greenstone, see Nadel and Langer [14].
2. For example, in 2004 ACEEE found that energy efficiency programs cost utilities an average of about 3 cents per kWh [15]. A similar study in 2009 found an average of 2.5 cents per kWh [77].

- Many utility regulators are supportive of utility energy efficiency investments, and regulated utilities generally prefer to satisfy regulators. In recent years, as discussed in the chapter by Cavanagh, many states have improved the business case for utility investments in energy efficiency, including program cost recovery, addressing impacts of lost sales on the ability to recover fixed costs, and providing shareholder incentives for meeting energy efficiency goals. Figure 2.6 shows the status of shareholder incentives as of April 2012. An analysis by Hayes et al. [18] found that utilities that received such incentives on average spend 2.1 to 5.0 times more on energy efficiency per customer, varying by year, than utilities without such incentives. These issues are described more fully by York et al. [19].
- Energy efficiency programs can typically be ramped up in a few years, much quicker than a new baseload power plant.
- To buy time until regulatory uncertainties are resolved. Presently there is a lot of uncertainty regarding future environmental regulations, particularly regarding greenhouse gas emissions. Some utilities are postponing major new power plant investments until these uncertainties are resolved. Energy efficiency can help get demand and supply in balance in the mean time.
- To help meet environmental rules. Many new environmental rules are being set, including for air toxics, nitrogen oxide levels in eastern states, coal ash, cooling water, and carbon dioxide emissions. Energy

Shareholder incentives for electric ans natural gas utilities

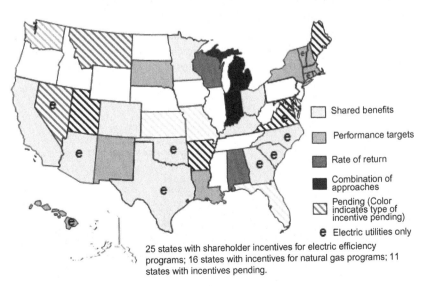

FIGURE 2.6   States with Shareholder Incentives for Energy Efficiency as of April, 2012. *Source: [17].*

efficiency has zero or very low emissions and can be part of a compliance plan with these rules. Hayes and Young [79] discuss these issues in more detail.

- Energy efficiency has been shown to work. As more and more states and utilities embrace and are successful with energy efficiency, states and utilities that were initially skeptical can become convinced that efficiency investments have a role to play.

However, not all utilities have embraced these rationales. In particular, some states have not addressed the business case for utility investments in energy efficiency and as a result, energy efficiency investments may be less profitable than investments in new power plants. Likewise, some utilities are accustomed to proposing new power plants and other "supply-side" investments without a careful analysis of energy efficiency and other "demand-side" alternatives, and some utility commissions are willing to approve these proposals.

While these rationales for energy efficiency apply to the United States in particular, many of them are likely to apply in many other countries.

4 RECENT EFFORTS IN LEADING STATES

As noted above, the average American utility is saving about 2 percent of kWh sales from energy efficiency programs [2]. Incremental annual savings of measures installed in 2009 averaged 0.37 percent of sales [1]. However, results vary widely from state to state, as is illustrated in Figure 2.7. In 2009, the last year for which complete data are available, five states were saving more than 1 percent of sales from efficiency measures installed that year under their programs. In addition, several states have recently made major commitments to expand their energy efficiency programs. In the sections below we discuss some of these leading states.

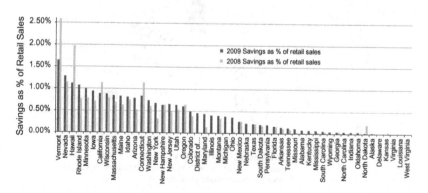

FIGURE 2.7 Electricity Savings from Utility sector Energy Efficiency Programs in 2008 and 2009. *Source: [1].*

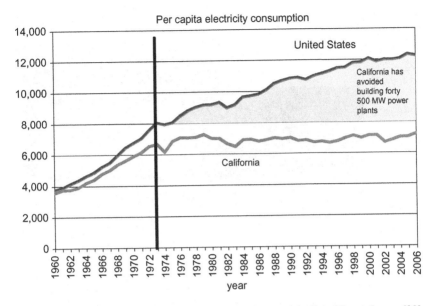

FIGURE 2.8   Per Capita Electricity Consumption in California and the United States. *Source: [21].*

4.1 California

California was one of the first states to aggressively pursue energy efficiency, and the combination of utility sector energy efficiency programs combined with equipment efficiency standards and building code improvements has stabilized electricity use per capita in the state, while electricity use per capita has grown in most other states (Figure 2.8). Since the state's population is growing slowly (e.g., about 1 percent per year from 2000 to 2010[3]), electricity use is growing as well, albeit slowly.

As shown in Figure 2.7, California utility programs have been averaging savings of about 0.9 percent of sales each year. As discussed in the sections that follow, several states are doing even more.

4.2 Pacific Northwest

The Pacific Northwest — Washington, Oregon, Idaho, and western Montana — is another region that has been pursuing utility energy efficiency investments since the 1980s. In the Northwest, power sector planning is driven by a regional council, the Northwest Power and Conservation Council (NWPCC). Every five years they prepare a regional power plan that guides activities by

3. http://quickfacts.census.gov/qfd/states/06000.html

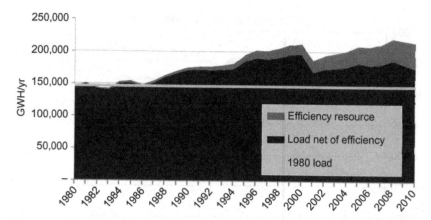

FIGURE 2.9　Role of Energy Efficiency in the Northwest Since 1980 Relative to Regional Electricity Sales. *Source: [22].*

the Bonneville Power Administration (a regional wholesale power provider owned by the federal government) and is also generally used by state utility commissions and many utilities. They also regularly summarize the role of energy efficiency in the region since 1980. A recent summary is presented in Figure 2.9 and shows that energy efficiency has "supplied" about half of the regional sales growth since 1980.

In their Sixth Power Plan, adopted in 2010, the NWPCC decided to significantly ramp up energy efficiency investment and savings because it was cost effective, was subject to less uncertainty than other new energy sources, and would also help meet environmental objectives such as reducing emissions of greenhouse gases. Specifically, they ramped up incremental efficiency savings to 1.5 percent of sales each year, essentially ending sales growth. This is illustrated in Figure 2.10. They estimate that energy efficiency (labeled "conservation") will meet 85 to 90 percent of their resource needs − some renewable energy and natural gas generation will be added to offset power plant retirements.

In fact, over the 2000−2010 period, electricity sales actually declined in Oregon and Washington by 9 percent and 6 percent, respectively [24] although some of this decline is likely due to the impact of the 2007−2009 "Great Recession" as well as the closure of some major aluminum and paper plants in the northwest.

4.3 New England

New England states have been operating energy efficiency programs since the late 1980s. Figure 2.4 shows energy efficiency savings in Vermont, which in the past few years have been greater than the underlying sales

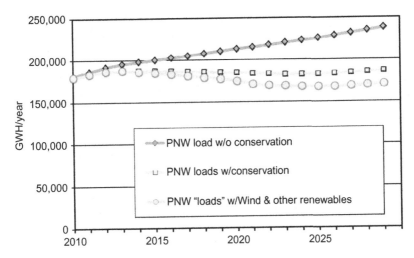

FIGURE 2.10   Role of Energy Efficiency in the 6th Northwest Power Plan. *Source: [23].*

growth of about 1.5 percent per year and therefore sales in Vermont are declining. Vermont has been a leader, setting aggressive energy savings goals that keep challenging them to do better, and then meeting these goals. As a result, Vermont leads the nation in electricity savings as a percent of sales (Figure 2.7). Vermont is also relatively unique in that they have established an "Efficiency Utility" to run programs statewide, with money for this utility provided by the individual electric utilities in the state. Program administration by utilities versus non-utilities is discussed later in this chapter in the section Issues to Address.

In southern New England − Connecticut, Massachusetts, and Rhode Island − efficiency programs have historically saved around 0.8 percent of sales each year (Figure 2.4). But in the past three years all three states have decided to approximately triple efficiency program spending and savings, driven by the lower cost of energy efficiency savings and a desire to help meet climate change goals. In all three states, legislation was enacted calling for utilities to acquire "all cost-effective energy efficiency." Advisory boards and collaborative program development efforts involving major stakeholders have developed plans to implement these directives, for review and approval by state utility commissions. As a result, all three states in 2012 or 2013 are targeting incremental energy efficiency savings of more than 2 percent per year, turning sales growth negative. Charts from the Massachusetts and Connecticut plans are shown in Figure 2.11. Over the 2000−2010 period, sales in Connecticut were essentially level, while Massachusetts and Rhode Island had modest growth −0.7 percent and 1.0 percent per year, respectively (EIA SEDS, 2012). With energy efficiency programs tripling in size, all three states are likely to "go negative" even when their economies recover.

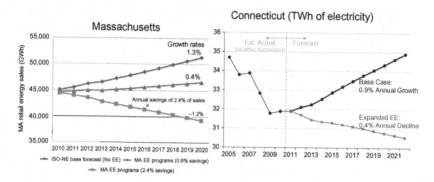

FIGURE 2.11 Estimated Energy Efficiency Savings and Sales Growth in Massachusetts and Connecticut. *Sources: [25] for Massachusetts, [26] for Connecticut Connecticut "Base Case" includes prior levels of energy efficiency programs.*

4.4 Midwest

Most of the midwestern states began or increased their utility energy efficiency program efforts in the 2007–2012 period. As shown in Figure 2.7, two midwestern states – Iowa and Minnesota – used efficiency programs to reduce electricity use by nearly 1 percent per year in 2009. Both states have set energy savings goals of about 1.5 percent of sales each year [20], and as of 2009 savings from these programs were still ramping up. However, these two states had above average growth in electricity use in the 2000–2010 decade (1.2 percent/year for Minnesota, 1.6 percent for Iowa) [24], and thus even with increases in energy savings, these states may still see very modest growth in electricity use. Many of the other Midwestern states also have energy savings goals of 1 to 2 percent per year including Illinois, Indiana, Michigan, and Ohio [20], but these programs typically began around 2009 and are still ramping up. Most of these states are growing slowly, if at all, and therefore it appears that using energy efficiency to end sales growth is possible (e.g., over the 2000–2010 period sales in Michigan and Ohio declined while sales in Indiana and Illinois grew by about 0.75 percent per year) [24]. Notably, Wisconsin also set similar energy efficiency goals as its neighbors, but following a change in ruling party in the 2010 election, reversed the decision to ramp up efficiency programs as part of efforts to cut spending across most areas affected by state government.

4.5 Southwest

Energy efficiency efforts have also surged in this region in the 2007–2012 period. Nevada is among the top states in Figure 2.7, and Arizona, Utah, and Colorado have been ramping up their efforts. Particularly notable is Arizona, which has established a savings goal for its utilities of 22 percent savings by 2020, including savings of more than 2 percent per year starting in 2014 [20]. However, despite the recent recession, electricity use in these states grew by an average of 1.8 to 2.1 percent per year over the 2000–2010 decade (EIA SEDS, 2012) and thus it will take 2 percent per year savings to stabilize sales. Arizona is planning for such an event but under current policy, modest sales growth is likely to prevail in the other states in the region.

4.6 Why the Regional Differences?

A key question is why have these five regions chosen to embrace energy efficiency much more than other regions of the United States? There are no absolute answers to this question, but there are some theories.

- First, many of these regions tend to be more environmentally concerned than other regions.
- Second, many of these regions tend to be more liberal than the nation as a whole, but there are also some fairly conservative states in this group such as Arizona, Utah, Idaho, Montana, Indiana, and New Hampshire.
- Third, some of these regions (California and New England) have above average electricity prices, although other regions are average (midwest) or even below average (southwest and northwest).

5 KEY LESSONS LEARNED

This brief review of past history and recent efforts helps to illustrate a number of key lessons that have been learned in the United States regarding utility energy efficiency programs over the past 30 years. Among the key lessons are the following:

- Large savings *can* be achieved. For example, Figure 2.4 illustrates how Vermont has used energy efficiency over the past decade to reduce energy use by about 14 percent relative to business-as-usual. Figure 2.9 illustrates how energy efficiency has met about half of sales growth in the northwest since 1980. In Vermont, as well as states such as Connecticut, Massachusetts, Washington, Oregon, Iowa, and Minnesota, energy efficiency has helped to dramatically reduce sales growth, and in some cases, even eliminate sales growth.
- Achieving large savings can be very *cost effective*. This is illustrated in Figure 2.5, which shows how the utility cost for energy efficiency savings has averaged about 2.5 cents per kWh, while power from new power plants costs two to ten times more. This is also shown in Figure 2.12,

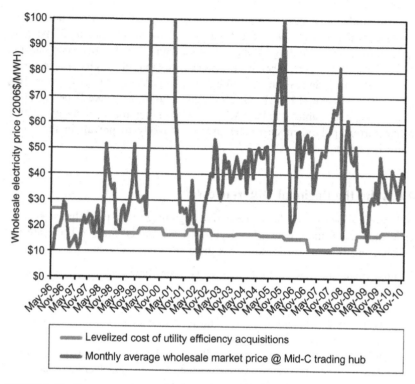

FIGURE 2.12 Comparison of Utility Cost of Energy Efficiency to Wholesale Power Costs in the Pacific Northwest, 1996–2010. *Source: [22].*

which shows financial savings from energy efficiency programs in the Northwest over the past 15 years.

• While some states and regions have embraced utility energy efficiency programs, others have not. This is illustrated in Figure 2.7. Of the kWh savings from utility sector programs in 2009, the top ten states accounted for 61 percent of the total savings [1].

• Effective policies can help drive utility energy efficiency investments. For example, of the 33 states with significant utility sector programs shaded in Figure 2.3, all but six have mandatory EERS in place. Likewise, as discussed previously, Hayes et al. [18] found much higher spending per capita on utility sector energy efficiency programs in states with shareholder incentives relative to states without such incentives. These findings are illustrated in Figure 2.13.

• It requires many different programs and program approaches to achieve large savings. Programs need to serve each of the major customer classes

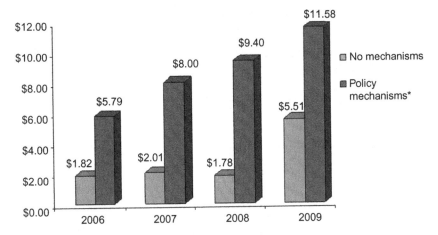

FIGURE 2.13 Comparison of Utility Spending Per Capita in State With and Without Shareholder Incentives. *Source: [18].*

(residential, commercial, and industrial), and even within these classes, each of the major submarkets (e.g., single-family and multi-family, space conditioning as well as lighting and appliances). There are no "silver bullets," but instead many "silver BBs."

- Achieving high participation rates requires both financial incentives and good information/marketing. Neither can do it alone. Earlier we discussed the results of a NYSEG refrigerator rebate program, but other examples are discussed by Geller [5]. The market transformation approach can be very effective in some markets, with high savings and low costs per kWh saved. For example, the Northwest Power and Conservation Council estimates that in 2010 the Northwest Energy Efficiency Alliance, a regional market transformation organization, captured energy savings at about 35 percent of the cost of other efficiency programs in the region [27]. Financing can also help, but so far participation rates in most financing programs have been modest (see [28] and the chapter by Hesser).

- Building energy codes and appliance and efficiency standards can also result in large energy savings. And in many cases, utility sector energy efficiency programs can lay the groundwork by building the market share for efficient equipment and practices and by helping to teach designers, builders, and developers about efficient construction techniques. Figure 2.14 illustrates the contribution that codes and standards have played as part of California's energy efficiency achievements. Utility efforts in California substantially contributed to the savings attributed in the graph to codes and standards.

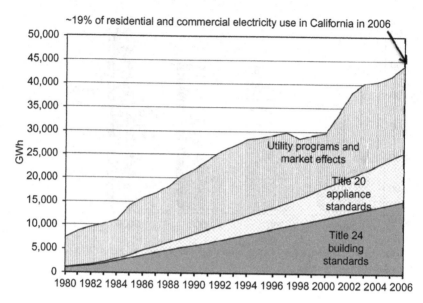

FIGURE 2.14 Efficiency Savings in California from Utility Programs, Codes and Standards. *Source: [21].*

6 LONG-TERM EFFICIENCY OPPORTUNITIES

This book addresses the end of demand growth. For demand growth to end, energy efficiency opportunities need to be available over many years and not just for a few years. In order to assess what may be possible going forward, it is useful to look at studies of available energy efficiency opportunities relative to established base case forecasts.

Probably the most widely used energy forecast is the Annual Energy Outlook (AEO), published each year by DOE's Energy Information Administration. The 2012 AEO projects that U.S. electricity use will increase by an average of 0.8 percent per year over the 2010–2035 period [29], slightly less than the 0.97 percent actual growth over the 2000–2010 period [30] but significantly slower than growth in earlier decades. The AEO forecast includes savings from federal energy efficiency standards and other federal policies that have already been adopted, but does not include any new federal standards nor savings from state policies such as state EERS. The EIA forecast is shown in Figure 2.15.

Several recent studies have projected business-as-usual energy use out to 2050 and then estimated how much this use could be reduced with cost-effective energy efficiency investments. The intent of these studies was to see if energy efficiency savings would be exhausted, or could be sustained

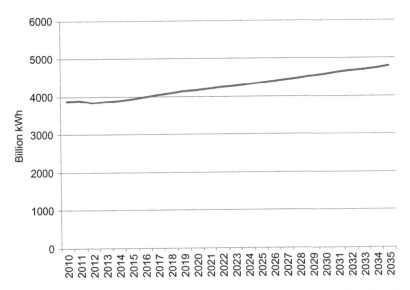

FIGURE 2.15   Energy Information Administration Forecast of Future Demand for Electricity. *Source: [29].*

over time, and also to look at opportunities for reducing greenhouse gas emissions in the long-term.

The first of these studies, entitled *California's Energy Future — the View to 2050* [31], charted paths to meet the state's objective of reducing greenhouse gas emissions by 80 percent by 2050. They found this goal could be met by a combination of improved energy efficiency, reducing energy use by about 40 percent overall, use of low-carbon fuels and electricity, and increased electrification. The latter figured into their strategy because there are more zero- or low-carbon electricity resources than low-carbon fuel resources. Their approach is illustrated in Figure 2.16, which shows how efficiency is squeezing demand from both sides while electrification is pushing this reduced demand to the right.

The second study was conducted by Amory Lovins and the Rocky Mountain Institute and published as a book titled *Reinventing Fire*. Their study sought to determine if the United States could wean itself off of oil and coal by 2050 and concluded, as illustrated in Figure 2.17, that such changes are achievable through a combination of reducing demand, better managing when demands occur, and shifting to low-carbon energy supplies. In terms of reducing demand, they estimate that end-use energy can be reduced by 46 percent relative to the EIA reference case. In terms of reducing and shifting demands, their analysis emphasizes looking at entire systems and not just individual technologies. Another key aspect of their study is to

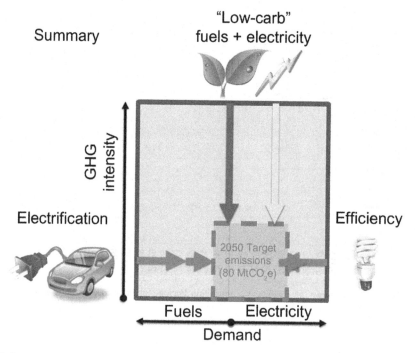

FIGURE 2.16　Summary of California's Energy Future in 2050. *Source: [32].*

identify opportunities for profitable business investments in making the transition to a society that does not rely on oil and coal [33].

Third, in early 2012, Laitner et al. at the American Council for an Energy-Efficient Economy (ACEEE) published an analysis of energy efficiency opportunities out to 2050 [34]. Like Lovins and RMI they examined savings by sector and their interactions. Also, like Lovins and RMI, there was a significant emphasis on savings from systems integration. They examined two scenarios — an advanced scenario that looked at just advanced technologies and practices and a "phoenix" scenario which included some changes in development patterns and existing infrastructure that further reduce energy requirements for buildings and mobility. Relative to the EIA reference case forecast, they found opportunities to use energy efficiency to reduce energy use by 42 percent (advanced case) to 59 percent (phoenix case). They estimate that pursuing these two paths will require $2.4 to 5.3 trillion dollars of public and private investment but will result in benefits of $15.0 to 23.7 trillion and will increase employment by a net 1.3 to 1.9 million jobs. Their results are summarized in Figure 2.18 and discussed in more detail in the chapter by Laitner et al.

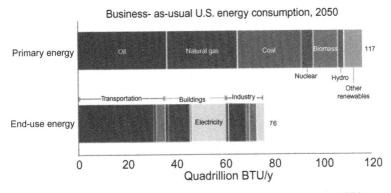

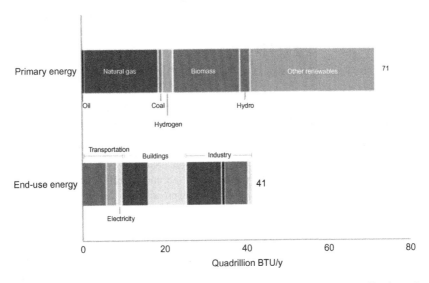

FIGURE 2.17 Estimated 2050 Energy Consumption Under Business-as-Usual and "Reinventing Fire" Scenarios. *Source: [33].*

All three of these studies found opportunities for at least 40 percent energy efficiency savings by 2050. Forty percent savings over 35 years (allowing some time for ramp-up) is about 1.45 percent per year, while 2 percent per year savings for 35 years means 51 percent savings. These findings suggest that energy efficiency programs should be able to find cost-effective opportunities to reduce electricity use by 1.5 to 2 percent per year for many decades.

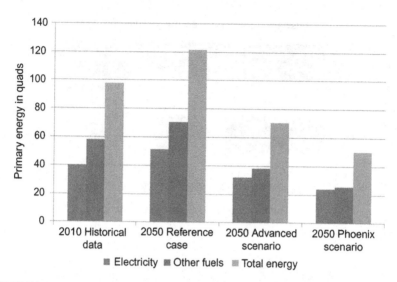

FIGURE 2.18   Summary of Savings Opportunities in ACEEE Study. *Source: [34].*

7 ISSUES TO ADDRESS

The discussion up to this point indicates large opportunities to use utility sector energy efficiency programs to reduce or eliminate sales growth, if a utility and state policymakers make serious commitments to energy efficiency. However, before such results can be achieved on a widespread basis, a number of issues need to be addressed. In this section we discuss many of the major ones.

7.1 Savings from Utility Programs Vs. from Codes and Standards

As discussed above, significant savings can be achieved from codes and standards. This has led some utilities and utility organizations to wonder if there will be much savings left to harvest with utility programs. For example, the Institute for Electric Efficiency [35] estimates that if codes and standards are set aggressively, these savings alone will be enough to eliminate sales growth in the United States overall. Their results are shown in Figure 2.19. In their moderate case, code and standard savings total 9 percent of 2025 electric sales; this climbs to 14 percent in their aggressive case. These estimates are more aggressive than other recent estimates. For example, Lowenburger et al. [36] estimate 7 percent electric savings from new equipment efficiency standards in 2025. As discussed above, utility programs can help lay the groundwork for achieving these savings. At times, utilities can even get credit for these savings [37]. Still, future increases in code and

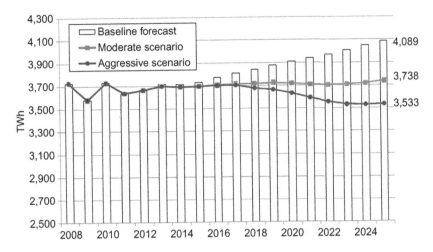

FIGURE 2.19   Estimated Impact of Code and Standard Energy Savings on Total US Electricity Consumption. *Source: [35].*

standard stringency mean that utilities will need to tap into other sources of savings to meet aggressive energy savings goals. As discussed in Section 6, system optimization is one such source of savings. Other new savings opportunities are discussed in Section 8.

7.2 Where Will Future Energy Savings Come from?

Many energy efficiency programs have relied extensively on a few specific energy efficiency measures to achieve large savings. These include compact fluorescent lamps and upgrades of commercial, industrial, and outdoor lighting systems. For example, Efficiency Vermont [38] reports that in 2010, 69 percent of the savings they achieved came from lighting. Eventually these "tried and true" measures will be exhausted. Also, some of these measures will be driven by minimum efficiency standards that have been set but not yet taken effect. Efficiency program operators wonder where they can get large savings once these tried and true measures have run their course. A few additional options are discussed by Platt et al.

As discussed in Section 6, substantial future savings will come from improved system efficiencies in residential and commercial buildings and in factories. Programs that spur energy saving behavior in homes and businesses are also growing – CEE [39] identified 105 such programs. New technologies will also provide significant savings, such as emerging LED lighting technologies [40], heat pump water heaters [41], variable speed air conditioning [42], and efficient electronic goods (to address growing loads in these end-uses). Some of these future savings targets are discussed further in Section 8.

7.3 Will Energy Efficiency Costs Increase or Decrease?

Standard economic theory says that markets will first acquire the lowest-cost resources, and as these are exhausted, costs per kWh from energy efficiency will go up. However, new energy efficiency opportunities continue to be invented and identified, adding new low-cost resources to the mix. Also, there can be economies of scale as programs ramp-up. As a result, an analysis by Takahashi and Nichols [43] estimates that the average price of a saved kWh has gone down, not up for utilities with high-savings programs (defined as annual savings of 1 percent or more of sales). Still, after many years of 2 percent per year efficiency savings, at some point costs may well go up, but when and by how much is very uncertain.

7.4 Evaluation, Measurement, and Verification

Energy efficiency savings are evaluated according to a variety of protocols, which can vary from state to state and even utility to utility and program to program. While some evaluation is very rigorous, other evaluation can be sloppy and can overestimate savings. In order to defer the need for new generation, transmission, and distribution infrastructure, the energy savings need to be real. While most states have evaluation protocols [80], some of these protocols can be improved. Increased standardization of protocols will also make it easier to compare and add together different evaluation results. Several regions have undertaken such efforts [44,45], and DOE is also beginning such an effort [46]. These efforts are important as they improve the credibility and accuracy of energy efficiency savings estimates, allowing systems planners to better count on energy efficiency as a system resource.

7.5 Sales Growth Versus Load Growth

Most of the discussion in this chapter is on total kWh sales. But utilities also need to have sufficient capacity to serve peak demand. In recent years, peak demand has been growing more than total sales. For example, over the 2000−2010 period U.S. electricity sales grew at an average annual rate of 0.97 percent [30], while peak demand grew at about 1.1 percent [47]. Going forward, the Energy Information Administration (EIA) projects electricity sales to grow at about 0.8 percent per year over the next decade [29], while the North American Electric Reliability Corporation (NERC) expects non-coincident peak demand to grow by about 1.1 percent per year [48]. Energy efficiency program portfolios tend to reduce loads during all hours and seasons, so that a 1 percent reduction in sales will, across an entire portfolio, result in roughly a 1 percent reduction in peak demand. In order to further reduce peak demand growth, demand response programs can be used that target peak demand reductions, including shifting loads from peak to off-peak and shoulder periods. A 2009 analysis for the Federal Energy Regulatory Commission [49] estimates that demand response programs can be used to reduce peak demand by 14 percent over a decade under an "achievable participation" scenario. Business-as-usual reductions (e.g., current efforts) save only 4 percent, indicating an opportunity for an additional 10 percent peak demand reduction from demand response.

7.6 Who Should Administer Utility Sector Energy efficiency Programs?

In much of the United States utilities take the lead on administering energy efficiency programs, but in some states non-profit organizations (Efficiency Vermont, Energy Trust of Oregon), state agencies (New York State Energy Research and Development Authority, New Jersey Board of Public Utilities), or private contractors selected via an RFP take the lead. Some of these alternative administrators have been very successful, such as Vermont, and others less so. For example, in New Jersey the governor has diverted money intended for energy efficiency to help cover the state budget. Other papers have discussed the pros and cons of utility vs. non-utility administration [50,51]. State administration can be problematic, as shown by the New Jersey example and in a case in Wisconsin more than a decade ago where a political appointee steered financial incentives to friends of the governor. On the other hand, state administration has worked well in New York as the administering agency has a structure that makes political interference more difficult. Utility administration varies, with some utilities doing excellent jobs and others not so well. A key for utility administration is whether utility management really wants to make the programs succeed and puts competent and creative staff in charge. A non-profit can also do a good job if they have the appropriate experience and staff. While a few states have recently hired third-party for-profit administrators, to our knowledge such an approach is too new to be evaluated.

7.7 Supporting Policies and Policy Stability

The majority of U.S. electricity is provided by investor-owned utilities that have a fiduciary responsibility to their shareholders. These companies will generally invest in energy efficiency only if it makes business sense to their customers and shareholders. Regulators and other policymakers need to establish a foundation for this business case. As discussed above, this involves cost recovery for programs, decoupling profits from sales in some manner, and providing a return to shareholders for successful achievement of energy efficiency goals. Essentially all states provide cost recovery and about half of the states are now addressing the other two criteria (see Figure 2.6). For utility energy efficiency programs to prosper nationwide, most of the remaining states will need to address these criteria.

In addition, top policymaker support is very useful, as is policy stability. In some states, policy support for utility energy efficiency programs has ebbed and flowed, particularly as political power changes from one administration to another, making it difficult for energy efficiency programs to prosper. Particularly troublesome can be "raids" on funds for energy efficiency programs, diverting funds to other purposes such as helping to balance state budgets. Policymakers need to recognize that energy efficiency programs operate best when they are allowed to "hit stride" and continue at a stable level. If changes are made, ramp-up or ramp-down time should be allowed. If efficiency programs are suddenly scaled back, the long-term relationships with trade allies such as retailers and equipment contractors are eroded, making it more difficult for programs to be successful in the future.

7.9 Potential Wild Card: Role of Electric Vehicles

One wild card regarding future electric demand is the role of electric vehicles. Currently, very little electricity is used for transportation. If electric vehicles become very common, this will be a major new load that will affect recent sales growth patterns. For example, in August 2011 the California Energy Commission forecast that peak demand for electric vehicles will be slightly above 250 MW in 2022. As noted by Kim et al. [52], "this is almost a rounding error in a state that hit a non-coincident peak load of 60,455 MW in 2010." Kim et al. go on to estimate peak loads in the United States for a level of 1 million electric vehicles and estimate peak demand could be as high as 1200 MW. This is about 1 percent of projected U.S. non-coincident peak demand for 2021 [48]. One million electric vehicles by 2021 might be optimistic[4] and thus electric vehicles are unlikely to be a major source of new sales over this next decade. But in the much longer term, electric vehicles could play a substantial role in electric demand. For example, a 2009 National Academy of Sciences study estimated 103 million plug-in vehicles in 2050 in a "probable penetration" scenario [54].

8 PROGRAM STRATEGIES FOR THE FUTURE

As noted in Section 7, some widely used energy efficiency measures will soon be mandated as part of new federal equipment efficiency standards,

---

4. For example, J.D. Power and Associates [53] estimated U.S. sales of about 100,000 battery electric vehicles in 2020 plus sales of about 1.7 million hybrid electric vehicles of which only some would be plug-ins.

while some other measures have been widely promoted and therefore available remaining savings will decline. To address these trends, while continuing to achieve savings at robust annual levels, program administrators will need to modify existing programs and in some cases add new programs. In this section we summarize some of the likely program strategies for the future. These include:

- *New construction programs* that promote savings substantially higher than required under new building codes. Circa 2011 versions of national model building codes will reduce energy use in new buildings by about 30 percent relative to prior codes [55]. Utilities can promote this level of savings to lay the groundwork for adoption of these codes and will then need to promote higher levels of savings, such as 50 percent savings relative to current codes. Notably, the 50 percent savings level is targeted by federal tax incentives enacted in 2005 [56], is targeted by the American Society of Heating, Refrigerating and Air-conditioning Engineers for their 2013 model code [57], and is also the subject of a set of building profiles [58]. In addition, programs should begin to focus on "zero net energy" as a long-term new construction target. Zero net energy buildings typically reduce energy use by roughly 75 percent relative to current codes and then generate the remaining power internally, at least on an annual average basis (e.g., they generate extra power some hours, less power other hours). Some programs are already beginning to target zero net energy [59]. The goal should be to eventually ratchet up building codes to achieve these levels of performance. Further information about zero net energy buildings can be found in several relevant chapters in this book.
- *Industrial process improvements and combined heat and power.* There are large savings available by modernizing industrial processes, particularly when production lines are reconfigured roughly every 20 years. Programs need to be in touch with customers to know when process refits will happen and have a cadre of expert consultants available who know about particular industries and can help with redesign. Programs can also encourage good operations and maintenance practices and can also encourage the incorporation of energy management into overall corporate management systems. A good example of the former is the Energy Trust of Oregon Industrial Efficiency Improvement program (see [76]). A good example of the latter is the Northwest Energy Efficiency Alliance Strategic Energy Management program (see [60]). In addition, programs should help identify good applications of combined heat and power (CHP) and waste energy recovery systems − systems that generate power and thermal energy for processes together, increasing efficiency relative to separate generating plants and steam boilers. The New York State Energy Research and Development Authority (NYSERDA) has run a successful program using rate-payer funds for several years [61].

- *Programs targeted at the largest customers and at cohesive commercial and industrial sectors.* Large savings can be achieved by building a long-term relationship with major customers where the customer and utility make multi-year commitments to work together on energy efficiency projects. For example, NStar Electric in Massachusetts has established several multiyear partnerships with its largest customers. To provide one example, in 2010 NStar and the Massachusetts Institute of Technology (MIT) instituted a $13 million three-year partnership designed to reduce MIT's energy use by 34 million kWh, about 15 percent of MIT's electricity use [62]. Likewise, Efficiency Vermont has established its Leadership Challenge program where it seeks commitments from its 300 largest customers to reduce energy use by 7.5 percent over 2 years. Efficiency Vermont has committed to helping each business create a comprehensive, long-term energy savings plan and is providing enhanced resources — both technical and financial — to help participants meet their goals [63]. There are also opportunities to work together with major cohesive business segments on efficiency projects involving an entire industry. For example, the Northeast Energy Efficiency Alliance has had successful programs with regional hospital associations [64] and with the Northwest Food Processors Association [65].
- *Deep retrofits* to substantially reduce energy use in existing buildings. Even in 2050 the majority of building floor area will be in buildings that are in existence today [29]. We need to reduce their energy use substantially through deep energy-saving retrofits at the time of major renovation. An example is the Empire State Building, which was recently retrofitted to reduce energy use by 38 percent [66]. Additional commercial building examples have been compiled by the New Buildings Institute [67]. Residential examples are being compiled by the 1000 Home Challenge, which is seeking to document 1000 deep retrofit examples [68]. Several program operators are now conducting pilot-deep retrofit programs such as National Grid [69].
- *Advanced technologies, including "intelligent efficiency."* There are many emerging technologies that efficiency programs can promote in the future, ranging from LED lighting (particularly once prices decline), variable speed air conditioners, heat pump water heaters, and "intelligent efficiency" strategies. *Intelligent efficiency* uses sensors, controls, software and information and communication technology to continually monitor and identify energy waste and opportunities to improve efficiency in building, manufacturing, and transportation systems and to provide this information to people or automatic controllers that can act on this information. *Intelligent efficiency* differs from component energy efficiency in that it is adaptive, anticipatory, and networked [70]. For all of these, market transformation strategies should be considered — multistage interventions that improve availability, reduce cost, and eventually make targeted

efficiency measures normal practice, such as leveraging codes and standards when possible. In the case of intelligent efficiency, work will also be needed on the best evaluation approaches so that savings can be documented.

- Strategies to reduce growing *"miscellaneous" energy uses.* DOE's Energy Information Administration estimates that by 2035, electronics and other miscellaneous energy uses will account for 50 percent of electricity use in the commercial sector and 39 percent in the residential sector [29]. These uses need to be better understood and strategies developed to reduce their energy use. For example, the New Buildings Institute and Ecova monitored these loads in a library and small office, installed a variety of no- and low-cost upgrades and reduced plug-load energy use by 17 percent in the library and 46 percent in the small office [71].

- *Smart grid and smart behavior.* The "smart grid" has received a lot of attention but thus far has not been widely used to help promote energy efficiency. There is a need to develop and test strategies for taking the information the smart grid makes available and providing this information to people and automatic controllers in ways that they can use to save energy. This will require learning by trial and error. For example, in the commercial sector, data on energy use can be monitored in real time and warnings given to building operators when systems are operating out of defined ranges [72]. Campaigns can be undertaken using the best social marketing and engagement techniques to encourage efficient workplace behaviors [73]. In the residential sector, feedback on energy use in real-time, presented in ways that catch attention and motivate action, can spur reductions in energy consumption. Foster and Mazur-Stommen [74] reviewed several pilot programs in this area and found they achieved an average of 4 percent energy savings, although results are preliminary and there is likely substantial room for improvement.

9 MOVING FORWARD

The proceeding discussion shows how the best utility sector programs are reducing electricity use by 1.5 to 2 percent per year and that available savings opportunities could allow this level of savings to continue for many decades. A few states have even "bent the curve" and electricity use is now declining rather than increasing.[5] A key question is whether this could happen for the entire United States. For the country as a whole, the answer is probably yes. EIA projects U.S. electricity use to grow only about 0.8

---

5. It should be noted that declines in electric sales were not a result of declining Gross State Prudct (GSP). GSP grew over the 2000–2010 period in all states, generally by 35 percent or more. The only states growing by less were Michigan (14 percent growth) and Ohio (25 percent growth) [75].

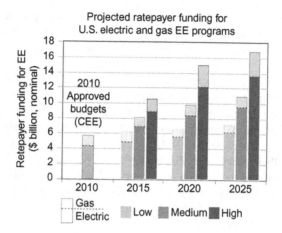

FIGURE 2.20   Note: "Low," "Medium," and "High" are scenarios for spending on programs. *Source: [78].*

percent per year over the 2012−2035 period [29]. This projection includes only limited utility programs − utility program savings are implicitly assumed to continue at historic levels, and the EIA projections do not include expected increases in utility program spending and savings in response to mandatory energy savings targets (EERS) and other drivers. The latest estimates from Lawrence Berkeley National Laboratory based on a detailed review of state laws and regulations is that utility sector energy efficiency spending will likely double relative to 2010 spending (see Figure 2.20) and could triple or even quadruple relative to 2009 spending (see Figure 2.1). If historic programs, as represented by 2009 savings, are saving about 0.4 percent of sales each year [1], then ramping-up average savings to 1.2 percent per year, commensurate with tripling spending, would eliminate overall sales growth.

This possibility is illustrated in Figure 2.21, which compares the EIA reference case forecast of electricity sales to a scenario in which energy efficiency savings are increased by 0.8 percent per year, with the increased savings phased in over four years. Under this scenario electricity use grows slightly in the early years and then levels off. We also include lines for 1.5 percent and 2 percent per year total efficiency savings (0.4 percent savings in base plus additional savings of 1.1 percent and 1.6 percent per year phased-in over six and eight years, respectively). The latter scenario results in a decline in total U.S. electricity use over the period of analysis.

However, sales growth is uneven, with some regions growing and some not. In order to get a rough picture of sales growth, electric sales in 2000 and 2010 by state can be compared, calculating the average annual percentage change over this period. These calculations are shown in Figure 2.22.

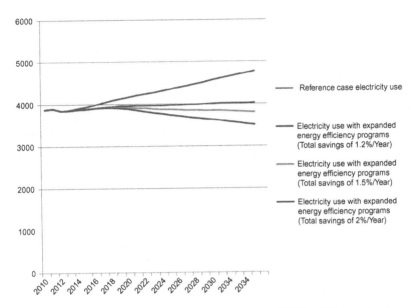

FIGURE 2.21 Comparison of EIA Reference Case to Cases with Higher Efficiency Savings.

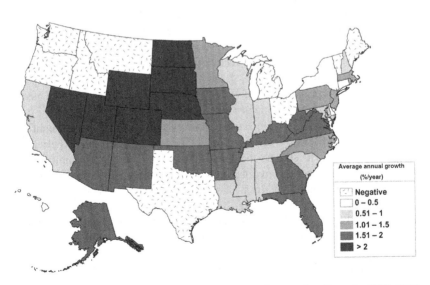

FIGURE 2.22 Average Annual Growth in Electricity Consumption Over the 2000–2010 Period by State. *Source: ACEEE analysis based on data in [2] and [30].*

Electricity sales in nine states actually declined over this period, driven by such factors as the Great Recession, changes in industry mix, and energy efficiency programs. Electricity sales in an additional 22 states grew by 1.5 percent per year or less and thus if these growth rates continue, expanded energy efficiency programs could eliminate sales growth. But in seven states, primarily in the mountain west and upper Great Plains, sales grew by more than 2 percent per year. In these states this rough analysis indicates that energy efficiency could reduce but not eliminate sales growth. And 12 states are on the cusp, with annual growth rates of 1.5 to 2 percent. This analysis is very rough, as past changes in sales are not necessarily indicative of the future. Still, this analysis does illustrate what might be possible and what is not.

10 CONCLUSIONS

Utility sector energy efficiency programs have a multi-decade history in the United States but have increased substantially in recent years. Leading states are using energy efficiency programs to reduce sales by 1.5 to 2.5 percent per year, and long-term analyses of energy efficiency potential indicate that savings of 1.5 to 2 percent per year could continue out to 2050. In the post-2030 period, growth in electric vehicles could have a substantial impact on sales growth, but until then, energy efficiency programs could eliminate sales growth in a substantial majority of states and overall.

However, achieving these savings will not be easy. We will need improved and new program approaches in order to move beyond lighting and other "tried and true" measures. We will need strong policy support, such as mandatory energy savings targets and a "three-legged stool" to support the business case for utility investments in energy efficiency – cost-recovery, decoupling/lost revenue recovery, and shareholder incentives. Most importantly, we will need smart, creative people to make this all happen. Energy efficiency programs involve both art and science, and we will need to be creative and learn by doing to make such large savings happen.

The benefits of such large energy savings are likely to be substantial. Even if energy efficiency savings double in cost from recent experience, energy efficiency will still be the low-cost resource, saving consumers and businesses money, thereby freeing up resources to spur a strong economy and meet other societal objectives.

This chapter has focused on the United States. However, many of these lessons are likely to apply to other countries while recognizing each country's unique characteristics. All countries for the most part have the same major uses of electricity and therefore similar efficiency opportunities. Where countries perhaps differ most is in their growth rates. For countries where electricity use is growing by about 2 percent per year or less, efficiency may be able to eliminate load growth. For countries growing more rapidly, efficiency opportunities may increase somewhat as they generally have a lot of new construction, and new construction presents particular opportunities to incorporate efficiency at modest cost. Still, for countries growing rapidly, efficiency investments are likely to reduce but not eliminate load growth.

REFERENCES

[1] Sciortino M, Neubauer M, Vaidyanathan S, Chittum A, Hayes S, Nowak S, Molina M. The 2011 state energy efficiency scorecard. Washington, D.C.: American Council for an Energy-Efficient Economy; 2011a Report E115, <http://aceee.org/research-report/e115>.

[2] [EIA] U.S. Energy Information Administration. Electric power annual. Washington D.C.: U. S. Energy Information Administration; 2011a, <http://www.eia.gov/electricity/annual/>.

[3] [EIA] U.S. Energy Information Administration. Annual energy review. Washington D.C.: U.S. Energy Information Administration; 2011b, <http://205.254.135.24/totalenergy/data/annual/index.cfm>.

[4] Collins N, Berry L, Braid R, Jones D, Kerley C, Schweitzer M, Soresen J. Past efforts and future directions for evaluating state energy conservation programs. ORNL-6113. Oak Ridge, Tenn: Oak Ridge National Laboratory; 1985.

[5] Geller H. Lessons from utility experimentation with appliance efficiency incentive programs. Proceedings of the 1988 ACEEE summer study on energy efficiency in buildings 1988;6.50–4.

[6] Nadel S. Lessons learned: a review of utility experience with conservation and load management programs for commercial and industrial consumers. Washington, D.C.: American Council for an Energy-Efficient Economy; 1990 Report U901, <http://aceee.org/research-report/u901>.

[7] Nadel S, Reid MW, Wolcott DR. Regulatory incentives for demand-side management. Washington, D.C.: American Council for an Energy-Efficient Economy; 1992, <http://aceee.org/ebook/regulatory-incentives-for-demand-side-management>.

[8] Kushler M, Witte P. Can we just "Rely on the Market" to provide energy efficiency? An examination of the role of private market actors in an era of electric utility restructuring. Washington, D.C.: American Council for an Energy-Efficient Economy; 2001, Report U011.

[9] Cain C, Lester JA. Retail rate comparisons and the electric restructuring debate bates. San Diego, Calif.: Bates White, LLC; 2008.

[10] Bacon D. California's disaster with electrical deregulation, <http://dbacon.igc.org/PJust/02ElecDereg.htm>; 2001.

[11] Taylor J, Van Doren P. Lights out on restructuring? <http://www.cato.org/publications/commentary/lights-out-electric-restructuring>; 1998.

[12] Nadel S, Amann J, Sachs H, Prindle B, Neal Elliott R. Market transformation: substantial progress from a decade of work. Washington, D.C.: American Council for an Energy-Efficient Economy; 2003.

[13] Efficiency Vermont. Year 2010 savings claim. Burlington, Vt.: Vermont Energy Investment Corporation; 2011, <http://www.efficiencyvermont.com>.

[14] Nadel S, Langer T. Comments on "Is There an Energy Efficiency Gap?". Washington, DC: American Council for an Energy-Efficient Economy; 2012.

[15] Kushler M, York D, Witte P. Five years in: an examination of the first half-decade of public benefits energy efficiency policies. Washington D.C.: American Council for

an Energy-Efficient Economy; 2004 Report U042, <http://aceee.org/research-report/u042>.

[16] Leuthauser F"Rick", Weaver E. Leveraging customer satisfaction through energy efficiency. Proceedings of the 2006 ACEEE summer study on energy efficiency in buildings. Washington, D.C.: American Council for an Energy-Efficient Economy; 2006, <http://www.aceee.org/proceedings-paper/ss06/panel05/paper20>.

[17] Sciortino M. Energy efficiency resource standards: definitions and state progress. Presentation to IEA. Washington, D.C.: American Council for an Energy-Efficient Economy; 2012.

[18] Hayes S, Nadel S, Kushler M, York D. Carrots for utilities: providing financial returns for utility investments in energy efficiency. Washington, D.C.: American Council for an Energy-Efficient Economy; 2011a Report U111.

[19] York D, Kushler M. The old model isn't working: creating the energy utility for the 21st century. Washington, D.C.: American Council for an Energy-Efficient Economy; 2011, <http://aceee.org/white-paper/the-old-model-isnt-working>.

[20] Sciortino M, Nowak S, Witte P, York D, Kushler M. Energy efficiency resource standards: a progress report on state experience. Washington, D.C.: American Council for an Energy-Efficient Economy; 2011b, <http://aceee.org/research-report/u112>.

[21] Rosenfeld A, Poskanzer D. A graph is worth a thousand gigawatt- hours. Innovations J 2009; <http//www.energy.ca.gov/commissioners/rosenfeld_docs/INNOVATIONS_Fall_2009_Rosenfeld-Poskanzer.pdf>.

[22] Eckman T. The value energy efficiency as a resource option: three decades of pnw experience. Presented at DOE, IEA and RAP workshop on policies for energy provider delivery of energy efficiency. Portland, Ore: Northwest Power and Conservation Council; 2012a.

[23] Eckman T. In the PNW we do more than plan!. Presented at ACEEE energy efficiency as a resource conference. Portland, Ore: Northwest Power and Conservation Council; 2011.

[24] [EIA] U.S. Energy Information Administration. State energy data system. Washington D.C.: U.S. Energy Information Administration; 2012b, <http://www.eia.gov/state/seds/#> [visited 21.04.12].

[25] Horowitz P, Schlegel J, Sherman M. Programs and strategies to achieve all available cost-effective energy efficiency: early report on bending the curve in Massachusetts. Proceedings of the 2010 ACEEE summer study on energy efficiency in buildings. Washington D.C.: American Council for an Energy-Efficient Economy; 2010, <http://eec.ucdavis.edu/ACEEE/2010/data/papers/2082.pdf>.

[26] [DEEP] Connecticut Department of Energy and Environmental Protection. Integrated resource plan − 2012. Hartford, Conn: Connecticut Department of Energy and Environmental Protection; 2012, < http://www.ct.gov/deep/cwp/view.asp?a=4120 &q=486946 >.

[27] Eckman T. Spreadsheet titled "regional conservation summary 1978−2009 adjusted for BPA co-funding and including line losses". Portland, Ore: Northwest Power and Conservation Council; 2012.

[28] Hayes S, Nadel S, Granda C, Hottel K. What have we learned from energy efficiency financing programs? Washington, D.C.: American Council for an Energy-Efficient Economy; 2011 Report U115, <http://www.aceee.org/research-report/u115>.

[29] [EIA] U.S. Energy Information Administration. AEO. Washington D.C.: U.S. Energy Information Administration; 2012.

[30] [EIA] U.S. Energy Information Administration. Electric power monthly, March. Washington D.C.: U.S. Energy Information Administration; 2012c, <http://www.eia.gov/electricity/monthly/>.

[31] [CCST] California Council on Science and Technology. California's energy future – the view to 2050, summary report. Sacramento, Calf: California Council on Science and Technology; 2011, <http://www.ccst.us/publications/2011/2011energy.pdf>.

[32] Long J. Chair's lecture: California energy futures study results, <http://www.arb.ca.gov/research/lectures/speakers/long.pdf>; 2011 [15.07.11].

[33] Lovins A, Rocky Mountain Institute. Reinventing fire: bold business solutions for the new energy era. White River Junction, Conn: Chelsea Green; 2011.

[34] Laitner JA"Skip", Nadel S, Neal Elliott R, Sachs H, Siddiq Khan A. The long-term energy efficiency potential: what the evidence suggests. Washington, D.C.: American Council for an Energy-Efficient Economy; 2012, Report E121, <http://aceee.org/research-report/e121>.

[35] [IEE] Institute for Electric Efficiency. Assessment of Electricity Savings in the U.S. Achievable through new appliance/equipment efficiency standards and building efficiency codes (2010–2025). Washington D.C.: Institute for Electric Efficiency; 2011, <http://www.edisonfoundation.net/iee/Documents/IEE_CodesandStandardsAssessment_2010-2025_UPDATE.pdf>.

[36] Lowenberger A, Mauer J, deLaski A, DiMascio M, Amann J, Nadel S. The efficiency boom: cashing in on the savings from appliance standards. Washington, D.C.: American Council for an Energy-Efficient Economy; 2012. Report A123, <http://aceee.org/research-report/a123>.

[37] [IEE] Institute for Electric Efficiency. Integrating codes and standards into electric utility energy efficiency portfolios. Washington D.C.: Institute for Electric Efficiency; 2011, <http://www.edisonfoundation.net/iee/Documents/IEE_IntegratingCSintoEEPortfolios_final.pdf>.

[38] Efficiency Vermont. Efficiency vermont annual plan 2012. Burlington, Vt.: Vermont Energy Investment Corporation; 2012, <http://www.efficiencyvermont.com/docs/about_efficiency_vermont/annual_plans/EVT_AnnualPlan2012.pdf>.

[39] [CEE] Consortium of Energy Efficiency. Consortium for energy efficiency 2012 behavior program summary. Boston, Mass: Consortium for Energy Efficiency; 2012.

[40] Navigant Consulting, Inc. Energy savings potential of solid-state lighting in general illumination applications. Washington, D.C.: Navigant Consulting, Inc; 2012, <http://apps1.eere.energy.gov/buildings/publications/pdfs/ssl/ssl_energy-savings-report_jan-2012.pdf>.

[41] Shapiro C, Puttagunta S, Owens D. Measure guideline: heat pump water heaters in new and existing homes. Washington, D.C.: U.S. Department of Energy; 2012, <http://www.nrel.gov/docs/fy12osti/53184.pdf>.

[42] [DOE] U.S. Department of Energy. Building technologies program. Washington D.C.: U.S. Department of Energy; 2011, <http://apps1.eere.energy.gov/buildings/publications/pdfs/alliances/techspec_rtus.pdf>.

[43] Takahashi K, Nichols D. The sustainability and costs of increasing efficiency impacts: evidence from experience to date. Proceedings of the 2008 ACEEE summer study on energy efficiency in buildings. Washington, D.C.: American Council for an Energy-Efficient Economy; 2008, <http://www.aceee.org/proceedings-paper/ss08/panel08/paper30>.

[44] [RTF] Regional Technical Forum. Guidelines for the development and maintenance of rtf savings estimation methods. Portland, Ore.: Northwest Power and Conservation Council; 2011, <http://www.nwcouncil.org/energy/rtf/subcommittees/deemed/Guidelines%20for%20RTF%20Savings%20Estimation%20Methods%20%28Release%206-1-11%29.pdf>.

[45] [NEEP] Northeast Energy Efficiency Partnerships. Northeast energy efficiency partnerships, EM&V forum. Portland, Ore.: Northwest Energy Efficiency Alliance; 2012, <http://neep.org/emv-forum>.

[46] Schiller S, Goldman C, Galawash E. National energy efficiency evaluation, measurement and verification (em&v) standard: scoping study of issues and implementation requirements. Berkeley, Calf.: Lawrence Berkeley National Laboratory; 2011, <http://eetd.lbl.gov/ea/EMP/reports/lbnl-4265e.pdf>.

[47] [NERC] North American Electric Reliability Corporation. Reliability assessment 2000–2009: the reliability of bulk electric systems in North America. Atlanta, Ga.: North American Electric Reliability Corp; 2000.

[48] [NERC] North American Electric Reliability Corporation. 2011 long-term electric reliability assessment. Atlanta, Ga.: North American Electric Reliability Corp; 2011, <http://www.nerc.com/files/2011%20LTRA_Final.pdf>.

[49] [FERC] Federal Energy Regulatory Commission. A national assessment of demand response potential. Washington, D.C.: U.S. Department of Energy; 2009.

[50] Eto J, Goldman C, Nadel S. Ratepayer-funded energy efficiency programs in a restructured electricity industry: issues and options for regulators and legislators. Berkeley, CA: Lawrence Berkeley National Laboratory; 1998 Report Number LBNL-41479.

[51] Harrington C. Who should deliver ratepayer funded energy efficiency programs? Montpelier, VT: Regulatory Assistance Project; 2003.

[52] Kim EL, Tabors RD, Stoddard RB, Allmendinger TE. Carbitrage: utility integration of electric vehicles and the smart grid. The Electricity J 2012;25(2):16–23.

[53] Power JD, Associates. Future global market demand for hybrid and battery electric vehicles may be over-hyped. Westlake Village, Calf: J.D. Power and Associates; 2010, <http://businesscenter.jdpower.com/news/pressrelease.aspx?ID=2010213>.

[54] [NAS] National Academy of Science. Transition to alternative transportation technologies – plug-in hybrid electric vehicles. Washington, D.C.: The National Academies; 2009.

[55] [DOE] U.S. Department of Energy. 2012 IECC final action hearing delivers DOE's 30 percent energy savings goals. Building energy codes program. Washington D.C.: U.S. Department of Energy; 2011b, <http://www.energycodes.gov/status/2012_Final.stm>.

[56] Nadel S. The federal energy policy act of 2005 and its implications for energy efficiency program efforts. Washington, D.C.: American Council for an Energy-Efficient Economy; 2005 Report E053

[57] [ASHRAE] American Society of Heating, Refrigerating and Air Conditioning Engineers. SSPC 90.1 2013 work plan. Atlanta, Ga: American Society of Heating, Refrigerating and Air Conditioning Engineers; 2011.

[58] Higgins C, Castillo G, Egnor T. Getting to 50: drivers and data of measured energy performance. Proceedings of the 2010 ACEEE summer study on energy efficiency in buildings. Washington D.C.: American Council for an Energy-Efficient Economy; 2010, <http://eec.ucdavis.edu/ACEEE/2010/data/papers/2008.pdf>.

[59] [NBI] New Buildings Institute and Preservation Green Lab. Getting to zero 2012 status update. Vancouver, Wash: New Buildings Institute; 2012, <http://www.newbuildings.org/getting-zero-2012-status-update-first-look-costs-and-features-zero-energy-commercial-buildings>.

[60] Jones T, Crossman K, Eskil J, Wallner J. The evolution of continuous energy improvement programs in the northwest: an example of regional collaboration. Consortium for energy efficiency. Proceedings of the 2011 ACEEE summer study on energy efficiency in buildings. Washington D.C.: American Council for an Energy-Efficient Economy; 2011.

[61] Chittum A, Kaufman N. Challenges facing combined heat and power today: a state-by-state assessment. Washington D.C.: American Council for an Energy-Efficient Economy; 2011 Report IE11, <http://aceee.org/research-report/ie111>.

[62] MIT. MIT efficiency forward exceeds electricity reduction goal. Cambridge, Mass: Massachusetts Institute of Technology; 2011 (press release), <http://web.mit.edu/press/2011/nstar-mit.html>.

[63] Edwards Bruce. Companies commit to major energy savings. Rutland Herald 2011; Decemeber:25, <http://www.efficiencyvermont.com/docs/about_efficiency_vermont/news_articles/RutlandHerald_122511.pdf>.

[64] [NEEA] Northwest Energy Efficiency Alliance. Undated1. NEEA Success story: commercial. Portland, Ore.: Northwest Energy Efficiency Alliance, <http://neea.org/successstories/docs/neea_success_story_commercial.pdf>.

[65] [NEEA] Northwest Energy Efficiency Alliance. Undated2. NEEA Success story: industrial. Portland, Ore.: Northwest Energy Efficiency Alliance, <http://neea.org/successstories/docs/neea_success_story_industrial.pdf>.

[66] Navarro M. Empire state building plans environmental retrofit. New York Times, <http://www.nytimes.com/2009/04/07/science/earth/07empire.html>; 2009.

[67] [NBI] New Buildings Institute and Preservation Green Lab. Deep retrofits. Vancouver, Wash: New Buildings Institute; 2011, <http://www.newbuildings.org/sites/default/files/11DeepSavingsEBCaseStudiesNBI.pdf>.

[68] Wigington L. Introduction to the thousand home challenge, <http://thousandhomechallenge.com/sites/thousandhomechallenge.com/files/user-files/Thousand-Home-Challenge-Intro_Webinar_10-16-2011.pdf>; 2011.

[69] Neuhauser K. Deep energy retrofit incentive programs: the national grid pilot. Presented at the residential building energy efficiency meeting. Somerville, Mass.: Building Science Consulting; 2010, <http://apps1.eere.energy.gov/buildings/publications/pdfs/building_america/ns/a24_deep_energy.pdf>.

[70] Elliott RN, Molina M, Trombley D. A defining framework for intelligent efficiency. Washington D.C.: American Council for an Energy-Efficient Economy; 2012 Report E125

[71] Mercier C, Moorefield L. Commercial office plug load savings and assessment: executive summary. Vancouver, Wash: New Buildings Institute; 2011, <http://www.newbuildings.org/sites/default/files/OfficePlugLoadAssessment_ExecutiveSummary.pdf>.

[72] Forsman K, Silver-Pell T. Energy information systems: a utility perspective. Presentation at the 2010 Market transformation symposium, <http://www.aceee.org/files/pdf/conferences/mt/2010/C2_Theda_Silver_Pell.pdf>; 2010.

[73] Shui Bin. Greening work styles: an analysis of energy behavior programs in the workplace. Washington, D.C.: American Council for an Energy-Efficient Economy; 2012 Report B121, <http://aceee.org/research-report/b121>.

[74] Foster B, Mazur-Stomman S. Results from recent real-time feedback studies. Washington D.C.: American Council for an Energy-Efficient Economy; 2012 Report B122, <http://www.aceee.org/research-report/b122>.

[75] Economy.com. 2012. Data buffet: gross state product, historical, by state. Downloaded May 2012.

[76] Crossman K, Brown D. Energy trust of oregon kaizen blitz pilot program. Proceedings of the 2009 ACEEE summer study on energy efficiency in industry. Washington D.C.: American Council for an Energy-Efficient Economy; 2009, <http://www.aceee.org/proceedings-paper/ss09/panel03/paper07>.

[77]   Friedrich K, Eldridge M, York D, Witte P, Kushler M. Saving energy cost-effectively: a national review of the cost of energy saved through utility-sector energy efficiency programs. Washington D.C.: American Council for an Energy-Efficient Economy; 2009 Report U092, <http://www.aceee.org/research-report/u092>.

[78]   Goldman CA, Barbose G, Hoffman IM, Bilingsley M. A rising tide for utility customer-funded energy efficiency. Proceedings of the 2012 ACEEE summer study on energy efficiency in buildings. Washington D.C.: American Council for an Energy-Efficient Economy; 2012 Lawrence Berkeley National Lab. 2012

[79]   Hayes S, Young R. Energy Efficiency: The Slip Switch to a New Track Toward Compliance with Federal Air Regulations. Washington, D.C.: American Council for an Energy-Efficient Economy; 2012, <https://www.aceee.org/research-report/e122>.

[80]   Kushler M, Nowak S, Witte P. A National Survey of State Policies and Practices for the Evaluation of Ratepayer-Funded Energy Efficiency Programs. Washington, D C. American Council for an Energy-Efficient Economy; 2012, < https://www.aceee.org/research-report/ u122 >.

[81]   Lazard, Ltd. Levelized cost of energy analysis – version 5.0. Energy 2011; (no. June):0−15.

Chapter 3

A Global Perspective on the Long-term Impact of Increased Energy Efficiency

Paul H.L. Nillesen[1†], Robert C.G. Haffner[1] and Fatih Cemil Ozbugday[2]

[1]PwC, The Netherlands, [2]Faculty of Economics & Business Administration, Department of Economics, Tilburg University, The Netherlands

1 INTRODUCTION

Energy efficiency has for long been perceived as the "holy grail" of energy policymakers due to a number of reasons. Firstly, energy efficiency directly reduces negative environmental effects associated with the consumption of fossil fuels and therefore directly contributes to achieving environmental policy objectives. More importantly, improvements in energy efficiency often have a positive business case, as saving energy means saving money. This contrasts starkly to many other (sometimes exotic) environmental protection measures considered by policymakers, companies, and consumers. As McKinsey [1] points out, energy efficiency has the potential to offer a vast, low-cost energy resource, which could yield gross energy savings worth more than $1.2 trillion for the U.S. economy alone. Figure 3.1 shows the significant potential offered by various measures aimed at improving energy efficiency as calculated by McKinsey [1]. As the study shows, many measures actually have negative abatement costs. Summarizing various studies on the potential of energy efficiency in the Australian residential building sector, PwC (2008) [2] reaches a similar conclusion, pointing out the negative levelized costs of many investment opportunities. An array of other studies point in the same direction.

For policymakers, investments in energy efficiency are particularly attractive as they offer both microeconomic and macroeconomic benefits. At a microeconomic level, energy-saving measures can have a positive impact

---

†Corresponding author. Any errors or omissions are the responsibility of the authors. The views expressed here are those of the authors and do not necessarily reflect the views of PricewaterhouseCoopers Advisory N.V. or any of the PricewaterhouseCoopers network of firms e-mail: paul.nillesen@nl.pwc.com.

Energy Efficiency. DOI: http://dx.doi.org/10.1016/B978-0-12-397879-0.00003-7

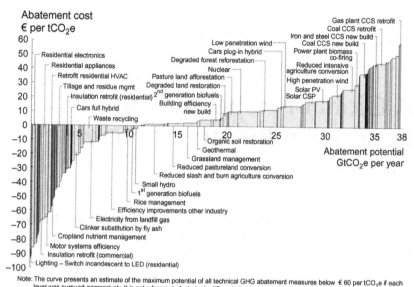

FIGURE 3.1   McKinsey Abatement cost of different technologies. *Source: McKinsey.*

on household incomes, reduce vulnerability of households to changes in energy prices, and also raise environmental awareness. The benefits at the sectoral and macroeconomic level are similar, especially for energy importing countries, as competitiveness is improved and exposure to energy price spikes is reduced with improved energy efficiency. The paradox is that given the significant benefits involved with investments in energy efficiency, in practice it has proved challenging to actually unlock this potential at the scale perceived by studies as of McKinsey [1]. The chapter by Allcott and Greenston [3] in this book is devoted to this question; part of the answer, lack of focus on consumers, is provided in the chapter authored by Laskey and Syler [4] as well as others in this volume.

However, at a macroeconomic level, improvements in energy efficiency have still been significant. Since the 1970s, economic growth has become significantly less energy intensive, not only in developed countries but also in non-OECD countries. According to the International Energy Agency (IEA) recent *World Energy Outlook*, energy intensity has declined by approximately 1.3 percent annually in OECD-countries and almost 1.5 percent in non-OECD countries in the period 1985–2009. For the future, the IEA has developed a number of scenarios. Interestingly, both under a "current policies" scenario and under alternative scenarios, global energy intensity is projected to decline in 2009–2035 by 36 to 44 percent depending

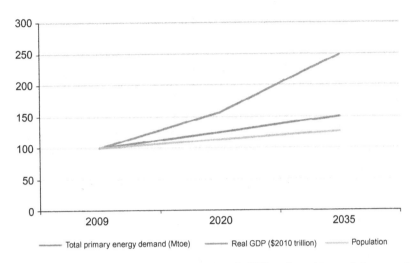

——— Total primary energy demand (Mtoe)   ——— Real GDP ($2010 trillion)   ······· Population

FIGURE 3.2 World energy demand, world real GDP and world population growth (2009 = 100). *Source: IEA, World Energy Outlook (2011), Annex A (Unchanged policies scenario).*

on the scenario. The extent to which energy intensity reduces in non-OECD countries, particularly China and, to a lesser extent, India and Russia, is actually the main change across the IEA scenarios that demonstrates the crucial importance of this variable.

Figure 3.2 summarizes IEA's current policies projections for energy demand, GDP growth, and population growth for the period 2009–2035. The figure shows that while GDP growth is expected to more than double, energy demand is increasing only by 50 percent, thus illustrating the expected fall in energy intensity. Population is expected to increase by 26 percent during this period, thereby indicating that at a per capita level more energy will be used in the future than today when policies are unchanged.

These projected improvements in energy efficiency on a global scale give the impression of a rather smooth and gradual improvement in energy efficiency. However, a lot of the underlying dynamics are hidden, as the expected improvements are the potential result of a number of diverse developments. Firstly, as indicated by the IEA, one would expect differences between developed and developing countries. Countries that are still at the beginning of their development may first see a significant reduction in energy efficiency as they develop from an agricultural society to a more industrial one. The industrialization phase is associated with energy growth and increasing energy intensity. After having completed the industrial phase, energy intensity may decline once again as countries become more services oriented.

This chapter's first question therefore is: what is the typical pattern observed in historical data with respect to energy intensity in different stages of development of countries? The second question is directly related to the first one. To what extent is the change in energy intensity caused by changes in the shares of different sectors of the economy such as agriculture, industry, and services? If changes in energy intensity are mainly caused by changes in the sectoral composition of economic growth, this is probably less the result of deliberate policies aimed at improving energy efficiency but more a "natural" economic development. Thirdly, what are the main determinants of changes in energy intensity? Based on an analysis of the determinants of historical developments in energy intensity, alternative forecasts are provided of expected future developments and compared to the projections provided by the IEA. The fourth and final research question therefore is: which changes in the level of energy intensity are likely based on historical evidence and alternative scenarios for the future? What do fore-casted energy intensity levels mean for the level of energy consumption in 2035?

The main value added of the chapter is to provide a comprehensive over-view of historical developments in energy intensity (and its determinants) on a *global* scale with data on 65 OECD and non-OECD countries over a relatively long time period compared to other studies. Based on this analysis, a projection of expected developments in energy consumption is given and policy conclusions based on this analysis are provided.

The plan of the chapter is as follows: The next section discusses the available economic literature on the development of energy efficiency at a macroeconomic and sectoral level. Section 3 then briefly describes the dataset and countries used to undertake the analysis and presents the method-ology. More details on the methodology can be found in the annex to this chapter. Section 4 presents the results of the analysis. The last section summarizes the conclusions.

2 LITERATURE REVIEW

There is broad literature on the costs, benefits, and potential of energy efficiency. This section focuses on the macroeconomic and sectoral analyzes as these are relevant to the research questions. Some of the analyses discussed assess the existence of an *environmental* "Kuznets curve." The traditional Kuznets curve is a graphical representation of the relationship between income inequality and income per capita. The hypothesis is that as a country develops, there is a natural cycle of economic inequality driven by market forces, which at first increases inequality and then decreases it after a certain average income is attained. As a result, the Kuznets curve has an inverted U-shape.

The environmental version of this curve follows a similar line of reasoning (see Figure 3.3). It basically states that the relationship between energy inten-sity and income levels (GDP per capita) has an inverted U-shape. Following

FIGURE 3.3   The environmental Kuznets curve.

Jaffe [5], the reasoning behind why energy intensity at first rises and then falls past some turning point as GDP per capita rises is based on three pillars.

Firstly, rising income levels are accompanied by structural changes in the economy, involving a move away from an industrialized economy with relatively high energy intensity per unit of value added, toward an economy that is increasingly dominated by the service sector or relatively cleaner industries.

Secondly, rising incomes raise the demand for environmental quality at a more-than-proportional rate. The reason is that the income elasticity of environmental quality is greater than one, indicating that energy intensity has characteristics of a luxury good. This would be due to the fact that rising income levels translate into political pressure for "greener" policies, such as tougher environmental policy standards and subsidies for energy conservation programs.

Finally, technological improvements in production processes may reduce the amount of energy needed to produce one unit of value added. High or rising energy prices may play a part here as the business case of investing in energy efficient technologies will improve with higher energy prices. Moreover, as a country gets richer it can afford to spend more on research and development, which can lead to more advanced and environmentally friendly production techniques. If there are increasing returns to abatement technology, this can also generate an inverted U-shape.

It is important to take into account that energy *intensity* does not say anything about how efficiently the energy is used. Energy use depends on many socioeconomic and environmental circumstances, such as comparative advantages for energy-intensive activity, the endowment with natural resources, but also population density and even climatological circumstances, such as extreme heat and extreme cold.

Energy *efficiency* is a measure of how resourcefully energy is used under these conditions (and given prices). Comparisons of energy intensity do not account for the different circumstances the various countries are facing as it combines differences in both energy efficiency and other conditions. For this reason it is useful to disentangle the developments in energy intensity and energy efficiency.

The following is a brief overview of the main (academic) studies and their findings on the development in energy intensity and energy efficiency over time. This chapter focuses on two types of studies. The first category analyzes the presence of convergence or divergence in energy intensity levels. The second decomposes the development in energy intensity in sub-categories to shed light on the main drivers of these developments. This provides the basis for the research presented in this chapter.

2.1 Convergence or Divergence in Energy Intensity

One of the early academic studies analyzing energy intensity trends is authored by Nilssen [6]. He focuses on 31 developed and developing countries in the time period 1950–1988. He notes that in about half the countries energy intensity decreases over time, despite falling prices, pointing toward a convergence in the energy intensities of developed and less developed countries. The author also points out that there is a large variation in trends and intensities between different countries. This reflects different economic structures and different development paths. Making predictions of energy intensity in the future based on highly aggregated data can be misleading as a result of these differences. Time series analyses of individual countries or groups of countries are more informative to discover trends than highly aggregated analyses. Moreover, cross-country comparisons of energy intensity should take into account differences in economic structure and climate. Finally, Nilssen notes that electricity intensities have on average been rising, with the exceptions of some of the high-income countries. Electricity appears to be the energy carrier of choice in many countries.

Mielnik and Goldenberg [7] also observe energy intensity trends of developing versus developed countries, but for the period 1971–1992. Their data shows convergence with both groups approaching a common pattern of energy use when using real GDP based on purchasing power parities. Energy intensity of developed countries generally decreases, whereas in developing countries intensity increases. The data show that 18 industrialized countries are on a decreasing trajectory and 23 developing countries follow an increasing energy intensity path. As a reference, the upper limit is given by the United States' energy intensity trajectory and the lower limit is the Indonesian energy intensity path over time. The authors conclude that the declining energy intensity in developed countries points to the availability of energy efficient technology. Developing countries could use this to fulfil Kyoto Protocol goals, while maintaining the pace of economic development.

Another paper reviewing convergence in energy intensity is by Sun [8]. He analyzes the presence of mean deviation in 27 OECD countries from 1971–1998. Results confirm the findings of other studies that convergence in energy intensity levels is taking place. He concludes that the mean deviation of energy intensity in OECD countries has decreased. Sun believes that his results provide strong indications of the presence of technological advancements and advancements in social economic structures, which are transferred from one country to the next. He expects this to be a strong driver of the patterns measured.

A special category of countries is studied by Cornillie and Fankhauser [9] who focus on energy intensity in transition economies, mainly Eastern Europe and the Commonwealth of Independent States. The analysis centers around 22 transition countries in six regional groups over the period of 1992−1998 using data from various sources. Intensity patterns vary strongly across different groups. However, energy intensity in most country groups has decreased. Note that energy intensity in these economies has traditionally been high compared to OECD countries but started to decrease since the beginning of transition towards a market economy. The improvement in energy efficiency reflects the reallocation of resources compared to its prior pricing systems, making production processes more efficient and reducing energy consumption. Therefore, this paper clearly shows the link between changes in economic structures and changes in energy intensity. The fact that energy intensity levels at the end of 1998 are still behind average OECD levels indicates that the transition process was not yet completed by 1998.

2.2 Decomposition Analysis

Metcalf [10] analyzes U.S. data regarding energy consumption at the state level for the period 1970−2001. He decomposes changes in energy intensity into two different effects, in the same way as is done in this chapter. The first is what he calls an efficiency effect. This is the change in energy intensity given a constant composition of economic activity. The composition of economic activity is measured by the share of residential, commercial, industrial, and transportation sectors in the state economies. The second is the activity effect, which is the change in energy intensity as a result of the change in the composition of economic activity. The decomposition suggests that roughly three quarters of the reduction in energy intensity is due the first effect: changes in energy efficiency. Metcalf concludes that this is evidence disproves concerns that the U.S. economy is moving energy-intensive industries outside and imports corresponding carbon-intensive consumer products. Metcalf also finds that energy intensity declines when per capita income rises. This conclusion is based on a regression analysis explaining the development of energy intensity over time across states.

Following up on the study of Metcalf, Mulder, and de Groot [11] illustrate that his approach is actually too crude when analyzing developments in energy intensity across countries. Most studies like Metcalf's that decompose the development in energy intensity into an efficiency effect and an activity effect only look at broad categories of sectors. Mulder and de Groot show that it is important not only to study changes in energy intensity between sectors but that changes *within* sectors can also be important. Using a high-quality sectoral dataset based on the EU KLEMS Growth and Productivity Accounts together with physical energy data from the IEA, these authors decompose changes in energy intensity in a structure effect and an efficiency effect. They cover 25 manufacturing sectors and 23 service sectors as well as the transport, agriculture, and construction sectors. They find that across countries energy intensity levels tend to increase in a fairly wide range of services sub-sectors, but decrease in most manufacturing sectors. Moreover, cross-country variations in aggregate energy intensity levels tend to decrease since 1995. Put differently, energy intensity levels in OECD countries tend to converge both in manufacturing and in service sectors. Interestingly, this convergence is almost exclusively caused by convergence of *within-sector* energy intensity levels, and not by convergence of the sectoral composition of economies.

3 DATA AND METHODOLOGY

In order to examine the potential from energy efficiency, a unique dataset for 65 countries has been collected for this chapter. Using this dataset and the decomposition methodology of Metcalf [10], the trends in energy intensity and energy efficiency are analysed separately for OECD countries and the rest of the world. Furthermore, the change in energy intensity is explained using regression analysis. Based on this regression analysis a forecast is made of energy demand for a sub-sample of countries for which the necessary data was available. This projection of energy intensity in 2035 is then compared to the IEA forecast.

However, from the perspective of global warming energy intensity may not be the most relevant policy variable as energy consumption may go up even when energy intensity is reduced. For this reason, the implications of the forecast for energy consumption are also discussed in this chapter. The chapters by Chang et al and Levine et al in this volume focus on energy efficiency trends in ASEAN countries and China, respectively.

The decomposition analysis disentangles the development in energy intensity into two important determinants: changes in energy efficiency and changes in economic activity. These are labelled *efficiency* and *activity* determinants in the remainder of the chapter. *Efficiency* indicates the decreased energy consumption per unit of economic activity, whereas *activity* refers to the shifts from energy-intensity economic activities to non-energy intensive economic activities holding efficiency constant.

When assessing changes in the composition of economic activity, a distinction is made between four different sectors: industry, services, transportation, and residential. For the energy use in the industry, services, and transportation sectors, value added in these sectors is used as the key driver of energy demand. As to residential energy consumption, household consumption expenditures are considered the key driver of residential energy demand.

Table 3.1 indicates the sectors and the measures of economic activity in those sectors that are used for the decomposition.

TABLE 3.1 Sectors for Decomposition Analysis at Country Level

Sector	Economic Activity Measure	Sectoral Energy Efficiency Measure
Residential	Household Consumption Expenditures ($2005)	Energy Consumption (in thousand tonnes of oil equivalent (ktoe) on a net calorific value basis) per dollar ($2005)
Services	Value Added in Services Sector ($2005)	Energy Consumption (in thousand tonnes of oil equivalent (ktoe) on a net calorific value basis) per dollar ($2005)
Industrial	Value Added in Industrial Sector ($2005)	Energy Consumption (in thousand tonnes of oil equivalent (ktoe) on a net calorific value basis) per dollar ($2005)
Transportation	Value Added in Transportation Sector ($2005)	Energy Consumption (in thousand tonnes of oil equivalent (ktoe) on a net calorific value basis) per dollar ($2005)
Total	GDP ($2005)	Energy Consumption (in thousand tonnes of oil equivalent (ktoe) on a net calorific value basis) per dollar ($2005)

Source: *Energy and electricity consumption data from the International Energy Agency. Economic data from the United Nations Statistics Division (UNSD). Data runs from 1971 to 2009.*

TABLE 3.2 List of Countries included in the Dataset

OECD Countries	Non-OECD Countries
Australia, Austria, Belgium, Canada, Chile, Denmark, Finland, France, Germany, Greece, Hungary, Iceland, Ireland, Israel, Italy, Japan, Mexico, Netherlands, New Zealand, Norway, Portugal, South Korea, Spain, Switzerland, Turkey, UK, U.S.	Argentina, Bangladesh, Bolivia, Brazil, China, Colombia, Costa Rica, Cote de Ivorie, Cuba, Cyprus, Ecuador, El Salvador, Ghana, Guatemala, India, Indonesia, Jamaica, Jordan, Kenya, Libya, Malaysia, Morocco, Nigeria, Pakistan, Panama, Paraguay, Peru, Philippines, Romania, Senegal, Singapore, South Africa, Syria, Thailand, Tunisia, Uruguay, Venezuela, Zimbabwe

Energy and electricity consumption data are obtained from energy balance databases of OECD and a selection of non-OECD countries (henceforth non-OECD countries). These databases are run by IEA. The datasets include 65 countries. The list of these countries is given in Table 3.2. Collectively, these 65 countries cover 94 percent of global output and

77 percent of the global population. Importantly, from an energy efficiency perspective, the analysis includes countries with large populations and strong economic growth, such as China, India, and Indonesia. Value-added and GDP data are obtained from the United Nations Statistics Division (UNSD). These figures are at 2005 prices and in U.S. dollars. To project energy consumption in 2035 the analysis relies on the GDP and population forecasts made by PwC in January 2011.

4 EMPIRICAL RESULTS

4.1 Overall Results

Figures 3.4 and 3.5 show the development of total energy demand and total electricity demand from 1971 to 2009 for the chapter's sample.

Total energy consumption in the OECD countries in the sample has increased from approximately 2.2mtoe to 3mtoe between 1971 and 2009 – an increase of 36 percent. The non-OECD countries have almost quadrupled energy demand in the same period, from 0.7mtoe to 2.5mtoe.

Electricity demand shows a similar pattern. However, for both OECD and non-OECD countries the increase in demand has been significantly larger than for total energy demand, confirming the switch to more electrical power as a share of energy consumption. The fall in electricity demand in OECD countries in the final years of the sample is also noteworthy, but it is still too early to say whether this signals a structural break as this development coincides with the global economic slowdown.

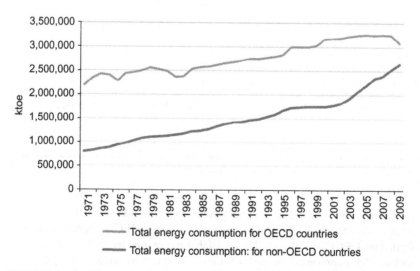

FIGURE 3.4   Total Energy Consumption: OECD Countries and non-OECD countries.

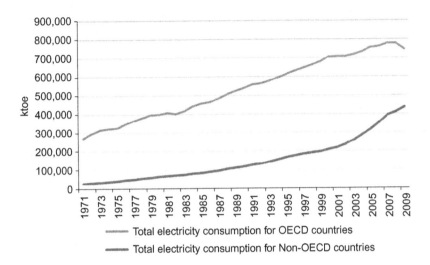

FIGURE 3.5   Total Electricity Consumption: OECD Countries and non-OECD countries.

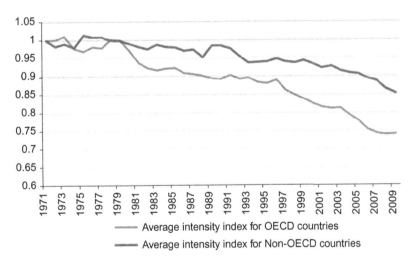

FIGURE 3.6   Average Intensity Index for Total Energy Consumption: OECD Countries and non-OECD countries (1971 = 100).

Figures 3.6 and 3.7 show the results of the decomposition analysis for OECD and non-OECD countries for total energy consumption. According to Figure 3.6 average aggregate energy intensities in OECD and non-OECD countries in 2009 are 74 and 85 percent, respectively, of their intensity levels in 1971. This implies that energy intensity has decreased in both OECD and non-OECD countries, suggesting a higher energy efficiency. The increase in

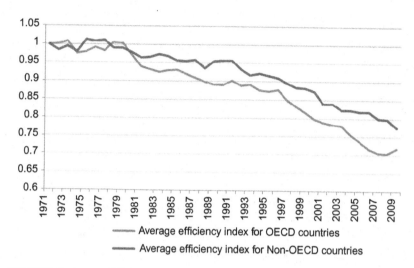

FIGURE 3.7   Average Efficiency Index for Total Energy Consumption: OECD Countries and non-OECD countries (1971 = 100).

efficiency has been more pronounced in OECD countries. Nevertheless, non-OECD countries have managed to reduce energy intensity, even though there was rapid economic development.

Part of the reduction in energy intensity may be due to a change in the composition of economic activity (e.g., a switch from industrial to services sectors). To analyze the impact of such changes, Figure 3.7 shows the development of the energy efficiency index assuming a constant composition of economic activity. The average efficiency index for OECD countries is 72 percent of its 1971 level. Stated differently, if the composition of economic activity had not changed between 1971 and 2009, energy intensity in OECD countries would have been 72 percent of its 1971 level. Comparing this to the energy intensity index (73 percent) shows that changes in the composition of economic activity − shifts *between* sectors − have *not* been a main driver of changes in energy intensity. Changes in economic structure have on balance contributed positively to the reduction in energy intensity.

For non-OECD countries, the constant structure efficiency index is approximately 77 percent of its 1971 level on average. Comparing this with the energy intensity index (85 percent) of Figure 3.6 demonstrates that changes in economic structure have been slightly unfavourable, counterbalancing the reduction in energy intensity. Overall, it is fair to say that the results for the non-OECD countries are more impressive given the degree of economic development during this period.

Figures 3.8 and 3.9 show the results of the decomposition analysis for OECD and non-OECD countries for *electricity* consumption. Electricity consumption per unit of GDP has increased in both sets of countries. For OECD

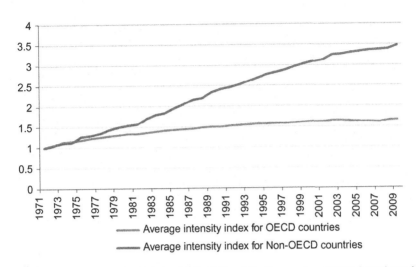

FIGURE 3.8 Average Intensity Index for Total Electricity Consumption: OECD Countries and non-OECD countries (1971 = 100).

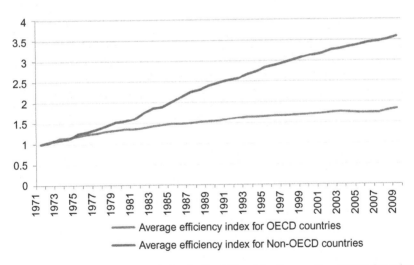

FIGURE 3.9 Average Efficiency Index for Total Electricity Consumption: OECD Countries and non-OECD countries.

countries, the intensity index has stagnated recently while it took off for non-OECD countries. Interestingly, this index suggests that electricity consumption has actually developed in line with GDP as no clear reduction in electricity consumption is witnessed in recent years. The intensity for non-OECD countries has almost quadrupled. As to the efficiency index, the development of this index is in line with the intensity index, suggesting that changes in economic structure have not been a major driver of changes in energy intensity. Comparing the two sets of indices suggests that changes in economic structure have had a slightly favorable impact on electricity intensity in both OECD countries and the rest of the world.

Overall, there is a pattern of decreasing energy consumption per unit of GDP in both OECD and the non-OECD countries and a clear increase in electricity consumption per unit of GDP. This suggests a switch to electricity as main form of energy consumption. At the same time the decomposition analysis shows that changes in economic structure have had a slight positive impact on energy intensity and electricity intensity in both sets of countries. However, the highly aggregated analysis above masks some interesting differences between countries that are discussed now.

4.2 Country-Specific Results

In the sample of 65 countries there are some countries that deserve specific attention. In this sub-section, the development of energy intensity in China, United States, India, Brazil, and Germany is given a closer look. These countries represent large populations and a large share in total global energy consumption.

Figure 3.10 shows the energy intensity expressed as tons of oil equivalent consumed per GDP for Germany, United States, Brazil, China, and India between 1971 and 2009. The figure is dominated by the sharp fall of China. China has managed, mainly as a result of rapid economic growth, to reduce its energy intensity. However, the level of energy intensity still remains higher

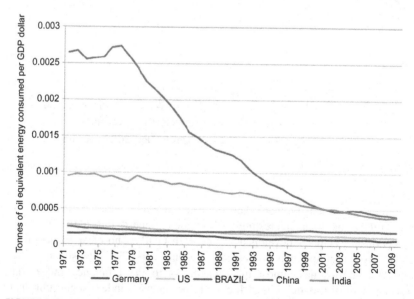

FIGURE 3.10   The Evolution of Energy Intensity over Time for the Selected Countries.

than that of Germany and the United States. India has also managed to reduce its energy intensity significantly, almost halving it over the 1971–2009 period. Even though this is not clearly visible in the figure because of the scaling of the y-axis, the energy intensity of Germany and the United States fell by more than 50 percent over this period. Thus, even though China and India are reducing their energy intensity rapidly, the developed markets (like the United States and Germany) are also managing to reduce their intensity over time. Interestingly, the reduction in intensity in India and China is largely driven by increased economic activity, which is a denominator effect.

In Germany and the United States the reduction is mainly the result of a slowdown of the growth in energy consumption in the United States and a stabilization in Germany, which is a numerator effect. This suggests that there remains a large potential to increase energy efficiency when countries like China and India start to focus on energy saving, especially given the still relatively large gap in the *levels* of energy efficiency between China and India and countries like Germany, the United States, and even Brazil.

The degree of convergence can be measured over time by examining the difference between the maximum and minimum energy intensity in the sample. Figure 3.11 shows the difference over time. The difference has halved over the period, with the notable decline between 1977 and 1991. From 1991 there seems to be relatively little convergence. From 2004 until the economic crisis in 2008 there is actually divergence. This is the result of developments in particular countries in our sample where economic growth has coincided with higher energy consumption. With the economic crisis impacting

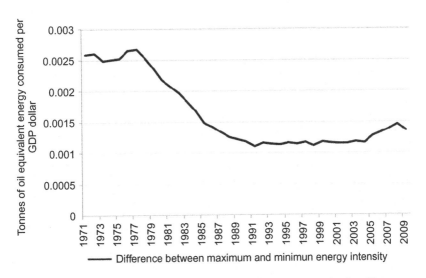

FIGURE 3.11   Difference between Maximum and Minimum Energy Intensity Over Time.

economic activity the energy intensity rates have started converging again recently.

4.3 Relationship between Energy Intensity and Income Levels

In the literature review the environmental Kuznets curve is discussed. This curve suggests an inverse U-shaped relationship between energy intensity and income levels. In particular, as income levels increase energy consumption and energy intensity initially increase, and then decrease as income levels continue to rise. The dataset of 65 countries covering nearly 40 years used in this chapter could help gain insight into whether a Kuznets-type relationship exists, and whether the shape is changing over time. Ideally, one would expect a contraction of the U-shape over time due to technological developments. This means that the increase in energy consumption becomes less pronounced when a country starts to develop economically and starts to decline earlier as incomes rise above a certain threshold.

In order to detect for the presence of a Kuznets curve, Figure 3.12 plots the energy intensity (expressed as total energy consumption divided by total GDP) versus the income per capita in 1971 and 2009 of the countries in the sample.

Visual inspection of Figure 3.12 does not suggest a clear inverted U-shaped relationship between income levels and energy intensity for 1971

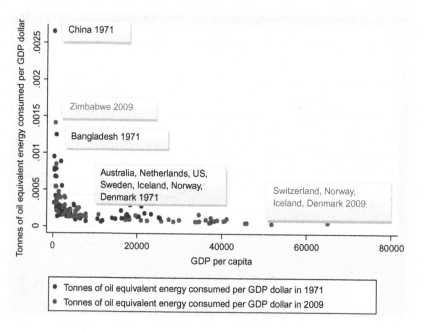

FIGURE 3.12   Energy Consumption per GDP versus GDP per capita in 1971 and 2009.

or 2009. The data suggest a rapid decrease in energy intensity as income levels increase initially – between 0 and U.S. $10,000 GDP per capita. The data also show a clear increase in the income levels between 1971 and 2009 without a pronounced U-shape, although in 1971 there seems to be an increase around U.S. $20,000 GDP per capita, and subsequent decline. However, for 2009 the data suggest a flat relationship between income and energy intensity.

In order to make more sense of the possible relationship between energy intensity and income levels, the sample is divided into four groups. These four groups essentially represent poor, low-income, middle-income, and high-income countries.[1] Subsequently the lagged value of energy consumption, population, population growth, GDP and GDP per capita growth is regressed on energy consumption, by performing a system of generalized methods of moments (GMM) estimation.[2] The reason for dividing the sample into four groups is that it is expected that the relationship between energy consumption and the explanatory variables differs between these groups. This is confirmed in statistical tests.

The regression results can be seen in Table 3.3. The coefficient of the lagged energy consumption (ENERGY CONSUMPTION (-1)) is highly statistically significant (at 1 percent level) for all categories. That is to say, energy consumption in the previous year strongly determines energy consumption in the subsequent year. Not surprisingly, increases in population and GDP are in most cases associated with higher energy consumption. Finally, increases in the rate of change of population and GDP in most cases raise energy consumption. Interestingly, these regressions confirm the absence of Kuznets-type relationships as the coefficient for GDP per capita growth is relatively constant for low- , middle- , and high-income countries.

4.4 Projecting Energy Demand to 2035

The results of the regression analysis make it possible to produce a forecast of future energy consumption, given forecasted values for GDP and population growth derived from PwC [14]. Forecasting based on extrapolating relationships, which were valid in the past, deserve a serious caveat. Due to technological developments, energy policies, and structural economic changes it cannot be expected that the future will follow the trend of the past, nor is it hoped for by policymakers. The forecasts presented in this section can therefore be seen as a base case scenario if no such structural developments occur. This provides an informative benchmark to compare with forecasts by other institutions such as the IEA.

---

1. China is excluded from the econometric analysis, since it is severely affects the outcome.
2. For more details on the GMM methodology, see Arellano and Bond [12] and Blundell and Bond [13].

TABLE 3.3 Dynamic Panel Data Regression Analysis for Total Energy Consumption

Energy Consumption	Poor Countries	Low Income Countries	Middle Income Countries	High Income Countries
Energy	0.9048***	0.9581***	0.8777***	0.9438***
	(0.034)	(0.024)	(0.018)	(0.012)
LN GDP	0.0512**	0.0332**	0.0817***	0.0100
	(0.025)	(0.015)	(0.018)	(0.006)
LN Population	0.0458**	0.0111	0.0407**	0.0433***
	(0.021)	(0.019)	(0.014)	(0.011)
GDP Per Capita Growth	0.0018**	0.0064***	0.0055***	0.0068***
	(0.001)	(0.001)	(0.001)	(0.001)
Population Growth	0.0025	0.0183**	0.0034	0.0174***
	(0.003)	(0.009)	(0.004)	(0.004)
Constant	−1.1092**	−0.6311**	−1.5501***	−0.3896***
	(0.493)	(0.269)	(0.236)	(0.115)
Observations	608	608	608	608
Number of countries	16	16	16	16
Arellano-Bond test for AR(1)	z = −2.55, p = 0.01	z = −3.02, p = 0.00	z = −2.21, p = 0.03	z = −3.51, p = 0.00
Arellano-Bond test for AR(2)	z = −0.01, p = 0.99	z = 1.45, p = 0.15	z = 0.97, p = 0.33	z = 1.43, p = 0.15

Notes: *: Significant at 10 percent level, **: significant at 5 percent level, ***: significant at 1 percent level. Poor Countries: Ghana, Kenya, Cote De Ivorie, China, Philippines, Zimbabwe, Paraguay, Nigeria, Pakistan, Indonesia, Syria, Bangladesh, Senegal, Morocco, India, Bolivia, Thailand.
Low-Income Countries: Panama, Peru, Uruguay, Malaysia, Guatemala, Jordan, Romania, Ecuador, Brazil, El Salvador, Turkey, Colombia, Tunisia, Cuba, Costa Rica, Korea.
Middle-Income Countries: Venezuela, Ireland, South Africa, Spain, Mexico, Portugal, Italy, Hungary, Finland, Argentina, Greece, Singapore, Jamaica, Chile, Israel, Cyprus.
High-Income Countries: Sweden, Denmark, Iceland, Japan, Belgium, Germany, Austria, Australia, Switzerland, Canada, New Zealand, France, U.S., Norway, Netherlands, UK.

For each category of countries (poor, low-income, middle-income, high-income) the forecast for energy consumption in 2035 is derived by using the coefficients obtained in the regression analysis and filling in the projections for GDP and population. As the PwC study only has GDP and population growth projections for a limited number of countries, the forecast is based on this more limited set of 18 countries.[3] As this sub-sample contains the largest economies, the results provide an indication of the global picture.

The results show that total energy consumption for the 18 countries is projected to increase by 72 percent between 2010 and 2035. The IEA forecasts an increase of world primary energy demand of 51 percent in the unchanged policies scenario in the period 2009−2035.[4] Our results also forecast a reduction in energy intensity by 39 percent between 2010 and 2035. Interestingly, the IEA arrives at the same expected reduction in energy intensity in the period 2009−2035.

The results show that the end of demand growth is not yet in sight. As a matter of fact, energy consumption is set to increase substantially; only energy intensity is set to decline over the next decades. Those countries with lower population growth and lower economic growth are expected to achieve a "flat-line" consumption pattern. However, the main challenge will be to further de-couple economic growth (driven by population growth) from energy demand growth in order to mitigate the expected growth in energy consumption.

While the forecasts of the IEA and this chapter's forecasts are in line with each other with respect to the forecasted energy consumption, it is important to note a number of differences, as follows:

- Firstly, the IEA forecast is for a much bigger sample of countries.
- Secondly, the IEA uses forward looking information in deriving this forecast, whereas this paper uses the estimated historical relationships.
- Finally, the underlying assumptions with respect to GDP growth and population growth differ.

The results show that although energy savings are implemented, there will still be a substantial increase in total energy consumption. The increase in demand is mainly driven by non-OECD countries, with a strong shift to more electrical power. Meeting this increased demand will require substantial investments. If this increase in demand is met through fossil-based sources, then there will be a substantial increase in CO_2 emissions. On the other hand, if this gap is met with renewable sources of energy, a substantial investment will also be required that is likely to be even larger given the current cost level difference between conventional and renewable sources of energy.

An estimate of the value of energy efficiency is obtained by examining the difference between the projected increase in energy demand and keeping

---

3. United States, India, Japan, Brazil, UK, Germany, France, Italy, Spain, Canada, Australia, Korea, Mexico, Indonesia, Turkey, Argentina, South Africa, and Nigeria.
4. In IEA's new policies scenario, energy demand still increases by 40 percent in this period.

current energy consumption constant. In essence, this is the value of avoiding an increase in energy consumption. The cumulative difference between energy consumption in 2010 and projected energy consumption until 2035 for the 17 countries[5] in the sample in 2035 is equivalent to approximately 225 billion barrels of oil, which corresponds to 9 bn barrels of oil per year (equivalent to 25 percent of annual consumption).

This suggests an average annual value of $900 bn, at an oil price of U.S. $100 per barrel, excluding the costs of CO_2 emissions rights and other environmental policies. While this number is not equal to the benefits of investing into energy saving technologies, it does point out the large potential cost savings that could be made over time if world energy consumption were to become more efficient.

5 CONCLUSIONS

Higher energy prices have led to improved business cases for energy efficiency. Yet, the historical data and this chapter's projections suggest that despite all the efforts to conserve energy, total energy consumption looks set to increase further in the next decade and beyond. At the same time both OECD and non-OECD countries have significantly increased the efficiency of energy use since 1971. Both rich countries, such as the United States and Germany, have managed to sharply increase the energy intensity of their economies, as well as emerging economies, such as China and India.

In emerging markets energy efficiency is predominantly driven by economic growth, whereas in the high-income countries it is relative reductions in energy consumption levels that have driven the improvements.

The evidence suggests both emerging and rich economies are able to achieve energy savings, either through greener growth or through reductions in consumption. At the same time, there seems little evidence of a relationship between income levels and energy intensity, suggesting a clear possibility to combine energy efficiency with additional growth.

The results from this chapter show that between 1971 and 2009 the countries in our sample needed 25 percent less energy to produce the same output. The projections also suggest that there are substantial gains to be realized if the expected growth in energy demand is either met with non-CO_2-emitting energy sources, or avoided through energy savings, or a combination of both. The increase in energy demand over the next 25 years in the 17 large economies of our sample is equivalent to 225 bn barrels of oil, or $900 bn avoidable costs per year. This suggests that the possible savings that can be achieved are large and tangible.

---

5. Note that China is excluded from the analysis due to its unique structure.

Descriptive Statistics for the Variables Used (for 2009)

	OECD Countries					Non-OECD Countries				
	Obs.	Mean	Std. Dev.	Min	Max	Obs.	Mean	Std. Dev.	Min	Max
Energy Consumption (in ktoe) in Industry Sector	28	28,691	51,913	1,273	273,225	37	32,117	117,974	297	709,952
Energy Consumption (in ktoe) in Services Sector	28	16,822	39,628	207	206,576	37	3,318	9,595	71	57,195
Energy Consumption (in ktoe) in Transportation Sector	28	40,806	107,235	290	577,759	37	12,735	28,493	359	160,799
Residential Energy Consumption (in ktoe)	28	24,309	49,438	526	262,069	37	22,786	63,127	314	348,755
Total Energy Consumption (in ktoe)	28	110,628	246,342	2,627	1,319,629	37	70,955	215,749	1,580	1,276,701
Electricity Consumption (in ktoe) in Industry Sector	28	8,571	13,263	776	68,720	37	7,752	30,420	51	183,161
Electricity Consumption (in ktoe) in Services Sector	28	8,704	21,536	88	113,813	37	1,476	2,819	51	14,486
Electricity Consumption (in ktoe) in Transportation Sector	28	323	443	0	1,671	37	128	487	0	2,803
Residential Electricity Consumption (in ktoe)	28	8,792	21,961	76	117,154	37	2,617	7,095	73	41,901
Total Electricity Consumption (in ktoe)	28	26,390	56,544	1,347	300,358	37	11,972	40,417	195	242,351
Value Added in Industrial Sector ($2005 million)	28	490,000	870,000	5,870	4,330,000	37	171,000	572,000	2,780	3,470,000

(Continued)

Descriptive Statistics for the Variables Used (for 2009) — (cont.)

	OECD Countries					Non-OECD Countries				
	Obs.	Mean	Std. Dev.	Min	Max	Obs.	Mean	Std. Dev.	Min	Max
Value Added in Services Sector ($2005 million)	28	832,000	1,800,000	9,360	9,390,000	37	95,500	231,000	2,470	1,290,000
Value Added in Transportation Sector ($2005 million)	28	82,200	148,000	936	770,000	37	16,900	34,500	216	179,000
Household Consumption Expenditures ($2005 million)	28	808,000	1,720,000	8,300	9,040,000	37	118,000	245,000	6,380	1,270,000
GDP ($2005 million)	28	1,280,000	2,440,000	17,100	12,600,000	37	232,000	607,000	5,930	3,520,000

Decomposition Analysis

Following Metcalf [14] energy intensity can be specified as a function of energy efficiency and economic activity. Stated more mathematically,

$$e_t \equiv \frac{E_t}{Y_t} = \sum_i \left(\frac{E_{it}}{Y_{it}}\right)\left(\frac{Y_{it}}{Y_t}\right) = \sum e_{it} s_{it}, \tag{3.1}$$

where E_t is aggregate energy (electricity) use in year t, Y_t is real GDP in year t, E_{it} is energy (electricity) consumption in sector i in year t, and Y_{it} is a measure of economic activity in sector i in year t[6]. Equation (3.1) indicates that aggregate energy intensity can be written as a function of sector-specific energy efficiency (e_{it}) and sectoral activity (s_{it}).

We first construct an energy intensity index as $I_t = e_t/e_0$, e_0 referring to the aggregate energy intensity for a base year. Metcalf [14] argues that a Fisher Ideal index yields a perfect decomposition of an energy intensity index (I_t) into economic efficiency (F_t^{eff}) and activity indexes (F_t^{act}) with no residual:

$$\frac{e_t}{e_0} \equiv I_t = F_t^{eff} F_t^{act} \tag{3.2}$$

To construct the Fisher Ideal index, we first construct Laspeyres and Paasche composition and efficiency indexes. The Laspeyres indexes are given as

$$L_t^{act} = \frac{\sum_i e_{i0} s_{it}}{\sum_i e_{i0} s_{i0}} \tag{3.3}$$

$$L_t^{eff} = \frac{\sum_i e_{it} s_{i0}}{\sum_i e_{i0} s_{i0}} \tag{3.4}$$

and the Paasche indexes are given as

$$P_t^{act} = \frac{\sum_i e_{it} s_{it}}{\sum_i e_{it} s_{i0}} \tag{3.5}$$

$$P_t^{eff} = \frac{\sum_i e_{it} s_{it}}{\sum_i e_{i0} s_{it}} \tag{3.6}$$

The Fisher Ideal indexes are then given by

$$F_t^{act} = \sqrt{L_t^{act} P_t^{act}} \tag{3.7}$$

$$F_t^{eff} = \sqrt{L_t^{eff} P_t^{eff}} \tag{3.8}$$

---

6. Energy (electricity) use in the sectors must sum to aggregate energy (electricity) use while the measures of economic activity need not add up to GDP. Furthermore, they do not necessarily have to be in the same units.

REFERENCES

[1] Mckinsey. Unlocking energy efficiency in the U.S. economy. Energy 2009.
[2] PwC. Review of energy efficiency policy options for the residential and commercial building sectors; 2008.
[3] Allcott H, Greenstone M. Is there an energy efficiency gap? Hunt Allcott and Michael Greenstone is there an energy efficiency gap? Energy institute at haas working paper 2012.
[4] Laskey A, Syler B. The Ultimate challenge: Getting consumers engaged in energy efficiency. Energy Efficiency, Chapter 23, this volume.
[5] Jaffe A, Newell R, Stavins R. Technological change and the environment. Handbook of Environmental Economics. North-Holland; 2003.
[6] Nilsson LJ. Energy intensity trends in 31 industrial and developing countries 1950–1988. Energy J 1993;18(4):309–22.
[7] Mielnik O, Goldemberg H. Converging to a common pattern of energy use in developing and industrialized countries. Energy Policy 2000;28(8):503–8.
[8] Sun JW. The decrease in the difference of energy intensities between OECD countries from 1971 to 1998. Energy Policy 2002;30(8):631–5.
[9] Cornillie J, Fankhauser S. The energy intensity of transition countries. Energy Economics 2004;26(3):283–95.
[10] Metcalf GE. An empirical analysis of energy intensity and its determinants at the state level. The Energy J 2008;29(3):1–26.
[11] Mulder P, de Groot HLF. Structural change and convergence of energy intensity across OECD countries, 1970–2005. SSRN Electronic J 2012;:10.2139/ssrn.2028990.
[12] Arellano M, Bond S. Some tests of specification for panel data: monte Carlo evidence and an application to employment equations. Review of Economic Studies 1991;58:277–97.
[13] Blundell R, Bond S. Initial conditions and moment restrictions in dynamic panel data models. J Econom 1998;87:115–43.
[14] PwC the world in 2050. 2011.

Chapter 4

Carpe Diem – Why Retail Electricity Pricing must Change Now

Allan Schurr[1] and Steven Hauser[2]

[1]IBM, [2]New West Technologies

1 INTRODUCTION

A few months before the U.S. Continental Congress approved the Declaration of Independence (July 4, 1776), a little known Scottish philosopher published a book titled *An Inquiry into the Nature and Causes of the Wealth of Nations*. This now famous book and its author, Adam Smith, described the concepts on which we have built our modern economy. Smith's statement "consumption is the sole end and purpose of all production" became the root of what we now commonly refer to as "supply and demand" where the science and mysteries of balancing the two drive successful businesses.

For the past century, no industry has embraced and exploited these concepts more than the electric utility industry, building a massive and complex infrastructure for delivering power to homes and businesses around the world. With rapidly increasing demand throughout most of that century, business models were created that assured adequate supply at a reasonable cost to consumers. Over the past decade, as the growth in demand has leveled and in some cases even decreased, utilities have begun to explore new ways of achieving the supply/demand balance.

Electric utilities, as is common for companies that depend on large capital investments, are a mostly fixed cost industry with a modestly rising cost structure for labor, materials, and compliance. Reductions, however, in expected demand and sales growth due to energy efficiency, self-generation, and economic dislocation threatens the core assumption of the historic regulated utility monopoly model. These reductions, while non-uniform, are nonetheless a secular trend in most of the OECD countries and will eventually result in fundamental policy and business model changes for this important infrastructure.

Energy Efficiency. DOI: http://dx.doi.org/10.1016/B978-0-12-397879-0.00004-9

Different jurisdictions have begun to experiment with changes in the traditional business models by passing along certain variable costs directly to consumers; for example, an automatic fuel adjustment as input fuel prices fluctuate. But progress has been inconsistent and too timid to balance long-term goals of security of supply and infrastructure adequacy, environmental compliance, and price competitiveness.

During the expected long transition to achieving this balance, the traditional utility operational model will in the meantime, result in strong pressure on managers to reduce operating costs and on regulators to approve prudent costs over fewer units of sales. This "business as usual" approach will slowly raise prices so that a dilemma emerges:

- Do utilities recover prudent costs in a highly fixed asset business by increasing variable prices, which drives lower sales growth though encouraging core efficiency and self-generation? or
- Do they restructure pricing to create more accurate market signals for the core utility components of energy production and delivery?

This chapter will argue that in the long run, utility leaders will make required production and delivery investments only if marginal prices are accurate and if customer decisions to reduce consumption are based on true avoided costs that minimize cost shifting. To achieve this balance, pricing reform is needed now.

The system today encourages energy efficiency and self-generation through a somewhat distorted pricing structure. For example, a solar system supplying 100 percent of the energy for a home or business, served via net metering arrangement, may have a zero cost bill, but still relies on the grid for off-hour energy and capacity. These "zero net energy" homes, as described by Rocky Mountain Institute's recent report on the subject, create complex value streams and cost shifting [1]. In the example here, all other customers end up paying for this solar customer's "free" grid support, but may still enjoy the environmental benefits. Likewise, a business using 20 percent less energy than her next-door competitor may nevertheless induce substantially the same delivery costs on the grid. This paradigm is not sustainable from a cost standpoint in the midst of rapid adoption of alternatives to traditional electricity supply of historic volumes.

In response, utilities may delay the onset of the most challenging regulatory decisions around cost allocation by pursuing dramatic cost reductions to maintain cash flows. Or they may find other stop-gap measures, like new product and service revenues, to postpone the time when financial pressures become too high to ignore the need for more fundamental reform. The long-term trends are inevitable, however, and indeed it is desirable now, due to unique economic and political factors, to address this revenue model challenge with regulators.

The timing to address such a shift in the cost recovery model is propitious, in the U.S. market particularly, inasmuch as any changes to consumers' bills over the next few years can be ameliorated by the historically low cost of natural gas. Utilities that act to resolve this dilemma now will enjoy comparative financial success over the long term as their peer group attempts what will surely be an increasingly challenging environment for structural change in the years to come.

This chapter will address these complex issues by providing an overview of the expected changes in the demand for electricity in sections 2 and 3. Section 4 provides specific examples of how these issues are being addressed by the industry. Section 5 outlines possible industry responses and suggests utilities take advantage of this window of opportunity to address structural pricing changes, since some options will merely postpone the inevitable financial impact of demand growth slowdown. Conclusions are drawn in section 6.

2 BREAKING THE ADDICTION TO ELECTRICITY

Many of us who were born in the 1950s or before remember the day that our mothers began using their first electric washing machine, maybe even the first refrigerator. Soon after came a television, a vacuum cleaner, a toaster, a range/oven, and so on. The marketing of new electric appliances was prolific in newspapers, magazines, radio and television even challenging us to "dream" about all the possible conveniences, as shown in the typical 1950 GE advertisement in Figure 4.1. Often local utilities were complicit with appliance manufacturers, supporting and even leading marketing campaigns. In most

FIGURE 4.1   GE Advertisement for Buying New Electrical Appliances.

cases it wasn't a hard sell, providing high value to customers in convenience, safety, health, and pleasure. "More time to play" was an often-used phrase promoting the time-saving nature of many of the devices. Implied, of course, but generally ignored or not understood, was how much electricity that was required to operate the device. Hans Rosling does an excellent job portraying this radical lifestyle change brought about by electricity.[1] This lifestyle is one that few of us would relinquish or even slightly modify today. While for many of us in developed countries these changes started more than a half century ago, for much of the world these changes are still yet to take place.

Our workplaces went through a similar transformation with large buildings requiring significant artificial lighting and increased amounts of cooling. Productivity grew rapidly starting with electric typewriters and evolving to massively complex computers and networks, moving information at enormously greater speeds, quantities, and distances. Overall, demand for electricity was growing at a fast pace and utilities were challenged to build more and more power plants and delivery infrastructure to keep pace with the demand.

Figure 4.2 shows the growth in electricity supply from 1950 forward. Federal and state regulations ensured that no matter where, when, and how much electricity consumers wanted, the utility infrastructure would be ready

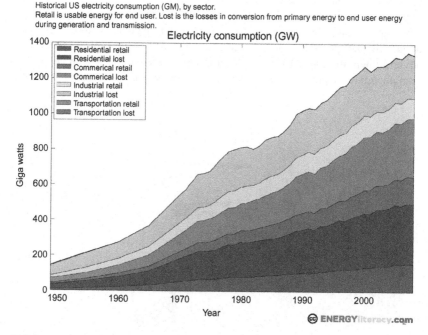

FIGURE 4.2 U.S. Historical Electricity Consumption.

1. See http://www.ted.com for Hans Rosling's presentation titled "The magic washing machine."

and able to supply it. Utilities built very successful businesses providing this electricity across the United States and in many other countries in the world. Both federal and state regulators worked closely with the industry to ensure reasonable business decisions resulting in electricity consumers could afford, driving our economy forward. When needed, they worked through major and minor disruptions, making adjustments to the successful model to minimize impacts to consumers.

By the early 1970s, this growth in the demand for electricity began to concern some because of the potential for huge environmental impacts as well as the growing need to import energy resources. As discussed by Sioshansi in the introduction to this book, several pioneers took on the challenge of both understanding the fundamentals of electricity use and began to develop ideas that showed how consumers could continue to avail themselves of the services that electricity provided while actually reducing the amount of electricity required for these services.

Figure 4.3 shows the historical and projected growth rates in electricity demand in the United States. While these rates were highest in the 1950s, technology improvements along with market saturation have continued to slow the growth in demand to the point where experts are now predicting less than 1 percent growth in electricity demand for the decades ahead. For example, technology improvements have substantially decreased the amount of electricity used in a typical refrigerator. In 1980, the typical household refrigerator was about 20 cu. ft. and consumed nearly 1300 Kwh/year. Today the typical refrigerator is over 22 cu. ft. and consumes less than 500 Kwh/year; the same or better "service" using less electricity. Nilleson et al further describe the global experiences in these advances.

The real question is whether growth will continue or whether a combination of technology and market forces will reduce existing demand sufficient to offset any new growth. Certainly examples abound, like those described in Australia by Smith. The Energy Information Administration compares the growth in electricity consumption compared to real GDP for the United

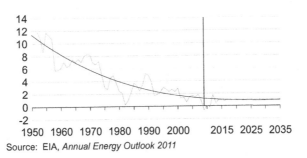

Source: EIA, *Annual Energy Outlook 2011*

FIGURE 4.3   Electricity Demand Growth Rates (Annual Percent Change) 1950–2035. *Source: EIA, Annual Energy Outlook 2011.*

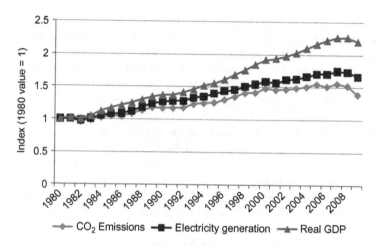

FIGURE 4.4 Growth Rates of CO_2 Emissions, Electricity Generation and Real GDP Compared. *Source: EIA, Annual Energy Outlook 2010.*

States from 1980 to 2010 (Figure 4.4). This shows that starting in the mid-1990s, growth in the U.S. economy is becoming less and less dependent on electricity production. Their projections in Figure 4.3 indicate an assumption that a small link between economic growth and electricity still exists, but the projected constant 1 percent growth in electricity demand is likely more based on assumption than fact.

A recent report by the American Council for an Energy Efficient Economy [2] examines the potential for a reduction in electricity consumption between 2010 and 2050. They assume continued strong economic growth of 2.8 percent per year resulting in a tripling of the U.S. GDP over the 40-year period. As a reference case they used the EIA growth projections in electricity consumption on a largely business as usual scenario resulting in an increase of about 25 percent in electricity demand in 2050. They then looked at two scenarios they called Advanced and Phoenix. The Advanced Scenario assumed that current and near commercial technologies available today are deployed throughout the residential, commercial, and industrial sectors. In other words, no technology R&D or breakthroughs are required. The resulting energy efficiency reduces the amount of electricity required by about 25 percent (translating to a 0.8 percent per year real decline in electricity demand).

The Phoenix scenario is a bit more aggressive, assuming, for example, that the average sq. ft. per home in the United States declines slightly between now and 2050. The resulting decrease in electricity demand in 2050 is over 40 percent. The reduction in this scenario is still based on available technologies but assumes faster and deeper penetration. Laitner et al also make similar arguments in their chapter.

Recent advances in smart grid technologies are rapidly becoming a key enabler of energy efficiency by engaging consumers in new and much more effective ways. Laitner presents an elegant analysis of the impact that semiconductor-based technologies have had on the U.S. economy in the last 40 years and projects an even greater impact going forward. He projects, for example, that between 2010 and 2030 that semiconductor-enabled efficiency can reduce electricity consumption by more than 20 percent [3].

While predicting the future is always difficult, the average growth in electricity demand in the United States is most likely to range between a 1 percent per year increase to a 1 percent per year decrease. Of course, this will vary significantly between regions, states, and even cities. Nadel summarizes the variations by state and highlights states where recent policy changes have accelerated the growth in energy efficiency programs and the resulting reduction in electricity demand. While a decade ago, energy efficiency was largely focused in a few states and utilities, programs now are sprinkled around the country. California has always been thought of as the leader in this endeavor, and they certainly can rightly claim the role of pioneer, but as Nadel shows, other states like Vermont, Nevada, Hawaii, Rhode Island, Minnesota, and Iowa are now ahead of California in the percentage of reduction in electricity due to energy efficiency programs. To make the point about the widespread nature of energy efficiency investments, even utilities like Oklahoma Gas and Electric have launched aggressive demand reduction programs that will help them defer the need to build new generation for the next decade or beyond.

3 SELF-GENERATION BECOMES THE NORM

Driven recently by renewable portfolio standards and generous incentives such as feed-in tariffs and net metering, rooftop solar systems for both residential and commercial applications have dramatically increased in count. This growth in the market for such systems combined with innovative manufacturing and reductions in the price of silicon has continued to push down the cost of solar panels. Figure 4.5 shows both the increase in sales and the decrease in cost, depicting the correlation between the two and indicating the continued potential for further cost reductions. These cost reductions are difficult to forecast, but still highly likely as described in the current Department of Energy "Sunshot" initiative where costs for PV systems are expected to reach grid parity (without subsidy) by the end of this decade [4].

The southwest U.S. is experiencing high growth rates in rooftop solar systems[2] with San Diego, Phoenix, and Las Vegas leading the way.

---

2. See http://www.openpv.nrel.gov

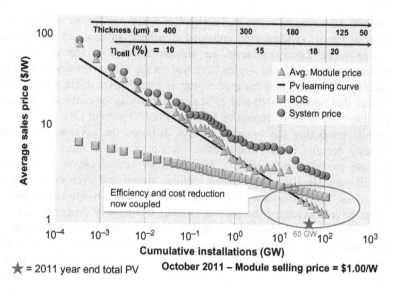

FIGURE 4.5   The Predictable Cost Reduction of Photovoltaic Systems.

Some newer housing developments have rooftop PV on every home, result-ing in 20 percent or more of the electricity for that neighborhood being supplied by self-generation. While extremely low natural gas prices may slow the growth, it is a trend that will continue especially as solar prices continue to fall.

A 2010 analysis by Arizona Public Service shows (Figure 4.6) that nearly half of the expected growth in electricity demand by 2025 will be met by energy efficiency (EE) and distributed energy (DE) resources. Germany has clearly demonstrated the feasibility of very large percentages of solar power reaching nearly 50 percent of the total electricity generated in a May 2012 two-day period.[3]

While electricity storage is still too expensive to allow consumers to totally disconnect from the grid in most situations, the cost of storage is projected to come down over the next several years motivating even more self-generation. Current grid-connected systems rely extensively on local distribution grids to supply power when the sun doesn't shine, at night and during maintenance events. Utilities are beginning to charge consumers addi-tional fees for "leaning" on the grid with their own generation. As these fees become more widespread, they will provide additional incentives for consu-mers to add storage and permanently disconnect.

This is the key issue: consumers use less grid-generated electricity while exerting increased stress on the grid to keep themselves supplied when

---

3. See Reuters, May 2012.

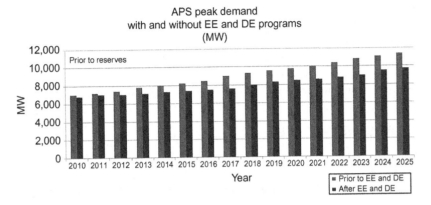

FIGURE 4.6   Arizona Public Service projections on demand growth through 2025.

they need it. Existing pricing structures result in cost shifting and underfunding of needed operations and investments.

Certain communities, especially where the price of utility-supplied electricity is very high, are focused on maximizing both energy efficiency and self-generation. The Vineyard Energy Project on Martha's Vineyard is an example where a variety of smart grid projects, energy efficiency, and local renewable energy will help them achieve their long-term goal of operating an independent power grid, fueled by local renewable energy.[4]

With natural gas prices in the United States at historic lows, many commercial and industrial consumers are considering generating their own power using small- and medium-sized high-efficiency turbines. These systems can provide local, reliable, and inexpensive electricity, especially when combined with productively using the waste heat, improving power quality and availability, and displacing utility-generated electricity that must be transported long distances or through congested transmission and distribution lines. Commonly referred to as cogeneration or combined heat and power (CHP), these systems are well understood and very dependable.

A report by the Oak Ridge National Laboratory [5] suggests that with pro-CHP policies, the United States could generate 240GW or 20 percent of its electricity needs from CHP systems by 2030. This equates to three times the current amount of CHP in the United States, but roughly equivalent by percentage to what is already done in Europe. Combined with new microgrid control systems that allow users to optimize the integration of multiple supply options (e.g., solar PV and CHP) with demand management, these self-generation options may become a preferred strategy for many companies.

---

4. See www.vineyardpower.com

If these new CHP systems were built and operated by local utilities, they could add instead of subtract from revenues, but this would require new business models for most utilities.

As a case study in what is occurring in practice, Portland General Electric is profiled as to their energy sales outlook, their business strategy regarding promoting energy efficiency, and their current revenue model.

3.1 Portland General Electric Viewpoint

Portland General Electric (PGE) is a vertically-integrated electric utility serving the Portland Oregon metropolitan area. It has approximately 822,000 customers and $1.7B in annual revenues. Its energy supply mix is comprised of coal, natural gas, hydroelectric, wind, and outside purchases. Retail rates range from $0.109/kWh for residential to $0.062/kWh for industrial customers.

According to Jim Piro, CEO, the combination of relatively strong economic growth, its low-cost generation portfolio, and appropriate regulatory recovery mechanisms means that PGE's customers and shareholders are well positioned even as energy efficiency grows its already strong role in the resource mix of this utility [7].

PGE estimates load growth of 2.3 percent, with energy efficiency offsetting about half of that in the near term (Figure 4.7). The expected load growth is dominated by industrial load growth from the technology sector

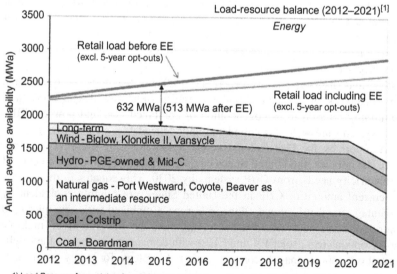

1) Load-Resource forecast data from 2011 integrated resource plan update, filed with the OPUC on 11/23/2011

FIGURE 4.7 Portland General Electric IRP Resource Outlook [6].

(e.g., Intel) in the PGE service area. Commercial and residential rooftop solar PV is growing but is seriously limited by a limited number of rooftops with good solar insolation. Portland is one of the leading cities for electric vehicle adoption, and PGE estimates that its system is capable of supporting up to 50 MW additional sales from plug-in electric vehicles over the next 10 years. The impact of self-generation is expected to be largely offset by new load.

Since Oregon regulators approved revenue decoupling for residential and small commercial energy sales, after adjustments for weather, the net revenue effect of energy efficiency driven sales and demand reductions is minimized. In addition, revenue adjustment for outside power purchases and fuel costs are addressed outside of general rate cases allowing for more timely collection of revenues with a mitigating effect on volume and price risk in wholesale generation.

Additionally, to ensure competitive financial returns and avoid unnecessary rate impacts, PGE is in the midst of an operations and maintenance (O&M) expense productivity program timed to coincide with its anticipated worker retirement over the next several years. All operating areas are first benchmarked against industry best practices and then investments are prioritized to target highest impact improvements. Even with this aggressive program, PGE projects approximately 1 percent/year growth in the cost of O&M.

As Jim Piro sees it, citing the impressive efficiency plans of a local grocery chain, energy efficiency is good for customers' competitiveness and an important element of the regional energy mix. PGE even collects funds for Energy Trust of Oregon (ETO) and other organizations, anticipated to be nearly $90 M in 2012, for use in implementing energy efficiency programs and other public purpose activities in the PGE service area. Energy efficiency is just part of the business landscape for PGE.

3.2 Raymond Gifford Viewpoint (Former Chairman of the Colorado Public Utility Commission)

Of the several possible regulatory responses, the experience of the 1980 and 1990s era telecommunications analog is instructive. Former Colorado PUC Chair Raymond Gifford offers his thoughts on applying this experience to energy. Chairman Gifford describes the critical need for regulators and utilities to come together in a new compact that encourages both cost and service innovations and pricing structures that do not shift costs to one group from another based solely on the former's lack of participation in energy efficiency or self-generation [8].

According to Gifford, one key to that result is a state regulatory model that saw great success in the telecom business. Known as "price cap" regulation, the construct is that a revenue requirement and resulting price is set for several years, often 7 or more, less a stipulated productivity factor that reduces the price each year from the prior year. The consumer sees reduced costs, and the utility can benefit from accelerated productivity investments. An additional feature of this scheme is that utilities also get authorization and pricing latitude for certain quasi-regulated services. In the telecom case, these were calling features like caller ID, voicemail, three-way calling, and call forwarding. All of these are considered optional services, but which can be marketed and delivered by the incumbent utility, thus increasing revenue and margin further.

Gifford is not very optimistic that such a new model applied to energy utilities will come easily. Regulators are not likely to lead in this regard, and utilities are not known for proposing innovative regulation until the existing model is not yielding favorable returns. Inasmuch as utilities are seeing only minor sales growth erosion and many have expectations that sales will return to pre-recession level soon, they are reluctant to propose something that does not yet have an experience curve in the energy utilities space.

4 INDUSTRY IMPLICATIONS

While some electricity sales reductions are operationally beneficial if they occur during the coincident peak, most mechanisms for sales growth reductions are not so well integrated into utility capacity planning as to result in much if any avoided cost. Except for utilizing the marginal peaking generation resources, substations have the same equipment, trees still fall across overhead pole lines and cause outages, consumption is measured, and bills are sent. Only reliable and persistent demand response resources have both an immediate and long-term cost avoidance opportunity. These phenomena come from the modern utility industry model, which was conceived and deployed around the three pillars of increasing load growth, declining marginal costs, and cost-of-service price regulation. When one or more of these three pillars erodes, the economic underpinnings of the system become imbalanced. In addition to the sales growth reduction from energy efficiency and self-generation, the other two pillars are under challenge too.

O&M costs have been rising faster than inflation for much of the past decade. Materials have been impacted by global economic growth, and commodity labor prices continue to rise. The Edison Electric Institute reports that O&M expenses for its investor-owned utility members grew an average of 4.6 percent from 2003 to 2011, with some volatility due to M&A accounting and a dip in 2008 due to the recession [9,10]. Generation construction prices have also increased during this time due to high costs of materials, permitting, and labor. Fortunately, in the United States, natural gas prices in summer and Fall of 2012 are between $2–3 MMBtu, which mitigates the total cost of energy in most markets. Many experts believe these low prices will persist for many years to come.

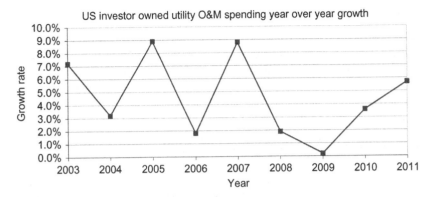

FIGURE 4.8   U.S. Investor-Owned Utility O&M Spending [9,10]. *Source: Edison Electric Institute.*

The regulated monopoly behind cost-of-service ratemaking is also beginning to unravel at the seams.

First, in many markets, social spending in the form of low-income service, renewable feed in tariffs, energy efficiency investments, and economic development programs all serve to increase the overall revenue requirement above the basic cost-of-service.

Secondly, rate design is yet another policy decision that can skew the price signal of marginal cost of energy and the associated investment return of energy efficiency or self-generation. For example, in markets like California, significantly differentiated tiered pricing places the prices for the highest tiers at multiples above the lowest tier to encourage such energy efficiency and self-generation options.

Finally, the presence of regulatory compensation or performance incentive schemes that reduce the harm from sales erosion on the investment and operating decisions made by utilities is a factor in the financial stability and investment attractiveness in utility companies.

In the long run, a sustainable energy infrastructure will only emerge from the reassessment and the replacement of these three elements with a coherent system of consumer price structures that reflect the short run and long run marginal price of a unit of electricity.

The need to achieve this sustainable financial structure is apparent. The positive feedback loops of high avoided unit prices, reduced cost of alternatives to utility supply, and supported social policy spending will most likely accelerate these trends toward more energy efficiency and self-generation. In the limit, non-participating consumers will pay more for the basic commodity service or utilities will see diminished financial returns that raise the cost of capital, and thus rates, as investors evaluate the prospects of earnings, cash flow, and risk.

5 POSSIBLE RESPONSES

Addressing the cycle in Figure 4.9 can take many forms. The best and most efficient way of addressing this cycle are structural pricing changes that would give the accurate market signals for both capacity and energy, but these may be the most politically difficult due to the elimination of certain intra-class subsidies and the resulting bill impacts.

Less structural in nature and less efficient in result, energy efficiency incentive and sales decoupling mechanisms, currently implemented in various states, can partially offset the financial downside to the utility of lost sales and revenues. Also, aggressive productivity and cost reduction programs can help maintain financial performance by mitigating some of the intra-class pricing increases from any structural price changes. And finally, some utilities are hedging these effects by creating plans for energy services offerings that can yield new revenue streams.

Tailored to local conditions, all these steps should be pursued as a portfolio of responses to the fundamental challenge to the financial integrity of utility providers and to support a long-term investment climate in low-cost and reliable energy service.

5.1 Regulatory

Given the strong role of regulation in the utility industry, it would seem that a response from local and national regulators to this slower sales growth condition would be logical. Mixed objectives, however, make these regulatory adjustments difficult to implement given the many stakeholder

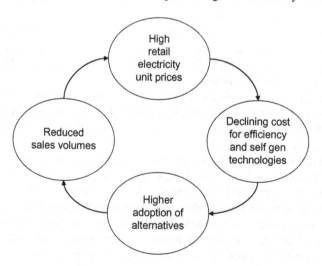

FIGURE 4.9   A Non-Virtuous Cycle.

positions. The most efficient approach in the long run is one where all parties – utilities, customers, and new entrants – make rational investment decisions based on avoided costs or expected gain. Parties may disagree as to the appeal of each, but there are three obvious and necessary regulatory adjustments.

Some background on traditional ratemaking is helpful. Utilities typically aggregate their cost-of-service from generation through delivery and customer accounts, including depreciation, taxes, and financing expenses. These costs are reviewed by regulators for prudency to establish an annual revenue requirement to build and operate the system going forward. In some cases, such as the example cited above for Portland General Electric, there are automatic adjustments for fuel costs. Once the revenue requirement is established the regulator performs class allocations to residential, commercial, and industrial customers based on their share of the costs incurred.

In its most simple form, prices are then set by taking this class revenue requirement and dividing by the forecasted sales to create a volumetric unit price. It is because this last step has been simplified significantly that leads to risk in cost recovery for utilities. Notably, fixed costs are recovered through volumetric charges, leaving the risk of under collection of the cost-of-service as sales decline due to high penetration of energy efficiency and self-generation.[5]

In the larger commercial and industrial classes there may be a demand charge component to address the maximum rate of electricity consumed and compensate for fixed costs like generation plant, cable and poles, transformers, and metering. But the smaller commercial and residential consumers are typically charged on an all-volumetric rate. Furthermore, the time of day or day of the week when energy is consumed is often not reflected in the pricing structure for any of the classes, even though there is substantial time variation in generation costs to adequately meet peak period demand.

Therefore, the typical rate structure often obscures the true cost-of-service, and decisions to reduce consumption through efficiency or self-generation may not reduce utility costs commensurate with the reduced electricity sales. This is especially true when "net metering" or "feed-in tariffs" are in place. These two mechanisms pay the retail price or greater for electricity produced from selected renewable energy sources like solar. Yet the residual cost to serve by the utility is not reduced proportionately with the revenue reduction. Zero net energy buildings, as described by Rajkovich et al and elsewhere in the book, merely demonstrate the point. Unless the building is entirely grid independent and disconnects, there is residual cost by the utility of providing back-up electricity service, but with no revenue resulting. All volumetric pricing to recover a significant fixed cost business creates false price signals for these measures.

---

5. See James Bonbright's classic reference "Principles of Public Utility Rates."

The three response mechanisms are marginal cost rate design, direct cost recovery, and energy efficiency performance incentives.

5.1.1 Marginal Cost Rate Design

The most economically direct mode of addressing the short-term and long-term revenue implication of sales growth erosion is utilizing cost-based pricing with rate design to reflect true marginal cost revenue collection. In other words, the marginal cost to produce and deliver a kWh is reflected in the marginal revenue collected for that same kWh. The following example is one way to address this imbalance of marginal cost-of-service matching the marginal revenue collected.

Liander in the Netherlands is a distribution system operator (DSO) in a competitive retail market structure. Energy consumers separately contract with competitive retailers for energy, usually at all volumetric pricing. But for the delivery portion of the bill, Liander has a fixed cost rate design to recover what are non-variable costs of running the distribution network. If a consumer is connected to the grid, they pay a fixed fee per month, regardless of the throughput, but as a function of amperage of the electric panel and whether it's single-phase or three-phase service. In the 2012 tariffs, a residential customer with three-phase service between 25 and 35 amps will pay approximately $100 per month in delivery and metering costs, including taxes [11].

The Liander example is a unique case in the utility industry. In other deregulated markets, the DSO charges are still collected in mostly volumetric pricing. Other industries, however, are similarly considering capacity-pricing models, such as the broadband Internet industry that may price for maximum bandwidth provided not minutes of use or megabits of data.

Some utilities in the United States are starting to move toward a hybrid of higher fixed monthly fees to recover a portion of the fixed cost of operating the assets along with a volumetric fee that is applied to the energy consumed at that delivery point. Sacramento Municipal Utility District (SMUD), for example, recently adopted a multi-year transition to a $25/month fixed customer fee, with offsetting reductions in unit electricity prices. But regulators are reluctant to make large changes in the rate structure for fixed cost recovery pricing due to adverse bill impacts on existing low-consumption users. Any such shifts will likely happen in stages, but failure to do so will increase the prices to those customers unable to participate in energy efficiency or self-generation strategies, whatever the reason.

5.1.2 Cost Recovery and Performance Incentives

A second alterative to mitigate this revenue erosion is incentive regulation that financially rewards the utility for encouraging energy efficiency or compensates the utility for lost revenue, thus making them indifferent to or even encouraging them to promote energy efficiency in their service area. According to the U.S. EPA, numerous states have implemented these mechanisms for direct cost recovery of energy efficiency program costs, fixed cost recovery of lost revenue, and performance incentives to further promote energy efficiency [12]. These mechanisms rarely target the negative revenue effects of self-generation, however, and as pointed out at the beginning of section 5, only delay the top-line financial effect from the "non-virtuous cycle." Indeed these incentive mechanisms result in higher costs, which create a positive feedback loop for price increases and more sales erosion.

California has had these three mechanisms in place since the 1990s. As described by Cavanaugh, all four investor owned utilities (e.g., PG&E, SCE, SDG&E, SoCal Gas) are incented to promote energy efficiency, get cost recovery for energy efficiency and distributed renewable generation programs, and are further recompensed for lost revenue through a decoupling mechanism that dates back to the 1980s.

As described by Laitner et al, Duke Energy has an approved innovative energy efficiency incentive program known as "Save a Watt" in the North Carolina market, which became effective in 2009. While controversy remains amongst various stakeholders as to the level of avoided costs, it represents a bold admission by Duke and its regulators that financial incentives matter if utility behavior is to change over a sustained period.

In anticipation of these significant changes to their sales outlook, utilities should concurrently work with regulators on implementing proper marginal costs signals and to put into place mechanisms that help recover lost short-term revenue from sales erosion. The energy efficiency incentive mechanisms have an added feature of building and maintaining utility energy end-use expertise and brand permission for future services by encouraging utilities to support energy efficiency programs directly.

5.2 Enterprise Model

Additional strategies to mitigate this lost margin that do not depend on new regulatory models should also be considered. These options, at a minimum, might reduce the financial impact due to adoption lags in these more favorable regulatory models, but also can boost returns in the short run and create greater customer value.

5.2.1 Cost Management and Productivity

Operating costs for utilities are always under the watchful eye of management. While certain back-office functions like finance, accounting, HR, and supply chain are addressed as part of an enterprise resource planning system (ERP), process standardization and new analytics are able to improve performance even further when applied to these common systems.

However, the long asset life and risk adverse culture of utilities can lead to slow adoption of other modern practices for cost management, especially in the asset management and front-office operations. Combine this under-exploited opportunity with the imperative to increase worker productivity as seasoned utilities workers face retirement and new workers take their place, and it is easy to see that cost management requires more focus than ever before. Many utilities in North America and in Europe have realized the urgency and are taking steps to re-engineer operations substantially across the value chain.

Cost management can take many forms at a utility. In the generation domain, fuel and maintenance costs are the primary opportunity areas for cost reduction. Many strategies are used to achieve these savings, including condition-based maintenance and asset management, described below, as well as fuel procurement optimization.

In the transmission and distribution assets and operations domain, a significant opportunity for both short-term and long-term cost reduction includes asset utilization, asset management, and compliance management. There are too many to mention here, but some of the more interesting that take advantage of new solutions and systems include:

- Mobile field workforce management is a common strategy to reduce unproductive work through better scheduling, materials and equipment integration, and work analytics.
- Condition-based maintenance, which reduces unneeded inspection functions and relies on remote monitoring of asset health and automatically triggers repair work when specific conditions are present or anticipated.
- Increased asset utilization to reduce unnecessary capital expense by understanding asset loading, leveraging new sensors and consumption data from advanced metering infrastructure (AMI), and then more dynamically operating the network through the use of remotely-operated switches and voltage control.

In the meter to cash domain, common cost reduction areas include meter reading, credit and collections processes, and commercial losses. While not as large as other business domains, its predominance of O&M expense makes savings here a direct and immediate impact on earnings and helps lower bills over the longer term.

For example, traditional manual meter reading is costly, and hard-to-access meters require appointments or estimated bills due to missed meter reads. Replacing electro-mechanical meters with radio accessible, solid-state meters, automated meter reading (AMR), or the more capable AMI will reduce the cost to read and improve accuracy through reduced exceptions and estimated reads.

Credit and collections is another candidate for cost efficiencies. Reducing bad debt and write-off is the primary goal by using improved credit analytics and collections process improvement. In addition, reducing the costs of establishing new service, or of managing repeat collections problems, often accompanies AMI investments via remote connect and disconnect of service.

Finally, self-service and transaction automation are emerging areas of focus. With the advent of new approaches for self-service that supplement typical utility web offerings, such as smart phones, tablets, and social media, utility customer business processes can benefit from both lower costs and higher satisfaction.

In sum, these productivity enhancements can address a moderately large spend category with cost reductions amounting to double-digit percentage improvements in overall costs.

5.2.2 New Revenues

New revenue sources are another option to mitigate the revenue loss from core utility services. Traditionally, utilities have been successful at finding opportunities in wholesale generation and trading, where excess generating or transmission capacity can be utilized to create new revenues through in-market wholesale power sales and transmission revenues.

But can new retail customer revenues streams also have a material effect in addressing sales erosion? At a minimum, there are hedging strategies that can be employed by utilities to allow them to participate in the same market trends that are slowing sales growth. For example:

In addition to the energy efficiency performance incentives described above, utilities in the United States and across Europe already have or are creating unregulated energy service companies aimed at profiting from energy efficiency and self-generation investments. For the most part, the track record in the US is either mixed or still unproven. In Europe, where market liberalization has unbundled energy retailing from the delivery function, many utility retail arms have offerings for energy efficiency, particularly targeted at larger non-residential customers. These entities will provide any of the several functions needed for market adoption such as design, procurement, installation, and financing. In the United States, there are fewer in operation, after a flurry in the later 1990s and early 2000s before many were divested or shut down. But advances in Internet-based energy monitoring and management hold promise for new penetration into the small commercial and residential market, which are being tested by many utilities against a backdrop of new offerings from telecommunication and cable TV firms.

Duke, once again, has shown a novel way to participate in the distributed generation market through a program in North Carolina that provides approximately $50 M in capital to purchase and install utility-owned distributed solar, while paying a lease fee to the building's owner for "roof rights." Duke earns a return on investment and avoids the loss or retail energy sales since the solar energy produced is viewed as utility generation.

Many utilities are currently considering additional participation plays including on-bill financing, combined heat and power (CHP) unregulated operations, and for-fee energy information products.

Finally, many utilities look to the growth in electrified transportation such as electric vehicles (EVs) as an offset to reduced sales. In certain markets, the forecasted energy sales to EVs can have a mitigating impact in the long run, particularly if increased sales occur in low-cost nighttime hours.

Each of these, if well executed, can serve to simultaneously hedge against sales erosion and improve the utility's brand perception and customer satisfaction. But they require new skills in marketing, product development, and business operation that are only partially available from an incumbent resource pool.

Most importantly, these are merely short-term mitigation steps to address a long-term flaw in marginal price structures that permeate the utility ratemaking process. Absent reforms in price structure, cost reduction, and revenue improvements are insufficient to maintain the investment stream of the grid itself.

6 CONCLUSIONS

Energy efficiency and efficient distributed generation are already cost-effective investments for many, and will become increasingly so as technology developments drive cost down and innovative business models overcome other barriers to customer adoption. These favorable economics are often significantly benefited by the existing volumetric biased utility pricing structures, even though a significant amount of fixed cost is collected in these volumetric prices. Combined with variable economic growth across geographies, future sales for most utilities are expected to be flat or negative.

As this trend becomes more visible and utilities adjust their prices to slower growing sales, prices will increase to all customers as remaining and growing fixed costs are spread across fewer sales. This situation, while favorable to efficiency and renewable energy adoption, leaves non-participants with even higher bills and creates an unsustainable situation for the utility business.

Forward-thinking utilities are considering several options and some are taking aggressive action. Some options address the fixed cost recovery problem directly, while the remainder either delay the effect described above or are only partial hedges against the financial effects of decreases in revenue. The most successful utilities will be those that address the long-term structural cost recovery problems head-on with their regulator. Their own long-term viability is at stake.

A bright spot in this dilemma is that natural gas prices are at historic lows in real terms and the resulting negative pressure on prices affords both regulators and utilities alike a short window of opportunity to put into motion structural fixes that address customer equity and ensures reliable infrastructure built on competitive investor returns.

REFERENCES

[1] Rocky Mountain Institute, Net Energy Metering, Zero Energy and the Distributed Energy Resource Future, March 2012.

[2] Laitner et al. The Long-term Energy Efficiency Potential: What the Evidence Suggests ACEEE Report number E121 January 2012.

[3] Laitner et al. Semiconductor Technologies: The Potential to Revolutionize U.S. Energy Productivity, ACEEE Report number E094 May 2009.

[4] Sunshot Vision Study DOE/GO-102012-3037 February 2012.

[5] Shipley et al. Combined Heat and Power: Effective Energy Solutions for a Sustainable Future. Oak Ridge National Laboratory report number TM-2008/224; December 2008.

[6] Portland General Electric, Investor Presentation, Williams Capital Group March 22, 2012.

[7] Interview, Jim Piro, CEO Portland General Electric; April 27, 2012.

[8] Interview, Raymond Gifford, Former chair Colorado Public Utilities Commission; May 2, 2012.

[9] Edison Electric Institute, Industry Financial Review; 2010.

[10] Edison Electric Institute, Industry Financial Review; 2011.

[11] Elektrciteit Tarieven, Stedin.net; 2012.

[12] Aligning Utility Incentives with Investment in Energy Efficiency, U.S. Environmental Protection Agency, Office and Air and Radiation, Climate Protection Partnerships Division; November 2007.

Chapter 5

Is There an Energy Efficiency Gap?

Hunt Allcott[1] and Michael Greenstone[2]

[1]*New York University,* [2]*Massachusetts Institute of Technology*

1 INTRODUCTION

Many analysts of the energy industry have long believed that energy efficiency offers an enormous "win-win" opportunity: through aggressive energy conservation policies, we can both save money and reduce negative externalities associated with energy use. In 1979, Pulitzer Prize-winning author Daniel Yergin and the Harvard Business School Energy Project made an early version of this argument in the book *Energy Future*:

If the United States were to make a serious commitment to conservation, it might well consume 30 to 40 percent less energy than it now does, and still enjoy the same or an even higher standard of living. Although some of the barriers are economic, they are in most cases institutional, political, and social. Overcoming them requires a government policy that champions conservation, that gives it a chance equal in the marketplace to that enjoyed by conventional sources of energy.

Thirty years later, consultancy McKinsey & Co. made a similar argument in its 2009 report, *Unlocking Energy Efficiency in the U.S. Economy*:

Energy efficiency offers a vast, low-cost energy resource for the U.S. economy — but only if the nation can craft a comprehensive and innovative approach to unlock it. Significant and persistent barriers will need to be addressed at multiple levels to stimulate demand for energy efficiency and manage its delivery. If executed at scale, a holistic approach would yield gross energy savings worth more than $1.2 trillion, well above the $520 billion needed through 2020 for upfront investment in efficiency measures (not including program costs). Such a program is estimated to reduce end-use energy consumption in 2020 by 9.1 quadrillion BTUs, roughly 23 percent of projected demand, potentially abating up to 1.1 gigatons of greenhouse gases annually.

In economic language, the "win-win" argument is that government intervention to encourage energy efficiency can improve welfare for two reasons. First, the consumption of fossil fuels, which comprise the bulk of our current

Energy Efficiency. DOI: http://dx.doi.org/10.1016/B978-0-12-397879-0.00005-0

energy sources, causes externalities such as harm to human health, climate change, and constraints on the foreign policy objectives of energy-importing countries. Second, other forces such as imperfect information may cause consumers and firms not to undertake privately profitable investments in energy efficiency. These forces, which we refer to as "investment inefficiencies," would create what is popularly called an energy efficiency gap: a wedge between the cost-minimizing level of energy efficiency and the level actually realized. Yergin, McKinsey & Co., and other analysts have argued that this gap represents a significant share of total energy use: in their view, the ground is littered with $20 bills that energy consumers have failed to pick up.

The energy efficiency policy debate often comingles these two types of market failures — energy use externalities and investment inefficiencies — causing imprecision in research questions and policy goals. In this paper, we distinguish between the two market failures and clarify their separate policy implications. If energy use externalities are the only market failure, it is well known that the social optimum is obtained with Pigouvian taxes or equivalent cap-and-trade programs that internalize these externalities into energy prices, and that substitute policies are often much less economically efficient. If investment inefficiencies also exist, the first-best policy is to address the inefficiency directly: for example, by providing information to imperfectly informed consumers. However, when these interventions are not fully effective and investment inefficiencies remain, policies that subsidize or mandate energy efficiency might increase welfare. The central question in this context is thus whether there are investment inefficiencies that a policy could correct — in other words, "Is there an Energy Efficiency Gap?"

This chapter examines two classes of evidence on the existence and magnitude of investment inefficiencies that could cause the energy efficiency gap. First, we examine choices made by consumers and firms, testing whether they fail to make investments that would increase utility or profits. Second, we focus on specific investment inefficiencies, testing for evidence consistent with each. After presenting the evidence, the chapter discusses policy implications.

Three key conclusions arise. First, although there is a long literature assessing investment inefficiencies related to energy efficiency, this body of evidence frequently does not meet modern standards for credibility. A basic problem is that much of the evidence on the energy cost savings from energy efficiency comes from engineering analyses or observational studies that can suffer from a set of well-known biases. Furthermore, even if the energy cost savings were known, energy efficiency investments often have other unobserved costs and benefits, making it difficult to assess welfare effects. This problem is general to other economic applications: in order to argue that an agent is not maximizing an objective function, the analyst must credibly observe that objective function in full. We believe that there is great potential for a new body of credible empirical work in this area, both because the questions are so important and because there are significant unexploited opportunities for randomized controlled trials and quasi-experimental designs that have advanced knowledge in other domains.

Second, when one tallies up the available empirical evidence from different contexts, it is difficult to substantiate claims of a pervasive energy efficiency gap. Some consumers appear to be imperfectly informed, and the evidence suggests that investment inefficiencies do cause an increase in energy use in various settings. However, the empirical magnitudes of the investment inefficiencies appear to be smaller, indeed substantially smaller, than the massive potential savings calculated in engineering analyses such as McKinsey & Co. [1].

Third, because consumers are quite heterogeneous in the degree of their investment inefficiencies, it is crucial to design targeted policies. Subsidizing energy efficient durables, for example, changes relative prices for all consumers. While this policy will increase welfare for some consumers, such benefits must be traded off against distortions to consumers not subject to inefficiencies. Policy evaluations must therefore consider not just how much a policy increases energy efficiency, but what types of consumers are induced to become more energy efficient. Welfare gains will be larger from a policy that preferentially affects the decisions of consumers subject to investment inefficiencies.

The chapter is organized as follows: Section 2 provides the context of the debate. Section 3 discusses tests of whether consumers and firms appear to leave money on the table by not being more energy efficient. Section 4 discusses tests of specific investment inefficiencies. Section 5 provides an assessment of policy implications of our findings. The chapter's conclusions are summarized in section 6. A model used to derive the results presented is provided as Appendix A.

2 BACKGROUND FACTS ON ENERGY DEMAND

2.1 Overview of Energy Demand and Energy Efficiency

Table 5.1 presents the breakdown of total energy demand across the sectors of the U.S. economy. Much of our discussion focuses on household energy use and personal transportation instead of commercial and industrial energy use, because these are areas where inefficiencies of imperfect information might be more severe. In 2007, the average U.S. household spent $2,400 on gasoline for their autos and another $1,900 on natural gas, electricity, and heating oil [5]. Of this latter figure, heating and cooling are the most significant end uses, which suggests that they may also be the areas where energy conservation could have the largest effect.

The smaller the variance in energy costs across products relative to the total purchase price, the more likely it is that consumers will choose to remain imperfectly informed about, or inattentive to, these costs [6]. Figure 5.1 shows the lifetime energy cost of a selection of energy-using durables, discounted at 6 percent over each good's typical lifetime, as well as the ratio of energy cost to the

TABLE 5.1 U.S. Energy Use

By Sector [3,4]	
Commercial	19%
Industrial	30%
Transport	29%
Residential	22%
Residential Categories [2]	
Refrigerators	5%
Air conditioning	8%
Water Heating	20%
Space Heating	41%
Other Appliances and Lighting	26%

Source: Data are from U.S. Energy Information Administration [2–4].

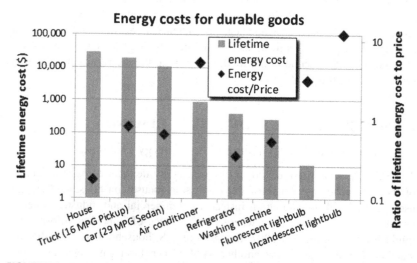

FIGURE 5.1 Energy Costs for Durable Goods. *Source: Authors.*

purchase price. For example, if gasoline costs $3 per gallon, lifetime gasoline costs are $19,000 for a typical pickup truck, or 83 percent of the purchase price, and $10,000 for a relatively energy efficient sedan, or about 66 percent of purchase price. Typical lifetime energy costs are five times greater than purchase prices for air conditioners and 12 times greater for incandescent light bulbs, but only about one-third of purchase price for a typical refrigerator.

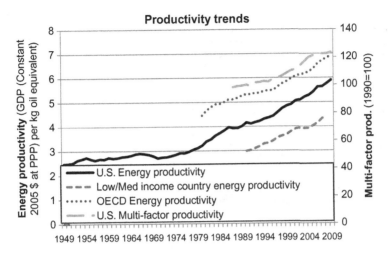

FIGURE 5.2   Productivity Trends. *Note: PPP is "purchasing power parity."* Multifactor *productivity index equals 100 in 1990.*

The most aggregate measure of energy efficiency is the ratio of GDP to total energy use, with different energy sources combined using common physical units. As shown in Figure 5.2, U.S. "energy productivity" per unit of GDP is 2.4 times higher than in 1949. Various factors drive this continual improvement, including compositional changes in the economy toward less-energy-intensive industries, energy efficiency policies, and other forces that drive total factor productivity growth. Energy prices also induce factor substitution and technical change: the figure suggests this effect, showing that the fastest improvements in energy productivity were in the 1970s and the most recent 15 years, both periods of relatively high energy prices. The figure also shows that U.S. energy productivity has grown faster than total factor productivity since the beginning of that data series in 1987, meaning that through some combination of directed technical change and factor substitution, the United States is economizing on energy faster than it is economizing on other factors. The U.S. economy is more energy intensive than other OECD countries, although it has improved more quickly since 1980, and less energy intensive than the set of low- and middle-income countries. In sum, the U.S. economy is progressively becoming less energy intensive, although this is uninformative about whether the United States is at or near the economically efficient level of energy efficiency.

2.2 Energy Efficiency Policy in the United States

The United States has enacted a wide array of policies to encourage energy efficiency, many of which were originally promulgated during the energy crises of the 1970s. Table 5.2 presents the most significant of these policies,

TABLE 5.2 Significant U.S. Energy Efficiency Policies

Name	Years	Magnitude
Corporate Average Fuel Economy Standards	1978–	$10 billion annual incremental cost form tightened 2012 rule [7]
Federal Hybrid Vehicle Tax Credit	2006–2010	$425 million total annual credit [8]
Gas guzzler tax	1980–	$200 millions annual revenues [8]
Federal Appliance energy efficiency standards	1990–	$2.9 billion annual incremental cost [9]
Residential and commercial building codes	1978–	
Electricity Demand-Side Management programs	1978–	$3.6 billion annual cost [10]
Weatherization Assistance Program (WAP)	1976–	$250 million annual cost [11]
2009 Economic Stimulus	2009–2011	$17 billion total [11]
Additional WAP funding		$5 billion
Recovery Through Retrofit		$454 million
State Energy Program		$3.1 billion
Energy Efficiency and Conservation Blocks Grants		$3.2 billion
Home Energy Efficiency Tax Credits		$5.8 total credit in 2009 [12]
Residential and Commercial Building Initiative		$346 million
Energy Efficient Appliance Rebate Program		$300 million
Autos Cash for Clunkers		$5 billion

Source: Authors.

along with some measure of their annual costs. Auto industry policies include: Corporate Average Fuel Economy (CAFE) standards, which require that the new cars and trucks sold by each auto manufacturer meet a minimum average rating based on miles-per-gallon; tax credits of up to $3,400 for hybrid vehicle buyers; and "gas guzzler taxes" ranging from $1,000 to $7,700 on the sale of passenger cars with low fuel economy. There are a series of national-level minimum energy efficiency standards for household appliances, such as refrigerators, air conditioners, and washing machines. Additionally, many states have building codes that encourage energy efficiency by, for example, stipulating minimum amounts of required insulation. Furthermore, electricity bill surcharges fund billions of dollars of utility-managed "demand-side management" programs, which include subsidized residential and commercial energy audits, energy efficiency information provision, and subsidies for energy efficient appliances and other capital investments.

"Weatherization" is frequently used as a general term for a set of residential energy efficiency investments primarily including wall and attic insulation, improved heating, ventilation and air conditioning systems, and "air-sealing," which reduces the leakage of hot or cold outside air. Through the Weatherization Assistance Program, the federal government transfers $250 million annually to state agencies to weatherize approximately 100,000 low-income homes. Weatherization funding grew significantly due to the 2009 American Recovery and Reinvestment Act. In total, that legislation and related economic stimulus bills included $17 billion in energy efficiency spending, including non-low-income weatherization programs, automobile and appliance cash-for-clunkers programs with energy efficiency requirements on new models, and other grants to state programs.

In this chapter, the phrase "energy efficiency policies" refers to this set of subsidies and standards that directly encourage investment in energy efficient capital stock but do not directly affect energy prices. Although gasoline taxes, cap-and-trade programs, or other policies that affect energy prices will of course also increase investment in energy efficient capital stock, these policies that act through energy prices are conceptually distinct in our policy analysis.

3 EVIDENCE ON RETURNS TO ENERGY EFFICIENCY INVESTMENTS

This section analyzes the evidence on whether consumers and firms leave profitable energy efficiency investments on the table. There are four categories of evidence: engineering estimates of returns to potential investments, empirical estimates of returns to observed investments, the cost effectiveness of energy conservation programs run by electric utilities, and estimated demand patterns for energy-using durables.

3.1 Engineering Estimates of Energy Conservation Cost Curves

While the McKinsey & Co. [1] study quoted in our introduction has garnered substantial attention, it is preceded by a long literature that uses engineering cost estimates to construct "supply curves" for energy efficiency (e.g., [13–21]). The basic approach in such studies is to calculate the net present value of a set of possible energy efficiency investments given assumed capital costs, energy prices, investment horizons, and discount rates.

Across many studies from different industries and sectors, a common theme seems to emerge: large fractions of energy can be conserved at *negative* net cost. That is, the studies conclude that consumers and firms are failing to exploit a massive amount of profitable investment opportunities in energy efficiency. For example, a meta-analysis by Rosenfeld, Atkinson, Koomey, Meier, Mowris, and Price [19] concludes that between 20 and 60 percent of total electricity use, depending on the study and the electricity cost assumption, can be conserved at negative cost. The McKinsey & Co. [1] analysis quoted in our introduction suggests that 23 percent of U.S. non-transportation energy demand can be eliminated at negative cost. These engineering studies are a large part of the basis for the claims about the energy efficiency gap.

However, it is difficult to take at face value the quantitative conclusions of the engineering analyses as they suffer from the empirical problems introduced in the previous section. First, engineering costs typically incorporate only upfront capital costs and omit opportunity costs or other unobserved factors (ξ in the model presented in the Appendix). For example, Anderson and Newell [22] analyze energy audits that the U.S. Department of Energy provides for free to small- and medium-sized enterprises. They find that nearly half of investments that engineering assessments showed would have short payback periods were not adopted due to unaccounted physical costs, risks, or opportunity costs, such as "lack of staff for analysis/implementation," "risk of inconvenience to personnel," or "suspected risk of problem with equipment."

Second, the engineering estimates of energy saved may be faulty. For example, in the context of home energy weatherization, Dubin, Miedema, and Chandran [23], Nadel and Keating [24], and others have documented that engineering estimates of energy savings can overstate true field returns, sometimes by a large amount. Even in the two decades since these studies, some engineering simulation models have still not been fully calibrated to approximate actual returns [25].

3.2 Empirical Estimates of Returns on Investment

Another approach to measuring the energy efficiency gap is to use empirical energy use data to estimate the average returns for the set of consumers that adopt an energy efficient technology. Most of the evidence in this category analyzes the costs and benefits of the Weatherization Assistance Program, which is intended to be both a transfer to low-income homeowners and an energy efficiency investment with positive net returns. The typical empirical analysis compares natural gas billing data in the first year after the weatherization work was done to the year before, using either a statistical correction for weather differences or a non-randomly selected control group of low-income households. Schweitzer [26] analyzes 38 separate empirical evaluations of weatherization projects from 19 states from between 1993 and 2005, re-weighting them to reflect the observable characteristics of the national Weatherization Assistance Program. The average weatherization job costs $2,600 and reduces natural gas use by 20 to 25 percent, or about $260 per year.

As evidence on the energy efficiency gap, such analyses again suffer from the problems introduced in the previous section. First, there are potentially substantial unobserved costs and benefits (the ξ in our model) from weatherization. Weatherization takes time, and for most people it is not highly enjoyable: the process requires one or sometimes two home energy audits, a contractor appointment to carry out the work, and sometimes additional follow-up visits and paperwork. Some benefits are also difficult to quantify: for example, weatherization typically makes homes more comfortable and less drafty. Furthermore, weatherization reduces the cost of energy services such as warmer indoor temperatures on a cold winter day, and this cost reduction causes people to increase their utilization of these services. (In the energy literature, this is called the "rebound effect.") Measuring the change in energy use from weatherization without accounting for the utility gain from an increase in utilization of energy services understates the welfare benefits.

Second, the net present value of energy cost reductions is unknown. The empirical estimates are based on short-term analyses, and the persistence of returns over many years is rarely assessed.[1] If the $260 annual savings from Schweitzer [26] are assumed to have a lifetime of 10 years or less, then weatherization does not pay back the $2,600 cost at any positive discount rate. At lifetimes of 15 or 20 years, the discount rate that equates future discounted benefits with current costs (the internal rate of return) is 5.6 or 7.8 percent, respectively. Furthermore, all of the estimates are non-experimental, and households that weatherize may also engage in other unobserved activities that affect energy use. This may be a larger concern with non-low-income weatherization programs, in which homeowners might be more likely to carry out renovations and energy efficiency work at the same time.

Third, the effects of weatherization on energy use are heterogeneous. For example, Metcalf and Hassett [28] estimate the distribution of returns to attic insulation in the U.S. population using a weather-adjusted difference estimator with nationally representative panel data. The estimated median and

---

1. Sumi and Coates [27] find 14 percent cumulative degradation of energy savings for Weatherization Assistance Program participants in Seattle after the first six years. Additional evidence using more recent data would be valuable, as would information on whether the rate of depreciation changes over time.

mean returns on investment are on the order of 10 percent, and one-quarter of households had returns greater than 13.5 percent. This heterogeneity means that while estimates of average returns for adopters could in principle be meaningful in evaluating the costs and benefits of an existing program, a simple selection model like the one above would imply that the net returns for adopters overstate the net returns for non-adopters. On net, the available evidence seems inconsistent with significant investment inefficiencies in the context of weatherization.

3.3 Cost Effectiveness of Energy Conservation Programs

Many electric utilities run "demand-side management" programs, which largely consist of subsidies to households and firms to purchase energy efficient appliances, air conditioning and heating systems, and other equipment. If these programs can reduce energy use at less than the cost of energy, the argument goes, then there were investment inefficiencies, and the programs should be viewed as welfare-enhancing.

The simplest example of this approach is to divide the annual spending on these programs by utilities' estimates of electricity savings, as in Gillingham, Newell, and Palmer [9]. These estimated savings are typically from engineering estimates of electricity savings or non-experimental comparisons of energy use between program participants and non-participants. For 2009, U.S. electric utilities reported $2.255 billion in direct costs and 76.9 terawatt-hours of savings for demand-side management programs, according to the 2009 *Electric Power Annual* ([10], Tables 9.6 and 9.7). Dividing these two figures gives a cost effectiveness of 2.9 cents per kilowatt-hour (kWh). Friedrich, Eldridge, York, Witte, and Kushler [29] also use utilities' estimates of electricity savings to calculate a cost effectiveness of 2.5 cents per kilowatt-hour.

Analyses such as these suffer from the same problems introduced in the previous section. First, the reported "costs" are typically costs to the utility, not including costs incurred by program participants, which may be almost as large [29–32]. Second, energy savings are estimated using engineering analyses or observational data, and it is difficult to establish a credible counterfactual level of energy use in the absence of the program. The most rigorous solution is to use randomized controlled experiments to evaluate demand-side management programs. The feasibility of this approach is demonstrated by recent experimental evaluations of programs that send letters that compare a household's energy use to that of their neighbors and provide energy conservation tips [33,34].

The most advanced estimate in this literature is by Arimura, Li, Newell, and Palmer [35], whose point estimates indicate that between 1992 and 2006, demand-side management conserved electricity at a program cost of 5.0 and 6.1 cents per kilowatt-hour, assuming discount rates of 5 and 7 percent, respectively.[2] If one further assumes, based on the analyses in the paragraph above, that additional costs to consumers might be 70 percent of program costs, one concludes that demand-side management programs have reduced energy use at an average cost of $5.0 \times (1 + .70) = 8.5$ cents/kWh or $6.1 \times (1 + .70) = 10.4$ cents/kWh, again using 5 or 7 percent discount rates, respectively. Comparing the investment cost per kWh conserved to the national average electricity price of 9.1 cents/kWh, the investments that occurred because of demand-side management programs were barely profitable at a discount rate of 5 percent, and barely unprofitable at a discount rate of 7 percent.[3] Arimura et al. [35] estimate that these programs reduced 1 to 2 percent of national electricity demand. Given that only a small percent of total electricity demand was reduced at nearly zero excess profits, this evidence on demand-side management energy conservation programs does not suggest a pervasive energy efficiency gap.

3.4 Tradeoffs between Durable Goods

The final way of determining whether there are profitable returns to energy efficiency investments involves estimating consumer demand for household appliances or automobiles. This approach typically uses a discrete choice model to estimate utility function coefficients on purchase price and on the present discounted value of energy costs. The estimated coefficient on energy cost should be the same as the estimated coefficient on price: that is, consumers should be indifferent between spending a dollar in present value on energy and a dollar in present value on purchase price. If the analyst's assumptions about discount rates, product utilization, and energy prices are correct, the ratio of these two coefficients is the γ in our model in the Appendix. (In that model, $\gamma = 1$ indicates no investment inefficiencies, and $\gamma < 1$ indicates that there are investment inefficiencies.)

In a seminal paper, Hausman [36] estimated a discrete choice model using 65 observations of consumer choices between air conditioner models, which vary in upfront cost and energy efficiency rating. Hausman framed his analysis as an estimate of an "implied discount rate" that rationalizes the demand system by assuming $\gamma = 1$. Hausman's paper, along with Dubin and McFadden's [37] analysis of households' choices between heating systems, was the state of the art in this literature for 30 years. Both papers find real

---

2. Arimura, Li, Newell, and Palmer [35] also calculate cost effectiveness of 3.0 and 4.1 cents per kilowatt-hour using discount rates of zero percent and three percent, respectively. The authors focus on their results for a five percent discount rate.

3. The text offers a back-of-the-envelope version of the more sophisticated calculation that should be done. One implicit assumption is that there are no consumers that are inframarginal to the demand-side management subsidies. If there are inframarginal consumers, then some of the program costs were in fact transfers, and the incremental investment costs induced by the demand-side management programs were smaller.

implied discount rates of 15 to 25 percent, which is higher than returns on stock market investments but not much different from real credit card interest rates, which were around 18 percent.

However, such analyses suffer from the problems introduced in the previous section. First, unobserved product attributes (which are analogous to ξ in our formal model) complicate the cross-sectional econometric approach. The coefficient on the present discounted value of energy costs is biased if energy efficient products have better or worse unobserved characteristics. For example, automobile prices actually *decrease* in fuel economy, as the more energy efficient vehicles are smaller and often have fewer luxury amenities. Furthermore, product prices will often be correlated with unobserved attributes, giving the usual simultaneity bias in estimating price elasticity. As in Berry, Levinsohn, and Pakes [38], this issue can potentially be addressed by using instrumental variables, but the instruments available may be dissatisfying.

Working papers by Allcott and Wozny [39], Busse, Knittel, and Zettelmeyer [40], and Sallee, West, and Fan [41] use an alternative approach to address the problem of unobserved attributes. These papers take a panel of used durable goods, condition on product fixed effects, and test how the relative prices of more energy efficient vs. less efficient products change as energy price expectations vary over time. As an intuitive example of the identification strategy, notice that as expected gasoline prices rise, we should expect to see the market price of a three-year-old used Honda Civic increase relative to the price of a three-year-old Honda Accord, because the Civic is more energy efficient than the Accord. If market prices are not very responsive, this approach suggests that γ is small.

Relative to the other categories of evidence on the energy efficiency gap, this approach is especially appealing because the fixed effects eliminate unobserved costs by construction. However, these analyses still suffer from the second problem, which is that they still require assumptions about the relevant discount rate, vehicle-miles traveled, and consumers' expectations of future gasoline prices (r, m, and p in our model in the Appendix), and other factors. Allcott and Wozny's [39] results tend to suggest that $\gamma < 1$, while Busse, Knittel, and Zettelmeyer's [40] results tend not to support the hypothesis that $\gamma < 1$. The two analyses do agree that even if there is some investment inefficiency, the welfare losses would be relatively small. Allcott and Wozny show how to use discrete choice data to calculate the private welfare loss. Their preferred estimate of γ suggests that investment inefficiencies in the auto market cause a welfare loss of about $1 billion per year and an increase in gasoline consumption of about 5 percent. Busse, Knittel, and Zettelmeyer's results tend not to suggest that there are investment inefficiencies, implying that there would be zero welfare losses. In either case, the welfare loss and gasoline consumption increase appear to be small relative to the total market size.

4 INVESTMENT INEFFICIENCIES THAT COULD CAUSE AN ENERGY EFFICIENCY GAP

The previous section examined evidence on whether consumers and firms fail to exploit profitable energy efficiency investments. This section reverses the perspective by specifying particular investment inefficiencies that might cause underinvestment in energy efficiency and assessing the empirical evidence on their magnitudes.

4.1 Imperfect Information

Imperfect information is perhaps the most important form of investment inefficiency that could cause an energy efficiency gap. Two basic models of imperfect information are most relevant. In one model, consumers and firms may be unaware of potential investments in energy efficiency. For example, homeowners may not know how poorly insulated their home is and may not be aware of the opportunity to weatherize. Similarly, factory managers may not know about a new type of machine that could reduce their energy costs.

An alternative model resembles Akerlof's [42] "lemons" model. Buyers know that different products, such as apartments, commercial buildings, or factory equipment, have different levels of energy efficiency, but these differences are costly to observe. Thus, they are not willing to pay more for goods that are in fact more energy efficient. For example, a renter evaluating a set of different apartments may be aware that there is a distribution of wall insulation quality and thus of resulting heating costs, but the renter will not be willing to pay more for a well-insulated apartment without taking the time to inspect the insulation.

There are three approaches to assessing the magnitude of imperfect information in the context of energy efficiency. The first approach is to test for market equilibria consistent with imperfect information. Several recent projects used this approach in the context of renter-occupied vs. owner-occupied housing units. The theory is that because imperfectly informed renters will not be willing to pay more for energy efficient apartments, landlords have reduced incentive to invest in energy efficiency. Homeowners, on the other hand, do capture the benefits of improved energy efficiency, at least until they sell the property. Such a "landlord–tenant" agency problem implies that rental properties are less energy efficient than would be socially optimal.

As an example of this approach, Davis [43] studies the market penetration of refrigerators, dishwashers, light bulbs, room air conditioners, and clothes washers that have earned the U.S. government's "Energy Star" designation, meaning that they are relatively energy efficient. Conditional on observable characteristics, renters are 1 to 10 percentage points less likely to report having Energy Star appliances. In percentage terms, these differences are large: they represent between 5.6 and 68 percent of the overall average

Energy Star saturation rate. But because non-renters are themselves not very likely to own Energy Star appliances and because appliances make up only one-quarter of residential energy use, the differences in appliance ownership do not add up to a large difference in energy use: the author calculates that if renters had the same energy efficient appliance ownership rates as owner-occupied homes, total energy bills in rental homes would be 0.5 percent lower.

Heating and cooling represent close to one-half of residential energy use, meaning that insulation is a more important investment that could be subject to the landlord–tenant agency problem. Gillingham, Harding, and Rapson [44] analyze data from California and show that when the resident pays for heating and cooling, owner-occupied dwellings are 13 to 20 percent more likely to have insulation than rentals, conditional on other observable characteristics of the property, occupant, and neighborhood. As in the Davis [43] analysis, large percent differences do not translate into large increases in energy use: under the optimistic assumption that insulation reduces total energy demand by 10 percent, we can see that rental properties would use 1.3 to 2.0 percent less energy if insulated at the same level as owner-occupied properties.

Additional research in this area would be important to address problems of both internal and external validity. As both Davis [43] and Gillingham, Harding, and Rapson [44] note, conditional differences in appliance ownership between owners and renters are not ironclad causal evidence of a market failure, because preferences could vary in unobservable ways. Furthermore, even if estimates from California are internally valid, differences in policies and housing stock make it difficult to generalize nationwide. For example, California was a relatively early adopter of building codes, which means that more buildings in the state may now be insulated, and thus the difference between rental and owner-occupied properties may be smaller than in other states.

If these estimates are assumed to be causal and generalizable, how big is the investment inefficiency from the landlord–tenant agency problem? The magnitude is the number of affected households times the extent of the reduced energy efficiency. Of U.S. households, 29 percent are rental units where the renter pays energy bills [45]. Multiplying this figure by several percent of total energy demand, to approximate the magnitude of the inefficiencies estimated above, implies that the landlord–tenant information problem might increase total residential energy use on the order of 1 percent. Thus, while the empirical evidence points to some inefficiency, it explains only a very small fraction of the purported energy efficiency gap.[4]

---

4. A related agency problem is that some landlords pay the utilities, while tenants set the utilization levels for appliances and heating and cooling equipment. This problem is a small, as only 4 percent of U.S. households are rentals with utilities included and demand for energy services is relatively inelastic. Levinson and Niemann [46] estimate that energy costs are 1 to 1.7 percent higher when utilities are included in rent.

An additional example of testing for imperfect information using equilibrium outcomes is to examine whether information disclosure increases the elasticity of energy-saving technical change with respect to energy prices. The idea is that consumers who are better informed about energy use will be more responsive to energy price changes when choosing between models of an energy-using durable. Therefore, firms with better-informed consumers will be more likely to offer more energy efficient models as energy prices rise. Newell, Jaffe, and Stavins [47] show that the mean energy efficiency of room air conditioners and water heaters was more responsive to energy prices after 1981 and 1977, respectively, the years when the federal government introduced energy efficiency labeling requirements for the two goods. While other factors might also have changed in these years, this finding suggests that the labeling requirements may have reduced the extent of imperfect information. However, this approach is not informative about the magnitude of any remaining investment inefficiency.

A second approach that can allow a direct assessment of the magnitude of imperfect information is to observe information sets through surveys. Turrentine and Kurani [48] and Larrick and Soll [49] use structured interviews and laboratory studies to show that consumers are not very good at calculating the gasoline costs for different automobiles. The 2010 Vehicle Ownership and Alternatives Survey [50,51] adds nationally representative evidence on how accurately consumers perceive the financial value of energy efficiency. The data suggest that consumers are indeed imperfectly informed: over half of Americans misestimate the gasoline cost differences between the vehicle they own and their "second choice vehicle" by more than 40 percent. However, the errors run in both directions. On average, consumers appear to either correctly estimate or slightly underestimate the energy cost savings from higher-fuel economy vehicles.

A third approach to assessing the magnitude of imperfect information is to test for the effects of information disclosure on purchase decisions. This approach has the benefit of being based on observed choices in the marketplace, instead of beliefs stated on a survey. We are not aware of any large-scale randomized evaluations of energy efficiency information disclosure.

4.2 Inattention

Interventions that resemble information disclosure might change the buying patterns of consumers who are already well-informed. For example, Chetty, Looney, and Kroft [52] find that despite the fact that consumers are well-informed about sales taxes, posting information about sales tax amounts in a supermarket changes buying patterns. This finding suggests the existence of another type of investment inefficiency, which behavioral economists call inattention.

The psychology of inattention starts by recognizing that choice problems have many different facets, and some of these facets are less salient at the time of choice even if they are potentially important to the utility that will later be experienced. When buying printers, for example, we might focus on the purchase price and fail to consider that replacement ink cartridges make up the bulk of the total cost. Inattentive consumers are misoptimizing: they fail to recognize opportunities to save money by choosing products with lower ancillary costs. Research in a variety of other non-energy settings is suggestive of inattention [53−55]. It seems possible that some consumers might be inattentive to energy efficiency when purchasing energy-using durable goods.

5 POLICY IMPLICATIONS

Our assessment is that the available evidence on the size of an energy efficiency gap is situation-specific, mixed, and often inconclusive. However, policymakers must make policy even in the absence of ironclad evidence.

If the goal is to achieve efficient economic outcomes, then market failures should be addressed as directly as possible. In response to energy use externalities, a Pigouvian tax gives the first-best outcome. If agents are imperfectly informed and the government has an inexpensive information disclosure technology, an information disclosure approach should be used. Formulating policy becomes more challenging if the first-best solutions are not possible − when information disclosure is not fully effective, or when a Pigouvian tax is not politically feasible because of aversion to new taxes or to policies that explicitly regulate greenhouse gases. This section examines the effects of energy efficiency policies, considered as a second-best alternative.

5.1 Energy Efficiency Subsidies and Standards as a Second-Best Approach to Pollution Abatement

Until now, we have set aside the uninternalized energy use externalities and focused on investment inefficiencies. We now examine the converse: imagine a setting where no investment inefficiencies exist, but energy is priced below social cost (in our model, $\varphi > 0$). If Pigouvian taxes or cap-and-trade programs are politically infeasible, would energy efficiency subsidies and standards be a relatively promising approach to pollution abatement?

When no investment inefficiencies exist, energy efficiency policies such as subsidies for energy efficient durable goods and minimum energy efficiency standards would have larger welfare costs per unit of pollution abated compared to the first-best Pigouvian tax for several reasons. First, subsidies and standards change relative prices for all consumers equally, while the Pigouvian tax provides a larger incentive for consumers with higher utilization to choose energy efficient capital stock. Second, the first-best policy must impose the right price on the utilization decision, which only the Pigouvian tax does. Third, it is difficult to calibrate the stringency of an energy efficiency standard or subsidy precisely, meaning that it will likely generate more or less carbon abatement than a Pigouvian tax set at the level of marginal damages. Energy efficiency policies in different sectors can also be miscalibrated against each other, causing inefficiency due to unequal marginal costs of abatement.

Of course, if these three theoretical factors were small in reality, then energy efficiency policies might be a reasonable second-best substitute for Pigouvian taxes. Several analyses have simulated the relative cost effectiveness of particular energy efficiency policies relative to Pigouvian taxes. Jacobsen [56], for example, simulates automobile supply and demand and shows that Corporate Average Fuel Economy (CAFE) standards have a welfare cost of $222 per metric ton of carbon dioxide abated, compared to $92 per ton for a gas tax that generates the same amount of abatement. Krupnick, Parry, Walls, Knowles, and Hayes [57] come to a similar qualitative conclusion. They compare the cap-and-trade provisions of the proposed Waxman—Markey climate change legislation to the legislation's energy efficiency provisions, which include standards for buildings, lighting, and appliances. The cap-and-trade, or an equivalent carbon tax, abates carbon dioxide at an upfront welfare cost of $12 per ton. Under the assumption that there are no investment inefficiencies, the energy efficiency standards are five times more costly, or $60 per ton. This significantly exceeds the United States government's estimated social cost of carbon dioxide emissions, which is about $21 [58].[5]

These results forcefully argue that Pigouvian taxes or cap-and-trade programs are the most efficient way to address energy use externalities. Energy efficiency subsidies, CAFE standards, and other energy efficiency policies can also reduce energy use externalities, but in any settings where there are no investment inefficiencies, such policies will often impose a significantly larger cost on the economy per unit of pollution reduction.

5.2 Energy Efficiency Subsidies and Standards as a Second-Best Approach to Correcting Investment Inefficiencies

The United States has long required energy-use information disclosure: for more than 30 years, retailers have been required to display fuel economy ratings for new vehicles and energy cost information for home appliances.

---

5. As we will argue in the next section, energy efficiency policies can increase welfare when there are investment inefficiencies. Krupnick, Parry, Walls, Knowles, and Hayes [57] quantify this, showing that under an assumed level of investment inefficiencies, energy efficiency standards abate carbon at a cost of $7 per ton.

However, consumers may not notice, understand, or pay attention to this information. If information disclosure or other direct solutions to market failures are not fully effective, how useful are energy efficiency subsidies and standards as a second-best approach to addressing investment inefficiencies?

Allcott, Mullainathan, and Taubinsky [59] analyze this question when consumers are inattentive to energy costs. As we discussed earlier, their model shows that energy efficiency subsidies can increase welfare and, when consumers are sufficiently homogeneous, the first-best can be obtained. The intuition is straightforward: if consumers and firms underinvest in energy efficiency, subsidizing or mandating them to invest more can increase welfare. However, any corrective policies must be properly calibrated. For example, the vast majority of benefits in the U.S. government's cost–benefit analysis of Corporate Average Fuel Economy Standards derive from the assumption that the regulation corrects consumers' inattention to energy efficiency when buying autos.[6] However, Allcott [51], Allcott and Wozny [39], Fischer, Harrington, and Parry [60], and Heutel [61] use different models to show that the current and proposed CAFE standards are much more stringent than can be justified by even worst-case estimates of investment inefficiencies.[7] Of course, if there are zero investment inefficiencies, then there are zero welfare benefits through this channel.

Heterogeneity in the investment inefficiency weakens the policy argument for subsidizing energy efficient goods, as [59] also show. For example, imagine that some consumers have no investment inefficiencies (i.e., have $\gamma = 1$) in our model in the Appendix, while others are subject to some investment inefficiency. A subsidy, while improving allocations for agents subject to investment inefficiencies, causes those not subject to investment inefficiencies to *over-consume* energy efficient goods. This offsets the welfare gains from improving allocations for the consumers subject to investment inefficiencies.[8] A key implication for policy analysis is that we must understand not just how *much* a policy increases sales of energy efficient goods,

6. For more background, see the federal government's Regulatory Impact Analysis of the 2012–2016 CAFE standards (National Highway Traffic Safety Administration, 2010) or the discussion in Allcott and Wozny [39].

7. Heutel calibrates the extent to which investment inefficiencies that cause consumers to undervalue gasoline costs by 30 percent decrease the average fuel economy of vehicles sold, relative to the optimum. An optimal policy response would therefore increase fuel economy by a corresponding amount. In Table 4, comparing column (1) to column (2) shows the effect of investment inefficiencies: 0.0439 gallons per mile (GPM) -0.0426 GPM $= 0.0013$ gallons per mile. Thus, an optimal policy response decreases GPM by 0.0013 gallons per mile, or equivalently increases fuel economy by slightly more than 0.5 MPG. In Heutel's model, CAFE standards would not be justified by his assumed level of investment inefficiency, given that they require fuel economy to increase by much more than 0.5 MPG.

8. In Figure 5.3 in the Appendix, imagine that some consumers are on demand curve D' with $\gamma = 1$, while others are on demand curve D with $\gamma < 1$. Now, a subsidy moves the high-γ agents to point z, where they consume more energy efficiency than they would in the social optimum.

but *who* are the people induced to buy these goods. For example, even in a setting where the average consumer is subject to some investment inefficiency, energy efficiency subsidies might decrease total welfare if they are largely taken up by environmentalists and homeowners, who are more likely to be well-informed about energy efficiency and are not subject to a "landlord—tenant" agency problem.

This discussion highlights that energy efficiency policies are more likely to increase welfare if they target agents subject to the largest investment inefficiencies. Some existing policies do appear well-targeted. For example, households that use more energy than other comparable households are more likely to have low-cost energy conservation opportunities of which they are unaware, and many U.S. utilities now target energy conservation information to these relatively heavy users [33,34]. "Smart meters" that record hourly consumption, which as described by Joskow (2012) are increasingly being deployed across the United States, also provide information useful for targeting. For example, utilities can now identify households that use more energy on afternoon hours of particularly hot days, suggesting that they have energy inefficient air conditioners, and send them information on new energy efficient models.

Aside from heterogeneity in the investment inefficiency, consumers and firms also have substantial heterogeneity in other factors that affect demand for energy and for energy efficient capital stock. For example, the mild climate of Los Angeles compared to the more extreme weather of Chicago means that there is substantial variation in utilization of air conditioners and heating equipment, and residential retail electricity prices vary across the country from 4 to 30 cents per kilowatt-hour. As a result, national-level minimum efficiency standards for home appliances seem likely to decrease welfare for subsets of consumers with low prices and utilization and could increase welfare for high-price and/or -utilization consumers with investment inefficiencies. Ideally, standards could vary geographically to take account of this, targeting consumers that may have the most to gain. For example, building codes in states with extreme weather often require more insulation than building codes in mild climates. On the other hand, home appliance standards are set at the national level, and appliance manufacturers and retailers operate nationwide. The benefits of heterogeneous standards must be weighed against the costs of regulatory complexity.

6 CONCLUSIONS

Since the energy crises of the 1970s, many have made the "win-win" argument for energy efficiency policy: subsidies and standards can both address investment inefficiencies in the purchase of energy-using durable goods and reduce externalities from energy use. However, a reliance on observational studies of variable credibility and the possibility of unobserved costs and benefits of energy efficiency make it difficult to assess the magnitude of the energy efficiency gap definitively. Nevertheless, the available evidence from empirical analyses of weatherization, demand-side management programs, automobile and appliance markets, the "landlord–tenant" agency problem, and information elicitation suggests that while investment inefficiencies do appear in various settings, the actual magnitude of the energy efficiency gap is small relative to the assessments from engineering analyses.

Furthermore, it appears likely that there is substantial heterogeneity in investment inefficiencies across the population. Thus, targeted policies have the potential to generate larger welfare gains than general subsidies or mandates. Given this heterogeneity, policy analyses need to do more than assess how much a policy affects energy efficiency: they must also identify what types of consumers are induced to be more energy efficient.

This area is ripe for rigorous empirical research. Future research should utilize randomized controlled trials and quasi-experimental techniques to estimate the impacts of energy efficiency programs on heterogeneous consumer types and to address the challenges posed by unobserved costs and benefits. The scientific insights from such research are potentially generalizable, and the policy implications are significant.

APPENDIX: A MODEL OF INVESTMENT IN ENERGY EFFICIENCY

The basic economics of energy efficiency are captured by a model in which an agent, either a profit-maximizing firm or utility-maximizing consumer, chooses between two different versions of an energy-using durable good such as an automobile, air conditioner, or light bulb.[9] This setup can also represent a choice of whether to improve the energy efficiency of an existing building, for example through weatherization. In the first period, the agent chooses and pays for capital investments. In the second period, the consumer uses the good and incurs energy costs.

The two different goods are denoted 0, for the energy inefficient baseline, and 1, for the energy efficient version. They have energy intensities e_0 and e_1, respectively, with $e_0 > e_1$. The energy efficient good has incremental upfront capital cost $c > 0$ and unobserved incremental opportunity cost or utility cost ξ. The variable ξ could either be positive (an unobserved cost) or negative (an unobserved benefit). The private cost of energy is p, and the risk-adjusted discount rate between the two periods is $r > 0$. The variable m depends

---

9. The model presented here is an adaptation of the model in Allcott, Mullainathan, and Taubinsky [59]. It resembles a generalized Roy model. It abstracts away from factors which may be relevant in some settings, including the irreversibility of some energy efficiency investments and uncertainty over energy costs [62,63] and explicit models of imperfect information in the purchase or resale of the good.

represents an agent's taste for usage of the durable good; a high m reflects an air conditioner user in a hot climate or a car owner who drives a long way to work. The variable m is implicitly a function of energy prices: as energy prices rise, the cost of utilization increases, so utilization decreases. We index m_i to explicitly recognize that it varies across agents, although in practice ξ and p will also vary.

In the basic case, an agent's willingness-to-pay for the energy efficient good is the discounted energy cost savings net of unobserved costs. Agent i will choose the energy efficient good if and only if willingness to pay outweighs the incremental capital costs:

$$pm_i(e_0 - e_1)/(1 + r) - \xi > c. \tag{5.1}$$

To capture the essence of the energy efficiency gap, we introduce the parameter γ, which is an implicit weight on the energy cost savings in the agent's decision. Now, the agent chooses the energy efficient good if and only if:

$$\gamma pm_i(e_0 - e_1)/(1 + r) - \xi > c. \tag{5.2}$$

For the purpose of determining the effects of subsidizing the energy efficient good, the γ parameter is a sufficient statistic for all investment inefficiencies. As we will discuss later in more detail, there are several distinct types of investment inefficiencies. First, agents may be unaware of, imperfectly informed about, or inattentive to energy cost savings. Second, agents may be themselves perfectly informed but unable to convey costlessly the energy intensity e_1 of an improved house or apartment they are selling or renting to others. Third, credit markets may be imperfect, meaning that agents may not have access to credit at the risk-adjusted discount rate r.[10] The γ parameter is conceptually related to what others have called an "implied discount rate," which is the discount rate that rationalizes the tradeoffs that agents make between upfront investment costs and future energy savings.

It is often asserted that $\gamma < 1$, meaning that investment inefficiencies cause agents to value discounted energy cost savings less than upfront costs. Notice that when this is the case, some agents do not choose the energy efficient good despite the fact that this would be profitable at current energy prices. Formally, asserting that there is an "energy efficiency gap" is exactly equivalent to asserting that there are investment inefficiencies and $\gamma < 1$. Of course, in some settings it might be that $\gamma > 1$.

Other than the investment inefficiencies captured by γ, the additional element of the "win-win argument" is that there are additional social costs from energy use that are not internalized into energy prices. We denote this

---

10. Credit constraints are a frequently discussed investment inefficiency. Although we note the issue in theory, there is not much empirical evidence in the context of energy efficiency, so we will not discuss it further.

uninternalized externality by φ. In the social optimum, the agent adopts the energy efficient good if:

$$(p + \varphi)m_i(e_0 - e_1)/(1 + r) - \xi > c. \tag{5.3}$$

The social optimum differs from the agent's choice in the previous equation for two reasons. First, the allocation accounts for the externality φ. Second, the allocation is not affected by investment inefficiencies, so $\gamma = 1$.

Figure 5.3 illustrates the three cases. The figure's horizontal axis represents the quantity of the energy efficient good that is purchased, while the vertical axis shows the incremental costs and benefits of purchasing that good. The height of a demand curve at each point reflects some individual agent's willingness-to-pay from the left-hand side of a corresponding equation above. The agents on the left side of the figure, with higher willingness-to-pay, tend to have high usage m, low unobserved cost ξ, and high energy price p.

The lowest demand curve, denoted D, reflects the case in the second equation with both investment inefficiencies ($\gamma < 1$) and uninternalized energy use externalities. In this case, the market equilibrium is at point a, the intersection of demand curve D with incremental cost c. Demand curve D' reflects the case in the first equation with no investment inefficiencies, but energy still priced below social cost. Demand curve D'' reflects the social optimum in the third equation, where there are no investment inefficiencies and energy prices

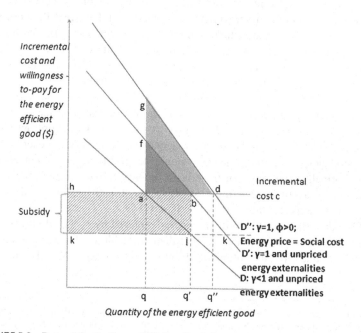

FIGURE 5.3 Demand for the Energy-Efficient Good. *Source: Authors.*

include externality φ. Adding a Pigouvian tax on energy consumption (based on the energy source's pollution content) increases willingness-to-pay more for the consumers on the left of the figure with higher utilization, so demand curve D'' rotates clockwise relative to demand curve D'. The first-best equilibrium is point d, where D'' intersects incremental cost c.

From a policy perspective, it is crucial to distinguish the two types of market failures, energy use externalities and investment inefficiencies. The reason derives from the general principle that policies should address market failures as directly as possible. If there are no investment inefficiencies but energy prices are below social cost due to uninternalized energy use externalities, demand is represented by D'. This causes a distortion both in the purchase and in the utilization of energy-using durables: for example, consumers buy too many gas guzzlers and drive them too much. A Pigouvian tax of amount φ on energy (on gas, in the example) would give both the socially optimal quantity demanded (q'') of the energy efficient good and the socially optimal utilization. By contrast, as long as utilization is not fully price-inelastic, a subsidy for the energy efficient good does not achieve the first best. While this could move quantity demanded to q'', consumers would not face the true social cost of energy when deciding how much to use the good: consumers would buy the right number of gas guzzlers but still drive them too much.

Many investment inefficiencies, on the other hand, distort purchases but not utilization. If there are investment inefficiencies but no uninternalized energy use externalities, the optimal corrective policy affects purchases, but not utilization. For example, Allcott, Mullainathan, and Taubinsky [59] show that when consumers have homogeneous $\gamma < 1$ and vary only in utilization m_i, the first-best policy involves a subsidy for the energy efficient good.[11] In Figure 5.3, that optimal subsidy would move quantity demanded from q to q'. Notice that an energy tax could potentially also correct the investment inefficiency, giving the same marginal consumer at q'. However, as long as utilization is not fully inelastic, an energy tax that gives price above social cost (to correct the investment inefficiency), would cause consumers to reduce utilization below the first-best level: consumers would buy the right number of gas guzzlers and then drive them too little.

Putting these arguments together, when there are distortions from both uninternalized energy use externalities and investment inefficiencies, the first-best policy involves both Pigouvian taxes on energy and a second mechanism to increase quantity demanded of the energy efficient good. This second mechanism may be a subsidy for the energy efficient good, although as we will discuss later in the paper, heterogeneity in the investment inefficiency γ makes subsidies potentially less desirable.

---

11. Heutel [61] obtains a comparable result using a different model of investment inefficiencies.

How can this framework be used for cost–benefit analysis? Consider first adding the subsidy in isolation, without any Pigouvian tax on energy. When there are investment inefficiencies, the original marginal consumer at quantity q gains amount af from being induced to buy the energy efficient good. In fact, there are allocative gains from inducing each of the consumers between q and q' to purchase the energy efficient good, as each of these consumers has benefits that are larger than incremental cost c. The total private welfare gains are illustrated by the triangle abf. If a Pigouvian tax on energy is added to this subsidy, then the total social welfare gain is illustrated by the triangle adg.

These benefits are then compared against the costs of the policy. A subsidy involves a transfer of public funds to consumers of amount $hbjk$, as illustrated by the shaded rectangle. If those funds could otherwise be used to lower labor taxes, the deadweight loss of these taxes would be included as a social cost, along with any other costs of administering the subsidy.[12] Similarly, an information program that moved demand from D to D' would also increase private welfare by abf, and this welfare gain would be traded off with the costs of implementation. For any policy, it will be an empirical question whether the costs exceed the benefits, and whether the net benefits are larger than alternative policies. This approach to assessing net welfare benefits is the appropriate test of whether energy efficiency policies are socially beneficial when Pigouvian taxes are also available to correct energy use externalities.

To summarize, this section has analyzed two forces that can cause behavior to differ from the social optimum: energy use externalities and investment inefficiencies. If there are energy use externalities but no investment inefficiencies, ideally only Pigouvian taxes would be used. If there are investment inefficiencies, energy efficiency policies such as subsidies for energy efficient capital stock might have benefits that outweigh their costs. If there are both investment inefficiencies and energy use externalities, then Pigouvian taxes should be used in combination with some welfare-improving energy efficiency policy. The central economic questions are thus whether there are investment inefficiencies, and if so, whether the benefits of a corrective policy outweigh its costs.

In the next section, we will examine choices by consumers and firms to adopt or not adopt energy efficient technologies and attempt to infer whether there is an energy efficiency gap. When there are no investment inefficiencies, agents' choices are governed by the first equation above, and unobserved factors such as costs ξ or utilization m can be inferred from their decisions. Some analysts have relied heavily on this framework in explaining away an apparent energy efficiency gap, with an argument along the lines that "agents are well-informed, so if they are not energy efficient, then

---

12. Analogously, a Pigouvian tax brings in public funds that can be used to lower labor taxes, which should be counted as an additional benefit [64].

it must be that the unobserved costs of energy efficiency are large." The analysis is more difficult when there might be investment inefficiencies. In that case, we now must know everything about agents' objective functions to estimate the size of γ.

Three types of problems will pervade the analyses we review in the next section. First, factors that are difficult to observe or quantify, as denoted by ξ in our model above, will be potentially very relevant. Second, estimates of the net present value of energy cost savings are often questionable. Depending on the setting, this could be because the analyst does not know the change in energy intensity ($e_0 - e_1$), the utilization m, or the appropriate discount rate r. Third, there is often substantial heterogeneity across consumers in utilization and unobserved costs, meaning that average returns for adopters might be uninformative about average returns for non-adopters or returns for the marginal adopter.

These empirical problems directly parallel other economic contexts. Consider, for example, the question of whether farmers in developing countries could profitably adopt agricultural technologies such as fertilizer and high-yielding variety seeds. These technologies have unobserved costs, such as increased labor inputs [65]. It is difficult to know the resulting increase in profits without randomized controlled trials, as in Duflo, Kremer, and Robinson [66]. Also, the substantial heterogeneity in costs and gross returns means that the fact that adopters have high returns does not imply that non-adopters are foregoing a profitable investment [67].

REFERENCES

[1] McKinsey & Co. Unlocking energy efficiency in the U.S. Economy, <http://www.mckinsey.com/clientservice/electricpowernaturalgas/downloads/US_energy_efficiency_full_report.pdf>; 2009.

[2] U.S. Energy Information Administration (EIA). Residential energy consumption survey: Table US12: total consumption by energy end uses, 2005, <http://www.eia.gov/emeu/recs/recs2005/c&e/summary/pdf/tableus12.pdf>; 2005.

[3] U.S. Energy Information Administration (EIA). Annual energy review Table 2.1a: Energy consumption estimates by sector, selected years, 1949−2010, <http://www.eia.gov/totalenergy/data/annual/pdf/sec2_6.pdf>; 2011a.

[4] U.S. Energy Information Administration (EIA). Residential energy consumption survey, <http://www.eia.gov/consumption/residential/>; 2011b.

[5] U.S. Bureau of Labor Statistics. Table 1. Quintiles of income before taxes: average annual expenditures and characteristics, consumer expenditure survey, 2007, <ftp://ftp.bls.gov/pub/special.requests/ce/standard/2007/quintile.txts>; 2007.

[6] Sallee J. Rational inattention and energy efficiency. Work in progress, University of Chicago; 2011;June.

[7] National Highway Traffic Safety Administration (NHTSA), Office of Regulatory Analysis and Evaluation. Corporate average fuel economy for MY 2012−MY 2016 passenger cars and light trucks. Final regulatory impact analysis. National highway traffic safety administration, U.S. Department of transportation; 2010;March.

[8] Sallee J. The taxation of fuel economy. NBER working paper 16466; 2010.
[9] Gillingham K, Newell R, Palmer K. Energy efficiency policies: a retrospective examination. Annual Review of Environment and Resources 2006;31:161−92.
[10] U.S. Energy Information Administration (EIA). 2009 Electric power annual. DOE/ EIA−0348(2009). Only available on the web at: <http://www.eia.gov/cneaf/electricity/ epa/epa_sum.html>; 2010.
[11] U.S. Department of Energy (DOE). American recovery & reinvestment act. Only Available on the web at: <http://www1.eere.energy.gov/recovery/>; 2011b.
[12] U.S. Internal Revenue Service (IRS). 2009 Estimated data line counts individual income tax returns. <http://www.irs.gov/pub/irs-soi/09inlinecount.pdf>; 2011.
[13] Meier A, Wright J, Rosenfeld AH. Supplying Energy through Greater Efficiency: The Potential for Conservation in California's Residential Sector. Berkeley, CA: University of California Press; 1983.
[14] American Council for an Energy Efficient Economy (ACEEE). The potential for electricity conservation in new york state. Prepared for the New York state energy RBD authority, Research report U891; 1989;september.
[15] Goldstein D, Mowris R, Davis B, Dolan K. Initiating least-cost planning in California: pre-liminary methodology and analysis, Natural resources defense council and the sierra club, prepared for the california energy commission docket no. 88-ER-8, revised 1990;May 10.
[16] Koomey JG, Atkinson C, Meier A, McMahon JE, Boghosian S, Atkinson B, et al. The potential for electricity efficiency improvements in the U.S. Residential sector. LBL−30477, Lawrence berkeley laboratory; 1991.
[17] Brown MA, Levine MD, Romm JP, Rosenfeld AH, Koomey JG. Engineering−economic studies of energy technologies to reduce greenhouse gas emissions: opportunities and challenges. Annual Review of Energy and the Environment 1998;23:287−385.
[18] National Academy of Sciences. Policy Implications of Greenhouse Warming: Mitigation, Adaptation, and the Science Base. Washington, DC: National Academy Press; 1992.
[19] Rosenfeld A, Atkinson C, Koomey J, Meier A, Mowris RJ, Price L. Conserved energy supply curves for U.S. buildings. Contemporary Policy Issues 1993;11(1):45−68.
[20] Stoft S. The economics of conserved-energy "Supply" curves. Program on workable energy regulation (POWER) Working paper 028; 1995;April.
[21] Blumstein C, Stoft SE. Technical efficiency, production functions and conservation supply curves. Energy Policy 1995;23(9):765−8.
[22] Anderson ST, Newell RG. Information programs for technology adoption: the case of energy-efficiency audits. Resource and Energy Economic 2004;26(1):27−50.
[23] Dubin J, Miedema A, Chandran R. Price effects of energy-efficient technologies: a study of residential demand for heating and cooling. RAND J Econ 1986;17(3):310−25.
[24] Nadel S, Keating K. Engineering estimates vs. Impact evaluation results: how do they compare and why? Research report U915, American council for an energy-efficient econ-omy. Available at: <http://www.aceee.org/research-report/u915>; 1991.
[25] Blasnik M. Energy models vs. Reality. Presentation. Southface lecture series 2010; January 26.
[26] Schweitzer M. Estimating the national effects of the U.S. Department of energy's weatherization assistance program with state-level data: a metaevaluation using studies from 1993 to 2005. Prepared for U.S. Department of energy. Working paper ORNL/CON−493; 2005;September.
[27] Sumi D, Coates B. Persistence of energy savings in seattle city light's residential weatherization program. Energy program evaluation: conservation and resource

management; Proceedings of the August 23−25, 1989 Conference, Chicago: Energy program evaluation conference; 1989. p. 311−316.

[28] Metcalf G, Hassett K. Measuring the energy savings from home improvement investments: evidence from monthly billing data. Review of Economics and Statistics 1999;81(3):516−28.

[29] Friedrich K, Eldridge M, York D, Witte P, Kushler M. Saving energy cost-effectively: a national review of the cost of energy saved through utility-sector energy efficiency programs. ACEEE report no. U092; 2009;September.

[30] Nadel S, Gelle H. Utility DSM: what have we learned? where are we going? Energy Policy 1996;24(4):289−302.

[31] Joskow P, Marron D. What does a negawatt really cost? Evidence from utility conservation programs. Energy J 1992;13(4):41−74.

[32] Eto J, Kito S, Shown L, Sonnenblick R. Where did the money go? The cost and performance of the largest commercial sector dsm programs. Lawrence Berkeley national laboratory working paper no. LBL-38201; 1995.

[33] Allcott H. Social norms and energy conservation. J Public Econ 2011;95(9−10):1082−95.

[34] Ayres I, Raseman S, Shih A. Evidence from two large field experiments that peer comparison feedback can reduce residential energy usage. NBER Working Paper 15386; 2009.

[35] Arimura T, Li S, Newell R, Palmer K. Cost-Effectiveness of electricity energy efficiency programs. Resources for the future discussion paper 09−48−REV; 2011;April.

[36] Hausman J. Individual discount rates and the purchase and utilization of energy-using durables. Bell J Econ 1979;10(1):33−54.

[37] Dubin J, McFadden D. An econometric analysis of residential electric appliance holdings and consumption. Econometrica 1984;52(2):345−62.

[38] Berry S, Levinsohn J, Pakes A. Automobile prices in market equilibrium. Econometrica 1995;63(4):841−90.

[39] Allcott H, Wozny N. Gasoline prices, fuel economy, and the energy paradox. Working Paper, MIT, <https://files.nyu.edu/ha32/public/research.html>; 2011;July.

[40] Busse M, Knittel C, Zettelmeyer F. Are consumers myopic? Evidence from new and used car purchases. American economic review, forthcoming. Available at Meghan Busse's website: <http://www.kellogg.northwestern.edu/faculty/directory/busse_meghan.aspx#research>; 2012.

[41] Sallee J, West S, Fan W. Do consumers recognize the value of fuel economy? Evidence from used car prices and gasoline price fluctuations. Work in progress, Macalester College; 2011;January.

[42] Akerlof GA. The market for 'lemons': quality uncertainty and the market mechanism. Q J Econ 1970;84(3):488−500.

[43] Davis LW. Evaluating the slow adoption of energy efficient investments: are renters less likely to have energy efficient appliances? Energy institute at haas working paper 205; 2010;June.

[44] Gillingham K, Harding M, Rapson D. Split incentives and household energy consumption. Energy Journal 2012;33(2):37−62.

[45] Murtishaw S, Sathaye J. Quantifying the effect of the principal−agent problem on US residential energy use. Lawrence Berkeley national laboratory working paper number 59773 Rev; 2006;August.

[46] Levinson A, Niemann S. Energy use by apartment tenants when landlords pay for utilities. Resource and Energy Economics 2004;26(1):51−75.

[47] Newell R, Jaffe A, Stavins R. The induced innovation hypothesis and energy-saving technological change. Quarterly J Econ 1999;114(3):941−75.

[48] Turrentine T, Kurani K. Car buyers and fuel economy? Energy Policy 2007;35(2): 1213−23.

[49] Larrick R, Soll J. The MPG illusion. Science 2008;320(5883):1593−4.

[50] Allcott H. Consumers' perceptions and misperceptions of energy costs. American Economic Review 2011;101(3):98−104.

[51] Allcott H. The welfare effects of misperceived product costs: data and calibrations from the automobile market. American Economic Journal: Economic Policy 2012; forthcoming. Available at: <https://files.nyu.edu/ha32/public/research.html>

[52] Chetty R, Looney A, Kroft K. Salience and taxation: theory and evidence. Am Econ Rev 2009;99(4):1145−77.

[53] Hossain T, Morgan J. Plus shipping and handling: revenue (non)equivalence in field experiments on eBay. Advances in Economic Analysis and Policy 2006;6(2):3 Article.

[54] Barber B, Odean T, Zheng L. Out of sight, out of mind: the effects of expenses on mutual fund flows. J Bus 2005;78(6):2095−120.

[55] Gabaix X, Laibson D. Shrouded attributes, consumer myopia, and information suppression in competitive markets. Q J Econ 2006;121(2):505−40.

[56] Jacobsen M. Evaluating U.S. Fuel economy standards in a model with producer and household heterogeneity, <http://econ.ucsd.edu/∼m3jacobs/Jacobsen_CAFE.pdf>; 2010; September.

[57] Krupnick A, Parry I, Walls M, Knowles T, Hayes K. Toward a new national energy policy: assessing the options. Resources for the Future Report, <http://www.energypolicyoptions. org/wp-content/uploads/reports/RFF-Rpt-NEPI_Tech_Manual_Final.pdf>; 2010;November.

[58] Greenstone M, Kopits E, Wolverton M. Estimating the social cost of carbon for use in U.S. Federal rulemakings: a summary and interpretation. MIT department of economics working paper no. 11−04; 2011;March.

[59] Allcott H, Mullainathan S, Taubinsky D. Energy Policy with Externalities and Internalities. NBER working paper 17977. Available at: <https://files.nyu.edu/ha32/public/research. html>; 2012; April.

[60] Fischer C, Harrington W, Parry I. Do market failures justify tightening corporate average fuel economy (CAFE) standards? Energy J. 2007;28(4):1−30.

[61] Heutel G. Optimal policy instruments for externality-producing durable goods under time inconsistency. NBER working paper 17083; 2011.

[62] Dixit A, Pindyck R. Investment under Uncertainty. Princeton, NJ: Princeton University Press; 1994.

[63] Hassett K, Metcalf G. Energy conservation investment: do consumers discount the future correctly? Energy Policy 1993;21(6):710−6.

[64] Bovenberg L, Goulder L. Optimal environmental taxation in the presence of other taxes: general equilibrium analyses. American Economic Review 1996;86(4):985−1000.

[65] Foster A, Rosenzweig M. Microeconomics of technology adoption. Yale university economic growth center discussion paper 984; 2010.

[66] Duflo E, Kremer M, Robinson J. Nudging farmers to use fertilizer: theory and experimental evidence from Kenya. American Economic Review 2011;101(6):2350−90.

[67] Suri T. Selection and comparative advantage in technology adoption. Econometrica 2011;79(1):159−209.

FURTHER READING

[1]   Auffhammer M, Blumstein C, Fowlie M. Demand-Side management and energy efficiency revisited. Energy J 2008;29(3):91−104.

[2]   Bollinger B, Leslie P, Sorensen A. Calorie posting in chain restaurants. American Economic Journal: Economic Policy 2011;3(1):91−128.

[3]   Davis L. Durable goods and residential demand for energy and water: evidence from a field trial. RAND J Econ 2008;39(2):530−46.

[4]   Davis L, Muehlegger E. Do americans consume too little natural gas? An empirical test of marginal cost pricing. RAND J Econ 2010;41(4):791−810.

[5]   DeCanio S, Watkins W. Investments in energy efficiency: do the characteristics of firms matter? Review of Economics and Statistics 1998;80(1):95−107.

[6]   Fowlie M, Greenstone M, Wolfram C. An experimental evaluation of the weatherization assistance program. Work in progress; 2011.

[7]   Gillingham K, Newell R, Palmer K. Energy efficiency economics and policy. Annual Review of Resource Economics 2009;1:597−620.

[8]   Hausman J, Joskow P. Evaluating the costs and benefits of appliance efficiency standards. American Economic Review 1982;72(2):220−5.

[9]   Jaffe A, Stavins R. The energy paradox and the diffusion of conservation technology. Resource and Energy Economics 1994;16(2):91−122.

[10]   Loughran D, Kulick J. Demand-side management and energy efficiency in the United States. Energy J 2004;25(1):19−43.

[11]   Meier A, Whittier J. Consumer discount rates implied by purchases of energy-efficient refrigerators. Energy 1983;8(12):957.

[12]   Meyers S, McMahon J, McNeil M. Realized and prospective impacts of U.S. Energy efficiency standards for residential appliances: 2004 update. Lawrence berkeley national laboratory working paper LBNL−56417; 2005.

[13]   Parry I, Evans D, Oates W. Are energy efficiency standards justified? Resources for the future discussion paper 10−59; 2010;November.

[14]   U.S. Department of Energy (DOE). Weatherization assistance program goals and metrics. Only Available on the web at: <http://www1.eere.energy.gov/wip/wap_goals.html>; 2011a.

[15]   Yergin D. Energy Future: the Report of the Energy Project at the Harvard Business School. New York: Random House; 1979.

Part II

The — Frustratingly Slow — Evolution of Energy Efficiency

Making Cost-Effective Energy Efficiency Fit Utility Business Models: Why has It Taken So Long?

Ralph Cavanagh[†]

Energy Program Co-Director, Natural Resources Defense Council

1 INTRODUCTION

Most utilities, whether publicly or privately owned, operate under a system of price regulation that treats them like commodity distribution businesses, whose principal rewards lie in boosting unit sales and reducing average unit costs. For electric utilities at least, this worked very well for much of the 20th century; from 1973–2000, for example, retail electricity sales more than doubled in the United States, while the population increased by only 33 percent.[1]

High confidence in rapid sales growth, coupled in many cases with automatic rate adjustments to ensure recovery of any increase in fuel costs, allowed many utilities to defer explicit requests for rate increases almost indefinitely; any escalation in non-fuel costs could be covered by rising electricity sales volumes. Customers, meanwhile, typically shouldered the risks and costs of any surges in fuel prices. Of course, this business model was inconsistent with any serious effort to promote energy efficiency, no matter how inexpensive it might be to achieve the savings.

As early as 1989, the National Association of Regulatory Utility Commissioners (NARUC) called on its members to encourage utilities to substitute cost-effective energy efficiency for more costly generation wherever feasible and to "ensure that the successful implementation of a utility's

[†] The author gratefully acknowledges perceptive comments and editorial guidance from Sheryl Carter, Sierra Martinez, Patricia Remick, and Devra Wang. Figures 6.1, 6.3 and 6.4 were prepared by Sierra Martinez.

1. Compare 2009 Statistical Abstract of the United States, p. 17 (Oct. 2008) (population) with U.S. Department of Energy, Energy Information Administration (May 2012) (electricity use).

Energy Efficiency. DOI: http://dx.doi.org/10.1016/B978-0-12-397879-0.00006-2

least-cost plan is its most profitable course of action."[2] This chapter discusses why that objective remains unrealized, even as the nation moves urgently to accelerate energy efficiency progress, expand its renewable energy base, and upgrade its transmission and distribution grids.

For more than two decades, debates have raged over proposals to break the linkage between utilities' financial health and their retail commodity sales, and to introduce earnings opportunities associated with energy efficiency gains in their service territories. In the United States, these decisions rest with state commissions (for shareholder-owned utilities) and local boards (for publicly-owned utilities).

Progress has been far too slow, in part because regulators and their staffs are protective of traditional practices, and in part because other parties, including the utilities themselves, have worried that change would introduce new risks or erode long-held advantages. Although California decoupled gas and electric utilities' financial health from sales more than 30 years ago, by 2008 only a dozen other states had taken similar action on behalf of at least one natural gas utility; for electric utilities the count stood at just four. Four years later, following a national campaign that prominently included the Natural Resources Defense Council (NRDC), the totals had reached 21 and 15, respectively.

The case for swifter progress becomes more urgent daily, given both the magnitude of the nation's unrealized efficiency potential, reinforced throughout this volume, and a widening consensus that robust growth in energy commodity sales is untenable on both economic and environmental grounds. From 2000–2011, for example, electricity sales lagged behind population growth for the first extended period in U.S. history (Figure 6.1); for natural gas utilities' retail sales, the same trend is much more firmly established.[3] Utilities across the nation face long-term prospects of at best modest increases in commodity sales, even as their needs to make and recover capital investments expand.

The remainder of this chapter explains why utilities need a new business model (in section 2), advocates revenue decoupling as a necessary but not sufficient element, and outlines (in section 3) the performance-based earnings opportunities that are essential to having utilities' profitability reflect their success in promoting cost-effective advances in energy efficiency and distributed generation.

---

2. D. Moskovitz, Profits and Progress Through Least-Cost Planning (NARUC: November 1989), Appendix C.

3. Ted Hesser's chapter notes that 2007 marked a particularly pronounced shift in electricity consumption. A comparison of the Statistical Abstract of the United States (2012) with the Monthly Energy Review of the U.S. Department of Energy (May 2012) indicates that U.S. population growth from 2000–2010 was 10 percent; retail electricity sales grew by less than 9 percent from 2000–2011. Natural gas sales in the mid-2000 s were about the same as those in 1973, and subsequent increases are mostly attributable to shifts from coal to natural gas for electricity generation, which do not affect sales by gas distribution utilities.

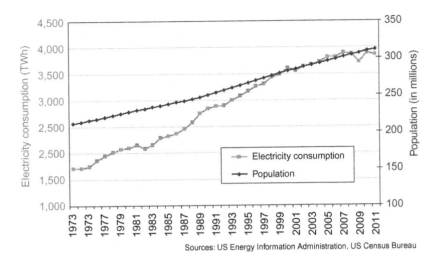

Sources: US Energy Information Administration, US Census Bureau

FIGURE 6.1 Trends in U.S. Electricity Consumption and Population, 1973–2011.[4]

2 A BROKEN BUSINESS MODEL

Given the sobering condition of federal and state governments' balance sheets, most new electricity infrastructure will have to be financed with utility revenues, in competition with alternative solutions to meeting energy service needs.[5] Moreover, much of the genius in such procurement is integration of diverse resources under ever-shifting real-time conditions. Demands for this kind of expertise are relatively new to an industry known mostly in decades past for managing giant construction projects. Consider, for example, one major electricity supplier's recent summary of its resource plan:

Most of [our] incremental energy needs for the next several years can be met by meeting [our] conservation targets. In addition to relying on conservation, [we] plan to continue to:

- Rely on short- and mid-term wholesale power market purchases.
- Facilitate the effective, efficient and reliable integration of renewable resources to [our] system.

---

4. Electricity consumption data are from U.S. Energy Information Administration, *Monthly Energy Review*, Table 7.6, Electricity End Use, "Electricity End Use, Total" (June 27, 2012). Population data are from U.S. Census Bureau, *Population Estimates*, Table 1 – Annual Estimates of the Population for the United States, Regions, States, and Puerto Rico (December 2011).

5. Matters are otherwise, of course, in places like China, France, and Russia, where national governments still routinely choose and finance electricity resources without any obvious reference to competitive considerations generally, or alternative energy efficiency opportunities in particular.

- Increase transmission grid operating flexibilities, develop smart grid technologies, and directly involve electricity users through demand response programs.
- Track, evaluate, and appropriately pursue availability of pumped storage and natural gas-fired resources for seasonal heavy load hour energy and/or balancing reserves.[6]

For utility systems with responsibilities like these – which is to say most of them, at least in the United States – clarity will be needed regarding both cost recovery and accountability. For example, when and on what terms may distribution utilities enter into long-term contracts with generation service providers? How will distribution utility responsibilities interact with the opportunities created for competitive retail suppliers in states with retail competition? Who has the responsibility for identifying needed enhancements to the transmission network? How will transmission providers be paid for securing them, and who will pay? What, if any, rewards will be earned by utilities that reduce the cost of balancing their systems' loads with variable-output generation from wind and solar plants?

Most nations rely at least in part on resource procurement and integration by regulated distribution companies. Even in countries and U.S. states (like Texas) that are inclined to view regulated utilities largely as managers of interconnected wires and pipes, with all or much resource procurement entrusted to other entities, utility-owned distribution systems increasingly need to address the challenge of integrating variable-output electricity production, and both electric and natural gas utilities' customers collectively would benefit from help in substituting low-cost energy efficiency for higher-cost alternatives.

For this purpose, energy efficiency should be treated as a utility system resource that is the functional equivalent of other ways to meet customers' aggregate demand. Regulators should aim to ensure an acquisition process open to all competitors, with results that minimize the life-cycle cost of reliable electricity service while meeting society's environmental goals. Within the for-profit sector, the theme of NARUC'S 1989 resolution remains compelling: utilities that effectively manage resource procurement and integration should be more profitable than inferior performers. And publicly-owned utilities should not confront an automatic deterioration in creditworthiness if their energy efficiency efforts cause sales to decline or grow more slowly.

2.1 The Wrong Path: Higher Fixed Charges

Since both for-profit and non-profit utilities recover most of their fixed costs of service through charges on electricity and natural gas use, increases or

---

6. The supplier in question is the Bonneville Power Administration, which sent out the quoted summary of its resource plan in a mass e-mail communication from John Taves, *BPA Issues Final Resource Plan* (Sept. 13, 2010).

Gas and Electric Decoupling in the US
JUNE 2012

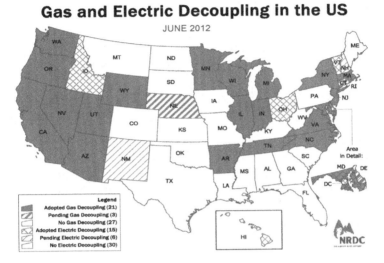

FIGURE 6.2   Gas and Electric Decoupling in the United States.

reductions in consumption will affect fixed cost recovery, even though the costs themselves do not change. Overcoming this problem means ensuring that fluctuations in sales (either up or down) do not result in over- or under-recovery of utilities' previously approved fixed costs.

The immediate temptation is to convert all or most fixed costs into fixed charges. This would indeed make the recovery of fixed costs independent of energy sales, but it would also significantly reduce customers' rewards for reducing energy use. That is a step in the direction of what might be termed "all you can eat" rates, which reduce or eliminate customers' rewards for saving energy by making their utility bills largely or wholly independent of total energy consumption. This may sound like an improbable caricature, but in 2012 it became a retail service option in Texas, where Reliant Energy Retail Services' "Predictable" rate plan was precisely that: residential customers who enrolled paid $150 per month regardless of electricity use.

What the world needs now to encourage more energy efficiency is not rate designs that encourage energy waste, but a strong embrace of inverted rates, where the rule is "the more you use, the more you pay." We need not and should not abandon reliance on commodity charges to recover the costs of electricity and natural gas service. Over the past 22 years, for example, as Figure 6.3 shows, U.S. electricity sales have declined by more than one percent only once (in 2009); in 18 of the 22 years, annual sales increased (although they have been increasing less rapidly than population since 2000). The changing electricity sales outlook does not portend sudden sharp declines that might force abandonment of volume-based cost recovery and the rewards it offers to those who use less. But utilities will legitimately

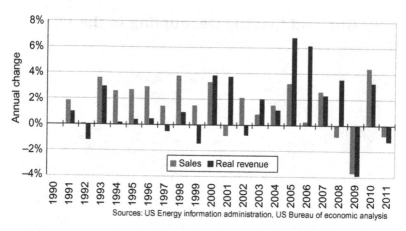

Sources: US Energy information administration, US Bureau of economic analysis

FIGURE 6.3   Annual Changes in Utilities' Electricity Revenues and Sales.[7]

question how the financial interests of their bondholders and shareholders can be squared either with rate designs that increase variability of revenue recovery, or with sustained recovery of largely fixed network costs over uncertain volumetric sales. Fortunately, there is a straightforward remedy called "revenue decoupling" that does not require changing rate designs or reducing rewards for customers who save energy.

2.2 The Revenue Decoupling Debate

If utilities continue relying on variable charges to recover all or most authorized fixed costs of service, disincentives for energy efficiency engagement will persist. Revenue decoupling uses small, regular rate adjustments to prevent over- or under-recovery of authorized costs. Under decoupling, a simple system of periodic true-ups in base rates either restores to the utility or gives back to customers the dollars that were under- or over-recovered because of fluctuations in retail sales. This corrects for disparities between the utility's actual fixed cost recovery and the fixed-cost revenue requirements approved by utility regulators.[8]

7. Sales and revenue data are from U.S. Energy Information Administration, *Form EIA-826 Data Monthly Electric Utility Sales and Revenue Data*, EIA-826 Sales and Revenue Spreadsheets (June 5, 2012). Revenues are presented in chained 2005 dollars. Nominal dollars of revenue were converted to chained dollars using U.S. Bureau of Economic Analysis, National Economic Accounts, Gross Domestic Product, Current-Dollar and "Real" GDP (June 28, 2012).
8. For a full explanation and numerical illustrations, see R. Cavanagh, "Graphs, Words and Deeds," *MIT Innovations* (Fall 2009), pp. 83−87.

Decoupling proposals were initially controversial among both utilities and consumer advocates, both of which were uncomfortable with a shift in the regulatory status quo and its allocation of revenue-based risks and opportunities. Electric utilities often were reluctant to surrender the upside in revenues associated with retail sales growth. For their part, many consumer advocates either opposed these proposals outright or sought to attach financial penalties in the form of reductions in utilities' authorized returns on equity. Utility managements were understandably unenthusiastic about absorbing an automatic upfront loss to shareholders as part of a regulatory reform that was supposed to remove a financial obstacle to utilities' promotion of energy efficiency gains.

In California, the state with by far the longest decoupling experience (starting in 1981),[9] it has been at least a decade since any party challenged the mechanism or sought to link it with any form of financial penalty to either gas or electric utilities. Elsewhere, revenue decoupling retains powerful opponents, although an urgent recent search for common ground is yielding promising results. For example, the online journal ElectricityPolicy.com recently published an extended exchange between the author (representing the Natural Resources Defense Council [NRDC]) and John Howat of the National Consumer Law Center.[10] NRDC is a long-time proponent of revenue decoupling, and NCLC is a leader among consumer advocates with a history of skepticism about the concept. An excerpt follows, which helps illuminate both the sources of resistance and the best ways to address them:

CAVANAGH [environmental perspective]: John, we've recently been in a hearing room together where, not for the first time, environmental and consumer advocates were at odds over whether to introduce revenue decoupling as part of a strategy for enhancing energy efficiency investment. What is your view here?

HOWAT [consumer perspective]: NCLC has on many occasions been critical of revenue decoupling mechanisms that blindly reward companies for reductions in sales for reasons that have nothing to do with utility-sponsored energy efficiency. But a well-structured decoupling mechanism is in my view far preferable to "straight-fixed variable" (SFV) design, for example, that penalizes low-volume utility consumers while removing volumetric pricing efficiency incentives for all utility customers. I urge colleagues to accept revenue decoupling that is directly tied to new investment in comprehensive, cost-effective energy efficiency programs and measures and that includes (1) rate increase collars that limit upside rate volatility, (2)

---

9. For more on the California PUC testimony that introduced revenue decoupling to the electricity sector in 1981, see R. Cavanagh, note 8 above, p. 89, n. 14.
10. R. Cavanagh & J. Howat, "Finding Common Ground Between Environmental and Consumer Advocates," ElectricityPolicy.com (April 2012).

explicit regulatory review and adjustment of return on equity to account for altered utility risk profiles (retrospective, but in a reasonable timeframe is fine with me), (3) review and adjustment of baseline utility cost structure assumptions including cost of capital on some regular basis, and (4) the "Tucson model" of implementing inclining block rates, where decoupling surcharges are tied to higher usage blocks and surcredits to the initial usage bock. Again, such a structure would, in my view, be far preferable to implementation of SFV in the name of promoting energy efficiency. Further, I've long agreed with you about the need to address the utility "throughput addiction," and that best-quality energy efficiency represents our most valuable energy resource.

CAVANAGH: Let's unpack this a bit, because I don't see anything here that should divide us. I agree on the need to pair revenue decoupling with enhanced energy efficiency performance and benefits, and we have supported rate increase collars of three percent for electric utilities and five percent for gas utilities (with no limit on rate reductions associated with decoupling). I supported the Tucson Electric proposal that you cite, which would apply any decoupling-related rate increases to the highest use block of consumption in a rate structure, and apply any reductions to the baseline block (so that any decoupling adjustments would amplify rather than mute the rewards for saving energy that inclining block rates provide to customers). So far so good?

HOWAT: Yes, there is plenty of room to work together here. We need to break the link between utility profits and sales, and design the decoupling mechanism in a way that makes sense for consumers interested in stable prices and an appropriate regulatory treatment of the utility cost structure and risk profile.

CAVANAGH: On cost of capital adjustments, the crucial phrase in your response is "retrospective, but in a reasonable timeframe." Our latest proposal, which you heard me defend before the Washington Commission, also reflects your call for "review and adjustment of baseline utility cost structure assumptions including cost of capital on some regular basis." We recommend that Commissions not link decoupling mechanisms with targeted prospective reductions in cost of capital, which may or may not materialize (and have yet to be documented empirically after three decades of experience), but we support continuous review of any changes in utilities' capital structure, whatever the cause, and full passthrough of any associated cost savings to customers. If, as authorities like the Regulatory Assistance Project maintain, decoupling should help establish a long-term foundation for consumer-friendly changes in capital structure, our proposal ensures prompt and full delivery of benefits if and when they appear.

HOWAT: I agree with you that the key, with respect to cost of capital adjustments, is in the assurance of periodic regulatory review. I was gratified to hear you state at the hearing in Washington that revenue decoupling should not be viewed as a means of doing away with regulatory process.[11] Rather, it is a means of re-aligning incentives to eliminate utility aversion to effectively promoting energy efficiency programs that work.

CAVANAGH: Finally, can we agree that revenue decoupling appropriately treats the "throughput addiction" to which you refer, in the simplest possible way, by avoiding efforts to adjudicate inevitably speculative causes of increases or reductions in sales, and simply ensuring instead that utilities' ability to recover fixed-cost revenue requirements is not affected by changes in retail sales that regulators did not anticipate when they set retail rates?

HOWAT: That is a great question that I frankly have struggled with over the years. Like many advocates, I have bristled at the prospect of "rewarding" utility companies for declining sales that have absolutely nothing to do with their efforts to enhance energy efficiency. After all, sales will decline in times of economic downturn, during mild weather conditions, when appliances become more efficient, when end-users invest in energy efficiency improvements on their own, and, in some instances, when fuel prices increase. However, because utilities inevitably file for rate increases anyway if revenues erode for any of the reasons listed above, and because revenue decoupling provides consumers with declining rates as sales increase for any reason, my thinking on this issue has evolved over time. I have come to agree that, as long as a utility company's return on equity is appropriately adjusted to reflect changes in the sales risk faced by that company through implementation of revenue decoupling, and the measures mentioned above are part of the design, it is appropriate to embrace a full – rather than partial – decoupling mechanism.

CAVANAGH: All of this is very helpful and I seek only one final clarification: can we agree that such regulatory adjustments should reflect observed changes in cost of capital once the mechanism has been adopted? To use your earlier phrase: "retrospective, but in a reasonable timeframe." We would support both regular reviews and immediate passthroughs of any savings; our objection is to imposing reductions in costs of capital prospectively, before there is evidence of whether and to what extent they have occurred.

HOWAT: I agree that as long as regulators retain full authority and responsibility to adjust return on equity to reflect changes in a company's risk profile, adjustments specifically related to a company's cost of capital may be made in a timely manner after evidence of actual increases or decreases is presented.

This dialogue suggests grounds for optimism about progress on revenue decoupling for electric and natural gas utilities. Figure 6.2 above shows that about half the states had embraced revenue decoupling for at least one of their investor-owned electricity and natural gas utilities by mid-2012, although decoupling mechanisms still covered well under half of total retail sales. Public power took its first strong step toward decoupling when the nation's largest municipal utility, the Los Angeles Department of Water and Power (LADWP), released a proposal to adopt it in May of 2012.[12]

Some have argued for an approach called "lost revenue recovery" as an alternative to revenue decoupling.[13] The superficially appealing rationale for such systems is that they focus solely on compensating utilities for lost revenues associated with their own energy efficiency programs, which allegedly removes any disincentive to help customers use less electricity and natural gas while leaving the allocation of other business risks unchanged.

But any such rationale is misleading in the extreme. Lost revenue recovery schemes leave intact automatic penalties, in the form of reduced fixed-cost recovery, for all cost-effective electricity savings not directly associated with a utility's own programs, even when the company by action or inaction could make a material difference in prospects for those savings. Examples include federal and state efficiency standards and programs administered by other parties, all of which can benefit significantly from utility cooperation and support. Small-scale "distributed" generation on customers' premises would produce the same kind of adverse balance sheet effects. Rewarding a utility for a limited category of adjudicated efficiency gains while penalizing it for all the rest is analogous to trying to drive with one foot on the brake and the other on the accelerator. Moreover, lost-revenue recovery would create a reason for utilities to promote programs that looked good on paper but delivered little or no savings in practice.[14] And it would ensure adversarial discord over every savings calculation, since significant financial stakes and rate impacts would hinge on the results. Finally, and most tellingly, rate adjustments keyed solely to adjudicated savings would mean automatic annual rate increases (unless the company was wholly ineffective), whereas decoupling adjustments can be either positive or negative.

Some have argued that revenue decoupling pays utilities for savings that they didn't help achieve. But decoupling mechanisms don't "pay" anyone any incremental amount for anything; they simply allow utilities to receive no more and no less than the fixed-cost revenue requirement that their regulators have reviewed and approved. This underscores the point that revenue decoupling, while a crucial step in the right direction,

---

12. LADWP, Power System Rate Proposal. FY 12/13 and FY 13/14, Summary and Supporting Information (May 3, 2012), p. 76. Publicly owned utilities often assume that their non-profit status and local control obviates the need for decoupling, overlooking the importance of assured fixed-cost recovery to financial ratings and the high transaction costs associated with formal rate adjustments in the absence of a pre-approved adjustment mechanism.

13. An example is a settlement proposal approved by the Arizona Corporations Commission for Arizona Public Service (over strong objections from NRDC and the Southwest Energy Efficiency Project) in Docket No. E-01345A-11-0224 (May 2012).

14. See, e.g., Arizona Corporations Commission, Docket No. G-01551A-10-0458, Decision No. 72723 (January 2012), pp. 39—40.

is not a panacea. It removes a powerful disincentive for energy efficiency progress, but it does not by itself create the prospect of reward for exemplary results. Both are needed. In the hope that the revenue decoupling debate has been outlined fully above, the final section of this chapter turns to what remains to be done, even assuming that decoupling becomes a regulatory norm for electric and natural gas distribution systems in the United States and abroad.

3 TOWARD A NEW UTILITY BUSINESS MODEL

To sustain their excellence in efficiency, the investor-owned utilities that deliver three-quarters of the nation's electricity and most of its natural gas will require more than just protection from instant pain. A dozen states have acted to assure that independently verified net energy efficiency savings to customers will also yield a reward for utility shareholders.[15] One option is to allow utilities to earn a rate of return on approved efficiency expenditures that is at least equal to the compensation afforded prudent generation or grid investments. A better alternative, however, is a compensation system tied to independently verified performance in delivering cost-effective savings to customers, rather than just "tonnage of capital committed."[16] Whether the issue is power plant construction or energy efficiency incentives, there is growing discomfort about shareholder rewards based on how much a utility spent to get the desired result.[17]

A balance must be struck between rewarding performance at a reasonable level and creating outsized "compensation" that invites endless discord over savings estimates that can never achieve complete precision (because one can never know with certainty what would have happened without utility intervention). The "lost revenue recovery" mechanisms described earlier are doomed to failure because any reasonable level of adjudicated lost revenue recovery dramatically raises the cost of every kilowatt-hour saved, given the fraction of the typical kilowatt-hour charge that fixed cost recovery represents (at least half). Utilities with appreciable annual savings would generate guaranteed, escalating annual rate increases under this model, with results

---

15. For a compendium of precedents, see the website of the Edison Electric Institute's Energy Efficiency Institute (http://www.edisonfoundation.net/iee). The key contrast between these mechanisms and the "lost revenue recovery" option criticized earlier is that the stakes per adjudicated unit of savings are much lower, which should ensure less discord and lower implementation costs.

16. I first heard this characteristically vivid comparison from Tom Page, then CEO of San Diego Gas & Electric.

17. Promising initial efforts have been launched by the Energy Foundation (under the leadership of former Colorado PSC Commissioners Ron Binz and Ron Lehr) and the Rocky Mountain Institute (where James Newcomb and Lena Hansen are heading a new Electricity Innovation Lab).

that would soon test even the most highly motivated regulators.[18] As explained above, revenue decoupling avoids any possibility of these adverse cumulative impacts, by using regular and modest rate adjustments that go in both directions, with an eye solely to preventing annual fixed-cost recovery from either exceeding or dropping below a previously approved target.[19]

On the other hand, it is no solution for regulators to overlook or deny earnings opportunities altogether, either for energy efficiency or for the broader array of functions associated with effective management of upgraded grids and diversified resource portfolios. Instead, utility business models should involve a sharing between utility customers and shareholders of long-term cost savings from an effectively managed and integrated portfolio of electricity and natural gas resources. The incentive should encourage diversified portfolios of long- and short-term investments, which insulate customers from excessive exposure to volatile spot markets. This requires reasonable, objective benchmarks against which portfolio performance can be measured, and consensus-based ways to evaluate that performance. An urgent and unresolved item on regulatory agendas involves fleshing out the concept of performance-based resource procurement and integration incentives, and devising specific proposals for utility regulators and managers to consider.

Some lessons are already clear from California's experience from 2006−2008 with performance-based energy efficiency incentives, which earned utilities more than $200 million but exacted a high price in terms of adversarial struggles and uncertainty (even though the sum in question represented less than two percent of utility profits over that period). Of particular importance is the now widely acknowledged point that performance benchmarks, once established, should not be subject to change in the middle of program implementation. Much of California's discord reflected efforts to make significant retroactive adjustments in such benchmarks, based on disputes over how much credit utilities should receive for undisputed installations of cost-effective measures. Prominent among those disputes,

---

18. Minnesota learned as much during the 1990 s, when the utility's annual lost revenue recovery request reached levels comparable to its annual energy efficiency expenditures, and the Commission responded by terminating the recoveries. Minnesota Public Utilities Commission, Order Disallowing Recovery of Lost Margins and Other Incentives, Docket No. E-002/M-99-419 (July 27, 1999) (lost revenue request was $26.9 million, while program expenditures totaled $33.3 million). Duke Power's widely publicized "Save-A-Watt" proposal attracted extensive opposition because of its reliance on generous payments to Duke for every kWh determined to have been saved through its programs; Duke opted recently in Ohio for revenue decoupling instead. Public Utility Commission of Ohio, Finding and Order, Case No. 5905-EL-RDR (May 30, 2012).

19. The success of U.S. revenue decoupling mechanisms in avoiding appreciable rate impacts is documented thoroughly in Pamela Morgan, "Rate Impacts and Key Design Elements of Gas and Electric Utility Decoupling: A Comprehensive Review," *The Electricity Journal* (October 2009).

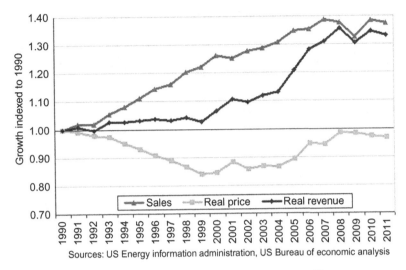

Sources: US Energy information administration, US Bureau of economic analysis

FIGURE 6.4   Trends in Utilities' Electricity Revenues, Sales, and Prices.[20]

for example, were whether and how much to alter initial estimates of how many utility customers would have installed compact fluorescent light bulbs if utilities had done nothing to promote them.

Even more important, it is past time to abandon ratemaking systems that rely on long- or short-term energy commodity rate caps as utilities' principal incentive mechanism. These are still common worldwide; they would make sense only if society's primary interest lay in minimizing commodity rates and maximizing sales, or if regulators thought it was practicable to achieve continuous reductions (or at worst a long-term freeze) in the delivered cost of kilowatt-hours and therms. Rate caps are flatly inconsistent with economic and environmental progress in a world of rising commodity costs, obvious infrastructure investment needs, and huge untapped energy efficiency opportunities that would raise commodity rates further by spreading fixed costs over smaller sales volumes, even though the energy efficiency improvements are overwhelmingly cost-effective (Figure 6.4).

[Cumulative growth in electric utilities' real revenue (1990–2011) was about equal to growth in consumption. Both were up about a third (sales at 37 percent, and revenue at 33 percent). Real average prices per kWh are now

---

20. Sales, revenues, and price data are from U.S. Energy Information Administration, *Form EIA-826 Data Monthly Electric Utility Sales and Revenue Data*, EIA-826 Sales and Revenue Spreadsheets (June 5, 2012). Revenues and prices are presented in chained 2005 dollars. Nominal dollars of revenue were converted to chained dollars using U.S. Bureau of Economic Analysis, National Economic Accounts, Gross Domestic Product, Current-Dollar and "Real" GDP (June 28, 2012).

about three percent less than they were in 1990, but with a clear rising trend since 1999.]

Instead, the principal aim of regulation should be to minimize the life-cycle costs of energy services. Utility managers themselves increasingly understand this and are emerging as advocates for change, which is among the best reasons for optimism about a prompt transition to a new business model. But success will require more engagement by all with a stake in reliable and affordable energy services.

4 CONCLUSIONS

Electricity and natural gas distribution no longer make sense as commodity businesses. No customer desires either fuel for its aesthetic qualities; it is reliable and affordable energy *services* that underpin healthy economies and improving standards of living. Everyone stands to benefit if utilities can help their customers find cost-effective strategies for getting more work out of less energy. At the same time, pressures are growing at all levels of society to reduce pollution emissions associated with fossil fuel use. If we were designing utility regulation for the first time, would anyone want to include potent rewards for promoting increased electricity and natural gas consumption? Yet that is the model to which most regulators and utilities remain tied, at least in part. Together, they now have an opportunity to invent the business anew, taking advantage of abundant experience in ways to chart a more productive course.

Chapter 7

The Evolution of Demand-Side Management in the United States

Frank A. Felder
Center for Energy, Economic and Environmental Policy, Rutgers University

1 INTRODUCTION

The expansionary history of demand-side management (DSM) is one of sub-stantial but by no means complete success. Among DSM proponents, there is no agreement on how to move forward, except to do so at an ever-accelerating rate. Many major organizational, programatic, and evaluation approaches are being questioned and disputed. Beneath this history is a vigorous debate regarding the efficacy of DSM programs, which is reflected in many chapters in this volume. At one end of this debate are DSM advocates who make strong claims regarding the significant amounts of energy efficiency that can be obtained at low and sometimes even negatives costs. At the other end of the spectrum are many economists who question the notion that there is the proverbial free DSM lunch. Granted, they argue, that there are market failures, such as imperfect information and not including the costs of negative external-ities (e.g., air emissions), for example, in energy prices, but nonetheless, the amount of cost-effective energy efficiency and moreover the ability of DSM programs to obtain achieve cost-effective outcomes is limited.

The history of DSM can be summed up in one word: expansion. Over the last four decades, DSM's scale, scope, and structure have grown. DSM has expanded from a response to the U.S. dependency on imported oil with its attenuate threat on national security, to a means of obtaining environmental improvements (starting with local and regional concerns and extending to global ones), and now as a means to advance economic growth. DSM began with programs that provided information. Now DSM contain numerous types of financial incentives and mechanisms from covering the cost of energy audits to giving rebates for energy efficiency measures, whole house and building efficiency programs, and cogeneration (also referred to as combined heat and power). DSM programs are also being integrated with wholesale

Energy Efficiency. DOI: http://dx.doi.org/10.1016/B978-0-12-397879-0.00007-4

179

and retail electricity markets including price-responsive components and being linked with renewable resources and distributed generation [1]. It has also expanded organizationally from programs run by utilities and their contractors to stand-alone energy efficiency utilities in some states and complex organizational and contractual structures in others. As program administrators run out of the low-hanging fruit, according to Michael Peeve, California regulators have moved upstream to building codes, net zero energy building, light-bulb phase-out, and net metering. Finally, DSM has expanded geographically across the United States and internationally.

Over time the term energy efficiency has started to replace the term DSM, although this chapter uses DSM to refer to energy efficiency programs undertaken in the electric sector. In the United States natural gas distribution utilities also have energy efficiency programs, but the term DSM is not usually applied to energy efficiency programs for natural gas. DSM has now become an umbrella term that refers to utility and government activities designed to change the amount and timing of electricity use for the collective benefit of DSM program participants, ratepayers, and utilities [2].

This chapter is organized as follows. The next section discusses the origins and evolution of DSM, focusing on the United States. Section 3 discusses the intellectual dispute regarding the energy gap and explores some hidden assumptions behind this debate. Section 4 discusses the future of DSM, and section 5 concludes.

2 ORIGINS AND EVOLUTION OF DEMAND-SIDE MANAGEMENT

DSM is not unique to the United States, but when it started in the United States it spread to many developed and emerging countries [2,3]. DSM analysts and commentators are in approximate agreement regarding the major phases of DSM in the United States (e.g., [2,3]). There have been five major waves of programs roughly corresponding to the periods of the 1970s, the 1980s, the early 1990s, the late 1990s, and beyond the year 2000 [2], each of which is discussed further below and indicated in Figure 7.1.

Since the beginnings of DSM, an important shift in the U.S. economy has been occurring. From the start of DSM, the U.S. economy continued its transition from a manufacturing-based economy to a more service-oriented one.

FIGURE 7.1   Five Major Waves of DSM in the United States.

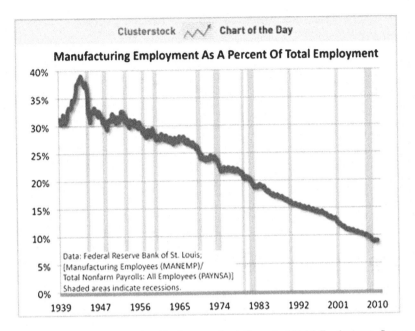

FIGURE 7.2 U.S. Manufacturing Employment As A Percent of Total Employment. *Source: Business   Insider   http://www.businessinsider.com/charts-of-the-week-manufacturing-2010-8#heres-the-trend-obama-is-fighting-if-he-wants-to-save-american-manufacturing-1.*

As indicated by Figure 7.2, in 1970, manufacturing accounted for about 25 percent of all of the jobs; in the year 2010, this was a little over 10 percent.[1] Needless to say, this continued transition is reflected in the United States' use of energy. Nillsen et al report and reflect on global trends in energy intensity and efficiency.

Correspondingly, U.S. imports of durable goods have increased, tracking the reduction in the United States annual growth rate of energy and with very high annual rate increases of energy consumption in emerging economies (Figure 7.3). Manufacturing is generally more energy intensive than other sectors of the economy, so changes in the economic structure of the U.S. economy must be considered when analyzing energy efficiency trends over time. Similar trends exist across other advanced economies.

Before describing the origins of DSM, a short description of the electric power industry's beginnings and characteristics is necessary. The electric power industry began in the 1878 with Thomas Edison producing and distributing electricity. Soon competitors entered the industry, and it was not until the 1930s or so that states started to regulate utilities. The economic argument for regulation is that the large economies of scale of the industry

---

1. http://www.policyalmanac.org/economic/archive/manufacturing_employment.shtml.

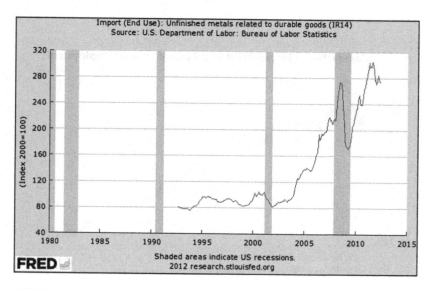

FIGURE 7.3   U.S. Imports of Durable Goods. *Source: Federal Reserve Bank of St. Louis, Economic Research http://research.stlouisfed.org/fred2/series/IR14?rid = 188&soid = 22.*

justified having a single vertically integrated utility regulated by the state. Regulated rates, based upon average cost, are necessary to prevent the utility from charging monopoly prices.

The 1930s to the early 1970s was the golden age of the industry: electric utilities continued to grow and expand rapidly, and improvements in technology and economies of scale led to lower costs and improved reliability, which resulted in more electricity customers. The U.S. Atomic Energy Act of 1954 permitted private development of commercial nuclear power. In short, the interests of the ratepayers, regulators, and utilities were aligned with increasing electricity demand. Both utilities and the federal government promoted the expansion of the use of electricity.

The power system pushed electricity from large, centralized power stations through the transmission and distribution systems to serve customers. For reliability reasons, particularly due to high growth in electricity demand, utilities had direct load control programs, such as the ability to cycle off residential appliances such as water heaters and air conditioners and interruptible and curtailable rates for commercial and industrial customers [3]. Early load control programs also existed in Europe and New Zealand [4]. For the most part, however, customer usage (amount and timing), was considered exogenous to the planning and operations of the utility [4].

The golden years came to an abrupt end in October 1973 with an oil embargo by Mideast producers of oil, although perhaps a harbinger of the end of the golden years was the 1966 large-scale blackout of the Northeast

TABLE 7.1 Percentage of U.S. Electricity Generated in 1973 versus 2011 by Top Five Fuel Types

Fuel	1973	2011
Coal	45.5	42.2
Natural Gas	18.3	24.8
Petroleum	16.9	0.7
Hydroelectric	14.8	7.8
Nuclear	4.5	19.2

Source: Nuclear Energy Institute: http://www.nei.org/resourcesandstats/documentlibrary/reliableandaffordableenergy/graphicsandcharts/uselectricitygenerationfuelshares/

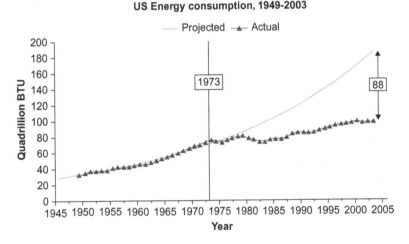

FIGURE 7.4   U.S. Energy Consumption, 1949–2003. *Source: Peter J Wilcoxen, The Maxwell School, Syracuse University http://wilcoxen.maxwell.insightworks.com/pages/804.html.*

United States and Canada. Fuel costs skyrocketed, the economy tanked, and inflation went through the roof. As Table 7.1 indicates, the industry reduced the percentage of petroleum it used from approximately 17% to less than 1 percent. The growth rate in electricity and other uses of energy dropped dramatically, and new power plants, transmission lines, and distribution facilities were no longer needed, as illustrated in Figure 7.4.

Nonetheless, given the regulatory structure of the industry, the costs of unnecessary facilities were for the most part passed on to consumers through higher electricity rates. Cost overruns also occurred on the order of 500 percent,

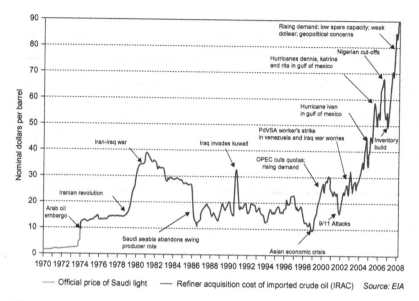

FIGURE 7.5   Oil Prices over Time. *Source: Penn State University https://www.e-education.psu.edu/engr312/node/163*

which were magnified by high inflation and therefore high borrowing costs, particularly in the construction of nuclear power plants. Orders for 65 nuclear power plants were cancelled. In addition, environmental concerns and associated policies, for example, the Clean Air Acts in 1963 and 1970, added costs and complicated utility investment decision-making. The fall of the Shah of Iran on January 16, 1979, and its associated increase in oil prices, also contributed to this higher energy cost trend along with the Three Mile Accident on March 28, 1979. All of these factors led analysts and policymakers to start questioning the fundamental premise of the industry, which was that a utility should build, maintain, and operate the electric power system taking customer demand and future demand growth as exogenous. Conservation and energy efficiency became the watchwords. In addition, the seeds were planted to restructure the industry and introduce competition into part of its supply chain. Figure 7.5 reports oil prices over time.

Following the 1973 oil embargo, a series of federal legislation was enacted: The Energy Policy and Conservation Action of 1975, the Energy Conservation and Production Act of 1976, the Public Utilities Regulatory Policies Act (PURPA) of 1978, and the National Energy Conservation Policy Act (NECPA) of 1978. The first two pieces of legislation established federal policies regarding energy efficiency standards. PURPA required utilities to purchase power from non-utility renewable and cogeneration facilities, referred to as qualifying facilities (QFs), at the utilities' avoided cost, and NECPA mandated that utilities provide residential customers energy audits.

FIGURE 7.6 President Jimmy Carter in His Cardigan Sweater Delivering His Energy Conservation Speech to the Nation.[2]

On April 18, 1977, President Carter addressed the nation in a televised speech proposing his energy policy in a cardigan sweater (Figure 7.6). He compared the difficult effort required to address the nation's energy problems as the "moral equivalent of war" [5]. Of the ten fundamental principles he articulated, five mentioned conservation including:

"The sixth principle, and the cornerstone of our policy, is to reduce the demand through conservation. Our emphasis on conservation is a clear difference between this plan and
others which merely encourage crash production efforts. Conservation is the quickest, cheapest, most practical source of energy. Conservation is the only way we can buy a barrel of oil for a few dollars" [5].

One of the original motivations for conservation, energy efficiency, and therefore DSM programs was the oil price shocks of the 1970s, but as discussed further below, the rapid rise in electricity prices, in part due to the oil crisis and the accompanying recession, and in part due to nuclear power plant cost overruns. Although the United States imports more oil both in absolute terms and as a percentage of its consumptions now than it did in the 1970s, its use of oil to generate electricity has declined tremendously. According to the U.S. Energy Information Administration in 2011, less than one percent of the United States' generation of electricity came from oil, and one percent of the nation's total consumption of oil was used to produce electricity. Dependence on oil imports is an important economic and national security problem, but at least in terms of electricity it is not. In the future, the electric power sector may be able to reduce materially U.S. oil consumption if electric vehicles become widespread.

---

2. http://www.weccusa.org/main/abouttimeline/title/Energy%20efficiency%20timeline

During this first wave of DSM programs, the term DSM had not yet been coined, and instead these programs were sometimes referred to as conservation and load management (C&LM) programs, indicating that conservation was added to existing load management efforts by utilities. The term demand-side management, or DSM, soon entered the industry's vocabulary. It was defined by Clark W. Gellings: "Demand side management is the planning, implementation, and monitoring of those utility activities designed to influence customer use of electricity in ways that will produce desired changes in the utility's load shape, i.e., changes in the time pattern and magnitude of a utility's load" [6] citing [7]).[3] Ironically, Gellings is featured in this volume examining energy efficiency gains on the supply-side of the industry.

DSM clearly expresses its core meaning: utilities would manage demand in order to reduce costs, or reduce the increase in costs, for both the utility and ratepayers and to improve the environment. DSM also includes load management, which is reducing demand during peak hours including by shifting it to off peak hours. Six specific DSM load shape changes are peak clipping, valley, filling, load shifting, load reduction or conservation, off-peak load growth, and enhancing flexibility to load [8]. DSM now has a broader meaning and includes energy efficiency and conservation in general.

DSM required a fundamentally rethinking of the industry, which occurred in a public policy fish bowl. With rising energy costs, a slowing economy, and heightened environmental concerns, politicians, regulators, industrial consumers, environmentalists, and ratepayer advocates all wanted to be involved in setting DSM policies and implementing programs. There seemed to be a competition among states and regions to be at the leading edge of DSM policy formation, particularly between the East and West coasts. These advisory and stakeholder processes continue today and are formally embedded in many programs. So in addition to the substantive policy objectives of DSM – lowering ratepayer costs, enhancing economic development, improving the environmental, and strengthening reliability – there are additional process objectives, such as allowing all interested parties to be able to participate and contribute to DSM program development.

Initially, DSM programs were hastily and perhaps sloppily designed and implemented. They included public appeals to conserve energy, as illustrated by President Carter's national address, informational campaigns, energy audits, weatherization programs, and fuel switching programs for larger customers [2,3]. According to the chapter by Nadel, information campaigns resulted in relatively small levels of energy savings. Some DSM programs, particularly earlier ones, also contained electrification components to increase electricity during off-peak or low-cost periods [2]. For example, in 1976 the utility Wisconsin Public Service

---

3. According to one source, the term DSM originated at a conference in the O'Hare Hilton in Chicago during a meeting of the Edison Electric Institute and the Electric Power Research Institute (EPRI) (Faruqui and Fox-Penner, 2011).

stopped selling appliances as it switched from encouraging energy consumption to conservation, and in 1978, initiated an energy audit program for its customers.[4]

The next wave of programs, the mid-1980s to the mid-1990s, consisted of cash rebates and low-interest financing to encourage customers to purchase more efficient appliances and build more efficient buildings and industrial facilities [3]. States that took the lead in reorienting their utilities to focus on the demand side of the business were California, Massachusetts, and Wisconsin, but they were by no means the only ones. Not surprisingly, these states along with the others that aggressively pursued these policies had higher electric rates and more aggressive environmentally policies than states with less aggressive or no DSM programs. In other states, DSM was needed because nuclear power plants were shut down, as in the case of Three Mile Island, or were cancelled, as in the Northwest (Nadel). In tandem with DSM programs, the federal and many state governments adopted appliance efficiency standards and building codes that set minimum levels of efficiency for new appliances and buildings.

Programmatically, DSM also evolved over time, starting with information and loans, followed by rebates on individual energy efficiency measures such as lighting and refrigeration, direct installation programs that offered a combination of measures, market transformation efforts in which the goal is to move the market for a specific appliance to a higher energy efficiency standard, to programs that fundamentally alter energy consumption such as whole or integrated building approaches and deep retro-fits [9]. In the early 1990s, 25 utilities pooled together $30 million for the Super Efficient Refrigerator Program, with a goal of 40 percent increase in efficiency. The chapter by LaRue, Cole, and Turnbull takes this concept one step further by discussing California's zero net energy goals for all new commercial and residential buildings. Stephen M. Wheeler and Robert B. Segar extend the concept of net zero energy buildings to communities with a case study of a neighborhood, University of California, Davis West Village. DSM is also converging with the penetration of renewable resources. Anderson et al discuss this issue.

LaRue et al in this volume summarize California's DSM history. As part of this history, the important concept of decoupling was developed and implemented, which is further discussed in chapters by Sioshansi, Cavanagh, and Hesser. The term refers to severing the link between a utility's revenues and throughput so that a utility no longer has the incentive to sell more in order to make more. While policymakers and DSM advocates were concerned that utilities may undercut the effectiveness of DSM programs if those programs threatened utility revenues and therefore profits, utilities were concerned about inadequate cost recovery. Decoupling is one way, but by no means the only one suggested, to address these twin problems. Another mechanism is for utilities to recover lost revenues due to DSM.

---

4. http://www.wisconsinpublicservice.com/company/history.aspx

The term negawatt was introduced by Amory Lovins in 1989 to capture the idea that not using electricity should be considered because it is typically less expensive than building a power plant to produce more electricity.[5] This term is still used today, and the concept is still alive in current references to energy efficiency as a resource. The fundamental idea is that since energy efficiency can displace supply-side resources, investments in energy efficiency should be integrated into utility resource planning. The terms least-cost planning (LCP), integrated resource planning (IRP), and integrated resource management (IRM) soon appeared in the industry's lexicon. Their intent was to expand the utility's focus to include not only the demand side, but to integrate DSM efforts with supply-side efforts in order to obtain the most cost savings and environmental benefits [2]. The National Association of Regulatory Utility Commissioners (NARUC) endorsed least-cost planning in 1989 and in doing so acknowledged that least-cost planning also had to be the utility's most profitable course [10]. IRP and IRM differed from least-cost planning by emphasizing a more active approach to identifying, developing, and incorporating into the portfolio of options both demand- and supply-side options for the utility to consider and adopt. Gas utilities also got into the act by adopting LCP [11].

The initial peak for DSM in the United States occurred in 1993: annual DSM expenditures reached $3.2 billion, representing 1.7 percent of utility revenues and DSM programs in place in 447 utilities [2]. In 1992, the Congress passed the Energy Policy Act, which, among other things, introduced created wholesale electricity markets. The generation of electricity was no longer thought to have sufficient economies of scale to warrant cost-of-service regulation. Instead, generation of electricity was subjected to competitive forces to improve its efficiency, while the transmission and distribution of electricity continued under cost-of-service regulation. In some regions of the country, such as California, New England, and New York, utilities sold off much if not all of their generation assets.

Many states also adopted legislation that created retail electricity markets in which third-party providers would procure and schedule electricity for retail customers. At the state level, as is commonly the case, California led the way and began restructuring its electric power sector in 1994. Many advocates of retail electricity markets argued that energy service companies (ESCos) would offer energy efficiency services, perhaps in conjunction with the provision of electricity, and that market participants would replace utility DSM programs. A period of DSM retrenchment occurred during the 1990s and early 2000s due to restructuring, but now DSM spending is increasing substantially and is forecasted to continue to increase according to Nadel's chapter in this volume (see also Figure 7.7).

The United States has two different architectures for electricity markets. One model is the centralized or organized Regional Transmission/Independent System Operator (RTO/ISO) markets (Figure 7.7) in California, the Midwest,

---

5. http://www.ccnr.org/amory.html

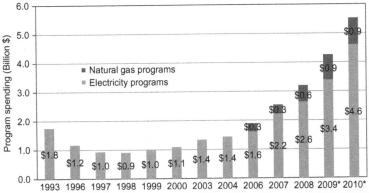

*All values are actual program spending except for 2009 and 2010, which are budgets. Notes: Includes ratepayer-funded programs. Natural gas efficiency program spending is not available for 1993–2004. Sources: Nadel et al. (2000); York and Kushler (2002), (2005); Eldridge (Molina) et al. (2008), (2009), (2010).

FIGURE 7.7   Annual Electricity and Natural Gas Energy Efficiency Program Spending or Budgets. *Source: ACEEE, http://www.investmentu.com/2011/October/energy-efficiency-spending-skyrocketing.html.*

New England, New York, the mid-Atlantic, and parts of the Midwest (PJM), and most of Texas (ERCOT). Although there are many important differences among these markets, the most important of which is the ERCOT market in Texas which does not have an installed capacity market, all of them contain real-time and day-ahead energy markets with hourly prices that vary by location (i.e., locational marginal prices) based upon dispatch and unit commitment and capacity markets. The other architecture is a decentralized or not organized market in which there is not structured real-time and day-ahead markets with transparent locational marginal prices. Instead, physical bilateral markets exist in which transactions are scheduled subject to available transmission. Other countries and regions, such as Australia, Europe, and elsewhere, have also restructured their electricity sector by developing wholesale and retail electricity markets.

The question is: how should DSM programs be structured to complement wholesale and electricity markets? DSM programs are developed in the context of state policies not federal ones. That being said, the Federal Energy Regulatory Commission (FERC), the agency that regulates the United States continental wholesale electricity markets except in Texas, in its Order No. 745 is pushing to increase DSM, particularly price-responsive programs, via organized wholesale electricity markets. Bresler et al describe how to improve demand response in RTO markets. A map of RTO markets is presented in Figure 7.8.

Some states, concerned that retail markets left to their own devices would not result in sufficient energy efficiency and conservation, adopted societal benefits charges, also referred to as public goods charges, that funded energy

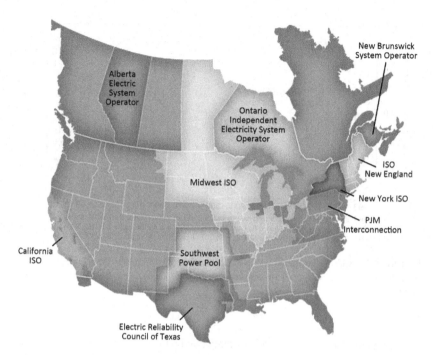

FIGURE 7.8   Map of North American Organized Wholesale Electricity Markets. *Source: California ISO, http://www.caiso.com/about/Pages/OurBusiness/UnderstandingtheISO/Opening-access.aspx.*

efficiency programs through retail rates [2]. These non-by-passable charges are levied on the sale of electricity and natural gas by distribution companies and have to be paid by end-use consumers regardless of who their provider of electricity or natural gas might be. New Jersey is one but not the only example. As part of the state's efforts to respond to federal policies promoting wholesale electricity markets, New Jersey adopted the Electric Discount and Energy Competition Act (EDCA) in 1999 that, among other initiatives, created a societal benefits charge (SBC), in essence a tax, to fund DSM initiatives along with other public policy programs. New Jersey has and continues to use multiple approaches to achieve energy efficiency goals for electricity and natural gas including building codes and standards, appliance standards, utility administered energy efficiency programs, and state administered ones overseen by the New Jersey Board of Public Utilities (BPU) and administered and implemented by third-party contractors.

Wisconsin is another example of a state that has a statewide public benefits program [12]. In 1999, Wisconsin adopted Act 9, which created such a program using third-party administrators, followed in 2005 by Wisconsin Act 141, which increased funding levels and allowed utilities to contract with these energy efficiency program administrators. Wisconsin uses a push-pull strategy to increase consumer demand for energy efficiency goods and services while motivating the supply chain to deliver them. The organizational structure is complex and consists of a Statewide Energy Efficiency and Renewable Administration (SEERA), a Program Administrator, a Fiscal Agent, an Evaluator, and a Compliance Agent. This structure is overseen by the Public Service Commission of Wisconsin.

In some cases, states adopted energy efficiency portfolio standards (EEPS), the energy efficiency analogue, but not equivalent because energy efficiency cannot be directly metered like renewable resources can be, to renewable portfolio standards (RPS). Market-based mechanisms have also been developed for DSM, although their use is not as widespread as renewable portfolio standards. To understand the motivation for EEPS, a little history is in order. The 1990 amendments to the U.S. Clean Air Act set up air emission allowance markets for sulfur dioxide and nitrogen oxides. The government capped these emissions from power plants greater than 25 megaWatts (MW) and allowed market participants to buy and sell the corresponding air emission allowances. A similar market-based mechanism showed up in installed wholesale electricity capacity markets as a floor-and-trade system to ensure generation adequacy [13]. With the adoption of renewable portfolio standards (sometimes referred to as green tags or green certificates), a floor-and-trade system to increase the amount of renewable resources used to generate electricity by approximately 30 U.S. states, the energy efficiency analogue developed referred to as white tags or white certificates [14]. The chapter by MacGill, Healy, and Passey in this volume further discusses markets for energy efficiency. Figures 7.9 and 7.10 present maps of U.S. states that have RPS and EEPS.

Over time, many types of DSM financing mechanisms have developed in addition to white tags. These include on-bill financing (OBF), property assessed clean energy (PACE), energy efficiency performance contracting (EEPC), energy service companies (ESCOs), third-party ownership, including by utilities, of customer-sited energy efficiency equipment, and third-party energy efficiency entities (EEU). The experience with these new forms of financing is mixed and limited. New Jersey is considering allowing commercial and industrial customers the choice of not paying their portion of the societal benefits charge so long as they use the same amount of money on approved energy efficiency measures. (See Hesser's chapter in this volume.)

The answer to the question of how best to organize the administrative structure to design, deliver, and evaluate programs has varied by state and has changed over time and by no means is a settled one. For instance, New Jersey has pursued a mixture of organizational strategies that have developed and evolved from originating with utility administered programs to ones that the BPU administered via contractors, to programs run by third

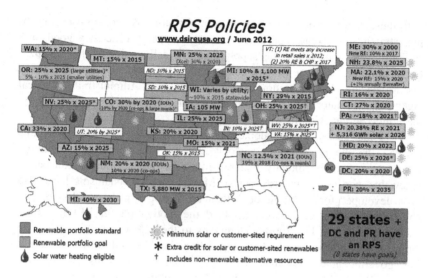

FIGURE 7.9   U.S. Renewable Portfolio Standard Policies. *Source: Database of state incentives for renewables and efficiency http://www.dsireusa.org/documents/summarymaps/RPS_map.pdf.*

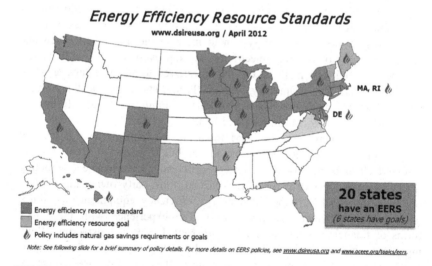

FIGURE 7.10   U.S. Energy Efficiency Resource Standards. *Source: Database of state incentives for renewables and efficiency, http://www.dsireusa.org/documents/summarymaps/EERS_map.pdf.*

parties but overseen by the BPU. Vermont, in contrast to New Jersey, developed an independent, non-profit, state-wide energy efficiency utility (EEU), Efficiency Vermont.[6] The benefits of a state-wide approach as opposed to a

---

6. http://www.veic.org/Implementation_Services/Project_Profiles/Efficiency_Vermont.aspx

utility-by-utility approach is that there is uniformity across the state as opposed to each utility having different programs, which creates some confusion in the marketplace and requires an extensive coordination effort, and to increase the availability of energy efficiency opportunities. In both states, stakeholders have the ability to review programs and provide input. Statewide approaches also have some disadvantages. They remove the utility from the picture, which may be the entity that has the best understanding of its customers. In addition, if there are major differences within customer classes across the state, uniform programs may not be as effective as ones tailored by geography.

3 THE INTELLECTUAL DISPUTE REGARDING THE ENERGY GAP AND ITS HIDDEN ASSUMPTIONS

The term energy gap is used to describe the paradox or apparent paradox between the amount of energy efficiency that is or is believed to be cost effective and the amount that energy consumers of all classes implement. There is a vigorous and unresolved debate between two perspectives. Jaffe et al. [15] lay out the differences in frameworks used by the two different camps: between technologists that emphasize engineering-economic calculations and economists that think about these issues from a market failure and social welfare perspective. The purpose of this section is not to resolve this debate, but instead by recapping some of the major points of this debate to unearth some important historical assumptions behind DSM.

This debate is reflected in this volume. For example, the chapter by Nadel finds that there is substantial energy savings to be had, energy efficiency is the lowest cost option, and strong policies, as opposed to leaving electricity consumers to their own devices, are necessary to achieve these energy savings.

Laitner et al build on these three claims linking the importance of energy efficiency to economic productivity and therefore growth and the need for not only aggressive energy efficiency policies but new business models as well.

These authors also reiterate the commonly made explanations for the energy gap including non-rational consumers who, for instance, prefer not to pay upfront costs even if the future savings appropriately discounted are larger than the initial outlay, the principal-agent problem. For example, landlords who do not invest in energy efficiency appliances because tenants bear costs of the corresponding higher utility bills, and the failure of energy prices to include all of their associated environmental costs. The response to such claims are advanced by Allcott and Greenstone in their chapter, as well as by Jaffe et al. [15] and numerous others. To answer the energy gap question requires empirical work, and Gillingham et al. [9] conduct a comprehensive and empirical retrospective examination of DSM policies.

As interesting and important as the energy gap debate is, it is just as critical to understand the underpinning assumptions behind the pro-DSM position. The DSM programmatic model consists of a centralized planner determining the amount and types of energy efficiency measures, which may be combined into various programs that target customer classes (e.g., residential, commercial and industrial), usage (e.g., lighting, pool pumps), and building types (e.g., new construction, retro-fits). The central planner has many decisions to make that require extensive amounts of data and information including the determination of what combination of information and incentives need to be provided to achieve program goals, presumably at low cost, organizational design and governance, program evaluation, and so forth.

The demands for data and information for DSM is reflected in many documents and budgets. For instance, the California Energy Efficiency Evaluation Manual is 487 pages long [16]. Evaluation budgets are millions of dollars, and in some cases they range from 3 to 5 percent of total DSM budgets, which can run into the hundreds of millions of dollars. A typical evaluation cycle consists of a series of studies, each of which can take many months if not longer to complete at costs of several hundreds of thousands of dollars if not more per study.

A DSM administrator first needs to establish a baseline of existing uses of electricity, such as the number, type, and energy consumption of appliances, levels of insulation and weatherization, feasibility of cogeneration, and so forth for all types of customers such as single and multi-family housing, apartments, small and large commercial facilities, and small and large industrial facilities, which can vary substantially by industrial application. This is no easy task because the baseline can be manipulated or gamed. Once the baseline is estimated, then a market potential study is needed to estimate the technical and market potential of implementing various energy efficiency measures. These measures number in the thousands and their level applicability and level of energy savings must be tailored to the climate of their location. All of these estimates are for a typical customer within a customer class or subclass or the average use of a measure, and variances among customers or use of measures are ignored.

The next analysis in this cycle is a DSM program design and implementation study. Presumably, a major factor in program design and implementation is achieving the most cost-effective program, but other objectives exist such as having broad program participation, allocating programs across rate classes commensurate with their funding contributions, serving low-income and disadvantaged communities, and satisfying process and procedural objectives.

After programs have been implemented, process assessments are conducted to ensure that program logic is solid and program delivery is optimized followed by impact evaluations required to estimate the actual energy savings. This has been long recognized as no easy task, which involves numerous tradeoffs [4]. One fundamental difficulty with estimating DSM energy savings is that savings cannot be directly measured [4]. Instead, some type of but for calculation is needed to assess what would have happened without a specific DSM program. Assigning causation, or in the language of DSM evaluation attribution, is extremely difficult and when it is combined with the fact that most DSM programs are cross-subsidized (i.e., funded by all ratepayers but only a relatively few number of ratepayers receive the direct benefits of DSM measures) the question of whether DSM programs benefit ratepayers and society in general becomes important and contentious.

To illustrate some aspects of this difficulty but certainly not covering all of the evaluation issues, consider the issues of spillover and free ridership [17]. Spillover refers to consumers adopting DSM measures even though they are not program participants or participate beyond the scope of a particular program. Free ridership, on the other hand, are program participants who would have adopted DSM without the program, but nonetheless participate in the program. For example, receiving a rebate or financial incentive even though they would have purchased the more efficient measure without the DSM program. Not surprisingly, determining if and the amount of each of these effects is difficult to do and subject to some uncertainty, and yet both spillover and free ridership must be known to accurately evaluate these programs. In the language of DSM evaluation, the combination of spillover and free ridership result, along with several other factors, in a net-to-gross (NTG) adjustment to determine the final or net amount of energy savings. Other NTG adjustments are needed to account for the fact that not all energy efficiency measures are implemented (e.g., unused compact fluorescent lights may be sitting in storage), measures may be installed incorrectly therefore reducing their effectiveness, and consumers may increase their use of energy because they can now afford to use more energy given that their usage is more efficient. This is referred to as take-back, snap-back, or the rebound effect.

After completing the necessary impact evaluations, programs need to be modified and updated, and the complete evaluation cycle needs to be repeated to account for changes in technologies, legislative and regulatory changes such as new appliance standards, changes in electricity prices, structural shifts in the economy, and so on. The above description does not do justice to the overwhelming informational needs of the central DSM planner and the numerous assumptions that must be made in order to make data collection practical.

The overwhelming amount of information that is needed by DSM program managers raises two important issues that have not received the attention that the energy-efficiency-gap debate has. First, there are substantial costs related to information collection. These costs of obtaining all of this information and the costs associated with its uncertainties, errors, and risks of being outdated need to be assessed and quantified in order to answer the question of whether these costs, which not only include the data collection costs but also the costs associated with poorly designed programs, are larger or smaller than the associated costs of not having DSM programs. Thus, having tracking and accounting systems that can be audited is also critical.

Second, DSM program managers may act in their own interests not necessarily in the public's interest. The information overload problem assumes that DSM program managers and the accompanying stakeholder processes act in the interests of end-use consumers. Perhaps they do, but there is little reason to believe that their self-interest is not also pursued in program design, implementation, and evaluation decisions. With the large informational requirements needed for DSM programs along with technical nature of the analyses that use this information, the principal-agent problem becomes a concern. Here, the DSM program manager is the agent that can pursue its own self-interest at the expense of the principal, in this case end-use consumers, given the asymmetric information between the two that would be costly for the principal to overcome. In short, DSM program managers are spending other people's money that it is costly for end-use consumers to monitor. The question then becomes under what conditions is it more efficient for end-use consumers to pursue energy efficiency measures on their own as opposed to having a central planner that has incomplete information and may be pursuing its own self-interest to a lesser or greater extent.

This review of the debate about the size of the energy efficiency gap and the costs and benefits of closing it has uncovered the question of how to close this gap, whether through planning, such as been the case historically with DSM programs, or through electricity markets with electricity prices that internalize environmental costs. Prior to the introduction of wholesale and retail electricity markets, there were no meaningful electricity prices, only electric rates based upon average not marginal costs, although it would have been possible to extract proxies for efficient prices from electricity rates. In addition, the recent and rapid reduction in the cost of smart meters that are able to track electricity consumption within an hour also facilitates transmitting and therefore using variable retail prices that reflect social marginal costs to end-use consumers. This is not to say that a pricing approach could not have been used since the 1970s, but instead that the introduction of electricity markets and the cost reductions in smart meters make this more of an option now than in the past. Thus, electricity markets in the United States have opened up another model, the market or pricing model, to achieve the public policy goals of DSM programs.

4 THE DSM PAST IS ENERGY EFFICIENCY'S PROLOGUE

With this brief history of DSM in mind, how will the energy efficiency debate evolve and how should the industry proceed? The forces pushing DSM's expansionary trend are strong. Even with the partial pause and rethinking of DSM that occurred during the formation of electricity markets, DSM has come back both in scale and scope. With its successful track record, which includes some hype and overpromises, DSM is an attractive policy with its allure of cost effectiveness, environmental benefits, and broader economic benefits. No doubt there are substantial opportunities for energy efficiency, conservation, and load management. This does not necessarily mean that government administered programs are the most cost-effective means of doing so. DSM is also enticing politically. The claims of large benefits at little or no cost are hard for the political process to resist. Finally, there is an active and strong DSM advocacy and analytical community that is continually pushing for more DSM and conducting studies that claim even more DSM should be done.

The historical model of ratepayers cross-subsidizing DSM programs with a relatively small percentage of ratepayers directly benefit from in a given year may be coming apart at current DSM funding levels and may not be scalable to achieve the amount of energy reductions that are envisioned as being necessary in response to climate change. The movement towards self-financing of DSM in which the major DSM beneficiary pays for the most of the associate costs is an encouraging response to this problem. Although DSM does provide benefits to non-program participants, quantifying these benefits is difficult. Non-participants may be skeptical that they are benefiting overall, particularly when low-cost measures, the proverbial low hanging fruit, are no longer available and more costly ones must be used.

Historically, DSM did not require well-defined property rights regarding electricity given the industry's vertically integrated and regulated structure. Implicit in a utility building its system to a specified level of reliability was that its customers, who paid for the system, could consume as much power as they would like whenever they desired. In order to induce customers to reduce their demand, for example during peak periods, some type of payment or compensation was necessary. In short, customers were paid for not consuming. With the adoption of wholesale and retail markets in the United States and many parts of the world, exactly who owns, if anyone, the right to sell back electricity and in what quantities and at what price has to be made clear and precise if electricity markets are to function efficiently.

DSM programs were developed in an industry that did not have prices, and therefore price signals could not be used to affect demand. This limitation is quickly being addressed with the strong emphasis on developing programs to increase price-responsive demand discussed by Faruqui in this volume. The advancement in various smart grids including smart meter technologies makes implementing price-responsive and load control programs easier. These technologies may fundamentally change the ability of end-use consumers to respond and react to all parts of the electric supply chain from generation, transmission, and distribution. Many important issues remain to be resolved, but nonetheless these technological advancements open up many more possibilities, particularly in the real- and near real-time response of DSM to system conditions [1]. An important element of smart grid technology is equity and including other DSM programs besides the ability of end-use customers to respond to changing prices can address some equity issues [18].

A final trend that DSM must respond to is the implications of shale gas. The vast resources of now low-cost natural gas from shale, combined with the current economic conditions, means that wholesale electricity pries are low, making energy efficiency less attractive. In many parts of the United States and during many hours, natural gas is the marginal fuel (i.e., it is the fuel that the most expensive power plant that is generating is using). In wholesale electricity markets, this marginal unit (or combination of units if there are binding transmission constraints) sets the wholesale electricity price. As natural gas prices rise and fall, so do wholesale electricity prices and eventually retail electricity prices. In addition to shale gas being inexpensive and available in large quantities, in many cases these deposits are located near load centers, reducing natural gas transportation costs. That being said, there are important and legitimate concerns with shale gas such as its environmental impact, questions regarding the actual amount of reserves, and to what extent it reduces, if at all, greenhouse gas emissions given methane leakage. In addition, questions as to whether natural gas generators will sign long-term natural gas contracts that are necessary to build new natural gas pipelines are arising in wholesale electricity markets in which utilities no longer have the obligation to serve retail customers.

5 CONCLUSIONS

Given these considerations, how should DSM proponents and analysts respond?

First, DSM proponents and analysts should think about how to once and for all substantially reduce if not outright eliminate the cross-subsidies that have been inherent in DSM programs. An important exception to this is low-income and disadvantaged populations. There needs to be a well-thought-out exit strategy so that the societal optimal level of energy efficiency occurs without cross-subsidies. Without such a strategy, DSM programs may not be financially viable, particularly if they are significantly expanded. To date, DSM proponents have identified numerous market failures, barriers, and irrational behavior to efficient levels of energy efficiency without DSM programs, but a lot less work has been done on quantifying these each of them and then specifically targeting them in DSM program design. Without this additional work, it will not be possible to address the fundamental problems that result in inefficient adoption of energy efficiency measures without DSM programs.

Second, DSM proponents need to engage the economic critique of DSM. Too many of the DSM evaluation reports ignore the fundamental challenges that many economists have been making of DSM. The two parallel and competing literatures on the efficacy of DSM do not seem to intersect. The purpose of recommending that these two opposing viewpoints be reconciled is not to promote an academic exercise of determining who is right. Instead, its purpose is to improve the efficient adoption of energy efficiency, conservation, and load management measures through the optimal combination of market mechanisms and DSM programs at both the wholesale and retail level.

To evaluate the efficacy of DSM programs, more randomized evaluations are necessary. Although some of these types of studies have been conducted, typical DSM impact studies do not use this methodology. Randomized evaluations, analogous to drug testing studies, could empirically answer many of the theoretical claims and counterclaims that are being made between the two camps. In addition, they could help answer the question of what is the most cost-effective means of obtaining efficient levels of DSM. The focus of current DSM evaluations is answering the question of whether a program is cost effective, not trying to determine the most cost-effective means of achieving the desired level of results.

Third, DSM proponents and analysts should recognize the implications of shale natural gas. The response should not be to find other benefits on paper of DSM to offset lower electricity costs, but to adjust and improve programs, and in some case eliminate them, given the anticipated but not guaranteed lower wholesale cost of electricity for many years to come. Another implication of low natural gas prices is that the difference in on-peak and non- and off-peak prices is likely to be lower on average than in the past, all else equal. The reason is that natural gas is likely to be the marginal fuel in both non-peak hours, set by a natural gas fired combined cycle unit, and during peak hours, set by a natural gas fired gas turbine. One way to think about shale gas from the perspective of DSM proponents is that it provides a backstop to DSM programs. The production of electricity from natural gas provides the conditions, economically and environmentally, that DSM programs must surpass.

This chapter's brief history of DSM in the United States is intended to provide some insights regarding the extent that energy efficiency can provide meaningful cost-effective and welfare-enhancing outcomes. This chapter reviewed this history and examined some of its important implications given current policies. Looking ahead, what is required is improving the economic content of DSM analysis including its objectivity and transparency. DSM has an important role to play in energy, economic, and environmental policy, and so over-the-top advocacy and associated claims are no longer needed. Instead, calm, dispassionate, and systematic analysis and critiques are required. Only then will society be able to achieve the socially optimal level of energy efficiency over time.

ACKNOWLEDGMENTS

The author would like to thank Mike Ambrosio, Jon Lowell, Bob Obetier, Jaclyn Trzaska, Yuemeng Zhang, and the editor of this volume for their input and comments.

REFERENCES

[1] Sioshansi FP, editor. Smart grid: integrating renewable, distributed & efficient energy. Elsevier Inc.; 2012.

[2] Charles River Associates (CRA). Primer on demand-side management: with an emphasis on price-responsive programs, February 2005.

[3] Faruqui A, Fox-Penner P. Energy efficiency and utility demand-side management programs, the brattle group, presentation to the world bank, July 14, 2011.

[4] Rau NS, Rose K, Costello KW, Hegary Y. The national regulatory research institute, methods to quantify energy savings from demand-side management programs, October 1991.

[5] Carter J. The president's proposed energy policy, April 18, 1977, Vital speeches of the day, Vol. XXXXIII, No. 14, May 1, 1977, p. 418–420. 1977.

[6] Broeer T, Djilali N. Defining demand side management, Pacific institute for climate solutions, June 2010.

[7] Gellings CW. The concept of demand-side management for electric utilities. Proceedings of the IEEE 1985;73(10).

[8] Gellings CW, Smith WM. Integrating demand-side management into utility planning. Proceedings of the IEEE 1989;77(6).

[9] Gillingham K, Newell RG, Palmer K. Retrospective examination of demand-side energy efficiency policies, resources for the future, RFF DP 04–19 REV, June 2004.

[10] National Association of Regulatory Utility Commissioners (NARUC). Committee on energy conservation, resolution in support of incentives for electric utility least-cost planning, July 27, 1989.

[11] Goldman CA, Hopkins ME, and the national association of regulatory utility commissioners, survey and analysis of state regulatory activities on least cost planning for gas utilities, LBL–30353, April 1991.

[12] Stemrich C. The wisconsin story, presentation, undated.

[13] Jaffe AB, Felder FA. Should capacity have a capacity requirement? If so, how should it be priced? The Electricity Journal 1996;9(10):52–60.

[14] Transue M, Felder FA. Comparison of energy efficiency incentive programs: rebates and white certificates. Utilities Policy 2010;18:103–11.

[15] Jaffe AB, Newell RG, Stavins RN. Energy-efficient technologies and climate change policies: issues and evidence, resources for the future, climate issue brief no. 19, December 1999.

[16] TecMarkt Works Framework Team, The california evaluation framework, June 2004.

[17] Nelson DJ, Hydro BC. Measurable spillover (free drivers): the search continues, american council for an energy-efficiency economy (ACEEE) Proceedings paper, panel 8 paper 16, 1994, available at <http://www.aceee.org/proceedings-paper/ss94/panel08/paper16>; 1994.

[18] Felder FA. The practical equity implications of advanced metering infrastructure. Electr J 2010;23(6):56–64.

[19] Felder FA. Examining electricity price suppression due to renewable resources and other grid investments. Electr J 2011;24:34–46.

Chapter 8

China: Energy Efficiency Where it *Really* Matters

Mark Levine, Nan Zhou, David Fridley, Lynn Price and Nina Zheng
Lawrence Berkeley National Laboratory

1 INTRODUCTION

As the world's most populous country and the second-largest economy, China's energy use and energy-related CO_2 emissions have been interlinked with its unprecedented economic growth and urbanization. Since economic reforms were first initiated in the late 1970s, China has continued to experience double-digit economic growth with current nominal GDP totaling nearly U.S. $7.3 trillion despite a relatively low per capita GDP of only U.S. $5,184 in 2011. Since China is still in the early stage of industrialization and modernization, the process of economic development will continue to drive China's energy demand and has already made China the world's largest CO_2 emitter in 2007 and largest energy consumer in 2010.

One sector that has clearly been impacted by China's rapid economic growth and urban development is the power sector, where electricity demand has grown at an annual average rate of 13 percent over the last decade. At the same time, total installed power generation capacity has tripled from 320 GW in 2000 to 960 GW in 2010. That is, electric capacity of more than 60 percent of the entire U.S. capacity was installed in a decade in China. China's power sector remains dominated by coal with an 80 percent share, which has important implications for both coal use and energy-related CO_2 emissions.

In recent years, China has taken serious actions to reduce its energy consumption and carbon emissions. This has included significant actions affecting the power sector. China's 11th Five Year Plan (FYP) goal of reducing energy consumption per unit of GDP by 20 percent between 2006 and 2010 was followed by extensive programs to support the realization of the goal. In November 2009, China also committed to reducing its carbon intensity (CO_2 per unit of GDP) by 40 percent to 45 percent percent below 2005 levels by 2020. Within the power sector, China's goal is to raise the share of alternative energy to 15 percent of total primary energy by 2020. Achieving the 2020 goals will require strengthening and expansion of energy efficiency policies in the end-use

Energy Efficiency. DOI: http://dx.doi.org/10.1016/B978-0-12-397879-0.00008-6

201

FIGURE 8.1   Map of China. *Source: U.S. Energy Information Administration, China Country Analysis.*

sectors that consume energy – industry, buildings, appliances, and motor vehicles – as well as further expansion of renewable and nuclear power capacity.

The past decade has seen the development of various scenarios describing long-term patterns of China's future energy and greenhouse gas emissions. In most of the models used to create the scenarios, however, a description of sectoral activity variables is missing. End-use sector-level results for buildings, industry, or transportation or analysis of adoption of particular technologies and policies are rarely if ever provided in global energy modeling efforts. This is a serious omission for energy analysts and policymakers, in some cases calling into question the very meaning of the scenarios. Energy consumption is driven by the diffusion of various types of equipment, and their performance, saturation, and utilization. These factors, not considered in most models used for scenario building in China, have a profound effect on energy demand. Policy analysts wishing to assess the impacts of efficiency, industry structure, and mitigation policies require a description of drivers that includes saturation of different types of energy-using equipment (and of measures affecting the thermal integrity of the building envelope) and the factors that are responsible for their level of use (or, in the case of the envelope, heat transfer).

Other chapters of this volume examine the various means, policies, and options to reduce the future growth of energy, particularly electricity, and are mostly focused on developed economies. Due to the sheer size of its population, projected growth of its economy, and magnitude of energy consumption, however, China's evolution in these regards will also have significant impact on global energy trends.

This chapter thus focuses on a China Energy Outlook through 2050, with 2020 and 2030 milestones that assesses the cross-sectoral roles of energy efficiency policies and structural change for transitioning China's economy to a lower emissions trajectory and examines the likelihood of meeting China's 2020 goals. This outlook is based on the Lawrence Berkeley National Laboratory (LBNL) China End-Use Energy Model, which addresses end-use energy demand characteristics including sectoral patterns of energy consumption, changes in subsectoral industrial output, trends in saturation and usage of energy-using equipment, technological changes including efficiency improvements, and links between economic growth and energy demand. Two scenarios are developed to evaluate the impact of different levels of energy efficiency and power sector policies on controlling energy demand growth and emission mitigation and progress towards meeting its 2020 goal [1].

Section 2 provides an overview of the LBNL China End-Use Energy Model and its underlying methodology and scenario assumptions. Section 3 covers aggregate energy and emissions modeling results. Sections 4 through 8 review the key sector-specific findings for the residential, commercial, industrial, transport, and power sectors, respectively. This is followed by a brief discussion of sensitivity analyses in section 9 and conclusions in section 10.

2 MODELING METHODOLOGY

The LBNL China End-Use Energy Model has been significantly enhanced since its establishment in 2005 and is based on levels of diffusion of end-use technologies and other drivers of energy demand on a sectoral basis and includes both demand and supply-side modules. Built using the Long-Range Energy Alternatives Planning (LEAP) modeling software developed by Stockholm Environmental Institute, this model enables detailed consideration of technological development − industrial production, equipment efficiency, residential appliance usage, vehicle ownership, power sector efficiency, lighting and heating usage − as a way to evaluate China's energy and emissions development path below the level of its macro-relationship to economic development. Within the energy consumption sector, key drivers of energy use include activity drivers (total population growth, urbanization, building and vehicle stock, commodity production), and economic drivers (total GDP, income), energy intensity trends (energy intensity of energy-using equipment and appliances). These factors are in turn driven by changes in consumer preferences, energy costs, settlement and infrastructure patterns, technical change, and overall economic conditions. From the supply side, the energy transformation sector includes an electricity sector module that can be adapted to reflect changes in generation dispatch algorithms, efficiency levels, fuel-switching, generation mix, installation of carbon capture and sequestration technology, and demand-side management.

This chapter presents two scenarios, Continued Improvement and Accelerated Improvement, further defined below, to represent distinct alternatives in long-term pathways given current trends, currently available and projected efficiency technologies, and policy choices and degree of successful implementation of the policies.

The Continued Improvement Scenario (CIS) assumes that the Chinese economy will continue on a path of lowering its energy intensity. Efficiency improvements in this scenario are consistent with trends in "market-based" improvement and successful implementation of policies and programs already undertaken, planned, or proposed by the Chinese government. We use the CIS as the reference case for evaluating energy savings and emissions reduction potential.

The Accelerated Improvement Scenario (AIS) assumes a much more aggressive trajectory toward current best practice and implementation of important alternative energy technologies as a result of more aggressive and far-reaching energy efficiency policies. Efficiency targets are considered at the level of end-use technologies, with Chinese sub-sector intensities being lowered by implementation of the best technically feasible products and processes in the short to medium term, taking into account the time necessary for these technologies to penetrate the stock of energy-consuming equipment.

The key energy efficiency policies driving different paces of efficiency improvements in the two scenarios are highlighted in Table 8.1.

TABLE 8.1 Key Assumptions of Continued Improvement and Accelerated Improvement Scenarios

	Policy Drivers	Continued Improvement	Accelerated Improvement
Macroeconomic Parameters			
Population in 2050	–	1.41 Billion	1.41 Billion
Urbanization Rate in 2050	–	79%	79%
GDP Growth	–		
2010–2020	–	7.7%	7.7%
2020–2030	–	5.9%	5.9%
2030–2050	–	3.4%	3.4%

(Continued)

TABLE 8.1 Key Assumptions of Continued Improvement and Accelerated Improvement Scenarios—(cont.)

	Policy Drivers	Continued Improvement	Accelerated Improvement
Residential Buildings			
Appliance Efficiency	Efficiency standards revision, strengthened enforcement of Energy Label	Moderate Efficiency Improvement (1/3 improvement relative to High Efficiency)	Moderate Improvement of new equipment in 2010– near Best Practice by 2020
Building Shell Improvements: Heating		Moderate Efficiency Improvement (1/3 improvement relative to High Efficiency)	50% improvement in new buildings by 2010–75% improvement in new buildings by 2020
Building Shell Improvements: Cooling		Moderate Efficiency Improvement (1/3 improvement relative to High Efficiency)	25% improvement in new buildings by 2010–37.5% improvement in new buildings by 2020
Commercial Buildings			
Heating Efficiency	Strengthening equipment efficiency standards, incentives for heat pump installation	Moderate Efficiency Improvement by 2020	Current International Best Practice by 2020
Cooling Efficiency	Strengthening equipment efficiency standards	Current International Best Practice by 2050	Current International Best Practice by 2020
Building Shell Improvements: Heating		50% improvement in fraction of new buildings growing by 1% per year	50% improvement in all new buildings by 2010, 75% improvement in all new buildings by 2025
Building Shell Improvements: Cooling		25% improvement in fraction of new buildings growing by 1% per year	25% improvement in all new buildings by 2010, 37.5% improvement in all new buildings by 2025

(Continued)

TABLE 8.1 Key Assumptions of Continued Improvement and Accelerated Improvement Scenarios—(cont.)

	Policy Drivers	Continued Improvement	Accelerated Improvement
Lighting and Equipment Efficiency	New commercial equipment efficiency standards, phase-out of inefficient lighting	18 % improvement relative to frozen efficiency by 2030	48 % improvement relative to frozen efficiency by 2030
Industrial Sector			
Cement	Continuation of Top 1000 Program, setting and enforcement of sector-specific energy intensity target such as the 11th FYP targets.	Current world best practice for Portland cement by ∼2025	Current world best practice for Portland cement by ∼2020
Iron & Steel		25% of production by electric arc furnace by 2050	40% of production by electric arc furnace by 2050
Aluminum		Moderate decline of energy intensity to 2050	Accelerated decline of energy intensity to world best practice levels before 2050
Paper		Moderate weighted average energy intensity reduction	Current world best practice energy intensity by 2030
Ammonia		Moderate energy intensity reductions without achieving all 11th FYP target	Achieve all 11th FYP targets through 2020 with continued decline thereafter
Ethylene		Meets 11th FYP energy intensity targets through 2020 and continuing reduction	Current world best practice by 2025 and continuing reduction through 2050
Glass		Moderate efficiency improvements	National average efficiency reach Shandong Top 1000 Program's best practice level by 2030
Transport Sector			
ICE Efficiency Improvements	Strengthening existing fuel economy standards for cars and trucks,	Moderate efficiency improvements in fuel economy of aircrafts,	Significant additional efficiency improvements in fuel

(Continued)

TABLE 8.1 Key Assumptions of Continued Improvement and Accelerated Improvement Scenarios—(cont.)

	Policy Drivers	Continued Improvement	Accelerated Improvement
	incentives or rebates for efficient car purchases, gasoline tax	buses, cars, and trucks through 2050	economy of buses through 2050
Electric Vehicle (EV) Penetration	EV mandates or targets for government fleet, economic incentives for private EV purchase	Electric vehicle penetration to 30% by 2050	Electric vehicle penetration to 70% by 2050
Rail Electrification	Public investment in upgrading and expanding rail network	Continued rail electrification to 70% by 2050	Accelerated rail electrification to 85% by 2050
Electricity Sector			
Thermal Efficiency Improvements	Mandate closure of small inefficient coal generation units, require	Coal heat rate drops from 357 to 290 grams coal equivalent per kilowatt-hour (gce/kWh) in 2050	Coal heat rate drops from 357 to 275 (gce/kWh) in 2050
Renewable Generation Growth	Renewable Portfolio Standard or Mandatory Market Share for renewables, feed-in tariff. Environmental dispatch order	Installed capacity of wind, solar, and biomass power grows from 2.3 GW in 2005 to 535 GW in 2050	Installed capacity of wind, solar, and biomass power grows from 2.3 GW in 2005 to 608 GW in 2050
Demand Side Management	Various demand-side efficiency programs and policies	Total electricity demand reaches 9100 TWh in 2050	Total electricity demand reaches 7,764 TWh in 2050

3 AGGREGATE ENERGY AND EMISSIONS MODELLING RESULTS

The LBNL modeling results show that by 2050 China's primary energy consumption will rise continuously in both scenarios, but approach a plateau starting in 2025 for the AIS and 2030 for the CIS (Figure 8.2).

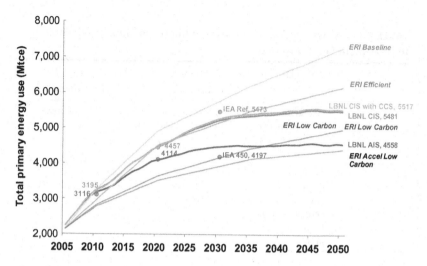

FIGURE 8.2 Primary Energy Consumption in Selected Scenarios. *Note: AIS is Accelerated Improvement Scenario; CIS is Continued Improvement Scenario.*

Energy demand grows from 2,250 million tons of coal equivalent (Mtce[1]) in 2005 to 5500 Mtce (161 EJ) in 2050 under the CIS. Energy demand is 900 Mtce lower – at 4,600 Mtce – in AIS in 2050, a cumulative energy reduction of 26 billion tons of coal equivalent from 2005 to 2050. If sufficient carbon capture and storage (CCS) capacity to capture and sequester 500 $MtCO_2$ by 2050 is implemented under the CIS, total primary energy use would increase to 5,520 Mtce in 2050 due to CCS energy requirements for carbon separation, pumping, and long-term storage, but carbon emissions would decline by 4 percent in 2050.

The notable difference between LBNL's scenarios and those produced by other research institutions is the shape of the energy and emissions trajectories over the long term. LBNL's projected energy consumption increases at approximately the same rate as shown in other models until 2025 or 2030 (depending on scenario), but diverges thereafter with a slow down or plateau. Most energy scenarios produced using other models exhibit an extrapolation of growth over the next two decades to 2050, as illustrated in Figure 8.2.

The ERI is China's Energy Research Institute, whose recent 2009 study results have been converted to IEA-equivalent figures given that ERI follows the convention of using power generation equivalent, rather than IEA and LBNL's use of calorific equivalent, to convert primary electricity [2]. This conversion of ERI results to the IEA/LBNL convention reduces the gross

---

1. Mtce is the standard energy unit used in China. 1 Mtce is approximately equivalent to 29.3 PJ.

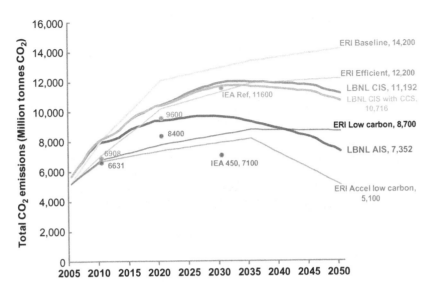

FIGURE 8.3 CO_2 Emissions in Selected Scenarios.

energy content of electricity generated from renewables and biomass by 66 percent. IEA results are taken from World Energy Outlook 2009 [4].

As seen in Figure 8.3, CO_2 emissions under both LBNL scenarios approach a plateau or peak in 2025 (AIS) and 2030 (CIS). The CIS reaches a plateau between 2030 and 2035 with 12 billion giga tons (Gt) in 2033, while the more aggressive energy efficiency improvement and faster decarbonization of the power supply under AIS result in a peak between 2025 and 2030 at 9.7 $GtCO_2$ in 2027. The alternative scenarios examined – including China's Energy Research Institute's low-carbon and accelerated low-carbon scenarios and the International Energy Agency's 450 scenario – rely heavily on application of CCS to reduce emissions growth. The LBNL scenarios are much more pessimistic regarding CCS than the other scenarios but nonetheless show growth in CO_2 emissions comparable to those of other analyses.

The CIS and AIS are both notable as two of the only three scenarios shown in Figure 8.3 that envision emissions peaks before 2050. In fact, the emissions peak is the earliest in the CIS and AIS, underscoring the important role that energy efficiency policies can play in carbon mitigation in the absence of carbon capture and sequestration.

3.1 Aggregate Results by Fuel and Sector

The share of coal in total primary energy use will be reduced from 74 percent in 2005 to about 47 percent by 2050 in the CIS, and could be further reduced to 30 percent in the AIS (Figure 8.4). Instead, more energy demand will be

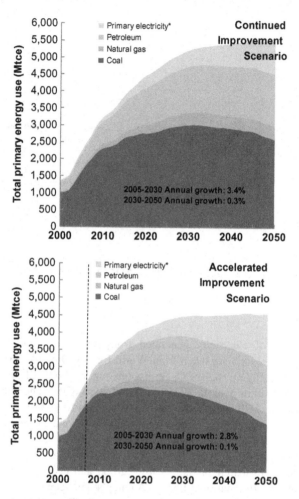

FIGURE 8.4   Total Primary Energy by Fuel Type, CIS and AIS. *Note: Primary electricity includes hydropower, wind, solar, and other renewables at calorific equivalent for conversion.*

met by primary electricity generated by renewables, hydro, and nuclear, which could reach 32 percent by 2050 with further decarbonization in the electricity sector under the AIS. Petroleum energy use will grow both in absolute terms and the relative share of overall energy consumption, as a result of an increase in vehicle ownership as well as freight turnover in the transportation sector.

The single largest emissions reduction potential among the energy end-use sectors is in the buildings sector, particularly commercial buildings, followed by the industrial sector, as illustrated in Figure 8.5. The industrial sector shows early achievement in emission reduction, but

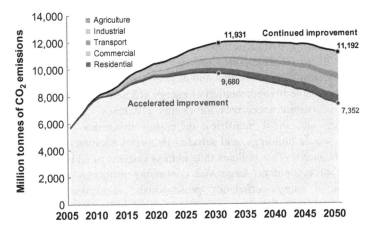

FIGURE 8.5 CO_2 Emissions Difference between Two Scenarios by Sector.

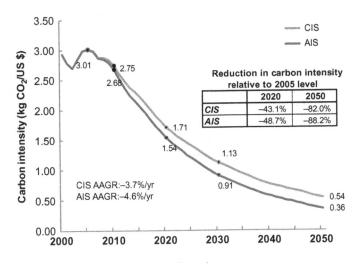

FIGURE 8.6   Carbon Intensity Reductions by Scenario.

within two decades the buildings sector achieves greater reductions. By 2050, more than half of the annual emissions reduction is in the buildings sector.

Overall, the growth of annual energy demand in China ranges from 3.4 percent (CIS) to 2.8 percent (AIS) between 2005 and 2030, and 0.3 percent to 0.1 percent between 2030 and 2050. In contrast, carbon intensity declines over this period, with annual average reductions of 3.7 percent for the CIS and 4.6 percent for the AIS from 2005 to 2050 (Figure 8.6). China will meet and even surpass its 2020 carbon intensity reduction goal of

40 percent to 45 percent under the CIS and AIS, respectively. Achieving such results in AIS will require strengthening or expanding energy efficiency policies in industry, buildings, appliances, and motor vehicles similar in scale and impact to the possible policy trajectories described in Table 8.1, as well as further expansion of renewable and nuclear power capacity.

With aggressive implementation of energy efficiency policies – including expansion of current incentives for energy efficiency and new incentives for buildings and small industries; increasing stringency of standards for appliances, whole buildings, and selected industrial equipment; development of standards and/or other policies that address systems in addition to individual pieces of equipment; large and continuing programs to enhance the capabilities of energy efficiency professionals; incorporation of energy efficiency as criteria for success in operation of industrial and agricultural enterprises, as well the performance of cities and counties (i.e., changing the evaluation system from purely economics to energy efficiency and environment); and decarbonization of the electricity sector under the AIS – China could reduce its 2005 carbon intensity by as much as 88 percent by 2050. Decarbonization of electricity will require continuation and expansion of the large-scale deployment of carbon neutral energy supply technologies (wind, solar, nuclear, zero-carbon biomass conversion, and hydropower) and using the experience to reduce implementation and operational costs of these systems. Successful energy efficiency policies depend not only on the promulgation of laws and regulations at the level of the national government but also effective implementation at the local level. A mix of incentives (for successful implementation) and disincentives (for lack of diligence) may be needed to improve implementation at the local level.

In light of the significant but differing energy savings and emissions reduction potential among the different end-use sectors, more detailed findings of projected energy – and electricity in particular – consumption trends and emissions implications for each end-use sector are presented in the sections below.

4 RESIDENTIAL BUILDINGS SECTOR FINDINGS

Although the ownership of many electricity-consuming appliances has become saturated in urban areas, new sales will remain strong over the period because of the rise in urbanization, with over 470 million additional people expected to become urban residents by 2050 (Figure 8.7).

As a result, electricity use from appliances will continue to grow rapidly. Electricity use reached 2.5 times 2000 levels in just ten years, growing from 170 TWh in 2000 to 430 TWh in 2010. Urban fuel consumption from space heating has also more than doubled in that decade and will continue its rapid growth in both the CIS and AIS due to increases in urban population and heating intensity. Rural electricity consumption will continue increasing in spite of the reduction in rural population due to increases in per household use of lighting and appliances. Biomass consumption will decrease considerably, with substitution of or conversion into commercial fuels.

Residential primary energy demand will grow rapidly until 2025 or 2030. In CIS, demand rises between 2005 and 2030 at an average annual rate of 2.8 percent. After 2030, it increases by only 0.6 percent per year. This slowing of growth is largely due to saturation effects, as the process of urbanization will be largely complete, most households will possess all major electricity-consuming appliances by 2030, and efficiency improvements in heat distribution will be largely complete. In spite of the saturation of major appliances, per household electricity consumption will experience some growth as rising incomes contribute to increased usage, larger refrigerators, more lighting, and more devices using standby power. Figure 8.8 shows the

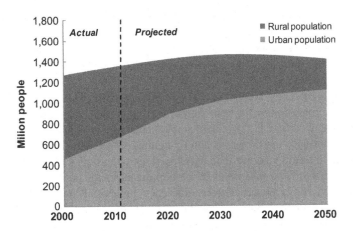

FIGURE 8.7 Historical and Projected Population and Urbanization Trends. *Source: Actual data from[3].*

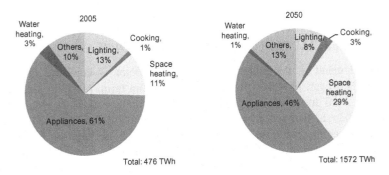

FIGURE 8.8 Residential Electricity Consumption Shares by End-Use, 2005 (left) and 2050 in the CIS (right).

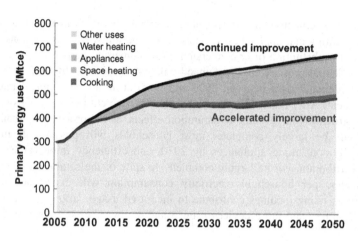

FIGURE 8.9   Residential Primary Energy Use and Potential Reductions by End-Use.

2005 and projected 2050 residential electricity consumption shares by major end-use under the CIS, with electricity consumption dominated by appliances and space heating.

The significant opportunity for reducing energy, and particularly electricity, consumption in households lies in two key areas: improvement of equipment efficiency through implementation of stringent efficiency standards and labeling programs and tightening of thermal shell of residential buildings. Figure 8.9 shows the energy savings opportunity in the residential sectors distributed across end-uses, with appliances and space heating having the largest savings potential from efficiency policies. From 2005 to 2050, accelerated adoption and implementation of more aggressive efficiency policies such as strengthened appliance efficiency standards and expansion of the China Energy Label to reach near international best practice by 2020 could lead to total cumulative CO_2 emissions reduction of 18.4 $GtCO_2$ under the AIS.

5 COMMERCIAL BUILDINGS SECTOR FINDINGS

Energy demand in the commercial buildings sector is currently growing rapidly. However, growth will slow in the medium term, reaching a plateau by about 2030 due to declining growth in total commercial sector employees and commercial floorspace approaching saturation. Total commercial building floorspace may saturate in the short term, but end-use intensity has much room to grow before reaching current levels in industrialized countries. In particular, the lighting, office equipment, and other electrical plug loads in commercial buildings will grow dramatically through 2030, but then level off thereafter in CIS.

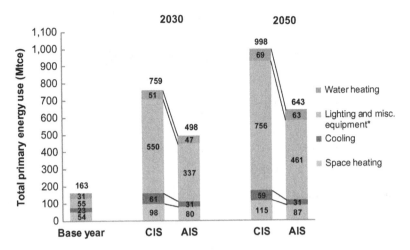

FIGURE 8.10 Commercial Primary Energy by End-Use. *Lighting and misc. equipment refers to electricity-consuming equipment such as computers, printers, audiovisual equipment, elevators, pumps, etc.*

The main dynamic of energy consumption in commercial buildings revealed by this study is that energy growth will be largely dominated by intensity increases, rather than overall increases in commercial floor area. The number of workers available to this sector will limit increases in commercial building space in China's future. Although the economic activity in this sector will continue to gain in significance, physical infrastructure will grow at a much slower rate than value-added GDP. In contrast, Chinese commercial energy use per square meter is still relatively low. For example, space cooling and equipment energy intensity is only a fraction of the current Japanese level. Thus, with rising commercial end-use energy intensity expected, building efficiency policies such as tighter building design codes, building efficiency and green building labeling programs, equipment efficiency standards, and phase-out of inefficient lighting will be important in controlling energy demand growth and CO_2 mitigation. Annual carbon mitigation under the AIS could reach 1180 $MtCO_2$ by 2030, or cumulative reduction of nearly 26 $GtCO_2$ emissions (Figure 8.10).

In final energy use terms, most of the electricity demand in commercial buildings will be driven by lighting and electric equipment (i.e., plug loads) usage and cooling to a smaller extent. As seen in Figure 8.11, electricity consumption by 2050 will be heavily dominated by lighting and equipment usage with much smaller shares of consumption by cooling, space heating, and water heating. Consequently, efficiency improvements in lighting and equipment will have the greatest electricity savings potential under the AIS, accounting for 90 percent of the 1044 TWh possible savings by 2050 followed by cooling efficiency improvements with 8 percent of savings.

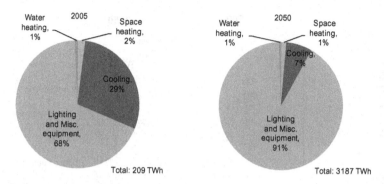

FIGURE 8.11   Commercial Electricity Consumption Shares by End-Use, 2005 (left) and 2050 in the CIS (right).

6 INDUSTRIAL SECTOR FINDINGS

Within industry, the energy consumption of the seven sectors singled out in China's long-term development plan for substantial energy efficiency improvements – cement, iron and steel, aluminum, ammonia, glass, paper, and ethylene – will gradually decline relative to other sectors, though still account for 47 percent of total energy consumption in 2050, down from 61 percent in 2005 in the CIS. Under both scenarios, a leveling in the output of cement and some chemicals is expected in the near term while others such as steel, aluminum, and glass production will increase at an annual average rate of 3 percent until 2020 and then start leveling off or declining. In the case of iron and steel and cement in particular, China's expected transition from rapid industrialization and infrastructure development to more intensive growth and expansion in the services sector after 2010 under-lies the slowdown and eventual decline in total iron and steel output and in the growth of the cement industry. Even so, China will remain the world's largest producer of these two commodities and will continue to produce enough steel and cement to meet domestic demand.

Among the industries in the "other industry" category, steady increases in energy consumption growth are expected from the refining sector, the coal mining and extraction sector, and the oil and gas exploration and production sector as the final energy intensiveness of fossil fuel extraction increases over time with declining energy return on energy invested as well as from manufacturing and other light industry.

Energy demand in China is currently dominated by a few energy-intensive sectors, particularly by the main construction inputs – cement and iron and steel. The recent explosion of construction in China has had a driv-ing role in these industries, and therefore Chinese energy demand as a whole. Because demand for energy-intensive industrial outputs such as cement, iron,

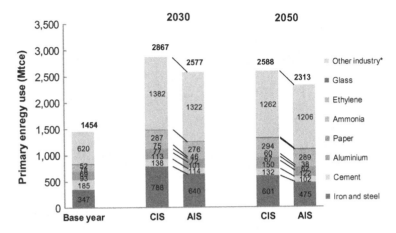

FIGURE 8.12 Industrial Primary Energy Use by Subsector. *Other Industry includes manufacturing, chemicals, light industry and all other small industrial subsectors.*

and steel, and aluminum are closely linked to construction, the slowing of this construction boom will therefore have a major impact as seen by the peaking of industrial primary energy use around 2030.

The energy use of each of these sub-sectors in absolute terms all decline modestly over time. The only exception is in energy use by the ethylene sub-sector, which grows notably from a 4 percent share of total industrial energy use in 2005 to 11 percent share in 2030 (Figure 8.12). The model results for projected CIS and AIS industrial energy use reflect key differences in only efficiency improvements, with a 290 Mtce reduction in energy use under the AIS scenario in 2030, and 274 Mtce in 2050. This translates into annual CO_2 emission reduction of 1550 $MtCO_2$ in 2050, or 40 percent of all emission reductions, and cumulative reduction of nearly 39 $GtCO_2$ by 2050.

Iron and steel and aluminum are the two single largest industries in terms of electricity consumption, although the "other industry" sub-sector representing manufacturing, chemicals, and other light industry consumes the most electricity in absolute terms (Figure 8.13). From 2010 through 2050, the relative share of electricity consumption by the iron and steel sub-sector will continue to grow with greater production from electric arc furnaces while the electricity consumption shares of cement and paper sub-sectors decline.

The more efficient AIS development trajectory has differing impacts on energy reduction in each of the seven industrial sub-sectors (Figure 8.14). Between 2005 and 2050, the iron and steel, other industry and cement sub-sectors comprise the largest energy reduction potential under both the CIS and AIS when compared to other sub-sectors. The ethylene sub-sector stands out as an exception with negligible energy savings under AIS, partly because energy consumption actually grows from 2010 through 2030 under both

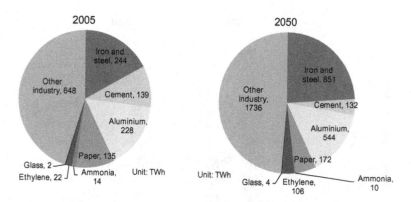

FIGURE 8.13   Industrial Electricity Consumption by Subsector, 2005 (left) and 2050 in the CIS (right).

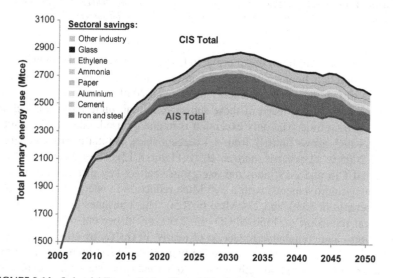

FIGURE 8.14   Industrial Energy Savings Potential by Subsector.

scenarios. Electricity savings through accelerated efficiency improvements under AIS will be realized in three key sub-sectors – iron and steel, cement, and aluminum. By 2050, each of these three sub-sectors will contribute nearly one-third of the total industrial electricity savings of 153 TWh under the AIS.

7 TRANSPORT SECTOR FINDINGS

The greatest growth for energy demand in the transport sector will be from passenger road transportation, with urban private car ownership expected to

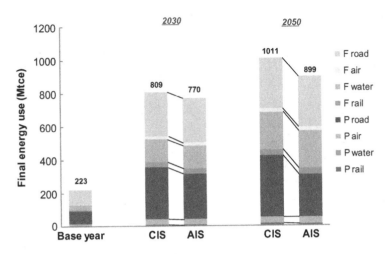

FIGURE 8.15 Transport Final Energy Consumption by Mode.

increase to over 356 million vehicles by 2050. Electrification of rail and vehicles will also increase electricity demand of the transport sector, while reducing gasoline and diesel demand. China has actively promoted transport electrification in recent years, setting specific targets for more efficient electric rail. In addition, subsidy programs have been launched to accelerate the market entry of electric vehicles and tap into their energy and environmental benefits (e.g., reducing heavy reliance on oil) and suitability for the predominantly short intra-city driving patterns in Chinese urban areas. Under CIS, total transport electricity demand will increase from 33 TWh in 2005 to 425 TWh in 2050, with a 65 percent share for electrified rail and 35 percent share for electric vehicles.

In primary energy terms, the impact of improved efficiency in motor vehicles and accelerated electrification of passenger cars and the national rail system will lower total transportation energy use in 2050 by 107 Mtce compared to the CIS (Figure 8.15). Increasing the 2050 proportion of electric cars from 30 percent in the CIS to 70 percent in the AIS reduces annual gasoline demand by 100 million tonnes of oil equivalent, but increases annual electricity demand by an additional 265 TWh in 2050 as a result of electric vehicles replacing more and more of gasoline cars and some hybrid cars over time. This produces the unintended result that China becomes a gasoline exporter, as demand for other oil products is not reduced commensurately.

Power decarbonization has important effects on the carbon mitigation potential of switching to electric cars technology. Greater transport electricity use under the AIS could result in net CO_2 reduction on the order of 5 to 10 $MtCO_2$ per year before 2030 and as much as 109 $MtCO_2$ by 2050 because

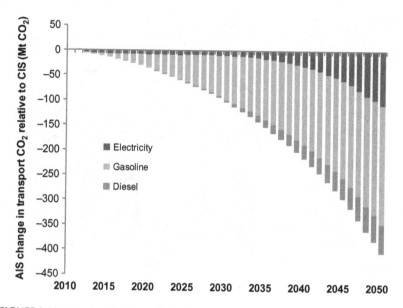

FIGURE 8.16   Transport CO_2 Emission Reduction under the AIS by Fuel Source.

the AIS power supply is less carbon intensive than CIS power supply (Figure 8.16). However, in the absence of any decarbonization in the power sector, electric vehicles will increase CO_2 emissions.

8 ELECTRICITY SECTOR FINDINGS

The electricity sector accounts for a large growing share of China's energy use and related carbon emissions. On the demand side, AIS results in 15 percent lower total electricity generation in 2050 than CIS. The net electricity reduction of 1285 TWh under AIS in 2050 includes a total reduction of 1540 TWh across the commercial, residential, and industrial sectors, offset by an increase of 255 TWh from electrification of the transport sector. The majority of the electricity savings will be from efficiency improvements in the commercial sector, followed by smaller savings in the residential and industrial sectors (Figure 8.17).

On the supply side, efficiency improvements and fuel substitutions bring the 2050 coal share of total electricity generation from 49 percent in the continued improvement scenario to 10 percent in the accelerated improvement scenario (Table 8.2).

Decarbonization also plays a significant role in carbon emission reduction in the electricity sector and substantially outweighs the potential impact of CCS. Besides the CIS and AIS scenarios of electricity sector development,

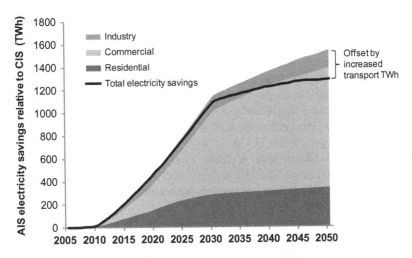

FIGURE 8.17 The AIS Net Total Electricity Savings and Savings Potential by End-Use Sector. *Note: Transport sector is not shown in this figure because it had negative savings (i.e., net increase) in electricity consumption under AIS. The net increase in transport electricity demand under AIS can be estimated by the difference between the sum of the residential, commercial and industry electricity savings and the total electricity savings.*

TABLE 8.2 Electricity Generation Shares by Technology, the CIS and AIS

	CIS			AIS	
	2005	2030	2050	2030	2050
Wind Power	0%	6%	13%	10%	16%
Nuclear Power	2%	13%	25%	19%	54%
NG Fired CC	1%	2%	2%	4%	2%
Hydropower	15%	12%	12%	17%	16%
Oil Fired Units	2%	0%	0%	0%	0%
Biomass and other Renew	0%	1%	1%	1%	1%
Solar	0%	1%	1%	1%	1%
Coal <100MW	21%	0%	0%	0%	0%
Coal 100−200 MW	11%	0%	0%	0%	0%
Coal 200−300 MW Subcritical Units	9%	0%	0%	0%	0%
Coal 300-600 MW Subcritical Units	35%	1%	0%	0%	0%
Coal 600−1000 MW Supercritical Units	2%	22%	8%	19%	0%
Coal >1000 MW Ultra Supercritical Units	0%	44%	41%	30%	9%
Total Electricity Generation (TWh)	*2620*	*7830*	*9100*	*6560*	*7760*

an additional scenario was added to represent the implementation of CCS to capture 500 MtCO$_2$ by 2050 under the CIS pathway of efficiency improvement and fuel shifting. Of the three scenarios, AIS requires the least primary energy and produces significantly lower energy-related power sector carbon dioxide emissions than either the CIS or CCS. In fact, AIS electricity sector emissions peak just below 3 GtCO$_2$ in 2019 and begin declining rapidly thereafter to 0.6 billion tons in 2050. The CCS base scenario results in 476 Mt fewer emissions in 2050 than the CIS scenario with a 1.4 percent increase in the total primary energy requirement for carbon capture, pumping, and sequestration.

Within the electricity sector, the greatest carbon emissions mitigation potential under AIS is from direct electricity demand reduction as a result of more aggressive policy-driven end-use efficiency improvements in the industrial, residential, commercial, and transport sectors. Figure 8.18 illustrates five wedges that lead to electricity sector emissions reductions of almost 3.5 GtCO$_2$ per year by 2030, where the solid wedges represent CO$_2$ savings from various electricity sector changes and the stripped wedge represents CO$_2$ savings from electricity demand reduction. One of the largest electricity sector mitigation potentials is from end-use efficiency improvements that lower final electricity demand and the related CO$_2$ emissions, which is about half of total CO$_2$ savings before 2030 and then one-third of total CO$_2$ savings by 2050. These results emphasize the significant role that energy efficiency improvements across all end-use sectors (see sections 4 through 7) will continue to play in carbon mitigation in the electricity sector (vis-à-vis lowering electricity demand), as efficiency improvements and can actually outweigh CO$_2$ savings from decarbonized

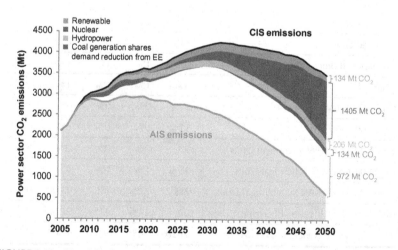

FIGURE 8.18   Electricity Sector CO$_2$ Emissions Reduction by Source in the AIS.

electricity supply through greater renewable and non-fossil fuel generation prior to 2030.

9 SENSITIVITY ANALYSES

Sensitivity analyses of drivers in the key economic sectors were conducted to evaluate uncertainties that exist in the model. In each sensitivity analysis scenario, a specific variable such as the urbanization level was tested for its impact on total primary energy use under the CIS, *ceteris paribus*. The results of the sensitivity parameters tested that had the highest level of uncertainties with changes of at least 300 Mtce (or 5 percent of total primary energy use) in 2050 are presented in Figure 8.19.

Among the different sensitivity analysis scenarios tested, variables in the industrial sector had the largest impact on total primary energy use, implying that there is a higher level of uncertainty surrounding these variables. For example, a 25 percent increase in the growth rate of other industry GDP that directly affects steel production for use in manufacturing can result in an increase of nearly 800 Mtce in total primary energy use by 2050. Likewise, uncertainties in the levels of heavy industrial output and in the energy intensity of "other industry" can result in changes in total primary energy use in the range of 300 to 700 Mtce in 2050.

The other variables included in the sensitivity analyses that resulted in medium (impact of greater than 50 Mtce or 1 percent) or low (impact of less than 50 Mtce) uncertainties are outlined in Table 8.3.

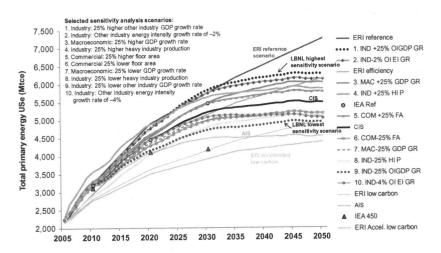

FIGURE 8.19 Sensitivity Analysis Scenario Results with Greatest Uncertainty.

TABLE 8.3 Parameters of Medium- and Low-Impact Sensitivity Analyses

Sensitivity Scenario Name	Sensitivity Scenario Description	Sensitivity Impact
MAC 67% Urban	Macroeconomic: 67% urbanization by 2050	Medium
RES +25% FA	Residential: 25% more floor area per capita	Medium
COM 25 Life	Commercial: 25 years building lifetime	Medium
COM 50 Life	Commercial: 50 years building lifetime	Medium
COM +25% LOI	Commercial: 25% higher lighting & other end-use intensity	Medium
COM −25% LOI	Commercial: 25% lower lighting & other end-use intensity	Medium
TRA 40% EV AIS*	Transport relative to the AIS*: 40% Electric Vehicle share of cars by 2050	Low
TRA 20% EV CIS	Transport: 20% Electric Vehicle Share of cars by 2050	Low
TRA −25% OFA	Transport: 25% lower Ocean Freight Activity	Low
IND 60% EAF	Industry: 60% EAF furnace penetration in steel production by 2050	Medium
IND 25% EAF	Industry: 25% EAF furnace penetration in steel production by 2050	Low

10 CONCLUSIONS

As China continues to pursue its social development goals, demand for energy services will continue to grow, presenting fundamental challenges as economic growth and projected rapid urbanization will significantly drive up energy demand and CO_2 emissions if there are not changes in energy efficiency and energy supply structure. This study evaluated how China can maintain its development trajectory and provide basic wealth to it citizens while being energy-sustaining, assessed the role of energy-efficiency for transitioning China's economy to a lower greenhouse gas trajectory, and evaluated China's long-term domestic energy supply in order to gauge the potential challenge China may face in meeting long-term demand.

By 2050, primary energy consumption will rise continuously in both scenarios but reach a plateau around 2040, with a cumulative energy reduction of 26 billion tons of coal equivalent under the AIS from 2005 to 2050. The plateau is reached in both scenarios due to saturation effects (e.g., most households will possess all major electricity-consuming appliances by 2030 and electrical plug loads in commercial buildings will level off around 2030) and efficiency policy effects (e.g., strengthened appliance and equipment efficiency standards). Future energy demand reduction potential is greatest in the industry sector in the earlier years and from the buildings sector in the long run. In terms of electricity consumption, the greatest potential for reduction under AIS is in the commercial sector, followed by the residential and industrial sectors.

CO_2 emissions under both scenarios could experience a plateau or peak around 2030, with AIS peaking slightly earlier at 9.7 $GtCO_2$ as a result of more aggressive energy efficiency improvement and faster decarbonization of the electricity supply, coupled with the electrification of passenger cars and the national rail system. The single largest end-use sector emission reduction potential could be seen in the buildings sector, particularly commercial buildings, followed by the industry sector. Further reduction of CO_2 under these scenario assumptions would require even higher levels of non-carbon-emitting electricity. The total national emissions mitigation potential of moving from a CIS to AIS trajectory of development is 3.8 $GtCO_2$ in 2050 with the power sector having the greatest mitigation potential.

Both the CIS and AIS scenarios demonstrate that with continuous improvement, the goal of 40 percent carbon intensity reduction by 2020 announced in 2009 is possible, but will require strengthening or expansion of energy efficiency policies in industry, buildings, appliances, and motor vehicles, as well as further expansion of renewable and nuclear power capacity. These results emphasize the significant role that energy efficiency policies and subsequent improvements will continue to play in decreasing the growth of energy demand and leading China on a lower carbon development pathway. The crucial impact of energy efficiency improvements on carbon mitigation is most readily apparent in the power sector (vis-à-vis lowering electricity demand), as efficiency improvements and can actually outweigh CO_2 savings from decarbonized power supply through greater renewable and non-fossil fuel generation prior to 2030.

REFERENCES

[1] Zhou N, Fridley D, McNeil M, Zheng N, Ke J, Levine MD. China's energy and carbon Emissions Outlook to 2050. Berkeley, CA: Lawrence Berkeley National Laboratory; <http://china.lbl.gov/publications/Energy-and-Carbon-Emissions-Outlook-of-China-in-2050;>; 2011.

[2] China Energy Research Institute (ERI). 2050 china energy and CO_2 emissions report (CEACER). Beijing: Science Press; 2009 [in Chinese].

[3] National Bureau of Statistics. Various years. China statistical yearbooks. Beijing: China Statistics Press.

[4] International Energy Agency (IEA). World energy outlook 2009. Paris: OECD Publishing; 2009.

Chapter 9

Rapid Growth at What Cost? Impact of Energy Efficiency Policies in Developing Economies

Youngho Chang and Yanfei Li
Nanyang Technological University, Singapore

1 INTRODUCTION

While many chapters of this volume focus on policies and measures to reduce demand growth in mature economies with low demand growth, this chapter examines the case of a number of developing economies from the Association of Southeast Asian Nations (ASEAN). This regional block for economic and political cooperation includes Brunei, Cambodia, Indonesia, Laos, Malaysia, the Philippines, Singapore, Thailand, and Vietnam.[1] This part of the world is experiencing fast economic growth and industrialization. Figure 9.1 illustrates the geographical location of this economic block and the average economic growth rates of the economies from 2001 to 2010.

Table 9.1 shows that the GDP of the ASEAN grew at an average rate of 5.2 percent annually over the period of 2000−2009. Over the same period, world average annual growth rate was 3.5 percent. At the same time, energy consumption of ASEAN economies grew at 4.8 percent annually, slightly slower than GDP expansion. World average energy consumption grew annually at 2.2 percent during this period. Electricity consumption grew faster than GDP does for both ASEAN economies and world average. The former grew at 6.6 percent annually, while the latter grew at 3.1 percent annually.

To better understand at what stage the ASEAN economies are, in terms of economic growth and growth in energy consumption as compared to the rest of the world, three more benchmark economies from the Asia-Pacific region is introduced. They are China, South Korea, and Japan. China is a typical fast-developing economy. Korea is a developed economy but still growing fast both in terms of GDP and energy consumption. Japan is a mature

---

1. Singapore is the only industrialized economy in ASEAN. Brunei has high GDP per capita thanking to its affluent crude oil and natural gas reserves.

Energy Efficiency. DOI: http://dx.doi.org/10.1016/B978-0-12-397879-0.00009-8

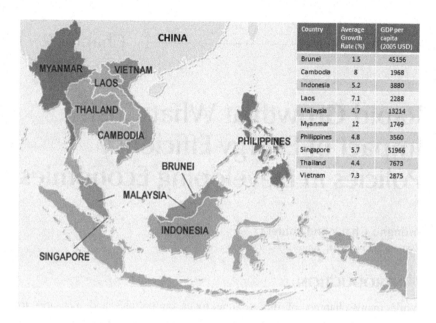

Country	Average Growth Rate (%)	GDP per capita (2005 USD)
Brunei	1.5	45156
Cambodia	8	1968
Indonesia	5.2	3880
Laos	7.1	2288
Malaysia	4.7	13214
Myanmar	12	1749
Philippines	4.8	3560
Singapore	5.7	51966
Thailand	4.4	7673
Vietnam	7.3	2875

FIGURE 9.1　Average Economic Growth Rates of ASEAN Economies from 2001 to 2010. *Source: Authors' estimation based on World Bank data.*

TABLE 9.1 Growth in GDP, Energy Consumption and Electricity Consumption – ASEAN compared to World Average in 2000–2009

	GDP	Energy Consumption	Electricity Consumption
ASEAN	5.2%	4.8%	6.6%
World Average	3.5%	2.2%	3.1%

industrialized economy and one of the most energy efficient economies in the world. It could serve as an example for ASEAN economies to follow.

Figure 9.2 provides an overview of the economic growth, energy consumption, and electricity consumption trends in these economies and compares them to the world average, during the period of 2000 to 2009. The case of China is covered in detail in a separate chapter of this volume.

Among the ASEAN plus three economies, only Brunei and Vietnam experienced faster growth in energy consumption than growth in GDP, which means lower energy efficiency for these two economies. All economies, except for Myanmar, Singapore, and Japan, experienced faster growth in electricity consumption than growth in GDP. This can be attributed to industrialization, shifting from agricultural sector to manufacturing and

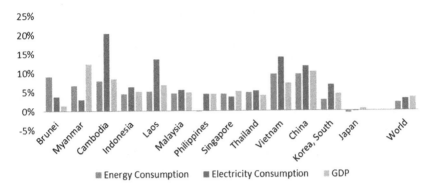

FIGURE 9.2   The Growth Rates of GDP and Electricity Consumption: 2000—2009. *Source: Authors' estimation based on World Bank and EIA data.*

commercial sectors, urbanization, as well as electrification substituting consumption of primary energy with consumption of electricity, all of which are taking place in these economies.

It is noted that world average growth in both energy consumption and electricity consumption are lower than that of GDP. ASEAN economies generally follow this trend in energy, but not in electricity consumption. In other words, growth in electricity consumption in ASEAN economies has been exceptionally high, which means it is reasonable to raise concern about the efficiency of electricity consumption.

In the rest of this chapter, section 2 analyzes the drivers of electricity consumption in the region. Section 3 estimates the extent of efficiency gap in electricity consumption, using Japan as a benchmark and provides the magnitude of electricity conservation potential in ASEAN economies by 2030. Section 4 discusses the details of possible policies and measures that ASEAN economies could adopt to achieve the efficiency potential in electricity consumption focusing on Singapore and Indonesia as case studies. Section 5 concludes the chapter's main findings.

2 DRIVERS OF ELECTRICITY CONSUMPTION IN ASEAN

Figure 9.3 shows the relative position of total electricity consumption levels of each country in the two groups as well as its changes over the period of 2000—2009. The empty box indicates a net increase in electricity consumption over the period. The figure shows that Indonesia, Malaysia, the Philippines, Singapore, Thailand, and Vietnam are the major contributors to growth in total electricity consumption in the region.

The data reported in Table 9.1, Figure 9.2, and Figure 9.3 suggests that electricity consumption in the ASEAN region is growing fast in terms of both growth rates and levels in kilowatt-hours (kWh). More importantly,

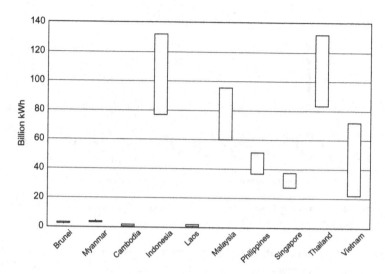

FIGURE 9.3   Trends of ASEAN Total Electricity Consumption in the Period of 2000–2009. *Source: Authors' estimation based on EIA data.*

the growth of electricity consumption in the region is generally faster than its growth of overall energy consumption, and much faster than the growth of GDP. The ensuing question is what the drivers of exceptionally high growth in electricity consumption are.

Electricity demand in the developing economies is primarily driven by GDP, prices (electricity tariffs), income, level and characteristics of economic activity, urbanization, and seasonal factors [1]. This chapter focuses on the two most important factors in determining electricity consumption in ASEAN economies, which are GDP and prices.

2.1 GDP and Electricity Consumption in ASEAN

It is noted in the empirical literature that the causality between electricity consumption and GDP appears to be bidirectional in the long run, but it goes from GDP to electricity consumption in the short run, and this is especially true in developing economies. These findings imply that developing economies usually have room for reducing electricity consumption without harming GDP in the short run [1].[2]

In the following figures, the historical electricity consumption of each ASEAN economy is presented together with GDP. Growth in GDP appears

---

2. The statement by Khanna and Rao [1] is based on literature review of around 30 empirical studies on the causal relation between economic growth and electricity consumption. We do note that Lee [2] finds that the causality runs from energy consumption to economic growth in both long run and short run in 18 developing economies investigated.

to move in tandem with growth in electricity consumption although with slightly different patterns in different economies.

Literature has provided insights into causality between the two for some of the ASEAN economies. For example, bidirectional causality between growth in GDP and growth in electricity consumption is found in the cases of Malaysia and Singapore, but uni-directional causality from growth in GDP to growth in electricity consumption in Indonesia and Thailand. These findings support the argument that growth in GDP is a key driver of growth in electricity consumption in the region. Meanwhile, they also imply the possibility of reducing electricity consumption without harming economic growth in the region. [3]

According to Figure 9.4 and Figure 9.5, with exception in the case of Myanmar, economies like Cambodia, Laos, and Vietnam, which are the low-income group of economies in ASEAN, tend to experience higher growth in electricity consumption than growth in GDP.

Economies like Indonesia, Malaysia, the Philippines, and Thailand could be treated as another group. These economies, which have higher GDP per capita as shown in Figure 9.1, tend to have electricity consumption and GDP growing at the same rate.

In the case of Singapore, with its GDP per capita so high that it is classified as an industrialized economy, the city-state generally experiences its

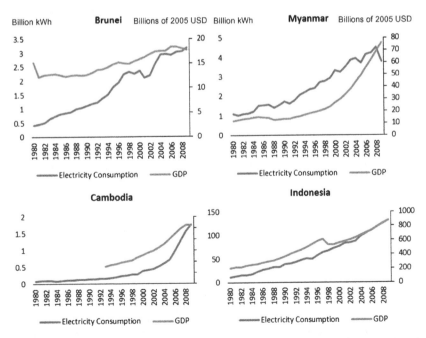

FIGURE 9.4   Electricity Consumption versus GDP of ASEAN Economies. *Source: EIA and World Bank data.*

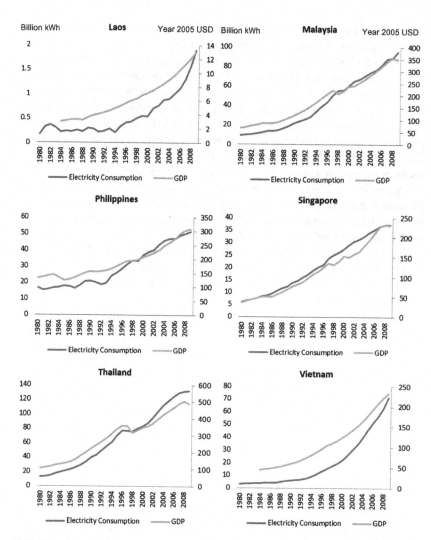

FIGURE 9.5   Electricity Consumption versus GDP of ASEAN Economies (Continued). *Source: EIA and World Bank data*

electricity consumption growing slower than GDP. Brunei is an exceptional case.[3]

---

3. The economy of Brunei has been dominated by the oil and gas sector, relying heavily on the export of the two. This feature renders the GDP of the economy vulnerable to international oil and gas prices. Its GDP peaked at 1980 because of the oil price increases of the 1970s, and then followed by a recession which lasted six years due to sharply lower oil prices. Since 1986, the economy started to pick up and it was until 1996 that its real GDP recovered to its 1980 level. Meanwhile, data show that its population has almost doubled by 2009 from the 1980 level, contributing to the decline in its real GDP per capita.

The above observations show that strong growth in GDP cannot be denied as a key driver to the growth of electricity consumption in ASEAN. Indeed, the figures show that strong and positive relation exists between growth in electricity consumption and growth in GDP, which imply positive income elasticity of electricity consumption.

Income elasticity is defined as the growth in electricity consumption in response to one percent growth in GDP. The income elasticity varies in different economies. Economies with high elasticity have more increases in electricity consumption over increases in GDP than economies with lower elasticity.

Appendix A gives estimates of the income elasticity of electricity consumption in ASEAN economies, which support the findings that there exist three groups among the ASEAN economies. The low-income group economies such as Cambodia, Laos, and Vietnam have income elasticity close to 2. The middle-income group economies such as Indonesia, Malaysia, the Philippines, and Thailand have lower income elasticity but still greater than 1. The high-income economy like Singapore has income elasticity less than 1.

It should be noted that the high- or low-income elasticity could be partially attributed to energy efficiency, and partially to the level of economic development and structure of the economy. That most ASEAN economies have relatively high income elasticity is partially due to low efficiency in electricity use and partially due to the fact that they are experiencing fast urbanization, fast electrification, and perhaps fast expansion of electricity-intensive industries.

2.2 Electricity Tariff and Electricity Consumption in ASEAN

It has been found in literature that electricity consumption unilaterally decreases in response to increases in electricity tariffs – negative price elasticity of electricity demand (see [1]). The finding is the theoretical basis for demand response (DR) schemes, which are being implemented in many developed economies such as the United States in pursuing the reduction of electricity consumption and is discussed in detail in other chapters of this volume including the chapter on PJM.

Table 9.2 and Table 9.3 list electricity tariffs of all ASEAN economies in 2011 for three different sectors – residential, commercial, and industry. The following observations could be made.

First, Brunei sets its electricity tariff to decrease as the amounts of electricity consumed increase – a typical declining block tariffs scheme.[4] It is therefore not surprising that Brunei's electricity consumption grows faster than its GDP. Moreover, the average tariff paid by the residential sector is

---

4. Many jurisdictions, such as California, have increasing block tariffs, which means higher consumption levels are charged at higher tariffs, to discourage high consumption.

TABLE 9.2 Electricity Tariff in ASEAN Economies in 2011

Country	Residential Tariff (USD/kWh)	Commercial Tariff (USD/kWh)	Industrial Tariff (USD/kWh)
Brunei	<10 kWh = > $0.195 <60 kWh = > $0.117 <100 kWh = > $0.078 >100 kWh = > $0.039 Average: $0.047	<10 kVA = > $0.156 <100 kVA = > $0.055 <200 kVA = > $0.047 >200 kVA = > $0.039 Average: $0.043	Average: $0.039
Cambodia	<50 kWh = > $0.096 <100 kWh = > $0.149 >100 kWh = > $0.176	Small Business = > Average tariff + $0.036 Medium Business = > Average tariff + $0.028 Large Business = > Average tariff + $0.024	Small Business = > Average tariff + $0.036 Medium Business = > Average tariff + $0.028 Large Business = > Average tariff + $0.024
Indonesia	<450 VA = > $0.043 <900 VA = > $0.063 <1300 VA = > $0.082 <2200 VA = > $0.083 3500–5500 VA = > $0.093 >6600 VA = > $0.138	<450 VA = > $0.056 <900 VA = > $0.066 <1300 VA = > $0.083 2200–5500 VA = > $0.094 >6600 VA = > $0.114	<450 VA = > $0.05 <900 VA = > $0.062 <1300 VA = > $0.08 <2200 VA = > $0.082 3500–14000 VA = > $0.095
Laos	<25 kWh = > $0.034 <150 kWh = > $0.04 >150 kWh = > $0.096	Low Volt 0.4 KV = > $0.104 Medium Volt 22 KV = > $0.088	Low Volt 0.4 KV = > $0.074 Medium Volt 22 KV = > $0.063
Malaysia	TNB = > $0.113 SESB = > $0.072 SESCO = > $0.10	TNB = > $0.096 SESB = > $0.109 SESCO = > $0.11	TNB = > $0.077 SESB = > $0.108 SESCO = > $0.102
Myanmar	$0.03	$0.06	$0.06
Philippines	Luzon = > $0.106 Visayas = > $0.092 Mindanao = > $0.067	N.A.	N.A.

Source: Estimated based on and adapted from http://talkenergy.wordpress.com/asean-electrical-tariff/.

TABLE 9.3 Electricity Tariff in ASEAN Economies in 2011 (Continued)

Country	Residential Tariff (USD/kWh)	Commercial Tariff (USD/kWh)	Industrial Tariff (USD/kWh)
Singapore	$0.201	High Tension Small (HTS) Peak $0.184 Off-peak $0.1131; High Tension Large (HTL) Peak $0.1828 Off-peak $0.113; Extra High Tension (EHT) Peak $0.1744 Off-peak $0.1116	High Tension Small (HTS) Peak $0.184 Off-peak $0.1131; High Tension Large (HTL) Peak $0.1828 Off-peak $0.113; Extra High Tension (EHT) Peak $0.1745 Off-peak $0.1116
Thailand	< 150 kWh = > $0.057 < 400 kWh = > $0.087 > 400 kWh = > $0.094	< 22 kV = > $0.054 22–33 kV = > $0.053 > 69 kV = > $0,053	N.A.
Vietnam	< 50 kWh = > $0.029 < 100 kWh = > $0.048 < 150 kWh = > $0.058 < 200 kWh = > $0.076 < 300 kWh = > $0.083 < 400 kWh = > $0.088 > 401 kWh = > $0.091	< 6 kV: Peak $0.153 Off-peak $0.089 Lower hours $0.051 < 22 kV: Peak $0.145 Off-peak $0.085 Lower hours $0.05 > 22 kV Peak $0.141 Off-peak $0.079 Lower hours $0.043	< 50 MVA: Peak $0.081 Off-peak $0.041 Lower hours $0.023 50–100 MVA: Peak $0.082 Off-peak $0.042 Lower hours $0.023 > 100 MVA: Peak $0.082 Off-peak $0.042 Lower hours $0.023

Source: Estimated based on and adapted from http://talkenergy.wordpress.com/asean-electrical-tariff/.

only $0.047/kWh, implying that the tariff system encourages residents to consume far more than 100 kWh per month.[5]

Second, Singapore has the highest electricity tariffs in the region, which is around $0.20/kWh.[6] It also has the highest efficiency in electricity use

---

5. In many European countries, by contrast, electricity tariffs are heavily taxed – rather than being subsidized.

6. Singapore electricity market has been deregulated since 2003 and electricity prices for most of the industry and commercial sector are determined in the deregulated electricity market. These tariffs are for residential consumers and light users in the industry and commercial sector.

among the ASEAN economies. The higher efficiency must be logically related to its high electricity tariffs. Generally speaking, electricity tariffs in Singapore for all sectors are about twice as high as tariffs in other ASEAN economies.

Third, many economies such as Indonesia, Laos, Malaysia, Myanmar, and Vietnam set electricity tariffs for the industry and commercial sector higher than that for the residential sector. This practice could again render a wrong incentive and incur higher costs to the industry and commercial sector. Considering the fact that the costs of supplying electricity to large industrial and commercial users could be lower than those to residential customers, this practice will further intensify the magnitude of inefficiency of electricity consumption.

The above observations show that electricity tariffs in most ASEAN economies, with the exception of Singapore, appear to give wrong incentives to consumers. They either encourage more consumption of electricity or discourage productive use of electricity. These facts about tariffs reflect the nature and structure of the power sector in these ASEAN economies. While Singapore has a liberalized power sector and electricity market, other economies do not. The power sector and electricity market in these economies are nationalized, vertically integrated, rigid, and serve social stability rather than cost minimization and energy efficiency in electricity use. Reforming the power sector and electricity market in these ASEAN economies will play a critical role to improve efficiency in electricity use.

2.3 Electric Productivity

An energy efficiency gap is an informative measure about how energy efficient an entity is and how much potential for conservation there is.[7] In this chapter, the electric productivity of an economy is applied to measure the efficiency in electricity consumption. The efficiency gap in electricity consumption is estimated as the difference in the electric productivity between economies. The definition of electric productivity is adopted from the one developed by the Rocky Mountain Institute [4]. Electric productivity is estimated as dollars of gross domestic product (GDP) per kWh consumed.

For ASEAN economies and the three reference economies namely China, Korea, and Japan, electricity consumption data from EIA and real GDP data

---

7. See detailed the discussion on energy efficiency gap in the first chapter by Sioshansi as well as the chapter by Alcott and Greenstone.

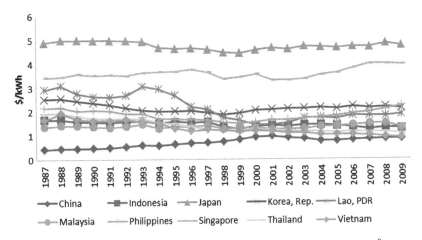

FIGURE 9.6 Electric Productivity of the ASEAN + 3 Economies from 1987 to 2009.[8] *Source: Authors' estimation based on EIA and World Bank data.*

from World Bank from 1987 to 2009 are used in estimating the electric productivity. Figure 9.6 presents the results.[9]

As Figure 9.6 shows, most ASEAN economies have electric productivity higher than that of China but lower than that of Korea. Japan has the highest electric productivity, implying high efficiency in electricity consumption. Singapore has the highest electric productivity among the ASEAN economies and is the second highest among all the economies in this sample.

Specifically, the Japanese economy produces about $5 per kWh of electricity consumed, Singapore about $4, Korea about $2, and China about $1. For all other ASEAN economies, it ranges between $1 and $2 for 2009.

The difference in the electric productivity is a direct measure of efficiency gap. Japan had arrived at about $5 per kWh as early as 1987 and stayed at this level afterwards. The question is, had ASEAN economies arrived at that level of economic development as Japan did in 1987, whether they could reach the same level of efficiency in electricity consumption of Japan as measured by electric productivity and narrow the efficiency gap.[10]

An optimistic answer is that the analysis of electric productivity presents the efficiency gap of ASEAN economies with Japan can be narrowed. There are three main reasons for the optimism. First, the level of the GDP per

---

8. The estimation of electric productivity has been adjusted for the share of electricity consumption in total primary energy consumption of ASEAN economies so as to remove the effect of low electrification rate in many of the ASEAN economies.

9. Brunei and Myanmar are not included in this sample due to the lack of valid data.

10. Although electric productivity of Singapore is high, the fact that it is a city-state without rural area and agriculture renders it inappropriate to be a benchmark of efficiency in electricity use for other ASEAN economies.

capita, urbanization, industrialization, and electrification that Japan has achieved in 1987 is expected for ASEAN economies to achieve in the next two decades, say, by 2030. Second, energy efficient technologies evolve all the time. By adopting existing as well as future energy efficient technologies, ASEAN economies are likely to fill much of the electricity efficiency gap with Japan's 1987 level. Third, most ASEAN economies, as shown in the case study of Indonesia, are yet to implement measures and policies to improve energy efficiency, including the efficiency in electricity consumption. The missing measures and policies imply there is significant room for improving the efficiency.

Economic development and evolution of technologies will do their part to contribute to narrowing the efficiency gap, but they are driven by forces beyond the scope of this discussion. Besides these two factors, the essential point of this study is to argue that had all ASEAN economies taken energy efficiency measures and policies as Japan did, the electric productivity of Japan is achievable. An estimation of how much electricity consumption could be reduced in 2030 is presented assuming that the efficiency gap could be narrowed. A detailed discussion on the specific measures and policies that ASEAN economies should adopt in order to narrow the efficiency gap follows the estimation.

3 PROJECTED ELECTRICITY CONSUMPTION AND CONSERVATION OF ASEAN ECONOMIES

The third ASEAN Energy Outlook (2009) [5] presents a projection on the electricity consumption of the ASEAN economies from 2010 to 2030 in a business as usual (BAU) scenario. The projection is taken as a benchmark scenario in this study. It serves to reflect the future electricity demand in these ASEAN economies if no radical measures or policies are taken to narrow the efficiency gap in electricity consumption. The projections are presented in Figure 9.7.

Figure 9.8 shows the projections of the electricity consumption in the scenario in which the efficiency gap is narrowed. The projection takes the same assumptions about projected GDP growth as those in the Third ASEAN Energy Outlook. Then it assumes that electric productivity of the ASEAN economies gradually increases and eventually arrives at the average electric productivity of Japan in the sample period (1987–2009), which is $4.75/kWh. The projected GDP is then translated into projected electricity consumption by applying the improved electric productivity.

Table 9.4 presents the potential in reducing electricity consumption in ASEAN economies by 2030 compared to BAU case. It is, of course, a hypothetical scenario, merely showing what the future could be if ASEAN economies were to evolve to the average efficiency level of Japan from 1987 to 2009.

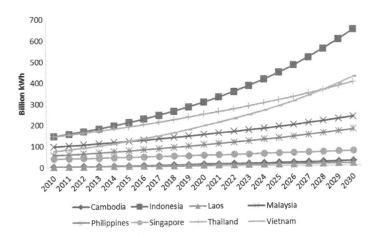

FIGURE 9.7   Projected Electricity Consumption of the Third ASEAN Energy Outlook (BAU Scenario). *Source: Adapted from the Third ASEAN Energy Outlook.*

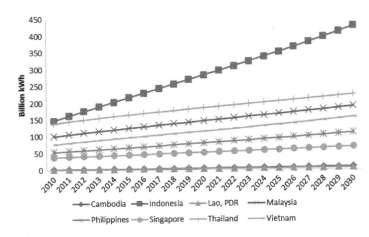

FIGURE 9.8   Projected Electricity Consumption with the Efficiency Gap Narrowed. *Source: Authors' estimation.*

Under this rather draconian scenario, the eight ASEAN economies can potentially reduce total electricity consumption by roughly 40 percent by 2030, equivalent to saving 782 billion kWh of electricity compared to BAU. It is worth noting that different economies would evolve along different paths with less-developed economies such as Cambodia, Laos, and Vietnam experiencing more than half of projected electricity consumption in the BAU scenario reduced. The reduction is mainly due to the fact that these economies are projected to grow at faster rates than others. However, if the

TABLE 9.4 Reduction of Electricity Consumption by 2030

	Conserved Electricity (Billion kWh)	Percentage of BAU Electricity Consumption Saved (percent)
Cambodia	14.2	63
Indonesia	216.6	34
Lao, PDR	6.0	52
Malaysia	43.3	19
Philippines	62.8	36
Singapore	2.1	31
Thailand	172.5	44
Vietnam	264.6	63
Total	782.2	40

Source: Authors' estimation.

efficiency gap is narrowed and electric productivity is improved, less growth in electricity consumption is required to support their fast-expanding GDP. Therefore, most of the growth in electricity consumption would be unnecessary and a certain amount of electricity consumption could be reduced in these economies.

4 MEASURES AND POLICIES TO NARROW THE EFFICIENCY GAP

The preceding discussion compared the BAU with a scenario in which the efficiency of ASEAN countries rises to match the average efficiency in electricity consumption of Japanese economy by 2030 – a rather ambitious assumption. This section describes, in broad terms, what sorts of policies may be required to achieve higher energy efficiency levels.

4.1 Measures and Policies towards Conservation

If the goal is to reduce electricity consumption by improving energy intensity, one must focus on two prime sectors of electricity demand – industrial and residential.

As described in the chapter by Levine et al, China has experienced surges in electricity consumption due to its fast economic growth during

1998−2007. It is found that the increasing scale of industrial activities is just one major factor driving the growth in electricity consumption. The second major factor is the structural change of industries, which has been moving from low energy-intensive to high energy-intensive industries as an inevitable process of industrialization. The third factor is the so-called "shift effect" in the energy consumption of industrial activities, which says electricity will see wider adoption to substitute other primary energy. Technological changes will offset growth in electricity consumption from the above sources to a certain extent [6]. This process that China has gone through will highly likely be followed by some ASEAN economies, such as Cambodia, Indonesia, Laos, Myanmar, the Philippines, and Vietnam, in the next decade or so.

Residential electricity consumption will also increase fast for the ASEAN economies that are experiencing high-income growth, which means higher living standard, and accelerated urbanization. It is also true that during this process, households will acquire electrical home appliances at exponential growth rates [7].

This section delves further into the key question of how to achieve the reduction of electricity consumption when the region is right in the position to go through the above processes that lead to massive increase in electricity consumption. Japan is one of the most energy efficient economies in the world. Knowing how Japan achieved its current electric productivity could suggest a path to achieve this.

For ASEAN economies to narrow the efficiency gap measured by electric productivity, the practices followed by Japanese government may be the best reference to policymakers of ASEAN economies. The key measures and policies that lead to the effective reduction of electricity consumption are legislation, incentive schemes, dissemination of technological know-how, management (operational improvement), and public awareness and information provision [7]. The following are the main success reasons Japan is efficient in electricity consumption.

First, formal legislation is fundamental, but the purposes of the law are well beyond imposing regulations to industries. The Japanese government had long realized the importance of reducing energy consumption. It has passed an Energy Conservation Law as early as 1979. The development procedure of the Energy Conservation Law served as a communication tool between the government and industries in the first place. Industry leaders first show conservation opportunities as well as best practices and then the government adopts them as guidelines and recommended practices. In this way, the law per se disseminates know-how to all companies in the industries.

Second, setting the correct incentives is critical. Improving energy efficiency usually requires additional investment and sometimes the financial burden incurred by such investment is the main barrier that makes industries hesitate. Incentive schemes should be provided by the government in the form of direct subsidies as well as tax alleviation to investments that aim at improving energy efficiency. Energy prices should reflect the true cost of energy to give end-users more incentive to conserve, and the energy market should be at work to allocate energy to the activities that create the most value-added. This implies the removal of fuel subsidies at the upstream of electricity supply chain and the removal of electricity subsidy at the down-stream. This is especially important to ASEAN economies as many of which provide subsidies to either fuel or electricity.

Third, energy efficient technologies and energy efficient operational improvements are equally important. Japanese government not only takes leadership in R&D of advanced and risky energy-related technologies but also provides technical services to SMEs, conducting energy audits and providing suggested solutions. In the meantime, the significant energy conservation opportunities that exist in operational processes should not be ignored. Japanese companies are enabled by their total quality control (TQC) system to achieve operational improvements and conserve energy. The implications to ASEAN economies are that policymakers need to find ways to help improving energy consumption management of industrial and commercial sector.

Fourth, energy efficiency labeling for home appliances and other con-sumer products is useful in both improving public awareness and disseminat-ing necessary information to consumers. Government could launch energy conservation campaigns from time to time. Improved public awareness on energy efficiency may even push industry associations to initiate voluntary energy efficiency programs to reinforce competitiveness of their products in this respect.

Table 9.5 summarizes implementable policies based on Japanese experi-ence. Broadly speaking, there are two major types of measures − one is *administrative* and the other is *market-oriented*. The administrative measures are based on legislation, and typically come with energy conservation law. The law would not only give obligations to industries, businesses, and indi-vidual consumers to conserve energy but also lay down the foundation to provide benchmarks, standards, guidance, subsidies, information, and techni-cal support. The market-oriented measures aim at providing incentives. Direct incentives include subsidies and tax concessions to investments by both industry and households and R&D activities to improve energy effi-ciency as well as removing subsidies and other distortions on fuel prices and electricity tariff to reflect the true costs of electricity. The indirect market-oriented means is to promote energy efficiency labeling, which induces com-panies to compete to provide energy efficient products to consumers. Successful examples include the Energy Star program of the United States and the Eco Points program of Japan [8,9].

To examine possible ways of implementing the measures and policies dis-cussed above in the ASEAN context, two specific cases, the developed econ-omy of Singapore and the rapidly developing case of Indonesia are reviewed.

TABLE 9.5 Recommendations on Implementable Policies

Means	Measures	Target Sectors	Actions
Administrative Means	Energy Conservation Law	– Manufacturing plants and business locations	Measures for energy consumption of machines and equipment
		– Transportation – Residential buildings and structures – Machinery and appliances	Energy audit, periodic report, and management
			Announce energy conservation criteria
			Providing information to the consuming public
Market-Oriented Means	Subsidies	Industry sector; households	Investments and R&D activities to improve energy efficiency
	Tax Incentives	Industry sector; households	Investments and R&D activities to improve energy efficiency
	Energy Prices and Electricity Tariff	All sectors and households	Removing subsidies to both fuel prices and electricity tariff, let energy prices reflect true costs
	Energy Efficiency Labeling	Home appliances, and other energy-consuming consumer products	Energy Star (U.S.); Eco Points (Japan)

4.2 Singapore Energy Efficiency Measures and Policies

The electric productivity of Singapore significantly picked up after 1999, as shown in Figure 9.6. The timing coincides with the release and implementation of major government legislation, policies, and measures aimed at improving energy efficiency of the economy.

Singapore started energy efficiency legislation in 1999 by passing the Environmental Protection and Management Act (EPMA). The Act implemented several energy efficiency standards such as the Mandatory Energy Labeling Scheme (MELS) and Minimum Energy Performance Standards (MEPS) for household appliances, and the Fuel Economy Labeling Scheme (FELS) for passenger cars and light goods vehicles.

In 2012, a new Energy Conservation Bill was passed by the Singapore Parliament. The Bill introduces mandatory energy management practices for large energy users in the industry and transport sectors. The Bill defines those that consume more than 15 GWh (gigawatt hour) of energy equivalent each year as large users and obliges these users to appoint an energy manager. Singapore government has intensively consulted large energy users in the industry and transport sectors on the energy management best practices to be mandated in the process of legislation formation.

Mandatory energy management in the industry sector turns out to be critical to improving efficiency in electricity consumption in Singapore. It is reported that the manufacturing industry in Singapore accounts for about 50 percent of the nation's electricity consumption. This is due to most manufacturing companies in Singapore consume energy in the form of electricity with the exception of the oil refining and petrochemical sub-sectors. This fact renders a major barrier to the adoption of energy efficient practices and technologies to the manufacturing processes, as there is fear of production disruptions. The most important factor to conquer this barrier is sufficient management support [10]. The latest Singapore legislation as mentioned above exactly tackles this barrier by making energy management compulsory.

Beyond legislation, Singapore's efforts on improving energy efficiency have been focused on raising awareness, building capabilities, and incentivizing the adoption of energy efficiency practices. For example, Singapore government provides no energy subsidies. Prices of energy, including wholesale electricity prices, are determined by energy markets to reflect the true cost.

To formally introduce incentives to improve energy efficiency, all Singapore government agencies have come together to establish an Energy Efficiency Programme Office (E2PO). The initiative is consisted of several energy efficiency programs targeting specifically at the private household sector, the industry sector, commercial buildings, and the transport sector. Table 9.8 in Appendix B provides a list of these incentive schemes and programs.

For Singapore, as a city-state without energy endowment, being efficient at not only energy consumption but other areas as well is the key to surviving and competing with other export-oriented economies. The Singapore experience in improving energy efficiency is successful as reflected in Figure 9.6. It re-emphasizes the importance of legislation, government incentive schemes, and awareness promotion. Along with having prices correct or unsubsidized, it shows the importance of understanding the ways that energy or electricity is consumed in each sector of the economy and then devising corresponding measures and policies.

4.3 Indonesia Energy Efficiency Measures and Policies

The Indonesia GDP growth rate increased from 5.7 percent in 2005 to 6.5 percent in 2011, and its population grew 1.1 percent in 2011, reaching more than 245 million. This altogether implies that GDP per capita of the economy increased by 5.4 percent in 2011. In such a fast-expanding economy, energy, including electricity consumption, inevitably increases fast.

TABLE 9.6 Paradigm Shift in Energy Policy of Indonesian Government and Targets

	Supply Side Management	Demand Side Management	Target
Demand	Lack of emphasis on energy efficiency: Inefficient sectoral energy use	Emphasis on energy conservation: Efficient sectoral energy demand	Reducing 33.85 percent of BAU energy consumption in 2025
Supply	Provide subsidized fossil fuel at any cost to meet the growing demand	Reduce reliance on fossil fuels	Reducing the share of fossil fuels to 75 percent of total energy supply
	Renewable energy development is not prioritized	Diversification and maximum utilization of renewable energy are promoted	Renewable energy to be 25 percent of total energy supply by 2025

To meet the fast-growing demand for energy and electricity, Indonesia government used to focus on supply-side management. The strategy was to meet energy demand through subsidies on fossil fuels regardless of the cost. Households, industrial, commercial, and transportation sectors therefore use energy in very wasteful manners.

More recently, the Indonesia government has adopted a paradigm shift towards demand-side management (DSM) (Table 9.6). The new policy regime emphasizes conservation and energy efficiency in all sectors by encouraging energy users to adopt more energy efficient behaviors and to start using more efficient technologies. The overall target is to reduce energy demand by 33.85 percent from the BAU scenario in 2025.[11]

Apart from promoting energy efficient behavior, the supply and use of renewable energy are prioritized and subsidized. The Indonesia government has adopted a strategy called "Vision 25/25," which outlined commitments to increase utilization of renewable energy to 25 percent by 2025.

Specifically in the household sector, the first step of energy efficiency measure is to improve energy awareness. The government has realized that the households should know the energy consumption of their home appliances such as air conditioners, heaters and water pumps, and electronic equipment. The household sector accounts for 40 percent of total national electricity consumption. Given the significant share, further energy efficiency measures are expected for this sector.

11. http://www.energyefficiencyindonesia.info/energy/indonesia

The industrial sector accounts for 44 percent of the total energy demand and 42 percent of total electricity consumption in the economy. It is noted that because the types of industry are very diverse, energy efficiency is highly dependent on the kinds of equipment and technology used for the production processes. To improve energy efficiency in this sector, emphasis has been given to two aspects: technology and management. First, encourage investment in energy efficient process technology. Second, implement energy audit and energy management to regulate and supervise the amount of energy consumed.

The commercial sector contributes to 4 percent of total national energy demand and 13 percent of total national electricity demand. Emphasis has been given to commercial buildings. Energy audit is required for the existing buildings, and Indonesian National Standards relating to energy conservation in buildings (lighting systems, air conditioning systems, and building envelopes) are applied to designing the new buildings.

The sector used to be monopolized by government-owned company, PLN, which controls 87 percent of power generation in the country and monopolizes power transmission and distribution networks. In the National Energy Management Blueprint, the Indonesia government has set agenda to introduce competition to the sector and to remove subsidies to electricity tariff. This move is also critical to improving efficiency in electricity use as it gives the right incentive to users to conserve and control costs.

The above-mentioned measures and policies introduced by the Indonesian government to improve energy efficiency focused on providing information to raise consumer awareness, energy auditing for better energy management, and power sector reform. The United Nations Environment Program (UNEP) asserts that there are four categories of barriers to energy efficiency: (1) lack of management support, (2) lack of information, (3) lack of financing, and (4) lack of policies and legislations [10]. In the case of Indonesia, the first two have been tackled to certain extent. However, legislations on energy efficiency are missing, and so are significant financial support schemes provided by the government [11]. The country has just begun its long journey to improve its energy efficiency. How effectively the released energy efficiency measures and policies achieve the intended goals remains to be seen.

5 CONCLUSIONS

Two key observations are made based on the preceding discussion:

- First, electricity consumption in the ASEAN region is growing faster than overall energy consumption and faster than GDP growth. This could be driven by three factors: growth due to increasing scale of industry activities, growth due to structural change of industries and shifting from low energy intensity industries to high energy intensity industries, and lastly the shift towards more electricity consumption to substitute primary energy consumption. The three factors are typical in fast-growing and industrializing economies like those of ASEAN.

- Second, ASEAN economies appear to fall into three groups at different stage of economic development. At a relatively low development level, like Cambodia, Laos, and Vietnam, economies tend to have higher growth of electricity consumption than the growth of GDP. At a middle level of development, like Indonesia, Malaysia, the Philippines, and Thailand, growth of electricity consumption appears to go hand-in-hand with GDP growth. And in economies with high development level, like Singapore, the growth of electricity consumption is slower than that of the GDP.

The price of electricity is equally critical in improving energy efficiency in ASEAN economies. ASEAN economies, except for Singapore, have low electricity tariffs and in a few cases declining block tariffs that deliver the wrong incentives. There is significant room for improvement in the efficiency of electricity consumption.

The region is wide and diverse, and some economies are already advanced and prices are cost-reflective, while others are lagging behind but rapidly developing. These economies have a lot of catching up to do if they are to achieve high levels of energy efficiency. The solution, in the end, is a combination of supportive policies and gradual adjustments in pricing. The case studies of Singapore and Indonesia present stark directions to improve electricity efficiency in ASEAN economies. The former shows that a systematic manner of promoting energy efficiency is the key to improve the efficiency. The latter sheds some light on where and how these ASEAN economies could start to improve.

APPENDIX A: INCOME ELASTICITY OF ELECTRICITY CONSUMPTION OF ASEAN ECONOMIES

To precisely measure how electricity consumption increases as response to economic growth in each country, the following univariate regression is run to obtain the income elasticity of electricity consumption:

$$lnE_t = C + \gamma * lnGDP_t + \mu_t$$

where E_t is total electricity consumption at time t, GDP_t is the GDP at time t, C is a constant, and μ_t is the residual. The estimated γ is the income elasticity of electricity consumption:

$$\gamma = \frac{\Delta E/E}{\Delta GDP/GDP}$$

If $\gamma > 1$, electricity consumption grows faster than economic growth, and if $\gamma < 1$, electricity consumption grows slower than economic growth.

Keeping in mind that economic structure, which is the combination of the agriculture, manufacturing, and service sectors in the overall economy, and that energy technologies evolve overtime, more recent data should be used in estimating the income elasticity of electricity consumption. In this study, the period of 1994–2008 is chosen. Table 9.7 presents the estimated elasticity of each country of the ASEAN + 3 economies.

The results are ordered from low elasticity to high elasticity. Myanmar appears to have extremely low elasticity, which may be attributed to its relatively low starting points in both electricity consumption and GDP as well as its exceptionally fast economic growth in the recent ten years, which averaged 11.6 percent per year.

Japan and Singapore have elasticity coefficients lower than 1 as one would expect. This is because both are industrialized and matured economy with high GDP per capita and high electricity consumption per capita. China appears to have lower elasticity as compared to Korea. This could be attributed to the fact that China experiences very strong economic growth in the period – close to 10 percent per year – and that China has made efforts during this period to bring itself down from an extremely high level of energy intensity.

TABLE 9.7 Income Elasticity of Electricity Consumption Estimated for Period 1994–2008

Country	Income Elasticity of Electricity Consumption
Myanmar	0.39
Japan	0.72
Singapore	0.90
China	1.07
Philippines	1.21
Malaysia	1.30
Korea, South	1.34
Thailand	1.48
Indonesia	1.72
Cambodia	1.80
Vietnam	1.94
Laos	1.95
Brunei	2.16

ASEAN economies, with exception in the cases of Myanmar and Singapore, all have higher elasticity greater than 1 and also greater than that of Japan and China. Except for Malaysia and the Philippines, these economies also have higher elasticity than Korea.

APPENDIX B: LIST OF RECENT SINGAPORE GOVERNMENT INCENTIVE SCHEMES AND PROGRAMS FOR ENERGY EFFICIENCY

TABLE 9.8 Recent Singapore Government Incentive Schemes and Programs for Energy Efficiency

Sector	Incentive/Program	Purpose
Household	10% Energy Challenge	Challenging households to reduce their energy use by 10% or more by practicing simple energy-saving habits
	The Energy SAVE Program	Providing solutions to existing residential buildings to 'modernized' and be more energy efficient
	Voluntary Agreement to Promote Energy Efficient Appliances	Labeling and giving support in kind to retailers and suppliers of household appliances which promote energy efficient appliances
Building	Green Mark Incentive Schemes	Labeling and subsidizing energy efficiency of buildings
	Building Retrofit Energy Efficiency Financing (BREEF) scheme	Financial support to energy efficiency of existing buildings
Industry	Design for Efficiency Scheme (DfE)	Encouraging energy efficiency in manufacturing development plans early in the design stage
	Energy Efficiency Improvement Assistance Scheme (EASe)	Funding detailed energy assessment performed by an Energy Services Company (ESCO)
	Grant for Energy Efficient Technologies (GREET)	Encouraging new and existing industrial facilities to invest in energy efficient equipment or technologies
	Singapore Certified Energy Manager (SCEM) Training Grant	Subsidizing the development the technical skills and competence for energy management

(Continued)

TABLE 9.8 Recent Singapore Government Incentive Schemes and Programs for Energy Efficiency—(cont.)

Sector	Incentive/Program	Purpose
Industry/ Commercial	Energy Efficiency National Partnership (EENP)	Introducing mandatory energy management requirements and recognizing excellent energy management practices
	ESCO (Energy Services Company) Accreditation Scheme	Accreditation to qualified ESCO
Public	Public Sector Taking the Lead in Environmental Sustainability (PSTLES)	Encourage Public sector agencies to implement energy savings projects and obtain long-term savings guarantees

REFERENCES

[1] Khanna M, Rao ND. Supply and demand of electricity in the developing world. Annual Reviews of Resource Economics 2009;1:567–95.

[2] Lee CC. Energy consumption and GDP in developing countries: a cointegrated panel analysis. Energy Economics 2005;27(3):415–27.

[3] Yoo SH. The causal relationship between electricity consumption and economic growth in the ASEAN countries. Energy Policy 2006;34:3573–82.

[4] Mims N, Bell M, Doig, S. Assessing the electric productivity gap and the U.S. efficiency opportunity. rocky mountain institute, January 2009. 2009.

[5] Institute of Energy Economics Japan, ASEAN centre for energy, and national ESSPA project team. The 3rd ASEAN energy outlook. 2011.

[6] Wang W, Mu H, Kang X, Song R, Ning Y. Changes in industrial electricity consumption in china from 1998 to 2007. Energy Policy 2010;38:3684–90.

[7] Thomson E, Chang Y, Lee JS. Energy conservation in East Asia. Singapore: World Scientific Publishing; 2011.

[8] Agency for Natural Resources and Energy, and Ministry of Economy, Trade and Industry, Energy conservation policies & measures in Japan. 2009.

[9] Ishihara A. Development of energy conservation policy, law, and management concept in Japan. Japan: The Energy Conservation Centre; 2008.

[10] Hin CK, Yeo C. Energy efficiency in Singapore's industrial sector. ESI Bulletin 2010;3(1): June 2010. http://esi.nus.edu.sg/publications/2012/02/02/energy-efficiency-in-singapore-s-industrial-sector

[11] Asia Pacific Energy Research Centre, APEC energy overview 2010. Tokyo, Japan. 2011.

Part III

Case Studies of Low-Energy Communities and Projects

Chapter 10

The Prospect of Zero Net Energy Buildings in the United States

Nicholas B. Rajkovich[1], William C. Miller[2] and Roland J. Risser[3]

[1]*PhD Candidate, University of Michigan, Urban and Regional Planning Program,* [2]*Program Manager, Lawrence Berkeley National Laboratory, Environmental Energy Technologies Division,* [3]*Director, U.S Department of Energy, Building Technologies Office*

1 INTRODUCTION

In the United States, a number of efforts are underway to achieve zero net energy (ZNE) in the built environment [1]. ZNE is generally defined as a highly energy efficient building or community where the balance of energy needs is supplied by renewable energy technologies.

Frequently described as the "Apollo" project for energy efficiency and renewable energy programs, ZNE buildings utilize cutting edge design techniques and technologies to exceed building energy codes by more than 60 percent [2]. This represents a paradigm shift in the design, construction, and operation of the built environment, implying an eventual end to electric load growth in buildings. While no national mandate to achieve ZNE currently exists, federal legislation (see section 3.1 of this chapter) has authorized initiatives to develop and disseminate technologies, practices, and policies to move the U.S. market toward ZNE.

This chapter is the first of four in this volume that discuss ZNE. LaRue, Cole, and Turnbull describe efforts by investor owned utilities in California to achieve ZNE in the residential and commercial building markets. Wheeler and Segar provide a case study of a residential community recently completed for the University of California, Davis. Finally, Anderson, Hauser, and Mooney discuss trends in the integration of energy efficiency and renewable energy systems by detailing achievements in one building type, the ZNE home.

This chapter broadly examines the push to ZNE. Section 2 begins with an overview of efforts to achieve a "steady state economy," and discusses how this concept and sustainable development influenced definitions of ZNE. Section 3 discusses the various organizations in the United States that are promoting a shift to ZNE. Section 4 discusses how the existing regulatory frameworks of

Energy Efficiency. DOI: http://dx.doi.org/10.1016/B978-0-12-397879-0.00010-4

energy utilities, in the United States an important way to implement ZNE, may impede progress to ZNE, and recommends changes to the rate case process to support experimental program design and rapid deployment of energy saving technologies. The final section of the chapter speculates what the potential impact of ZNE might be on the design, construction, and operation of the built environment. Although this chapter focuses exclusively on efforts in the United States, this is not to imply that the topic is limited to North America; several other countries are currently pursuing ZNE at a variety of scales.

2 FROM A STEADY STATE ECONOMY TO ZERO NET ENERGY

In the late 1960s, concerns about ecosystem degradation, overpopulation, and consumption dominated the economics, population biology, and popular literatures. Responding to an image from the NASA Apollo program showing Earth as a "fragile" blue dot in the blackness of outer space [3], economists such as Boulding and Daly reframed the global economy as a closed system using the "spaceship" as their metaphor. In this sealed and fragile system, natural resources like the atmosphere provided invaluable services; their degradation would lead to global collapse [4].

Under Boulding's definition of a closed system, energy was a necessary input to convert matter, such as ores, from their natural state into a usable form. Renewable energy was preferable because it was continually replenished by inputs from the sun. Energy derived from fossil fuels represented the depletion of a "capital" stock of stored-up sunshine. Therefore, the combustion of fossil fuels increased the entropy of the environment and the economy; a transition to a low- or no-entropy system was necessary to indefinitely sustain future generations.

Daly [5] argued that the growth paradigm had outlived its usefulness. To him, "The world is finite, the ecosystem is a steady state. The human economy is a subsystem of the steady-state ecosystem. Therefore at some level and over time period the subsystem must also become a steady state, at least in its physical dimensions of people and physical wealth." The steady-state economy was, "an economy in which the total population and the total stock of physical wealth are maintained constant at some desired levels by a "minimal" rate of maintenance." Under Daly's model, quotas on resource consumption would be set by the government. By reducing the number of quotas issued each year, and learning to "live off of our solar income," we would achieve a quality of life that was sustainable[1] .

---

1. Echoing these themes, Peevey argues in this volume that California has already achieved zero per capita growth in energy; he believes that a statewide push to ZNE will help stabilize future energy expenditures and help to stabilize the global climate. Faruqui states that the *Limits to Growth*, another influential text related to the zero growth economy, heavily influenced energy efficiency policy in the 1970s.

It was in this closed system, limited growth, and conservation mindset that the World Commission on Environment and Development (WCED) drafted its definition of sustainable development in 1987 in a report entitled *Our Common Future*. Defined as, "development that meets the needs of the present without compromising the ability of future generations to meet their own needs," the WCED report encouraged countries to promote economic development while protecting environmental systems [6]. Under their model, energy systems would co-evolve with economic growth, and progressive transitions to cleaner energy sources and renewable energy would occur as affluence increased and economies could afford better pollution control or emerging technologies.

The WCED definition of sustainable development is the foundation for much of the United Nations energy policy. In 1997, recognizing the impact of energy production on greenhouse gas emissions, 37 industrialized countries and Europe signed the Kyoto Protocol to voluntarily reduce their greenhouse gas emissions by 5 percent below 1990 levels [7]. Using three market-based mechanisms – an emissions trading market, the clean development mechanism (CDM), and joint implementation (JI) – the goal was to allow developing countries to continue to expand their economies while industrialized countries invested in clean technologies to reduce carbon emissions.

While *Our Common Future* and the Kyoto Protocol outlined what appeared to be a politically palatable solution, their lack of specificity has made it difficult to implement greenhouse gas emissions reductions at the local level. To actualize sustainable development, a number of organizations have attempted to simplify the concept of sustainable development into something for practitioners, such as architects, engineers, and planners to implement. These initiatives are discussed in detail in section 3 of this chapter.

ZNE is a conceptual descendant of Daly's steady state economy. Eliminating per capita growth in energy usage (and the associated greenhouse gas emissions), and relying heavily on renewable energy are first steps toward stabilization of the Earth's atmosphere. Within Daly's framework, ZNE buildings can be seen as a microcosm of society's transition to a low carbon future. Representing small steady-state systems themselves, they balance the amount of energy used with the amount of energy received from renewable sources. Because building energy use consumed 28.8 percent of the total primary energy used in the United States in 2010 [8], efforts to reduce energy use from the built environment have the potential for significant energy savings and greenhouse gas emissions reductions. As developing countries continue to urbanize, the need to limit energy demand and reduce greenhouse gas emissions will only increase in importance.

2.1 Definitions of Zero Net Energy

Torcellini et al. [9] broadly define ZNE buildings as *residential or commercial buildings with greatly reduced energy needs through efficiency gains such that the balance of energy needs can be supplied with renewable technologies.* In their paper, they attempt to detail all of the possible definitions of ZNE, showing how issues of scale, different types of renewable energy, and accessing renewables from various locations may affect energy usage and associated greenhouse gas emissions (Figure 10.1).

To illustrate the possible configurations of ZNE, their paper uses a sample of six low-energy buildings[2] to illustrate how ZNE is framed in each project, why clear and measurable definitions are needed, and how we had progressed toward the goal of ZNE in the United States by 2006. Four ways to define a ZNE building are discussed [9]:

- *Zero Net Site Energy* – A site ZNE building produces at least as much energy as it uses in a year, when accounted for at the site. Measurement of energy use is typically done at the electricity or gas meters. This is the most common accounting method used for ZNE buildings in the United States.
- *Zero Net Source Energy* – A source ZNE building produces at least as much energy as it uses in a year, when accounted for at the generation source. Source energy refers to the primary energy used to generate and deliver the energy to the site. To calculate a building's total source energy, imported and exported energy measured at the meter is then multiplied by an appropriate site-to-source conversion factor; these values vary by region and utility. Using source accounting is less common in the United States, though several organizations have advocated that this is the only way to account for upstream and downstream impacts of energy production and consumption.
- *Zero Net Energy Costs* – In a cost ZNE building, the amount of money the utility pays the building owner for the energy the building exports to the grid is at least equal to the amount the owner pays the utility for the energy services and energy used over the year. In states with net metering for renewable energy, this is a common approach because utilities are often only required to zero out utility bills.
- *Zero Net Energy Emissions* – A ZNE emissions building produces at least as much emissions-free renewable energy as it uses from emissions-producing energy sources. For example, renewable energy is used to offset carbon dioxide emissions from natural gas furnaces or other appliances.

In addition to the differences in accounting these definitions imply, there are also several ways to supply renewable energy to the building, as listed in Table 10.1.

---

2. The six buildings discussed are the Visitor Center at Zion National Park; the National Renewable Energy Laboratory Thermal Test Facility, the Chesapeake Bay Foundation's Merrill Center, the BigHorn Home Improvement Center; the Cambria DEP Office Building; and the Oberlin College Adam Joseph Lewis Center. Details of each building are discussed in reports issued by the National Renewable Energy Laboratory (e.g., [10,11]).

TABLE 10.1 ZNE Renewable Energy Supply Options

Location	Description	Technology Examples	Example Projects
Prerequisite to installation of renewable energy system (i.e., energy efficiency as a fuel source)	Reduce site energy use through demand side management technologies or operational improvements	Daylighting, solar hot water heating, high-efficiency lighting, high-efficiency HVAC equipment, natural ventilation, etc.	All ZNE community and building projects generally maximize energy efficiency to reduce renewable energy requirements and to control first cost
On-site renewable energy system	Use renewable energy sources within the building footprint	Building integrated photovoltaics, building mounted wind turbines, biomass pellet furnace, etc.	Aldo Leopold Legacy Center [12]
	Use renewable energy sources within the project or community site	Photovoltaics, wind turbines, biomass, hydroelectric, biogas digesters, etc. Definitions of renewable energy vary state to state, for example California excludes hydroelectric facilities above 30MW.	Adam Joseph Lewis Center for Environmental Studies at Oberlin College [13,14]
Off-site renewable energy system	Use renewable energy sources off-site (owned by project owner)		Cornell University Lake Source Cooling Project and Hydroelectric Facilities [15,16]
	Use renewable energy sources off-site (owned by others)		Renewable energy certificates (RECs) and/or power purchase agreements (PPAs)

Note: Table format adapted from [9].

A number of organizations in the United States have adopted their own definition of ZNE. The U.S. Department of Energy (DOE) Building Technologies Multi-Year Program Plan was completed in August 2005 and updated in January 2007. The plan describes the planned research, development, and demonstration activities for building technologies through 2012 and represents peer reviews, rigorous internal evaluations, as well as examines key opportunities offered by DOE external partners. The Plan is a living document, and is a part of a continuous planning process.

The Building Technologies Multi-Year Program Plan defined a ZNE building as "a residential or commercial building with greatly reduced needs for energy through efficiency gains, with the balance of energy needs supplied by renewable technologies [48]." The focus on deep energy efficiency gains is intentional; most buildings in the United States are far below their potential for efficiency. Focusing on energy efficiency also supports existing efforts such as technology and software development and allows for future integration of renewable energy technologies. This approach is called the "ZNE-ready" building by DOE, and is the current concept promoted by the Building Technologies Program.

The idea behind a ZNE-ready building is that if residences and buildings are designed and constructed to optimize their energy use while addressing occupant comfort and safety, the owner has the option in the future to offset the remaining energy loads with a renewable energy system. By definition, these homes and buildings are "ready" for future ZNE performance. This is because it is critical to lock in energy efficiency measures that cannot be easily upgraded once construction is complete, such as wall cavity insulation or building orientation. This also helps to ensure maximum energy performance before investing in renewable energy systems; these systems continue to decrease in cost as technologies and manufacturing processes improve.

FIGURE 10.1 The NREL Research Support Facility (RSF) in Golden, Colorado is a recently completed ZNE building. This aerial photo shows the extensive amount of photovoltaics required to offset the energy use of a multistory commercial building. For more information about the RSF, see http://www.nrel.gov/sustainable_nrel/rsf.html. *Photo Credit: Dennis Schroeder, DOE/ NREL.*

In addition to promoting building energy efficiency, DOE also currently supports a collaborative effort, called Sunshot, which will enable widespread large-scale adoption of solar energy by making solar energy systems cost competitive with other forms of energy by 2020. The Sunshot Initiative aims to reduce the total installed cost of solar energy systems by 75 percent (to about $0.06 per kilowatt-hour). This will enable solar-generated power to account for 15 to 18 percent of America's generation by 2030, including solar systems on homes and buildings. This initiative will be accomplished by reducing solar technology costs, reducing grid integration costs, and accelerating solar deployment. When combined with energy efficient buildings under construction today, we may see significant increases in the number of ZNE buildings in the future.

In an article entitled "Getting to Zero: The Frontier of Low-Energy Buildings," Malin and Boehland [17] describe several of the definitions of ZNE buildings as they are listed above, but extend their discussion beyond the building site. They argue that transportation energy and variations in the local transmission and distribution infrastructure are key considerations and should be considered as part of a comprehensive plan to reduce energy use and greenhouse gas emissions.

To date, ZNE building projects usually only focus on the design and operations of the building itself, ignoring the building's effect on transportation demand. However, a building located within walking distance of other amenities or along public transportation routes can be reached with minimal transportation energy, reducing total greenhouse gas emissions associated with the project. The amount of energy used for transportation to and from a building can exceed the amount of energy used in the building, especially in low-density cities or places with moderate heating and cooling loads.

If a project is attempting to reduce energy use and greenhouse gas emissions from all sources transportation becomes an important consideration. Two notable projects where transportation energy use was considered as part of the design are University of California, Davis West Village [18] and the Aldo Leopold Legacy Center [12]. With additional electrification of the passenger vehicle fleet, seen with the introduction of vehicles like the Chevrolet Volt, personal vehicle travel may be considered a plug load associated with a home or business in the very near future.

Second, emissions from energy generation are not uniform and constant but vary by location, season, and time of day. Utility companies often engage inefficient, more polluting generators to meet demand when power usage is at its peak, and can have more power than they can sell during the night. As a result, many utilities offer pricing structures with a higher rate for energy used at peak times, during the day, than at times of low demand. This is often called time-dependent valuation, or TDV [19]. Only the cost or zero net emissions definitions of ZNE account for the benefits of shifting loads; if projects use other definitions they may miss an important approach to meet grid peak demand and to lower greenhouse gas emissions.

2.2 Importance of the Definition of ZNE

Definitions are an important starting point for any discussion of ZNE because it affects the choices a design team might take to achieve the goal, how the system interacts with the local utility, energy load profiles, or whether or not the project can claim that they achieved the ZNE goal. For example, a definition could emphasize demand-side energy efficiency, various supply strategies such as on-site renewables, white certificates,[3] or whether fuel switching is appropriate to meet the ZNE goal.

In California, a number of organizations recently met to settle on a single definition of ZNE for use by the investor-owned utilities and state agencies [20]. Defining ZNE or its equivalent was deemed to be important for the formulation of energy codes, for the design of energy efficiency and renewable energy programs, and for providing guidance to developers, architects, engineers, and contractors. After several collaborative meetings that included representatives from the investor-owned utilities, state agencies, and engineering firms, no consensus on a single definition could be reached [21]. This could lead to ambiguity in achieving the ZNE goal and leaves open the possibility that regulating agencies in the state will adopt contrasting or conflicting definitions. This may complicate the evaluation, measurement, and verification (EM&V) process, or discourage design teams from attempting to achieve ZNE on their project. In addition, because other states also tend to follow the lead of California in adopting energy efficiency standards, it may also cause different approaches to ZNE to be pursued across the United States.

The scope of ZNE is also a point of contention, with urban and energy planners arguing for larger-scale interventions that incorporate transportation energy usage and attempt to achieve economies of scale by building utility scale renewable energy projects. This contrasts with efforts by architects and building services engineers to define ZNE at the building or site. Currently, these two different scales of development coexist without issue, but if regulators attempt to tightly define the scale of ZNE development this may change in the future.

Internationally, literature reviews of the definition of ZNE from other countries reveal similar conflict surrounding the definition of ZNE (e.g., [22]). As is the case with green building in general, it is unlikely that we will ever settle on a single approach or definition that is acceptable to all parties [23,24]. One policy option may be to set goals based on energy use intensity (watts/square meter of floor area) or carbon intensity (tons CO_2E square meter of floor area), as the overarching goal of most ZNE projects is to limit greenhouse gas emissions.

---

3. White certificates represent energy efficiency that has occurred at a particular location, but credit for the energy efficiency, usually for compliance purposes, is transferred via the ownership of the white certificate. It is similar in concept to renewable energy certificates, or "RECs."

3 ZERO NET ENERGY INITIATIVES IN THE UNITED STATES

Although the downturn in the U.S. economy slowed the pace of new construction, a recent report by the New Buildings Institute, [2] finds that a handful of ZNE or ZNE-capable buildings and communities are nearing completion in the United States. While the majority of the projects are in the mild climates of the west coast, several have been constructed in states such as Wisconsin [12] and Ohio [14] where heating loads dominate annual energy consumption (Figure 10.2).

In total, 15 commercial buildings have validated energy data documenting that they have achieved ZNE; another six buildings are currently commissioning their structures to achieve the goal. Another 39 projects in the United States are working toward a ZNE goal; at the time of the publication not enough data on these projects was available to state that they had definitively achieved their ZNE goal.

NBI found that the majority of projects are small, single-story buildings. They are often collegiate buildings or environmental centers, but projects on the boards include large office buildings and K-8 schools. Almost all projects use photovoltaics to supply the renewable energy to the project,

FIGURE 10.2 The Adam Joseph Lewis Center in Oberlin, Ohio, was one of the first ZNE buildings to be constructed in the United States. It is representative of most ZNE buildings as a small collegiate office and classroom building. However, while most ZNE buildings have been constructed in mild climates, such as California, the building at Oberlin has achieved ZNE status in the cold climate of northern Ohio. *Photo Credit: Nicholas B. Rajkovich.*

and increases in first cost are between 3 and 18 percent above standard levels [2]. Simple payback periods are generally 11 years or less, though the exact return on investment varies with differences in electrical rates, net metering structure, and local incentives for energy efficiency and renewable energy.

The characteristics of these first ZNE buildings align with a potential study conducted by the National Renewable Energy Laboratory (NREL). Entitled "Assessment of the Technical Potential for Achieving Zero-Energy Commercial Buildings" [49], the report looks at the technical feasibility of ZNE commercial using 15-minute annual simulations based on 5,375 buildings in the 1999 Commercial Buildings Energy Consumption Survey (CBECS) Public Use Database. The report determined that many single-story or low energy-intensity buildings could achieve ZNE in all climates of the United States, and using today's technologies and practices up to 64 percent of the commercial buildings in United States could be ZNE buildings.[4] There are no projections for the rate of adoption for ZNE buildings, but current DOE efforts attempt to reduce energy use in buildings by 50 percent, representing a 21 quadrillion BTU (Quad) reduction in primary energy usage in the United States.

3.1 Zero Net Energy Programmatic Efforts in the United States

Because many architects and engineers feel it is possible to design and construct ZNE projects, and a number of early adopting clients would like to pursue the goal to further their environmental objectives, a number of agencies are promoting the achievement of ZNE by 2030. The goal of ZNE projects by 2030 is driven by concerns about "tipping points" in the climate and because the majority of the U.S. building stock will overturn within the next 30 years [26]. This provides a major opportunity to reduce greenhouse gas emissions from the U.S. building stock.

One organization, Architecture 2030, has challenged design teams to immediately implement a 60 percent reduction in fossil fuel based energy consumption for new and renovated buildings and infrastructure. They go on to challenge design teams to further reduce their carbon dioxide emissions by 70 percent in 2015, 80 percent in 2020, 90 percent in 2025, and to be carbon-neutral by 2030 (Figure 10.3). Carbon neutrality is defined as using no fossil fuels for operations or construction of the project. These targets

---

4. Residential buildings (both single family and low-rise residential) are not included in the NREL study because numerous ZNE homes already exist. There is greater potential to achieve ZNE in residential structures because of their inherent low energy use intensity and because there are larger roof areas for the installation of photovoltaics [1,25]. However, Anderson, Hauser and Mooney discuss in this volume the particular challenges of balancing energy efficiency with renewable energy production.

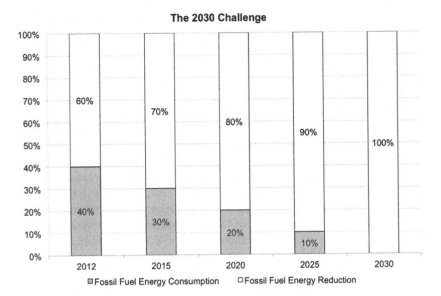

FIGURE 10.3   The 2030 Challenge encourages design teams to reduce the amount of energy their buildings draw from fossil fuel sources. These targets may be accomplished by implementing innovative sustainable design strategies, generating on-site renewable power and/or purchasing (20 percent maximum) renewable energy. *Note: Image adapted from http://architecture2030. org/2030_challenge/the_2030_challenge.*

may be accomplished by implementing innovative sustainable design strategies, generating on-site renewable power, and/or purchasing up to 20 percent renewable energy from outside organizations [27].

As one of the first organizations pushing for ZNE buildings and communities, the "2030 Challenge" has had a profound impact on the architecture/engineering/construction community. The American Institute of Architects (AIA) adopted the 2030 Challenge, and recently created a number of online tools to help design firms achieve their low carbon intensity goals [28]. In addition, AIA's commitment has led to a number of educational activities designed to provide architects with the necessary specific skills to design ZNE buildings. As of July 2010, 73 percent of the 30 largest U.S. Architecture/Engineering firms, have adopted and are implementing the 2030 Challenge. In total, approximately 41 percent of all U.S. architecture firms have adopted the Challenge. Most of the 15 extant ZNE buildings in the United States were built in response to this challenge.

Building on the success of the 2030 Challenge, organizations such as the Cascadia Green Building Council and the International Living Building Institute are already encouraging projects to go beyond ZNE, even though only a handful of projects have achieved ZNE status. Through their program entitled the "Living Building Challenge," they encourage project teams to consider additional issues such as the siting of buildings, water consumption, embodied energy, and social equity [29]. As a prerequisite to participation in the program the building must achieve ZNE status, defined as 100 percent of the project's energy needs supplied by on-site renewable energy on a net annual basis.

At a broader scale, the California Public Utilities Commission (CPUC) developed a strategic plan for energy efficiency in California in 2007 to guide energy efficiency programs through 2020 by focusing on longer-term, more visionary possibilities for improving efficiency, including ZNE.[5] The CPUC focused on this long-term strategy to align with California's implementation of Assembly Bill 32 (AB 32), the "California Global Warming Solutions Act of 2006," which requires the state's emissions of greenhouse gases to be reduced to 1990 levels by 2020, and to set the strategic directions that could produce significant reductions in energy use, and therefore greenhouse gas emissions, over the coming decade ([30], [47]. Specifically, the CPUC considered and approved a series of ZNE building goals to address climate challenges.

The goal and concept of ZNE was central to many of the findings of the strategic plan. Attaining these goals will cut the growth in California emissions significantly, as it pursues absolute declines through energy efficiency and other means. Currently, California expects that roughly 15 percent of its greenhouse gas reductions needed to meet the 2020 target will come from energy efficiency. As the keystone to this effort, the CPUC adopted the California Long-Term Energy Efficiency Strategic Plan (CEESP or Strategic Plan) in September 2008.

The Strategic Plan sets forth a state-wide roadmap to maximize the achievement of energy efficiency in California's electricity and natural gas sectors. The Plan also sets a number of specific goals for the years 2009 through 2030. Over 40 public workshops and meetings were convened by the CPUC and the investor-owned utilities (IOUs) leading to the adoption of the Strategic Plan [30] and updated in 2011 [31].

With the CEESP, the state of California has created a roadmap for a scaling-up of state-wide energy efficiency efforts. While the policies of the past three decades have been successful in raising public awareness of energy issues and laying the groundwork for large-scale efficiency efforts, the savings achieved through those policies have come through specific programs with targeted market impacts. The objective of the Strategic Plan is to sustain market transformation, moving California beyond its historic reliance on a near-term replacement of less efficient technology with more efficient

---

5. The California Public Utilities Commission (CPUC) is a regulatory agency, which regulates privately owned public utilities in the state of California, including electric power, telecommunications, natural gas and water companies. The CPUC is responsible for the regulation of investor-owned energy utilities including Pacific Gas and Electric (PG&E), Southern California Edison (SCE), Southern California Gas (SCG), and San Diego Gas and Electric (SDG&E).

technology and toward long-term, deeper savings achievable only through programs with broader, longer-term impact. This shift is in part to align with the goals of AB 32, but is also reflective of an evolution in program design as a result of a broader set of stakeholders engaged in energy efficiency planning.

Key to the Strategic Plan's success is four programmatic goals, widely viewed as ambitious, high-impact efforts. These goals, called the "Big, Bold Energy Efficiency Strategies" (BBEES), are listed in Table 10.2. The goals were selected not only for their potential impact, but also for their transparency and ability to galvanize market players.

For each of these measures, the Strategic Plan provides recommendations for coordinated action among the state, its utilities, the private sector, and other market players. The recommendations take advantage of the wide variety of stakeholder expertise engaged in the strategic planning process. While specific energy savings have not been assigned to each of the goals, the structure of each utility-administered program has been redesigned to address meeting the BBEES. LaRue, Cole, and Turnbull in this volume describe recent efforts by one investor owned utility, PG&E, to achieve these goals.

While no national mandate currently exists for ZNE, federal legislation such as the Energy Policy Act of 2005 (EPAct 2005, Public Law No. 109−58) and the Energy Independence and Security Act of 2007 (EISAct 2007, Public Law No. 110−140) have authorized initiatives that will develop and disseminate technologies, practices, and policies to move the U.S. market toward ZNE buildings [32].

TABLE 10.2 California Long-Term Energy Efficiency Strategic Plan Programmatic Goals, or the "Big, Bold Energy Efficiency Strategies"

Sector/Market	Goal
New Residential Construction	All new residential construction in California will be zero net energy by 2020.
New Commercial Construction	All new commercial construction in California will be zero net energy by 2030.
Heating, Ventilation, and Air Conditioning (HVAC) Industry	HVAC industry and market will be transformed to ensure that its energy performance is optimal for California's climate.
Low-Income Customers	All eligible low-income customers will be given the opportunity to participate in low-income energy efficiency programs by 2020.

Note: Table format adapted from the Strategic Plan, 2008.

These federal laws also initiate a path that federal buildings are to follow achieving ZNE by 2030. These were reinforced by Executive Order 13423 (signed January 24, 2007). Under this guidance, all new federal buildings must be designed to achieve "zero net energy" by fiscal year 2030, starting in 2020. Further, even in fiscal year 2010, all new or renovated federal agency building designs must reduce fossil fuel-generated energy consumption by 55 percent compared to a fiscal year 2003 baseline. Although projections of energy savings have not been calculated, these rules have already encouraged federal agencies to revise their design, construction, and operations standards.

3.2 Wider Reaching Zero Net Energy Educational Activities in the United States

There have been several efforts to communicate the concept of ZNE homes to broader audiences. One major effort by the DOE to support a push to ZNE is the Solar Decathlon. The program challenges collegiate teams to design, build, and operate ZNE houses that are cost-effective, energy-efficient, and attractive (Figure 10.4). The winner of the competition is the team that best blends affordability, consumer appeal, and design excellence with optimal energy production and maximum efficiency. The first Solar Decathlon was held in 2002; the competition has since occurred in 2005, 2007, 2009, and 2011. The next event in the United States will take place in October of 2013, at Orange County Great Park in Irvine, California [33].

Since 2002, the Solar Decathlon has inspired 92 collegiate teams to pursue multidisciplinary course curricula to design and build the solar-powered houses for the competition. Over 70 of the houses are still in use as homes or research facilities, establishing a worldwide reputation as a successful educational program and workforce development opportunity. Over 15,000 collegiate participants have worked on one or more of the homes, and recently the DOE expanded its outreach to K-12 students by inviting schools in the Washington, D.C. area to visit on class tours.

Each year the competition has more than 100,000 people visit the homes, and recently the competition was expanded to include competitions in Europe and China. Over time the program is expected to drive the demand for energy efficiency and renewable energy technologies by giving students the opportunity to practice ZNE design and construction prior to graduating from college [33]. It will also expose the public to new energy efficiency and renewable energy technologies, potentially driving demand for these new products.

Other educational efforts in the United States include an effort by the Society for Building Science Educators (SBSE) to establish a national ZNE curriculum. Since the early 1990s, the SBSE has worked to engage students (in their role as future market actors) to advance building performance as a

FIGURE 10.4   Southwest view of the Team Massachusetts House. The home incorporates a number of energy efficiency and renewable energy features and was open to the public during the 2011 competition in Washington, D.C. Two significant shifts from earlier iterations of the competition was a focus on grid interconnection and affordability. For more information visit http://www.solardecathlon.gov. *Photo Credit: Richard King, U.S. Department of Energy Solar Decathlon.*

consideration in all designs for the built environment. The curriculum project they are currently working on targets these actors, both in school and as they make the transition into practice. The proposed project is in its early planning stages, but builds on the success of three predecessor educational initiatives including the Vital Signs Project, the Agents of Change Project, and the SBSE Carbon Neutral Design Project.

SBSE proposes to engage schools of architecture, engineering, and planning around issues of integrated design, construction, and the management of a ZNE built environment. The target audience will be urban planners, architects, engineers, and construction managers in upper division university courses or in the first three years of practice. The proposal provides a near-term strategy to improve ZNE education through resources and strategies and projects mid- to long-term results that will change practice and the design and construction of the built environment. These efforts work in concert with the Solar Decathlon and efforts by professional organizations such as the AIA and the U.S. Green Building Council.

4 PLANNING FOR ZERO NET ENERGY

As described in the previous sections of this chapter, ZNE is a conceptual descendent of the notion of the steady state economy, can be defined in a number of ways, and is currently promoted by a diverse set of organizations. Though there are only a handful of ZNE projects currently in the United States, as the number of ZNE projects grows or ZNE becomes a public mandate in more states, it has the potential become a highly contested concept.

In any democratic process dealing with environmental issues, conflicts routinely emerge between those who value the long-term ecological viability of natural systems and those who value nature as the supplier of a commodity necessary for growth [34]. In addition, sustainable development projects can have an equity deficit; they promote environmental conservation and economic development over the needs of marginalized communities.

Agyeman [35] argues that it is only when issues of income distribution, civil liberties, and political rights are made equal to environmental conservation and growth will we be able to achieve a lasting, *just* form of sustainable development. As a subset of broader sustainable development efforts, ZNE is not immune from this criticism. Because it currently costs above market rate to build a ZNE project it puts the opportunity to live or work in a ZNE building outside of the reach of low-income communities; efforts to achieve ZNE may also divert funding from weatherization or other low-income energy efficiency programs. It is for these reasons the DOE Building Technologies Program continues to focus on new, less costly technologies to push more buildings to ZNE cost-effectiveness.

In addition, and as has been demonstrated in California [21], because ZNE will always be framed by a wide variety of discourses, no single definition will ever be able to "solve" all problems of energy production and consumption because it is a function of the local actors involved. Program planners can attempt to operate at the center of these conflicts – what Campbell [36] calls the planning triangle – to attempt to satisfy the competing logics of environment, economic development, and equity. Planners can also take a radical position by stepping outside of the policy process and call attention to deficiencies and unequal power relationships in the policy process. In either case, the goal is to achieve an acceptable, practical solution to competing and even conflicting policy goals.

This stands in stark contrast to the utility rate-setting process and physical-technical-economic model (PTEM) that dominates the utility and energy industries [37] and through which utility programs supporting ZNE are determined. The energy planning process promoted by energy utilities and their regulators sits within a command and control structure where competing logics of the environment, economic development, and equity are hashed out by attorneys in a litigation process. Environmental concerns, such as the amount of renewable energy required for a ZNE project, are argued in front of an administrative law judge; each side presents research from experts to justify their position. The rate case process uses highly technical language and acronyms that reduces or excludes the public's opportunity to participate, and some cases the regulatory agency may limit the number of actors participating in the process. The resulting programs produced are frequently criticized for their highly politicized nature and lack of community input [38]. The regulatory debate focuses on the prudency of the expenditure of ratepayer originated funds using quantitative metrics with the wider, future benefits of ZNE buildings disadvantaged in these calculations. This handicaps one of the significant mechanisms of ZNE building deployment.

As contributors either as staff or on behalf of the formal participants in the rate case process, building scientists focus on the control of environmental phenomena that affect building structure and function. Through empirical testing, economic analysis, and computer modeling, the goal of building science is to provide the architecture/ engineering/ construction and policy communities with data to: (1) predict and minimize energy use, (2) prevent system failures, and (3) protect human health, safety, and welfare [39].

Within the building sciences, contributions from the social sciences are limited to evaluating behavior or undertaking surveys to inform marketing campaigns. This work is often called the "behavioral" or "human" dimension of building energy efficiency. While this emerging area of research draws on concepts from sociology and environmental psychology, it does little to challenge the PTEM or the rate case process. A number of authors (e.g., [40,41]) have advocated for a robust inclusion of the social sciences in formulating energy efficiency policy and a push to ZNE.

Engaging multiple discourses may best be achieved through a collaborative planning process. Collaborative planning has been effective in solving other policy problems, most notably in Sacramento, California where the Sacramento Water Forum spent five years in an intensive consensus building process to manage their limited water supply [42]. In contrast to litigation, a collaborative group was convened that included state and federal agencies with jurisdiction over California's water supply. The collaborative planning process allowed for the creation of collective decisions that were defensible to the larger public. New relationships between former adversaries allowed for creative problem-solving. While there are concerns over the amount of resources participants would need to contribute to such a process, the concept may be transferrable to the energy sector [43].

True collaborative planning has been unusual in the past as it relates to energy efficiency and renewable energy. Even though collaboration has been highly recommended state-level activities that consider, authorize, and fund utility energy efficiency programs are predominantly quasi-legal regulatory processes. In such processes, collaboration has begun to play a role, but often as a precursor to a quasi-legal process. These have their own dynamic, in which the compromises inherent in collaboration can fall victim to a participant's perception of an opportunity to achieve more of its goals, sometimes called the prisoner's dilemma.

Still, engaging in a collaborative effort would be timely because several state public utility commissions are planning to leverage billions of dollars in public goods charge funding to support energy efficiency and renewable energy programs. By engaging in collaborative planning we might allow energy efficiency programs to truly support ZNE that could not have occurred in the past [44]. The result might be communities that are socially inclusive, and climate neutral; this would be a solution that would be truly sustainable in the long-term [45,46].

5 CONCLUSIONS

Although achieving ZNE in a single building is conceptually clear, and progress toward realization of this goal is being made, the path to ZNE for the entire built environment is far more complex. As discussed in this chapter, it involves a number of stakeholders, definitional issues, and significant limitations in the policy process. Although the route is unclear, a push to ZNE has the potential to engage communities of practitioners that have not coordinated in the past. In addition, because building energy use consumed 28.8 percent of the total primary energy used in the United States in 2010 [8], and the majority of the U.S. building stock will be remodeled or rebuilt within the next 30 years [26] these efforts will be the front line in the fight against climate change.

Although the concept of no growth dates to the 1970s, ZNE practice and policies are still early in their evolution. The unintended effects of governmental regulatory agencies adopting one definition of ZNE over another is an important issue, and constitutes a fertile area for future research.

With the introduction of several databases of building energy performance, including the USGBC Green Building Information Gateway (http://gbig.org), the DOE Renewable Ready Buildings database (http://buildingdata.energy. gov/renewable_ready), and DOE's National Building Performance Database (http://www1.eere.energy.gov/buildings/buildingsperformance),[6] the performance of certain buildings can be evaluated against their peers, and energy efficiency upgrades, including addition of photovoltaics (PV), could be analyzed for energy and financial performance across a large sample size. After controlling for variables such as the building type, climate, and delivery method, comparisons of key metrics such as energy consumption and financial results might reveal that those project teams that pursued ZNE voluntarily as opposed to a mandate (e.g., California) had a higher level of performance. In addition, teams with better collaborative practices might also achieve higher levels of performance. Qualitative analysis of project documents and

---

6. This database is in the pilot phase, and will be released in 2012.

interviews with project participants might reveal differences in framing environmental issues or approaches to energy efficiency. This research could help to inform future building energy codes, public policy, or improvements in the design and construction process, all required to achieve the deep energy reductions and increase in renewable energy promised by ZNE.

Significant questions emerge over the need for a new "systems integrator" profession that may be necessary to advance ZNE buildings. Since the introduction of the Leadership in Energy and Environmental Design (LEED) program by the U.S. Green Building Council, a number of state and federal agencies have promoted the development of LEED Accredited Professionals in a similar role for green buildings. But their knowledge and skills must go beyond technical aspects of green buildings and administrative aspects of the LEED process to achieve low carbon buildings. For example, these integrators must develop skills to engage in collaborative practices, especially with agencies such as utilities that are unaccustomed to an open process. This indicates the need to reshape post-secondary design and construction education and for new items to be added to the architecture, engineering, and planning curricula, such as the basics of generation, transmission, and distribution infrastructure.

ACKNOWLEDGMENTS

Portions of this work were supported by the National Science Foundation Graduate Research Fellowship under Grant No. DGE 0718128. Additional support was provided by the Michigan Memorial Phoenix Energy Institute and the Graham Environmental Sustainability Institute, both at the University of Michigan.

We would like to thank Richard King, of the U.S. Department of Energy, for the permission to reprint his photograph of the Team Massachusetts house from the 2012 Solar Decathlon competition. We would also like to thank Bill Burke, of the PG&E Pacific Energy Center, and Rick Diamond, of Lawrence Berkeley National Laboratory, for assistance with a conference paper in 2010 that influenced many of the findings of this chapter.

The views and opinions expressed in this chapter are wholly those of the authors and do not necessarily reflect the official policy or position of the University of Michigan, Lawrence Berkeley National Laboratory, or the U.S. Department of Energy. William C. Miller, PhD is currently employed by Lawrence Berkeley National Laboratory and provides support to the U.S. Department of Energy.

REFERENCES

[1] Rajkovich NB, Miller WC, LaRue AM. Zeroing in on zero net energy. In: Sioshansi FP, editor. Energy, sustainability and the environment: technology, incentives, behavior. Oxford: Butterworth-Heinemann; 2011. p. 497–517.

[2] New Buildings Institute (NBI). Getting to zero 2012 status update: a first look at the costs and features of zero energy commercial buildings. Vancouver, Washington: NBI; 2012.

[3] Hajer MA. The new environmental conflict. The politics of environmental discourse: ecological modernization and the policy process. Oxford: Clarendon Press; 1995.

[4] Boulding KE. The economics of the coming spaceship earth. In: Daly HE, editor. Toward a steady-state economy. San Francisco: W.H. Freeman and Company; 1973. p. 121–32.

[5] Daly HE. The steady-state economy: toward a political economy of biophysical equilibrium and moral growth. In: Daly HE, editor. Toward a steady-state economy. San Francisco: W.H. Freeman and Company; 1973. p. 149–73.

[6] World Commission on Environment and Development. Our common future. In: Wheeler SM, Beatley T, editors. The sustainable urban development reader. 2nd ed. New York: Routledge; 1987. p. 59–63.

[7] United Nations Framework Convention on Climate Change. Kyoto protocol. Retrieved August 22, 2011, from, <http://unfccc.int/kyoto_protocol/items/2830.php>; 1997.

[8] DOE. AEO2012 early release overview. Retrieved June 23, 2012, from, <http://205.254.135.7/forecasts/aeo/er/early_consumption.cfm>; 2012.

[9] Torcellini P, Pless S, Deru M, Crawley D. Zero energy buildings: a critical look at the definition. Paper presented at the American Council for an Energy-Efficient Economy (ACEEE) Summer Study on Energy Efficiency in Buildings; 2006.

[10] Torcellini P, Long N, Pless S, Judkoff R. Evaluation of the low-energy design and energy performance of the Zion National Park visitors center. Golden, Colorado: National Renewable Energy Laboratory (NREL); 2005.

[11] Torcellini PA, Deru M, Griffith B, Long N, Pless S, Judkoff R, et al. Lessons learned from field evaluation of six high-performance buildings. Paper presented at the American Council for an Energy-Efficient Economy (ACEEE) Summer Study on Energy Efficiency in Buildings; 2004.

[12] Utzinger M, Swenson S. The Aldo Leopold Legacy Center: expanding the definition of "community" in carbon management. Planning Theory and Practice 2012;13(1):156–66.

[13] Pless S, Torcellini P. Energy performance evaluation of an educational facility: the Adam Joseph Lewis Center for Environmental Studies, Oberlin College, Oberlin, Ohio. Golden, Colorado: National Renewable Energy Laboratory (NREL); 2004.

[14] Pless S, Torcellini P, Petersen J. Oberlin College Lewis Center for environmental studies: a low-energy academic building. Paper presented at the world renewable energy congress VIII and expo; 2004.

[15] Cornell University. Hydroelectric plant. Retrieved May 14, 2012, from, <http://energyandsustainability.fs.cornell.edu/util/electricity/production/hydroplant.cfm>; 2005a.

[16] Cornell University. Lake source cooling. Retrieved May 14, 2012, from, <http://energyandsustainability.fs.cornell.edu/util/cooling/production/lsc/default.cfm>; 2005b.

[17] Malin N, Boehland J. Getting to zero: the frontier of low-energy buildings. Environmental building news 2005;October.

[18] Wheeler SM, Segar RB. West village: development of a new ecological neighborhood in Davis, California. Planning Theory and Practice 2012;13(1):145–56.

[19] California Energy Commission. Time-dependent valuation (TDV). Retrieved May 14, 2012, from, <http://www.energy.ca.gov/title24/2005standards/archive/rulemaking/documents/tdv/index.html>; 2005.

[20] Diamond R. A zero net energy definition for residential and commercial buildings. (Personal communication). 2012.

[21] Diamond R, Hopewell B. A zero net energy definition for residential and commercial buildings. Sacramento, CA; 2011.

[22] Marszal AJ, Heiselberg P. A literature review of zero energy buildings (ZEB) definitions. Aalborg, Denmark: Aalborg University; 2009.

[23]   Guy S, Farmer G. Reinterpreting sustainable architecture: the place of technology. J Archit Educ 2001;54(3):140−8.

[24]   Guy S, Moore SA. Sustainable architecture and the pluralist imagination. J Archit Educ 2007;60(4):15−23.

[25]   Gray M, Zarnikau J. Getting to zero: green building and net zero energy homes. In: Sioshansi FP, editor. Energy, sustainability and the environment: technology, incentives, behavior. Oxford: Butterworth-Heinemann; 2011. p. 231−72.

[26]   Zhai J, LeClaire N, Bendewald M. Deep energy retrofit of commercial buildings: a key pathway toward low-carbon cities. Carbon Management 2011;2(4):425−30.

[27]   Architecture 2030. Architecture 2030: the 2030 challenge for planning. Retrieved May 14, 2012, from, <http://architecture2030.org/2030_challenge/2030_challenge_planning>; 2012.

[28]   American Institute of Architects (AIA). AIA 2030 commitment home page, programs and initiatives. Retrieved May 14, 2012, from, <http://www.aia.org/about/initiatives/AIAB079458>; 2012.

[29]   International Living Future Institute, and Cascadia Green Building Council. Living building challenge 2.1; 2012.

[30]   California Public Utilities Commission (CPUC). California energy efficiency. Retrieved April 13, 2010, from, <www.californiaenergyefficiency.com>; 2008.

[31]   California Public Utilities Commission (CPUC). Retrieved June 30, 2012 from, <http://www.cpuc.ca.gov/NR/rdonlyres/A54B59C2-D571-440D-9477-3363726F573A/0/CAEnergyEfficiencyStrategicPlan_Jan2011.pdf>; 2011.

[32]   National Science and Technology Council, and Subcommittee on Building Technology Research and Development. Federal research and development agenda for net-zero energy, high-performance green buildings. Washington, DC; 2008.

[33]   DOE. Solar decathlon. Retrieved May 14, 2012, from, <http://www.solardecathlon.gov/>; 2013.

[34]   Gandy M. Water, space, and power concrete and clay: reworking nature in New York City. Cambridge, Mass: MIT Press; 2002. pp. 19−76

[35]   Agyeman J. Toward a 'just' sustainability? Continuum: Journal of Media and Cultural Studies 2008;22(6):751−6.

[36]   Campbell S. Green cities, growing cities, just cities? Urban planning and the contradictions of sustainable development. J Am Plann Assoc 1996;62(3):296−312.

[37]   Lutzenhiser L. Social and behavioral aspects of energy use. Annual Review of Energy and Environment 1993;18:247−89.

[38]   Throgmorton JA. Survey research as rhetorical trope: electric power planning arguments in Chicago. In: Fischer F, Forester J, editors. The argumentative turn in policy analysis and planning. Durham: Duke University Press; 1993. p. 117−44.

[39]   Guy S, Shove E. The sociology of energy, buildings and the environment: constructing knowledge, designing practice. London; New York: Routledge; 2000.

[40]   Janda KB. Improving efficiency: a socio-technical approach. In: Jamison A, Rohracher H, editors. Technology studies and sustainable development. Munich and Vienna: Profil Verlag GmbH; 2002. p. 343−64.

[41]   Strengers Y, Maller C. Integrating health, housing and energy policies: social practices of cooling. Building Research and Information 2011;39(2):154−68.

[42]   Innes JE, Booher DE. Collaborative policymaking: governance through dialogue. In: Hajer MA, Wagenaar H, editors. Deliberative policy analysis: understanding governance in the network society. Cambridge: Cambridge University Press; 2003. p. 33−60.

[43] Innes JE, Booher DE. Reframing public participation: strategies for the 21st century. Planning Theory and Practice 2004;5(4):419–36.

[44] Rajkovich NB, Diamond R, Burke B. Zero net energy myths and modes of thought. Paper presented at the American Council on an Energy-Efficient Economy (ACEEE) Summer Study; 2010.

[45] Blanco H, Alberti M, Forsyth A, Krizek K, Rodriguez D, Talen E, et al. Hot, congested, crowded and diverse: emerging research agendas in planning. Progress in Planning 2009;71:153–205.

[46] Blanco H, Alberti M, Olshansky R, Chang S, Wheeler SM, Randolph J, et al. Shaken, shrinking, hot, impoverished and informal: Emerging research agendas in planning. Progress in Planning 2009;72:195–250.

[47] Decision Approving 2010 to 2012 energy efficiency portfolio and budgets, 08-07-031 C.F.R.; 2008.

[48] DOE. Building technologies program planning summary. Retrieved May 14, 2012, from, <http://www1.eere.energy.gov/buildings/plans_implementation_results.html>; 2010.

[49] Griffith B, Long N, Torcellini P, Judkoff R, Crawley D, Ryan J. Assessment of the technical potential for achieving net zero-energy buildings in the commercial sector. Golden, Colorado: National Renewable Energy Laboratory (NREL); 2007.

Chapter 11

What If This Actually Works? Implementing California's Zero Net Energy Goals

Anna M. LaRue[1†], Noelle C. Cole[1] and Peter W. Turnbull[2]
[1]*Principal, Resource Refocus LLC,* [2]*Principal Program Manager, Zero Net Energy Pilot Program, Pacific Gas and Electric Company*

1 INTRODUCTION

In the 2008 California Long Term Energy Efficiency Strategic Plan (Strategic Plan), the California Public Utilities Commission (CPUC) put forward four "Big Bold" goals to define California's energy future. Two of these goals focus on zero net energy (ZNE) construction: that all new residential construction be ZNE by 2020 and that all new commercial construction be ZNE by 2030. These ZNE goals are supported by policy initiatives of the California Energy Commission (CEC) and the California Air Resources Board (ARB). California investor-owned utilities also support these ZNE goals through a range of programs.

The goal of ZNE buildings has been a catalyst among building designers, forward-thinking building owners, and policymakers. At a high level, a "zero net energy" building is a highly energy efficient, grid-tied building that produces as much clean, renewable energy as it consumes, when measured over a calendar year. However, as elegant as ZNE has proven to be in concept, the reality involves a complex set of challenges at the intersection of energy policy, economics, and physics, all related to the electricity grid.

This chapter is the second of four chapters in this volume that discuss ZNE. Rajkovich et al discuss ZNE on a national scale. This chapter examines the status of ZNE in California and what will happen if the California ZNE goals for 2020 and 2030 are met. Wheeler et al focus on a case study of the ZNE community of West Village at the University of California, Davis. Anderson et al discuss the integration of energy efficiency and renewable energy systems in homes.

---

†Corresponding author. The views and opinions expressed in this chapter are wholly those of the authors and do not necessarily reflect the official policy or position of Pacific Gas and Electric Company.

Energy Efficiency. DOI: http://dx.doi.org/10.1016/B978-0-12-397879-0.00011-6

This chapter begins with an overview of energy efficiency and renewable policy progress to date in California. The chapter then discusses the current state of ZNE in California. It discusses how having a coordinated effort to drive California's building energy efficiency standards towards ZNE and to encourage investor-owned utilities and other regulatory stakeholders to incentivize ZNE development will result in an increasing number of ZNE-compliant buildings and potentially communities connected to the grid. The chapter goes on to examine how the simultaneous reduction in demand and increase in distributed generation associated with achieving widespread ZNE has the potential to increase utility rates, cause localized pressure on the electricity grid, and significantly change the relationship between utilities and their customers. Finally, the chapter discusses potential revisions to the ZNE definition to be used for California policy and building energy efficiency standards and ways in which stakeholders are considering implementing ZNE at different scales.

2 AN OVERVIEW OF CALIFORNIA ZERO NET ENERGY POLICY

There are three main agencies of state government in California that are engaged in energy issues related to ZNE and distributed generation – the CPUC, the CEC, and the ARB. The CPUC is the main regulator over most investor-owned utility (IOU) business issues, setting rates and utility service rules for customers and small generators. The CEC is responsible for assuring the state has an adequate supply of energy. Among other activities, the CEC administers the energy efficiency component of the state's building code. The ARB is the chief regulatory agency for air emissions-related activity, which comprises transportation and much electric power generation.

2.1 The Warren–Alquist Act and the California Energy Commission

By the 1960s, it was recognized that California faced explosive population growth in the decades to come. In fact, the population of California did explode, going from just under 20 million in 1970 to just under 34 million in 2000. This growth is larger than the relocation of an entire large state – such as Ohio, with a population of 11.4 million – to California over that time period. The state needed to add all of the services and physical infrastructure – roads, water systems, and power grid – required to serve that population in just 30 years. California utilities planned to meet the higher electricity demand growth with a series of nuclear power stations and an expansion of the transmission and distribution system. The environmental community instead suggested conservation and efficiency – or finding new ways to functionally use less – as a way to dampen the growth in demand. Politically, things were brought to a head by the energy crisis created by the "Arab Oil Embargo" of 1973–1974. Long lines for gasoline made headlines, but utility electricity supplies — which depended substantially on fuel oil at that time — were also of great concern.

The Warren–Alquist Act (WAA) of 1974 emerged as a compromise solution among the parties.[1] It authorized establishment of the CEC and designated two key functions to it: the authority to site power plants and to establish energy efficiency codes and standards to be part of the state building code.[2] The first set of building efficiency standards was issued in 1978. Per the WAA, these standards are updated on a regular basis every 3 to 4 years with revisions that are designed to ratchet up efficiency levels in buildings and appliances as technologies improve, lessening the need to acquire and build new energy supply and infrastructure. The CEC's role in establishing building and appliance energy efficiency standards is crucial to the current discussion of ZNE goals in California. The WAA has been a tremendous success. While the demand for power has risen substantially over the past four decades, and new infrastructure has in fact been built, on a *per capita* basis, growth has been flat.[3]

2.2 Rates and Ratemaking in California

Utility rates, ratemaking, and service rules have also shaped the energy landscape. In the next few sections, the rate policy mechanisms highlighted in Table 11.1 will be discussed. The chapter will later focus on the impact of these rate policy mechanisms as the state moves towards the 2020 and 2030 ZNE goals.

---

1. One of the key legislative findings of the Warren–Alquist Act includes the following language: "The Legislature further finds and declares that the present rapid rate of growth in demand for electric energy is in part due to wasteful, uneconomic, inefficient, and unnecessary uses of power and a continuation of this trend will result in serious depletion or irreversible commitment of energy, land and water resources, and potential threats to the state's environmental quality."
2. Nearly all parties agreed with the principles behind the law — the need for adequate energy supply and a reasoned, environmentally-sensitive approach to achieving it— but the WAA was opposed by Southern California Edison, General Electric, Pacific Gas and Electric Company (PG&E), the California Manufacturers Association and even the CPUC. Before signing the act in 1974 (post-oil-embargo), then-Governor Reagan had vetoed nearly identical legislation in 1973 (pre-embargo). Despite general agreement on the bill's objectives, some opposed the WAA as having an "anti-nuclear bias" (in the words of General Electric), and as an inappropriate overreach by the state to exercise authority over power plant siting. PG&E eventually sued over the power plant siting authority the act provides to the CEC, arguing to the US Supreme Court that the Atomic Energy Commission act of 1954 effectively pre-empted a state agency from exercising siting authority. The Supreme Court disagreed in a 1984 ruling, concluding that states had the right to control nuclear power station siting within their borders.
3. See the "Rosenfeld Curve" — Figure 7 in "A Graph is Worth a Thousand Gigawatt-Hours: How California Came to Lead the United States in Energy Efficiency" — located online at http://www.energy.ca.gov/commissioners/rosenfeld_docs/INNOVATIONS_Fall_2009_Rosenfeld-Poskanzer.pdf

TABLE 11.1 California Rate Policy Mechanisms

Rate Policy Mechanism	What It Does	Impact on ZNE and Customer-sited Renewables
Revenue Requirement	Establishes the amount of revenue utilities collect through rates in order to build, maintain, and operate all aspects of the electrical system.	Requires that the costs required to operate, upgrade, and maintain the grid be recovered in rates.
Revenue Decoupling	Separates commodity sales volume from the revenue requirement. Makes the utility indifferent to the commodity sales volume.	Requires commodity sales rates to go up or down to collect shortfall or surplus of revenue requirement.
Volumetric Rates	Establishes a rate structure where customers are billed by commodity unit (kWh for electricity) consumed.	Imperfect mechanism for collecting fixed costs (costs that tend not to vary by volume). Under scenarios involving reduced kWh sales, kWh rates will rise sharply to collect the revenue requirement.
Inverted Block Rates	Creates a rate structure with progressively higher charges per kWh for progressively higher levels of consumption.	Strongly discourages high consumption; makes efficiency and/or DG very attractive to high-use customers.

2.2.1 Revenue Requirement, Volumetric Rates, and the Utility Business Model

It is important to understand how electricity rates in California are structured in order to understand the pressure vectors on utility rates if the ZNE goals are achieved. A brief primer on the core utility business model is included to provide context.

The basic pact made between utilities and their regulators throughout the United States is that in exchange for an exclusive, stable arrangement to serve a given geographic market, a utility serves the territory on a *cost-of-service* basis. The utility is permitted to recover its costs to operate the system plus a modest rate of return. An obligation to serve comes with the territory to prevent the utility from selectively deciding who to serve based on a profit opportunity. The cost-of-service is adjudicated in a regulatory proceeding, and the utility sets rates to collect revenue equal to the agreed values based on a forecast of unit commodity (kWh) sales. As an example, if the cost-of-service for a fictional utility were determined to be $1 million dollars, and the kWh sales were forecast to be 5 million kWh, the rate would be set at $0.20 per kWh. Of course, in the real world of utilities, the numbers are much larger and the commercial expression of this basic pact contains myriad complexities and permutations.

For rate-making purposes, customers are divided into classes, typically commercial, residential, and industrial. A cost-of-service is determined overall for the utility and then parceled out to each class, to be collected by rates designed to recover the assigned costs by class. Seasonal and time-of-use parameters are included as well, based on the seasonal and diurnal differences in the cost of generating and delivering electricity. For the residential class, rate design is based almost entirely on volumetric, or cost per kWh, charges. For large commercial and industrial customers, while rate structures contain elements designed to collect some of the fixed costs of infrastructure, such as peak demand charges. These charges reflect the size of the wires, transformers, and other infrastructure needed to serve that customer. However, the volumetric components still dominate. Rate design is strongly biased toward volumetric charges for all classes; this bias suggests that utility costs are much more variable than fixed, which has never been true or close to true. Although the volumetric pricing rubric works mathematically for cost recovery, the reality is that fixed costs, which are mainly infrastructure-related, and variable costs, which are energy production related, are both very substantial; at an overview level, it is reasonable to think of them as being roughly equal. But today, utility rates are nonetheless "as if" almost all costs are variable costs.

2.2.2 Revenue Decoupling

At the same time the WAA was being implemented, California decoupled the commodity unit sales (the variable costs) from the cost-of-service, or revenue requirement, required to run the system. The goal was to remove the utility's inherent incentives – absent decoupling – to overbuild the infrastructure and sell more and more energy to customers. Under decoupling, the IOUs in California design rates to collect the required revenues based on forecast commodity sales. However, since forecasting is inevitably imperfect, the decoupling mechanism allows for balancing accounts, so that if commodity sales turn out to be higher or lower than forecast, the over- or under-collection of revenue is subtracted from or added to future rates. Commodity sales of electricity tend to be very stable, so forecasts are rarely more than a few percent different than actual sales. Decoupling is covered in greater detail in the chapter by Cavanagh.

Referring back to the earlier example, if the fictional utility sold 5.1 million kWh instead of the predicted 5.0 million, or 2 percent more than forecast, it would collect 2 percent more revenue than authorized by the regulators for that period of time, or a $20,000 overcollection. The overcollection would be recorded in a balancing account and refunded to customers in the form of a slightly reduced rate in the following year.

2.2.3 Inverted Block Rates

Another prominent feature of California residential rates is the inverted block structure, a rate structure with four rate blocks with increasing per unit prices designed to send sharp price signals to reduce incremental power consumption. The inverted blocks provide a modest amount of electricity to each residential customer each month at a low rate known as Tier 1 or the baseline allowance. After reaching the baseline amount, customers are charged a somewhat higher rate for Tier 2 usage. Tiers 3 and 4,[4] however, feature very sharp increases compared to the first two tiers, up to 300 percent more than Tier 1. An example of these tiered rates for certain residential PG&E customers is shown in Table 11.2. The tier allowances vary to account for California's 16 climate zones, such that customers in the warmer regions of the state are allowed larger amounts of power in Tiers 1 and 2 to account for higher power requirements associated with air conditioning.

One practical impact of this structure is that average per kWh rates for individual residential customers vary widely from the average kWh rate overall, with low-use customers paying an average rate substantially below the system average while high use customers pay substantially more. However, the revenue requirement – and allowed revenue recovery – for the class does not change. The tier structure is characterized as revenue neutral; although intended to reduce consumption and hold down costs long term, in the near term the rate structure is designed to recover the costs incurred by the utility and approved by the regulators as reasonable for the class.

2.3 Background on California Investor-owned Utilities, Energy Efficiency, and Renewables

California's customer-focused energy efficiency efforts, implemented by utilities under CPUC regulation, have been popular and effective. California IOUs have offered energy efficiency programs for over 30 years, but in that time program

TABLE 11.2 PG&E Residential Electric Rate Schedule E-1 (July 2012)

Tier 1 (Baseline)	Tier 2 (101–130% of Baseline)	Tier 3 (131–200% Baseline)	Tier 4 (201–300% of Baseline)	Tier 5 (Over 300% of Baseline)
$0.12845/ kWh	$0.14602/kWh	$0.29561/kWh	$0.33561/kWh	$0.33561/ kWh

Source: PG&E website - http://www.pge.com/nots/rates/tariffs/electric.shtml.

4. The Tier 5 rate is currently equal to the Tier 4 rate, so there are only four distinct rate blocks.

and utility goals, funding levels, target energy savings, and targeted customer seg-ments have changed significantly [1]. Since 1980, IOU spending on energy effi-ciency programs has been at least $200 million per year (in constant 2002 dollars) every single year, averaging over $400 million annually. Much of this funding has historically taken the form of cash rebates and incentives to help defer the incremental cost of new, more efficient equipment for customers. Much effort has likewise been devoted to energy efficiency training, information, and public outreach. The programs are funded by all ratepayers. Collectively, the funding does add more than 1 percent to the cost of an average kWh in CA. However, the programs have been deemed to save consumers more money than they cost.

In the last decade, California has also started to push aggressive distrib-uted generation goals. A Renewables Portfolio Standard (RPS) target that designates that a certain amount of a California utility's portfolio come from clean, renewable source energy was established in 2002 at a 20 percent target from renewable sources by 2017, and was later expanded in 2011 to be 33 percent by 2020.[5] California has additional targets for distributed genera-tion, or small-scale power generation located close to electricity loads. These targets, with associated aggressive incentive programs, have had a significant impact on the number of solar installations in California.

According to the National Renewable Energy Lab's (NREL) "Open Solar Project," California far outstrips other states in the number of solar installa-tions. At the time of writing, California solar installations are shown as exceeding the installations of the runner-up state, New Jersey, by more than a factor of 16, with Southern California counties, Fresno County, and Bay Area counties as the hottest spots. California clearly has a size advantage in such a comparison, but the uptake of solar in the state has also been spurred on by a number of state-wide and local efforts.

One such effort, "Go Solar California" (GSC), is an umbrella initiative by the CPUC, CEC, and the IOUs that includes the California Solar Initiative, the New Solar Homes Partnership, and a range of other programs.[6] The joint campaign goal is "to encourage Californians to install 3,000 megawatts of solar energy systems on homes and businesses by the end of 2016, making renewable energy an everyday reality" and represents a ratepayer funded investment of over 3.3 billion dollars between the years of 2007 and 2017 (Go Solar California) [16]. The GSC campaign aims to do this through policies that reduce the costs and smooth the pathway to solar installations "through four inter-related state policies: rebates, net energy metering (NEM), interconnection policies, and rate structures (e.g., tiered rates, time of use rates)" [2]. The cam-paign has had enormous early success. In particular, the California Solar

5. Detailed information on the California Renewables Portfolio Standard can be found on the CPUC website: http://www.cpuc.ca.gov/PUC/energy/Renewables/
6. More information can be found on the Go Solar California web portal at: http://www.gosolar-california.org/

Initiative reported that as of January 2011, the CSI program had installed over 61 percent of all California solar. Even as the market costs of solar have gone down and the rebates have accordingly decreased – the program began with rebates of $2.50 per watt in 2007 and stepped down in phases to $.35 per watt in 2011 due to market demand – the program continued to grow dramatically, with January of 2011 reporting a record number of installations [2].

The ZNE goals adopted by the CPUC, CEC, and ARB bring together the goal of increasingly aggressive energy efficiency in buildings and the goal of having a significant amount of grid-connected distributed generation into one concise policy target.

2.4 California Zero Net Energy Policy in Context

In 2006, the California legislature passed Assembly Bill 32 (AB 32), the California Global Warming Solutions Act, which requires California to reduce its greenhouse gas emissions to the 1990 level by 2020.[7] This reduction is to take place in large part through a number of coordinated state-wide efforts overseen by the ARB. The 2008 Climate Change Scoping Plan[8] provides a roadmap for reducing greenhouse gas emissions in California through a variety of direct regulated and voluntary actions that include a cap and trade program, a low carbon fuel standard, and targets for transportation related greenhouse gas emissions, among other measures.[9] Commercial buildings

---

7. Assembly Bill No. 32, CHAPTER 488, An act to add Division 25.5 (commencing with Section 38500) to the Health and Safety Code, relating to air pollution. Approved by Governor September 27, 2006. Filed with Secretary of State September 27, 2006.

8. The Climate Change Scoping Plan is available on the ca.gov website at: http://www.arb.ca.gov/cc/scopingplan/document/adopted_scoping_plan.pdf

9. A related piece of legislation, SB375 (Steinberg, 2008) requires a new coordinated regional planning process that links land use planning decisions to transportation emissions for the first time, as an additional means of achieving AB 32 goals [11]. SB 375 organizes state agencies into 18 Metropolitan Planning Organizations (MPOs) to produce regional strategies for the targeted reduction of vehicle emissions, giving priority to transportation and land use strategies that encourage compact development and intensive uses around transportation corridors, and discourage sprawl development [17]. Under this process the MPOs must engage in a coordinated land use and transportation planning process that will demonstrate the MPOs' ability to reach specified 2020 and 2035 emissions reduction targets. This new process works with and subsumes the already established Regional Transportation Plan (RTP), Regional Housing Needs Assessment (RHNA), and the California Environmental Quality Act (CEQA) processes. Further, it requires consistency and coordination between regions. It should be noted that local governments are not required to comply with the land use plans that come out the new regional planning processes; however, this is germane to the ZNE conversation because by its nature SB375 and the new regional planning processes will *encourage* densification and development patterns that will result in reduced vehicle miles traveled (VMT) through transportation funding incentives, and streamlined CEQA processes, while dis-incentivizing non-compliance with the SCS ,such as the loss of transportation dollars. If the ZNE 2020 and 2030 targets (also non- mandatory) are layered upon this, it can be inferred that already urbanized areas will be *incentivized* to densify further under a compact growth scenario and perhaps receive a substantial share of new (ZNE) construction.

account for 8 percent of California's greenhouse gas emissions and residential buildings account for 14 percent. Energy efficiency in new and existing residential and commercial construction is therefore a crucial component in realizing substantial reductions in greenhouse gas emissions in California [3].

To help tackle California's ambitious greenhouse gas reduction goals, the CPUC recognized the need for new long-term planning around energy efficiency. In September 2008, the CPUC adopted the California Long-Term Energy Efficiency Strategic Plan (Strategic Plan), which details a future vision for California energy efficiency, including four "Big Bold" goals [15].[10] Two of the Big Bold goals put forth in the Strategic Plan focus on ZNE − that all new residential construction in California be ZNE by 2020 and that all new commercial construction be ZNE by 2030. These goals have also been adopted by the CEC and the ARB.[11] Although the ZNE goals are not part of the mandatory and voluntary greenhouse gas reduction activities overseen by ARB, they are nonetheless aligned in purpose with the state-wide push to comply with AB 32. There have been several failed legislative attempts to mandate the 2020 and/or 2030 goals.[12] While many California stakeholders have roles that can support the 2020 and 2030 ZNE goals, the CEC's role in developing building and appliance energy efficiency standards that push new construction towards ZNE is crucial.

2.5 The Role of California Energy Efficiency Standards

Under the WAA, the building energy efficiency standards set by the CEC must be cost-effective and must not cause unreasonable disruption to industry compared to the amount of energy saved.[13] In the 2009 Integrated Energy Policy Report, the CEC stated as the first priority action that "The Energy Commission will adopt and enforce building and appliance

---

10. The "Big Bold Energy Efficiency Strategies" from the California Long Term Energy Efficiency Strategic Plan: 1) All new residential construction in California will be zero net energy by 2020; 2) All new commercial construction in California will be zero net energy by 2030; 3) Heating, Ventilation, and Air Conditioning (HVAC) will be transformed to ensure that its energy performance is optimal for California's climate; and 4) All eligible low-income customers will be given the opportunity to participate in the low income energy efficiency program by 2020.

11. CARB adopted the ZNE goals through the *2008 Energy Action Plan*, and the CEC adopted the goals through *2007 IEPR*.

12. The failed legislative attempts to mandate ZNE include: 1) AB 2112 (introduced by Saldana and Lieu in 2008) which was originally for ZNE residential by 2020 and ZNE commercial by 2030 but was amended to just cover residential; 2) AB 2030 (introduced by Lieu in 2008) was for commercial ZNE buildings; and 3) AB 212 (introduced by Saldana and Lieu in 2009) covering zero net energy residential buildings.

13. Section 25008 of the Warren−Alquist Act (May 2012) states, "It is further the policy of the state and the intent of the Legislature to promote all feasible means of energy and water conservation and all feasible uses of alternative energy and water supply sources."

standards that put California on the path to zero net energy residential build-ings by 2020 and zero net energy commercial buildings by 2030." The CEC updates the California Energy Efficiency Standards for Residential and Nonresidential Buildings, also referred to as Title 24, every three years, so there are several opportunities to push building codes to ZNE by 2020 and 2030. The CEC plans to update the building efficiency standards with energy efficiency measures that will increase the required energy efficiency at 20 to 30 percent per code cycle [12]. The schedule for future code cycles is shown in Figure 11.1. ZNE is a performance-based metric. Using the exist-ing code-compliance framework, which is based on the building assets, not the building's long-term performance, Title 24 will determine if a building is capable of being ZNE before it is occupied.

The most recent update for Title 24 was approved by the CEC on May 31, 2012, and will go into effect in January 2014, following approval by the California Building Standards Commission. The current update to Title 24 improves the efficiency of residential construction by 25 percent and non-residential construction by 30 percent over the previous standards [13]. Because ZNE buildings also include a generation component, the new standards will require that both new residential and new commercial build-ings are equipped with "solar ready" rooftop connections. This Title 24 update specifically addresses hotwater pipe insulation and air conditioning system installation verification for residential buildings and addresses light-ing controls and efficient process equipment for commercial buildings. The new standards also make recommendations for building envelope and insula-tion, higher performance windows, and cool roofing materials. Figure 11.2 illustrates certain new required and voluntary measures for residential and commercial buildings under the updated standards.

According to the CEC, buildings constructed after the new building energy efficiency standards are in effect are expected to:

- Save $1.60 billion in energy over a 30-year life;
- Save 200 million gallons of water per year; and
- Avoid more than 155 thousand metric tons of greenhouse gas emissions per year [4].

The CEC is also working to update appliance standards to include personal electronics and other plug loads [12]. Plug loads have a special significance to ZNE buildings because they are typically not addressed from the architecture-side; they are the add-on electrical uses from appliances, computers, and other personal equipment and devices that come in with the building occupants, after the building is complete. However, it is becoming increasingly apparent that once mechanical system and lighting loads are greatly reduced through integrated design, plug loads become a significant percentage of the overall building energy use. Since plug loads are currently unregulated and usually not addressed in the design phases it will be an

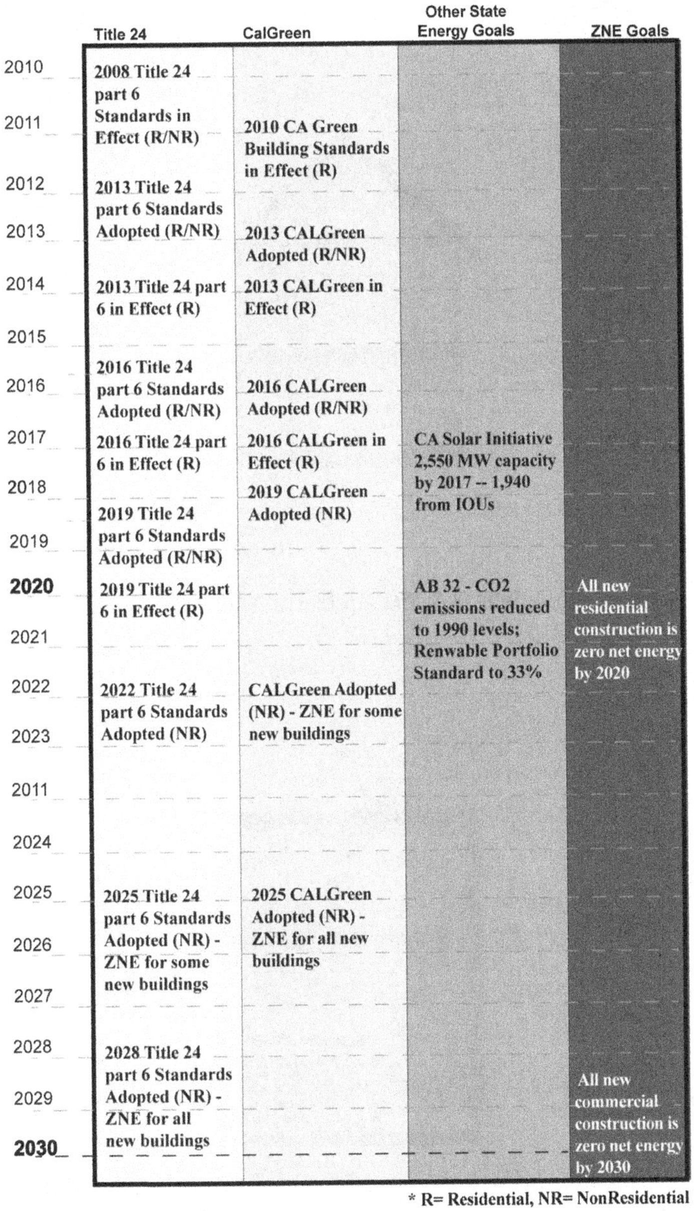

Year	Title 24	CalGreen	Other State Energy Goals	ZNE Goals
2010	2008 Title 24 part 6 Standards in Effect (R/NR)	2010 CA Green Building Standards in Effect (R)		
2011				
2012	2013 Title 24 part 6 Standards Adopted (R/NR)			
2013		2013 CALGreen Adopted (R/NR)		
2014	2013 Title 24 part 6 in Effect (R)	2013 CALGreen in Effect (R)		
2015				
2016	2016 Title 24 part 6 Standards Adopted (R/NR)	2016 CALGreen Adopted (R/NR)		
2017	2016 Title 24 part 6 in Effect (R)	2016 CALGreen in Effect (R)	CA Solar Initiative 2,550 MW capacity by 2017 -- 1,940 from IOUs	
2018	2019 Title 24 part 6 Standards Adopted (R/NR)	2019 CALGreen Adopted (NR)		
2019				
2020	2019 Title 24 part 6 in Effect (R)		AB 32 - CO2 emissions reduced to 1990 levels; Renwable Portfolio Standard to 33%	All new residential construction is zero net energy by 2020
2021				
2022	2022 Title 24 part 6 Standards Adopted (NR)	CALGreen Adopted (NR) - ZNE for some new buildings		
2023				
2011				
2024				
2025	2025 Title 24 part 6 Standards Adopted (NR) - ZNE for some new buildings	2025 CALGreen Adopted (NR) - ZNE for all new buildings		
2026				
2027				
2028	2028 Title 24 part 6 Standards Adopted (NR) - ZNE for all new buildings			All new commercial construction is zero net energy by 2030
2029				
2030				

* R= Residential, NR= NonResidential

FIGURE 11.1   Building Energy Efficiency Standards Update Schedule.[14]

14. CALGreen is a state-wide green building code with both voluntary measures that go beyond the energy efficiency required by Title 24.

CALIFORNIA'S 2013 - RESIDENTIAL BUILDING ENERGY EFFICIENCY STANDARDS
RECOMMENDED   REQUIRED

 whole house fan

Displaces warm air with cool outside air on cool summer nights.

 improved windows

Improved windows keep the sun's heat out of your home during hot summer months and keep warm air during winter months imporving comfort and reducing energy consumption.

 insulated walls

Better insulation reduces heating and cooling costs while improving comfort at home.

 solar-ready roof

Makes space available on the rooftops for easier installation of optional photovoltaic or solar thermal panels at a future date.

 hot water pipe insulation

Pipe insulation improves the overall efficiency of a home's hot water system.

 verify air conditioner installation

Improper installation of your cooling system reduces its efficiency. Having the installation verified by an independent inspection guarantees your air conditioner will operate as designed.

CALIFORNIA'S 2013 - NONRESIDENTIAL BUILDING ENERGY EFFICIENCY STANDARDS
RECOMMENDED   REQUIRED

 cool roof

Lighter colored roofing material reflects more of the sun's heat energy away from the building. This reduces a building's electricity bill by decreasing the amount of air conditioning required.

 high efficiency heating and cooling

To improve indoor comfort and reduce energy use, variable speed HVAC systems efficiently match heating and cooling requirements to a building's electricity budget.

 improved windows

Improved windows keep the sun's heat out of the building during hot summer monthe and keep warm air in during winter months. This improves comfort and reduces the building's energy use.

 solar-ready roof

Makes space available on the rooftops for easier installation of optional photovoltaic or solar thermal panels at a future date.

 lighting controls

Sensor-based lighting controls for fixtures located near windows adjust the lighting by taking advantage of available natural light.

 efficient process equipment

Improved technology offers significant savings by providing more efficient refrigeration equipment for supermarkets, computer data centers, and commercial kitchens.

FIGURE 11.2   Select Measures Included in the Building Energy Efficiency Standards. *Source: Adapted from Title 24 2013 infographics, CEC website: http://www.energy.ca.gov/title24/ 2013standards/, accessed 6/25/12.*

important area to address to ensure ongoing ZNE building performance. There will still be certain loads that are only regulated by the federal government and other loads that are not regulated at all.

2.6 Zero Net Energy Efforts from the Utilities

The primary role of gas and electric utilities is to deliver safe, reliable energy. Since the 1970s, the CA IOUs have the additional role of delivering energy efficiency expertise and information to the public. Both roles are relevant to the discussion in this chapter. All of the California IOUs have ZNE-focused program activities as part of their energy efficiency portfolios; however, the spending on ZNE activities is a small effort compared to the overall spending on other energy efficiency efforts.

PG&E Zero Net Energy Pilot Program

The PG&E Zero Net Energy Pilot Program was approved by the CPUC in 2010 and supports the Strategic Plan by offering design and technical assistance for new residential, commercial, and community-scale projects with ZNE goals, completing assessments of relevant building practices and technologies, and engaging in a range of education and outreach activities.[15] As a utility, PG&E can affect the future energy efficiency of buildings by either providing financial incentives to building designers and owners that are directly tied to energy performance or by addressing specific barriers to the design, construction, and operation of high-performance buildings. There are three main areas in which the ZNE Pilot Program activities are focused:

- Design and technical assistance to selected residential, commercial, and community-scale new construction projects;
- Technical assessments and research, including a roadmap to ZNE in California and a study on the technical potential for ZNE in commercial and residential new construction in California; and
- Outreach and education, including workshops, stakeholder forums, development of best practice design guidelines, and an annual design competition, Architecture at Zero.

The ZNE goals adopted by the CEC that are being incorporated into Title 24, and championed by utility ZNE programs and other stakeholders, focus on new construction. However, it is important to clarify that this focus does not mean

---

15. For the purpose of the program, "zero net energy" is defined as "the implementation of a combination of building energy efficiency design features and on-site clean distributed generation that result in no net purchases from the electricity or gas grid, at the level of a single 'project' seeking development entitlements and building code permits" per CPUC Decision 07-10-032, October 18, 2007. Projects must be grid-tied.

that only brand-new construction projects will be subject to ZNE performance goals. Rather, most buildings undergo significant renovations over their life-cycles, and in general, new construction building codes apply to these significant renovations. In this light, the ZNE goals for new residential and new commercial construction in California will touch a significant number of buildings in the coming years.

3 PATHWAYS TO THE 2020 AND 2030 GOALS

When the CPUC put forward the Strategic Plan and its "Big Bold" ZNE goals in 2008, they did so with the explicit intention that the finer-grained details of the implementation pathway would need to be developed and established over time. In fact, the pathway to ZNE in new construction by 2020 and 2030 is not without obstacles. It is useful to group these barriers into three main areas: (1) physical and technical, (2) financial and economic, and (3) policy. These issues are deeply interrelated and not neatly separable; for example, the policy set to define ZNE affects the technology choices used to achieve ZNE as well as the economic criteria used to evaluate these choices. In some respects, the barriers can be seen in terms of stakeholder and regulatory inertia; various components of state policy around rates and ratemaking have been highly impactful in reducing the state's per capita footprint, but some of those same policies are now potential barriers to the ZNE goals.

3.1 Physical and Technical Barriers to Zero Net Energy

Physical barriers to ZNE fall into two main categories: the challenges associated with getting individual buildings to ZNE and the challenges associated with the existing power grid's ability to absorb and distribute the large amounts of customer-sited photovoltaics (PV) required to achieve ZNE.

As noted above, a ZNE building is fundamentally a very energy efficient building with on-site renewable generation. As a practical matter, a ZNE building needs to have a low energy density (or kBtu footprint) and a relatively large roof compared to its overall square footage — in other words, it needs to be a low-rise building. Although there are plenty of exceptions, somewhere in the range of 75 percent of all buildings by square footage are already low-rise with modest kBtu footprints. These include such buildings as schools, warehouses, small office buildings, strip malls, single-family housing, and small apartment and condominium complexes. Driven by new technologies and increasingly stringent building codes, it is not difficult to foresee a future where the majority of residential and commercial square footage can have sufficiently low kBtu footprints such that the available rooftop area can generate the power needed to offset the electrical usage.

To provide a quantitative overview, and considering the electric load only for the time being, California's Commercial End Use Survey (CEUS) reports that the average commercial building in California consumes about 13 kWh per square foot annually (2006 data). A number of building types are below that amount. Residences, for instance, consume far less. For building types near this average figure, efficiency improvements in new construction would make it possible to reduce loads to about six kWh per square foot or below. Existing warehouses are already below five kWh per square foot. PV arrays can generate in the range of 16 kWh per square foot of panel, although the roof area available for such arrays is often in the range of 30 to 50 percent of gross roof area, reducing the output proportionately to about 5 to 8 kWh per gross square foot of roof area annually. This is not to say that it is a *fait accompli* that ZNE will occur for the majority of building stock, only that there do not appear to be insurmountable physical or technical barriers for much of the square footage to be built as ZNE. In fact, for nearly all residential buildings and for more than half of commercial buildings, it could be said that the ZNE goal is visible on the horizon from a technical standpoint within the 2020 and 2030 timeframes. The technically challenging buildings with respect to the ZNE goals would be urban high-rises, buildings with location-related solar access issues, and buildings with dense energy footprints such as data centers, restaurants, supermarkets, laboratories, and miscellaneous other buildings. Over time, high-efficiency solutions to serve the loads in such buildings will likely emerge. However, a strict prescription that *all* new buildings individually will be ZNE by 2030 or at any time in the foreseeable future is likely to remain an aspirational goal.

The second category of technical challenges involves the grid and integration of ZNE buildings into grid operations. The physical and technical issues with high-penetration scenarios involving ZNE and PVs are not well known. It is recognized that there will be potentially significant impacts on the grid from the high penetration rate of PV that will necessarily occur to the extent that the ZNE initiative succeeds. For example, one significant research project is under way to characterize and model circuits with high penetrations of distributed generation in order to be able to simulate and evaluate PV penetration levels for impacts on the grid and potential value [5]. One of the conclusions so far — and this point has been made before — is that solar PV provides kWh that can help meet daytime load, but the PV misses the evening peak. Many have pointed out that the distribution grid and its components have been designed and constructed for one-way current flow, from the substation to the customer. Under high penetration ZNE scenarios, this would no longer be the case. What happens over time to all of the distribution transformers and distribution control devices on the grid under these new conditions? What if localized peak grid demand becomes dominated by "outbound" PV production instead of by the consumption of the building?

The technical solutions to these and related problems are not likely to be particularly difficult, but they will need to be deployed at enormous scale for California to reach the ZNE goals of the Strategic Plan. It is essential that the issues be well understood so that effective, least-cost solutions can be developed. Although there is great societal value to the carbon-free energy that PVs produce, it does not appear likely that ZNE and PV deployment "at scale" will reduce distribution-level costs for grid operators; it will almost certainly increase those costs. Suffice it to say, the potential problems are not yet characterized in sufficient detail to determine the right mix of solutions — and what they might cost.

3.2 Financial and Economic Barriers to Zero Net Energy: California Rate Mechanisms and Net Energy Metering

As noted at the outset of this section, the lines distinguishing technical barriers from economic barriers or policy barriers are not necessarily clean and distinct. The two classes of physical barriers described above may ultimately prove to be economic barriers — that is, once the technical paths become clear, it is only a matter of determining how to pay for them. Other barriers that can be described in economic terms share a border with policy issues. Issues around rates and ratemaking fall into that realm.

In section 2.2, several fundamental rate policy mechanisms were described: the concept of a utility revenue requirement, of revenue decoupling, of volumetric rates, and of inverted block rates. In this section, the economic impact of these rate mechanisms on ZNE and DG, together with the concept of net energy metering (NEM), will be examined.

Residential IOU customers in California with PV systems are billed under "net energy metering" (NEM) arrangements. Under this tariff, the utility meter at the site effectively "spins backwards" whenever the output of the solar system exceeds the building load. Clearly, this phenomenon occurs during daytime hours in the summer; customers still rely on the power grid for electricity at night, and during most of the winter. For a residence to achieve ZNE on a site basis, the kWhs generated must equal or exceed those consumed on an annual basis.

Net energy metering[16] definitely motivates customers, particularly those with large amounts of expensive Tier 3 and Tier 4 usage, to install PV; customers with high usage and NEM are displacing very expensive electricity. The tension arises due to the revenue requirement by customer class discussed earlier. Once the revenue requirement is established for the class, the utility sets rates to collect that revenue from all customers in the class. If a given set of customers under NEM pays no net revenue to the grid operator, the remaining customers in the class must make up the shortfall through higher rates.

At relatively low rates of market penetration (say 1−2 percent), the bill impact on the "non-participants" is minimal. Since the PV participants had been, on average, relatively high users, they had previously paid somewhat

---

16. More information on net energy metering can be found on the CPUC website: http://www. cpuc.ca.gov/PUC/energy/DistGen/netmetering.htm

more than "average" rates owing to the tier structure. Simply for illustrative purposes, if 2 percent of the residential class installed PV systems, the revenue shortfall (which must be compensated for by the other class members) would be closer to 2.7 percent than 2 percent since expensive, high-cost Tier 3 and 4 electricity was displaced along with the associated revenue. As penetration rates of ZNE and PVs escalate, so does this phenomenon.

If the PV installations actually reduced the cost to operate the system by an amount equal to the revenue reduction, the math would all work out. Under the rate-making process, a future revenue requirement proceeding would reflect (again, illustratively) that 2.7 percent reduction in the cost to serve the residential class in the example above. A balancing account would be used to adjust rates to account for the value of the PV purchased, and all would be well. Although revenue would be down, costs would be down by an equal amount, and the grid operator and the non-participating customers would be made whole. However, few believe that the revenue avoided by NEM customers, which is effectively revenue paid to NEM customers, correlates well with the value of the grid savings attributable to their PV systems.

As the NEM system currently exists, a substantial revenue shortfall − with no clear offsetting savings − must be made up by the non-NEM-participating customers in the class. One holdover provision from California's energy crisis at the turn of the last century now comes into play. Californians are still paying for that crisis, and will continue to do so for years to come. One compromise reached to settle the issue provided that strict limits be placed on the amount by which Tiers 1 and 2 electricity rates could be allowed to increase; this was done to protect customers who use relatively small amounts of energy (and who, on average, are less well-off financially than those who use more). In short, increases in the revenue requirement beyond modest inflationary increases must be assigned to Tiers 3 and 4. This prescription widened the gap between lower cost Tiers 1 and 2 and higher cost Tiers 3 and 4.

What this means in an NEM context is that Tiers 3 and 4 must absorb the bulk of the revenue shortfall which could develop under rapid deployment of PVs. Customers staying mainly within Tiers 1 and 2 would not see dramatic impacts from NEM shortfalls, but customers that already have substantial Tier 3 and 4 usage − those customers that already have the highest bills − could see sizable new increases, especially during the warmest months. Although the utility would be made whole (mathematically) by the balancing account mechanism, it would need to face the prospect of increasing numbers of cost-sensitive, dissatisfied customers facing significant bill increases.

The market has taken note; as of summer 2012, solar vendors in northern California are marketing solar leasing programs that can provide solar installations to customers at little or no out-of-pocket cost to the customer with energy bills of a certain size. In effect, the solar company installs the system on its dime and the customer pays a monthly bill to the solar company, a bill "just like your utility bill, only lower" in the words of a current advertisement. This arrangement has the best economic feasibility for customers whose bills are dominated by expensive Tier 3 and 4 usage, and solar company advertising appears to reinforce this notion.

The problem with this approach becomes apparent by considering the following hypothetical case: if it were somehow possible for all residential customers to install PV systems and become ZNE tomorrow under today's tariff arrangements, the utility would be left with *zero revenue to serve the entire residential class*. This scenario would not be unique to IOUs; it would occur with any entity, including a municipally owned utility, that operated with the rate features described in this section. In theory, the grid operator would be charged with operating the system while receiving no net revenue. Yet the grid operator must still supply nighttime and wintertime electricity, which must be generated, transmitted, distributed and so on, all through an extensive power grid that must be operated, maintained, fixed in emergencies, and periodically upgraded. Clearly, the revenue requirement to do all of this grid operation and upkeep would not be zero or anything close to zero. Even allowing for the substantial societal cost reductions abundant PV power could create, a rate allowing the utility, or other grid operator, to collect zero revenue to operate the grid is clearly a nonstarter, irrespective of grid ownership and operation structures.

Understanding the Scale of the Power Grid: California's Two Largest IOUs

PG&E serves about 15 million people, which is roughly 40 percent of California's population, or just under 5 percent of the United States' population. PG&E reports over 141,000 circuit miles of distribution lines – San Francisco to New York, roundtrip, about 28 times – and 18,000 miles of higher voltage transmission lines. It has nearly 1,000,000 distribution transformers in its 70,000 square mile service territory.

Southern California Edison serves about 14 million people within a 50,000 square mile area and has slightly less transmission and distribution (T&D) infrastructure. Together, the utilities employ about 36,000 people. Despite the high societal value that large-scale deployment of ZNE and the accompanying PVs would produce, the need for the basic grid infrastructure and human resources required to operate and maintain it would not be reduced.

The solar industry and its advocates argue, correctly, that this scenario is obviously not going to happen tomorrow or anytime soon, and therefore NEM is needed to keep the industry moving forward and contributing to a sustainable future. There is certainly merit to this argument and time to adjust the relevant rate mechanisms and business models. Nonetheless, some urge caution even today in continuing down a path where the PV business case depends upon and requires the offset of high-cost electricity, which for policy reasons

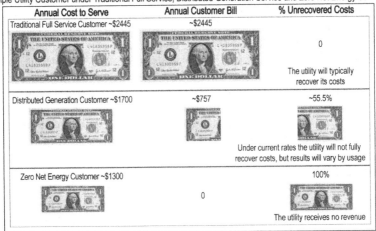

Sample Utility Customer under Traditional Full Service, Distributed Generation Service and Zero Net Energy Service*

Annual Cost to Serve	Annual Customer Bill	% Unrecovered Costs
Traditional Full Service Customer ~$2445	~$2445	0
		The utility will typically recover its costs
Distributed Generation Customer ~$1700	~$757	~55.5%
		Under current rates the utility will not fully recover costs, but results will vary by usage
Zero Net Energy Customer ~$1300	0	100%
		The utility receives no revenue

*Sample numbers from the Rocky Mountain Institute Report, Net Energy Metering, Zero Net Energy and the Distributed Energy Resource Future, p. 30

FIGURE 11.3   The Cost of Providing Service to Residential Customers.

is priced far above average or market rates. In the larger future vision — if anything close to the ZNE vision is to be achieved — some fundamental structural changes are needed to utility business and cost recovery models.

Summarizing, what have become traditional rate-making features in California — volumetric rates, inverted block rates, revenue decoupling — have the potential, when coupled with NEM, to create a perverse impact that could harm the industry it is designed to support. At low market penetration rates, these features strongly encourage the development of customer-sited small PV systems. But, at scale — which is the ultimate objective of the ZNE policy — these same features create an unsustainable marketplace, one that requires increasing numbers of PV installations be paid for by fewer and fewer customers.

Figure 11.3 details the utility costs to serve and the electricity bills for a sample traditional, full-service customer, a distributed generation customer, and a ZNE customer under the current tiered, volumetric rate structure, with values from the Rocky Mountain Institute (RMI) study.

3.3 Policy Barriers to Zero Net Energy: ZNE Definitions and Their Impact

With respect to the development of the Title 24 building energy efficiency standards, the CEC has adopted and is advocating a "societal value" definition of ZNE where "the societal value of energy consumed by the building over the course of a typical year is less than or equal to the societal value of the on-site renewable energy generated" [6]. In this case, societal value is a variation of time dependent valuation such that it is "the long-term projected cost of energy including the cost of serving peak demand and varies from hour-to-hour to account for peak demand and other fluctuating costs including projected costs for carbon emissions; for example, the time dependent valuation (TDV) of energy" [6]. The CEC argues that by using a metric that accounts for the societal value of energy, it will include the "critical impact of avoiding peak demand and the value of avoided carbon emissions," as well as other system costs [12]. This approach by the CEC in incorporating the societal value of energy is consistent with the time dependent valuation of energy used in current California efficiency standards for buildings.[17]

On the building side of the meter, TDV appears to work well; the higher valuations associated with peak energy savings in fact reduce the need for new generation and reduce what would otherwise be needed in terms of T&D infrastructure. TDV encourages measures such as peak-hour lighting reductions through daylighting controls, air conditioners that work efficiently during heat storms, and air conditioning load reduction measures such as cool roofs. However, applied to power exported from the customer to the grid, the logic and calculation mechanics break down. The TDV system simply does not provide information about avoided distribution grid costs in situations where power is exported to the grid. Under high penetration of ZNE and PVs, a completely new type of peak congestion issue could emerge; in short, instead of needing to size the distribution system to handle peak air conditioning loads requiring delivery of energy to customer, the system sizing criteria could become dominated, especially in localized areas, by power exports. This is not necessarily a doomsday scenario in any sense, but the TDV system, if applied in reverse, would simply not provide reliable information about avoided incremental costs under this scenario, nor was it designed to do so.

In sum, the concept of minimizing societal costs is a robust concept and completely consistent with the foundation established by the Warren–Alquist Act. Although existing TDV metrics are well established on the demand side of the equation, TDV provides no information or guidance about the cost (or savings) of power exported by the customer to the grid at the distribution level. Much work remains to be done in this area.

Without a consensus definition of ZNE, there is also no agreement on what successful implementation of ZNE will look like. However ZNE is ultimately defined for California, it is useful to think of the definition as being necessarily comprised of at least two parts. One part would be the metric that is used to define the energy accounting at the site, source, or grid

---

17. Detailed information on time dependent valuation can be found on the CEC website: http://www.energy.ca.gov/title24/2005standards/archive/rulemaking/documents/tdv/

level (i.e., TDV, "societal value," or other variant) and another that would define exactly what it is that is being measured (i.e., building-by-building ZNE, multi-building ZNE, community or regional-scale ZNE, or some combination of those).

If ZNE "equivalency" is included in the building energy efficiency standards, the CEC will need to establish a clear framework for determination and enforcement. The current code framework puts the onus on building owners and designers to create an energy efficient, code-compliant building. This is fundamentally an asset rating where the building is assumed to be capable of a certain level of performance. However, since ZNE is a measure of actual performance over the course of a year, what happens if the asset rating for a building is zero but the building does not perform accordingly? Where does accountability rest? Because ZNE is fundamentally a performance metric and not a standard to be met on a one-time basis, what is the appropriate interval for compliance under a ZNE scenario? How would corrections be made and assessed? Questions of this nature remain to be answered in code discussions.

4 IMAGINING ZERO NET ENERGY AT SCALE

Even with the policy, financial, and physical barriers that exist for ZNE in California, there is significant momentum among stakeholders that support the goals. There is interest from the design community, progress in the implementation of building energy efficiency standards in the direction of requiring ZNE capable buildings, and discussion around looking at the ZNE goals at a campus or community scale rather than building by building. However, a major concern among stakeholders remains that there is not a clear business model for utilities or other potential investors around widespread implementation of ZNE.

4.1 Current Status and Near Term of Zero Net Energy in California

There is momentum behind the ZNE goals. On the policy side, the 2013 version of the Title 24 building efficiency standards was recently approved by the CEC, and a recent decision by the CPUC broadened the definition of NEM, expanding availability. One of the reasons that there is so much momentum behind the ZNE goals in California is that unlike goals positioned relative to the changing baseline of building energy efficiency standards (i.e., 35 percent better than Title 24 2008), the concept of a ZNE building where "energy in equals energy out" seems comparatively straightforward and more easily quantifiable. Perhaps it is this reason that has allowed "zero net energy" to capture the imagination of builders, environmentalists, and the public in a way that more modest energy efficiency and renewable energy goals have not. California already has more ZNE buildings than any other state, although the number is not large and they are generally self-proclaimed to be ZNE.[18]

However, at this point there is no single, consensus definition of ZNE that includes a clear, well-defined metric for evaluating building performance with regard to ZNE. There are a series of metric-related challenges around site and source accounting (which vary regionally and over time). There are also issues around the proper valuation of energy saved at the building versus energy exported from the building to the grid. Two nominally identical buildings that are ZNE but that "achieve" ZNE performance with different strategies might easily impose substantially different costs or savings upon the grid. Ideally, building codes and utility rates would both encourage buildings to reach ZNE along a pathway that considers the full societal value and cost represented by the building and its performance. Determining which definition of ZNE[19] to use for policy and energy efficiency standards, and which metrics to use for evaluation of code compliance or ongoing building performance, will require major coordination among state agencies and buy-in from the IOUs and other stakeholders. Several areas are of concern with respect to metrics and definitions, as follows:

- *Valuation metrics on PV power exported to the grid.* Systems and metrics are needed to evaluate the true cost (and value) of electric power exported to the grid at the distribution level considering "at scale" deployment of ZNE and PVs. What happens when the critical capacity factor for a given distribution circuit becomes dominated by exported power rather than power delivered to customers? What kinds of control features are needed on the circuit to properly control power flow? Is there a point at which the cost of needed infrastructure upgrades exceeds the value of the exported power?

- *Site/source metrics as applied to appliances in buildings.* Better knowledge, measurement, and metrics are needed to guide the "site/source" discussion wherein the losses associated with electric power generation, together with the greenhouse gases production, are properly accounted for once they reach the site. In most of the United States, these losses are dominated by the combustion losses at the power plants. However, these losses can vary considerably according to the fuel mix, average heat rate, and amount of utility-scale renewable generation in a given region. For those

18. That California has more ZNE buildings than other states is supported formally by references in IEPR 2011 and the NBI report, and informally by increasing anecdotal evidence from building designers and researchers.

19. There are several definitions of ZNE. For more background on ZNE definitions, see "Zero Energy Buildings: A Critical Look at the Definition" by Paul Torcellini, Shanti Pless, and Michael Deru from the National Renewable Energy Laboratory, 2006. These definitions are also discussed in the previous chapter.

areas with higher average heat rates (or less efficient generation overall), source energy calculations would favor natural gas appliances over electric appliances at the building level. Without proper accounting for the site/source mix, a result might be inefficient uses of electricity requiring larger and more expensive than otherwise necessary PV installations. One way or the other, all ratepayers end up footing the bill if society encourages the use of expensive power sources to serve inefficient utilization.

Unless otherwise specified, many stakeholders in California discussing a ZNE building are generally using a version of the "zero net site energy" definition. While this definition may not be especially useful as a grid-level least cost planning tool, it is at once energizing, clear, and useful from the point of view of individual building design — it is a concept that has truly captured the imagination of many in the building design community. A commonly cited rule of thumb is that ZNE site buildings in California should have a target energy use of at least 50 percent below levels required by Title 24 2008. A major research project being conducted in 2012 by Arup and coordinated by the CA IOUs and CPUC Energy Division is modeling the technical potential for achieving ZNE commercial and residential buildings in California. One of the results of the study will be specific energy use intensity (EUI) targets (in kBtu/sf) for a range of building types, including medium and large offices, grocery stores, restaurants, retail, lodging, schools, and warehouses, for several representative California climate zones.

While the goal of getting all new commercial and residential new construction to be ZNE may seem futuristic, the buildings that have achieved quantifiable ZNE performance have all done so with building strategies and equipment that are "commercially available today" [7]. In fact, for all buildings in the United States, Griffith, et al., have estimated that "based on projections of future performance levels. . .62 percent of buildings could reach net zero" and "47 percent of commercial building floor area could reach the ZEB goal" [8]. One of the main reasons that buildings in that study were not able to get to ZNE is the roof area did not allow for enough solar PV to produce the energy required to offset the building loads (estimated based on high-performance building practices and technologies at the time). Significant improvements in energy efficient or generation technologies could help tip the balance so that more buildings could achieve ZNE performance. Since most of California generally has a fairly mild climate, relatively low heating and cooling loads, the percentage of buildings that could reasonably achieve ZNE with existing technology is likely higher in California than for the United States in general. There are fledgling efforts underway by the California IOUs to develop incentive programs to encourage high-efficiency and ZNE buildings.

The early ZNE commercial buildings have been a few consistent building types — primarily academic buildings and environmental centers — and with a few exceptions are relatively small buildings [7]. In the near future, specific types of buildings, such as warehouses and schools, will achieve ZNE performance earlier than others due to their lower EUIs and typically low-rise forms with large roof areas. Single-story and low-rise buildings, in general, have the best chance of achieving ZNE because they have lower EUIs and a roof area to floor area ratio that allows for enough for the on-site renewables needed to offset the building loads on an annual basis. Other building and building types are more difficult to design and operate for ZNE performance due to the size, occupancy pattern, the roof area to floor area ratio, or the building location, especially if it is in a dense urban area or other site that may not have sufficient solar or wind access for renewable energy generation. The Griffith study boiled it all down to four characteristics that determine whether a building is able to achieve ZNE: "(1) number of stories; (2) plug and process loads; (3) principal building activity; and (4) location" [8].

The current strategy of moving ZNE requirements for new residential and new commercial construction through building efficiency standards is a building-by-building approach. The Griffith study notes that with aggressive energy efficiency measures, many of the buildings modeled could produce more energy than they consume on site. If these "net positive" buildings are grouped with buildings that are not able to meet ZNE due to a reason such as building type, location, occupancy pattern, or insufficient solar access, the group may be able to achieve ZNE performance in aggregate.

There are other potential benefits to looking at ZNE at a larger scale. Buildings might be able to share larger energy efficiency or energy generation infrastructure that is higher-performance than a system at the individual scale, such as a district heating or cooling system. A study for the California Solar Initiative by E3 suggested an approach to reduce costs in reaching solar goals for both customers and the utility:

"...might be to promote a shift toward larger, community-based solar PV systems through a 'solar shares' or 'virtual net metering' approach. Larger systems have lower costs and better TRC results than rooftop solar, particularly as compared to smaller residential systems. Lowering costs through virtual net metering would improve TRC results and improve participant economics for smaller residential customers. Virtual net metering (VNM) installations would be located near customers, but would be larger systems (for example up to 5 MW or up to 20 MW), of which customers would 'purchase' a share. Such an arrangement would offer the economies of scale necessary to reduce systems costs, while maintaining a sense of customer ownership, as customers would see the installations in their communities. The systems could be utility-owned, community-owned, or third-party-owned" [9].

A number of communities in California have expressed interest in ZNE goals at a community scale as they wrestle with new targets to reduce greenhouse gases under AB 32.

While there may be many opportunities and potential models that present themselves when thinking about achieving ZNE at the aggregate level, pursuing large-scale ZNE would require a major feat of coordination among policymakers, utilities, and state energy agencies and industry. Currently, there are some preliminary efforts and programs underway for larger deployments of renewable energy systems in California, but there still remain many regulatory/logistical/permitting hurdles, high upfront costs, and cost recovery disincentives for developers who may wish to pursue district-scale energy efficiency and renewable systems for ZNE communities.

4.2 Business Model Issues: Who Would Invest in This?

As utilities and other stakeholders look towards a future with more and more ZNE buildings, there are growing concerns about how these ZNE customers are being essentially subsidized, and how that subsidization may eventually drive more and more customers to add distributed generation [14]. Rocky Mountain Institute advises "revising net metering rules and utility rates" to "reduce the conflicts created by the cross-subsidy implicit in the current rate systems which requires that customers without solar systems pay part of the cost of the utility services to those customers with solar" [10]. There have been attempts to revise net metering rules and rates, but it is very controversial, as demonstrated by the following San Diego Gas and Electric Company (SDG&E) example from the RMI study:

"As part of its General Rate Case in October 2011, SDG&E proposed modifying its residential electric rates to include a 'Network Use Charge,' which would bill customers for the costs associated with network use... SDG&E proposed to measure a customer's absolute demand as a basis for the charge, which would account for the fact that when a customer-generator has a negative demand − therefore exporting power − the customer is still in fact using the utility's network (NEM vs. NUC). Proponents of the Network Use Charge note that it would allow SDG&E to ensure that NEM customers pay for their fair share of distribution costs when exporting power... However, the measure met with fierce opposition from the solar industry, consumer advocates, environmentalists, and NEM customers. These groups argue that the Network Use Charge does not account for the benefits that DG systems provide to the network, that it runs contrary to California's renewable energy goals by discouraging solar, and that it does not send price signals that encourage reduction in coincident peak demand − rather, it pushed PV owners to shift their demand to times when their system is producing, i.e. midday" [10].

One of the issues in trying to revise net metering rules and rates is that there is not a good process by which stakeholders can determine the value created by ZNE buildings and the associated distributed generation systems when they are implemented at scale. Once the potential for value creation is understood and quantified, rates can be adjusted to encourage the desired behavior.

A potential approach is to adjust rates and other price signals to shift how customers optimize efficiency and generation choices and behavior, which may include such things as price incentives for shifting loads or shifting the time of electricity generation, such as by changing the orientation of solar to optimize for variables other than maximum production. Customers will increasingly be making individual decisions to deploy energy infrastructure and will be basing their choices on rates, incentives, and system availability that may or may not be set up to optimize the overall public good or to minimize impact on the grid. However, an organization, whether a utility or other investor, will be needed to monitor loads and maintain the grid and will need to be able to make a profit while doing so. Ultimately, investors need to have a compelling business case, and it is the regulatory community that sets the framework to allow that to happen.

What does all of the above mean for the changing role of utilities in California? Achieving the 2020 and 2030 ZNE goals would result in a different role for utilities, changing from having "control over investment decisions and operational management for most electricity system assets" to "coordinating a vast array of supply- and demand-side resources owned and operated by... independent actors" [10]. This is not idle speculation about the future of utilities in California. Up to "one fourth of the total new investments in generating capacity in PG&E's service territory between 2012 and 2020 could come on the customer's side of the meter," primarily as rooftop solar PV systems [10]. If California is successful in meeting these ZNE goals, which will require aggressive and unprecedented energy efficiency in California new construction as the first step, then declining electricity demand per building will occur in lockstep with increased penetration of distributed generation, especially solar generation.

If California meets the 2020 and 2030 ZNE goals, utilities will be less focused on selling units of energy and more focused on selling ways of using or of generating energy in an optimal way, whether for the building owner or occupant, the grid, the community, society at large, or the utility itself.

5 CONCLUSIONS

The utility grid was built to generate, transmit, and distribute electricity from a network of large central stations to individual customers. An elaborate policy and business framework supports that model — one that has served well. In a ZNE future, a framework that supports the *exchange of value* among many players — not "one way" power delivery — will be needed.

There is significant momentum behind the 2020 and 2030 ZNE goals in California. Policymakers have taken action to move building energy efficiency standards and rules related to distributed generation towards ZNE performance. Many enthusiastic building designers and local leaders have been galvanized by the seemingly straightforward goal and are beginning to think about how best to implement ZNE buildings for different building types and at different scales.

In the near term, achieving stakeholder agreement around a single ZNE definition for use in policy decisions — especially around building code issues — is needed, and soon. Included in the key issues that must be addressed are:

- Clarification of "site/source" metrics that address fuel switching and the use of natural gas
- Adoption of a robust system *within the building code* that accounts for and properly values the power to be exported by the building to the grid

The issue of a strict policy for building-by-building ZNE, implying PVs on every rooftop, bears reexamination. Does building-by-building ZNE represent a least-cost solution? Does the building-by-building approach supplant opportunities for district-level solutions that appear to have the potential to offer higher levels of efficiency? The solution would necessarily be complex, requiring changes to building codes, business models for property developers, and, of course, utility rules. The topic is ripe for further discussion.

Longer term, the role of purely volumetric rates as the core model of value exchange bears reexamination. Traditional volumetric rates inherently devalue infrastructure components of the grid — and it is new investment in exactly these components that will be required for society to capture the full potential value of ZNE buildings. Demand response approaches may hold promise in this respect. However, these tend to be superimposed on top of the core volumetric rate structure and are widely regarded by many customers as punitive in nature rather than as an opportunity for value exchange.

All of these issues lead into what may be the biggest issue of all: how can the proper investment incentives be placed before grid operators? What should those incentives be? If district-level solutions are determined to have great potential, who will build them — that is, where will the investment opportunity be placed? It appears clear that in the coming decades, substantially fewer electrons will be flowing from the power plant, through the transmission system, through distribution circuits, and on to the customer's building. High-efficiency buildings and communities themselves will require substantially less electric power on an annual basis, but they will need to be served by a grid that is at least as sturdy and robust as it is today. As society moves toward significant numbers of ZNE buildings and PV deployment, traditional grid operators will face a daunting dilemma: a modernized grid that can support high-performance ZNE buildings will still require new investment. New business models, with new — and robust — investment incentives will be required to realize the vision.

ACKNOWLEDGMENTS

We would like to thank Matt Heling and David Schoenberg of PG&E for directing us to specific research and data sources.

REFERENCES

[1]   Rajkovich NB. Zeroing in on zero net energy. In: Shioshansi F, editor. Energy, Sustainability, and the Environment: Technology, Incentives, Behavior. Oxford: Butterworth-Heinemann; 2011. p. 497–517.

[2]   Sterkel M. Update on the california solar initiative. Presentation. Retrieved from <www.cpuc.ca.gov/PUC/energy/solar>; 2011.

[3]   California Air Resources Board. (CEC). Climate change scoping plan: a framework for change. Retrieved June 22, 2012, from <http://www.arb.ca.gov/cc/scopingplan/document/adopted_scoping_plan.pdf>; 2008.

[4]   California Energy Commission. (CEC). State of california state energy resources conservation and development commission docket no. 12-BSTD-1 - [Proposed] order adopting proposed regulations and negative declaration. Retrieved June 25, 2012, from <http://www.energy.ca.gov/title24/2013standards/rulemaking/notices/2012-05-31_Draft_Adoption_Order.pdf>; 2012, June 25.

[5]   Sison-Lebrilla E. Seeing impacts of high penetration PV. Presentation. 2011.

[6]   Diamond R. A zero net nergy definition for residential and commercial buildings. Memo 2011 (draft).

[7]   New Buildings Institute. (NBI) Getting to zero 2012 status update: a first look at the costs and features of zero energy commercial buildings. Research report. 2012.

[8]   Griffith BL. Assessment of the technical potential for achieving net zero-energy buildings in the commercial sector. National renewable energy laboratory. 2007.

[9]   Energy and Environmental Economics, Inc. California solar initiative cost-effectiveness evaluation. San Francisco; 2011.

[10] Rocky Mountain Institute. (RMI) Net energy metering, zero net energy and the distributed energy resource future: adapting electric utility business models for the 21st century. snowmass, Co. 2012.

[11] Adams TE. Communities tackle global warming: a guide to SB 375. Retrieved from <http://www.nrdc.org/globalwarming/sb375/files/sb375.pdf>; June 2009.

[12] California Energy Commission. (CEC) Integrated energy policy report. California energy commission 2011.

[13] California Energy Commission. (CEC) (n.d.). Retrieved 06 25, 2012 <http://www.energy.ca.gov/releases/2012_releases/2012-05-31_energy_commission_approves_more_efficient_buildings_nr.html>.

[14] Cardwell D. Solar panel payments set off a fairness debate. NY Times 2012, June 4; Retrieved June 26, 2012, from <http://www.nytimes.com/2012/06/05/business/solar-payments-set-off-a-fairness-debate.html?_r=1&hp>.

[15]   CPUC. California long term energy efficiency strategic plan: achieving maximum energy savings in california for 2009 and beyond. San Francisco: CPUC; 2008.

[16]   Go Solar California. (n.d.). Retrieved June 22, 2012 <http://gosolarcalifornia.org/>.

[17]   Higgins B. Technical overview of SB 375. League of california cities. Retrieved from <http://www.calapa.org/attachments/wysiwyg/5360/SB375TechOV.pdf>; 2008.

Chapter 12

Zero Net Energy At A Community Scale: UC Davis West Village

Stephen M. Wheeler and Robert B. Segar

University of California, Davis

1 INTRODUCTION

If societies are to reduce greenhouse gas (GHG) emissions by 80 percent or more by mid-century, zero net energy (ZNE) and positive energy developments will be required strategies for success. The California Public Utilities Commission (CPUC) has in fact called for all new residential construction in the state to be ZNE by 2020 [1]. Building of this sort generally combines extreme energy efficiency with active generation of renewable energy (see chapters by Rajkovich et al and LaRue et al). Undertaking such development at a neighborhood scale is particularly desirable in order to bring a large number of ZNE buildings online quickly and to take advantage of district-wide systems design and economies of scale. Yet few such neighborhoods currently exist anywhere in the world. The University of California, Davis' West Village, a new neighborhood for 4,200 students, faculty, and staff that opened its first phase in 2011, is intended to be the first large-scale ZNE development in the United States. It is a landmark project that, while still in the early stages, serves as an important model for community-scale ZNE development.

The number of ZNE neighborhood-scale developments in the world is still very small, and there has been relatively little systematic evaluation of their performance. The first and perhaps best-known example is the Beddington Zero-Energy Development (BedZED) in the London borough of Sutton. Designed by architect Bill Dunster and the BioRegional Development Group, BedZED opened in 2002 as a mixed-use, mixed-income development of 82 housing units along with commercial space, a daycare center, and exhibit space. The project combined highly energy-efficient construction with passive solar design, a small-scale combined heat

Energy Efficiency. DOI: http://dx.doi.org/10.1016/B978-0-12-397879-0.00012-8

and power plant initially burning wood from municipal tree-trimming (considered to be a carbon neutral fuel), and a green transportation plan emphasizing walking, cycling, and use of public transport. An initial evaluation in 2003 found that BedZED housing units used 88 percent less energy than the British average for space-heating and that residents used 25 percent less electricity than average [2].

Other examples of neighborhood-scale, very-low-energy development include Hammarby Sjostad, a development in Stockholm expected to include 10,000 apartments housing 25,000 people when completed in 2016. Although not ZNE, the project aims to produce about half the energy it uses through photovoltaic panels and an efficient district heating system burning waste to produce energy. Vauban, a new neighborhood for 5,000 residents on the site of a former French military base in the German city of Freiburg, includes 100 units built to "passive house" standards, with no active heating or cooling systems, and 59 "plus energy" homes. The Kronsberg district in the German city of Hannover also includes 32 passive houses.

Masdar City, a new community for up to 50,000 people currently under construction near Abu Dhabi in the United Arab Emirates, is perhaps the world's most ambitious project aiming for ZNE status. In addition to energy efficiency measures, the project is expected to include large-scale solar power plants, wind farms, and geothermal energy. In the United States, a number of ZNE buildings have been built, but the largest neighborhood-scale ZNE development previous to West Village appears to be the partially built Green Acres subdivision of 25 upscale homes in the Hudson River Valley town of New Paltz some 85 miles north of New York City.

At UC Davis, West Village planners did not initially have a ZNE goal in mind. This objective emerged in the middle of the design process in part due to rapidly growing concern about climate change in California at the time. The zero net outcome was made possible by a creative public-private partnership, the forward-looking policy context in the state, institutional leadership at the University of California, and grant assistance from state and federal sources. Through its evolution and current operation, West Village helps illustrate the opportunities for ZNE development at the district scale, as well as some of the challenges facing such projects.

Section 2 discusses the background of the project and the process that led to its ZNE configuration, emphasizing procedural, institutional, and geographic factors that contributed to the project's development. Section 3 describes the energy efficiency measures that made West Village more than twice as efficient as required by California's already-stringent Title 24 building code. Section 4 describes the development's renewable energy strategies, principally employing PV, and section 5 discusses the project's other sustainability elements followed by the conclusions and what can be learned from the West Village example.

2 BACKGROUND AND CONTEXT

West Village is a new neighborhood for an eventual 4,200 residents on the existing campus of the University of California, Davis. This college town of about 65,000 people is located approximately 15 miles west of Sacramento in California's Central Valley (Figure 12.1). Initially opened with apartment housing for 800 students in 2011, the entire first phase of West Village will consist of apartment housing for approximately 1,980 students. Student apartments are typically designed as two and four bedroom units in three-story buildings, with additional apartments on the upper floors of four-story, mixed-use buildings surrounding a central village square. The first floor of the these mixed-use buildings at the village square is lined with approximately 45,000 square feet of dining, retail, and office space.

As the urban activity hub of West Village, the village square also includes a student community center with recreation and study spaces, and a community college center – the first of its kind on any University of California campus. Rounding out the mixed-use community will be 343 two-story, single-family homes for sale to UC Davis faculty and staff, operating on a model of capped appreciation intended to keep the housing affordable for future generations. A second phase of West Village is expected to include additional apartment units and single-family homes (Figure 12.2).

FIGURE 12.1 Davis CA Location Map.

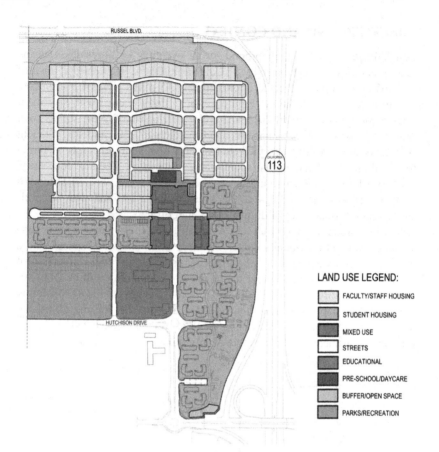

FIGURE 12.2   Site Plan of U.C. Davis West Village. The East-West orientation of streets and buildings in West Village makes possible passive solar orientation of buildings. *Source: Carmel Partners. U.C. Davis Office of Campus Planning and Community Resources.*

The layout of West Village is an integration of three main planning strategies. In line with principles of the New Urbanism, the community includes a grid-like network of streets and mixed-use public spaces designed to optimize sociability, walkability, and effective transportation. Secondly, as an extension of the local planning model in the City of Davis, the community includes an extensive greenway system that doubles as a stormwater drainage network and a comprehensive off-street and on-street bicycle network. Davis prides itself for being the most bicycle-friendly small city in the United States. And third, to take advantage of local climate factors, the layout of West Village utilizes long east/west streets to optimize solar access and capture summer cooling breezes from the

Sacramento River delta to the southwest of the site. Parking is provided at a ratio of three spots for every four apartment residents, located on the periphery of the site to allow uninterrupted bicycle and pedestrian access to apartment courtyards. Single-family homes are served by alley garages and streetfront visitor parking.

Planning for West Village began in 2000, after citizens of Davis approved a ballot measure preventing any future development on agricultural land or open space without voter approval. This severe growth constraint meant that private developers were unlikely to be able to provide substantial additional housing in Davis for UC Davis anticipated growth in faculty, staff, and students. The university was also concerned at the time about rapidly escalating housing prices. As part of a planning process to forecast campus needs, university planners put forward a variety of growth scenarios. Several scenarios included building a substantial new community on university-owned land, while others left staff, faculty, and students to find market-built housing in other cities and commute to Davis. The university held a series of workshops on such options with members of the Davis community in the early 2000s, engaged consultants such as well-known ecological designer Bill McDonough to help the community imagine a positive framework for growth, and prepared a long-range development plan (LRDP) and associated environmental impact report (EIR) by 2003. In addition to accommodating substantial growth in the teaching, research, and student services programs of the university, project goals included providing affordable housing for students, staff, and faculty. By choosing to develop a substantial community on university land, the campus reduced the negative environmental effects on air quality and transportation systems caused by the commute scenarios, and created a diverse, walkable community close to existing development on campus.

Energy objectives became more central to the planning of West Village in the mid-2000s due to state actions, rising public and professional concern about climate change, and the opening of multiple energy institutes on campus. Specifically, the signing of Executive Order S-03-05 by Governor Arnold Schwartzenegger in 2005 – committing the state to reducing greenhouse gas emissions 80 percent by 2050 – and legislative passage of AB 32 in 2006 – setting a 2020 target of reducing emissions to 1990 levels – provided major catalysts to public and professional interest in climate change. With California committed to becoming a global leader on climate change by the mid-2000s, officials at institutions such as the University of California began considering how they might assist this effort and develop green technologies and skills on their own campuses.

Although energy research had long been a strength of units within UC Davis such as the Department of Civil and Environmental Engineering and the Institute of Transportation Studies, the university established a number of new energy-related centers in the 2000s that became important resources for West Village planning. These centers included the following:

- The California Lighting Technology Center (established in 2003)
- The Sustainable Transportation Center (established in 2005)
- The Energy Efficiency Center (established in 2006)
- The Biogas Energy Project (established in 2006)
- The Western Cooling Efficiency Center (established in 2007)
- The Center for Water Energy Efficiency (established in 2009)

The initial technological and financial strategies for dramatically reducing energy use at West Village arose from a project advisory group convened by the newly formed UC Davis Energy Efficiency Center in 2006, with additional consulting assistance provided by the Davis Energy Group. This effort produced a set of increasingly aggressive energy efficiency 'packages' to lower demand. Encouraged by the prospect of deep reductions in energy demand through investments in efficiency, the project team began to pursue strategies for supplying all remaining energy demand through renewable sources. Modeling by the Energy Efficiency Center estimated base-case electricity consumption for the multi-family buildings at 13.8 kWh/sq ft/yr if built to 2008 California Title 24 standards. The modeling shows the energy efficiency package proposed by the center then reducing that amount to 5.6 kWh/sq ft/yr, a 58 percent reduction. For the single-family houses, modeling showed base case consumption to be 16.2 kWh/sq ft/yr, falling to 5.7 kWh/sq ft/yr with the deep energy conservation measures, a 65 percent reduction. The analysis indicated that total electricity use for the entire first phase of the project, including commercial spaces, the community college, and common area lighting, would be 23,295,000 kWh/yr, which could be reduced to 9.803,600 with aggressive energy efficiency improvements, a reduction of 58 percent [3]. Such data gave project planners optimism that renewable sources could then meet the remaining electricity demand, including up to 5 MW of PV, fuel cells run on biogas from a biodigester producing up to 300 kW, and a 1 MW battery (Figure 12.3).

During this time planners also secured grant funding to reduce project energy consumption from multiple sources. Grant sources included the following:

- The U.S. Department of Energy's Community Renewable Energy Deployment program ($2.5 million)
- The California Public Utility Commission (CPUC) California Solar Initiative ($2.495 million)
- The California Energy Commission's (CEC) Public Interest Energy Research (PIER)
- Renewable-Based Energy Secure Communities program ($1.94 million)

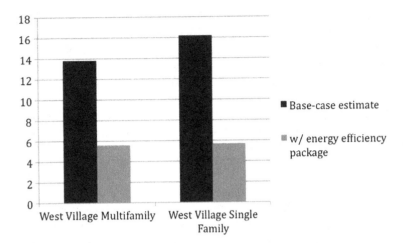

FIGURE 12.3 Modeled Energy Savings with West Village Efficiency Package (kWh/sq ft/yr). A package of added insulation, radiant barrier roof sheathing, solar reflective roofing, and high efficiency lighting, appliances, and HVAC units is expected to lead to a decrease of more than 50 percent in West Village energy use beyond California's already-strict Title 24 standards.

Pacific Gas & Electric (PG&E) and Chevron Energy Solutions provided assistance during this period as well. Actual construction of West Village began in 2008 with initial work on apartment housing for about 800 students in 315 apartment units, plus the community college facility serving about 2,400 students, and the village square mixed-use buildings. The university desired to build this central core of the neighborhood in the initial phase to create a heart to the community and a home for community-serving facilities. After this portion of Phase 1 opened in 2011, apartments rented quickly, and construction began on additional student housing units. The real estate market crash beginning in 2008 delayed development of the for-sale single-family homes; it is hoped that construction of that phase of West Village will begin as soon as the market recovers.

UC Davis did not have the capacity to construct the project from university funds, allocating most capital funds toward educational rather than residential buildings. The West Village community was conceived from the onset as a partnership with a private developer in order to finance and construct the project. Selected in 2004 through a request-for-qualifications process, West Village Community Partners (WVCP) is a collaboration of Denver-based Urban Villages, which has extensive experience in constructing walkable communities of similar scale to West Village, and San Francisco-based Carmel Partners, with significant experience financing and managing multi-family housing communities. UC Davis expended approximately $17 million to bring utilities to the border of the site, an investment that will be recouped through land lease payments from the developer and future homeowners. With a 65-year ground lease agreement with the university, WVCP anticipates a total investment of approximately $280 million to build out Phase One of the project. The financial model for the project was designed to place no financial burden on residents of West Village. In practice, the additional costs for energy efficiency investments and solar panels are recovered by the residents' "energy bill." The bill, in actuality, is not for power consumed, but to pay back the cost of efficiency and green power installations. The West Village breakthrough is to accomplish this redirection of funds towards SNE at no greater cost to the resident than the "business as usual" scenario.

Like other projects West Village is located within a specific geographical, political, and institutional context. Understanding this context and the development process is essential in order to determine its replicability and lessons for other mixed-use communities. First of all, the relatively mild climate of California's Central Valley greatly affects building energy consumption, especially for natural gas (far and away the primary heating fuel in California). Since winter temperatures rarely touch freezing, energy needs for heating are far less than in many other parts of the world. Winter is California's rainy season, but average annual precipitation is only 19 inches, and abundant sunshine makes passive solar building architecture relatively effective at further reducing heating needs. Summers are hot, with daily high temperatures averaging 94 and 93 degrees, respectively, in July and August, leading to high demand for air conditioning and correspondingly high electricity load on summer afternoons. Generally clear summer skies facilitate photovoltaic production during those hours, making Davis (like much of the American Southwest) an ideal environment for photovoltaics (PV).

Also, Davis is located in the part of the Central Valley closest to the Golden Gate, the natural gap in the state's coastal ranges leading to San Francisco Bay. Although Davis is some 70 air miles from the Golden Gate, cooling evening breezes from the Pacific Ocean often reach the town in warm weather – the "Delta breeze" – further reducing air conditioning needs especially if buildings are designed to capture those cool breezes through effective window placement, the provision of operable windows, and ventilation fans. So West Village has utilized many design and energy efficiency strategies to take advantage of its geographical context. The watchword of the project became "First Reduce, then Produce" – optimizing benefits from community design, passive solar orientation, active energy efficiency investments, and only then, renewable energy production on-site to cover the radically reduced energy demand. The political, cultural, and institutional contexts of West Village are important as well. As previously mentioned, the acceleration of state, regional, and local climate change planning within California beginning in the mid-2000s established an atmosphere in which institutions actively explored strategies to reduce greenhouse gas emissions.

Even before, student activism had led the UC Board of Regents to adopt a Green Building Policy in 2003 requiring that campus buildings at least meet the equivalent of a LEED "Certified" rating. In practice, many recent UC buildings have exceeded this standard, achieving certification at Silver, Gold, or Platinum levels. The City of Davis also had a tradition of ecological development dating back to the 1970s, epitomized in particular by the Village Homes development that pioneered neighborhood-scale passive solar design, albeit in a relatively low-density, primarily residential format (e.g., [4]). The city is also known for some of the most extensive bicycle infrastructure in North America, for a very extensive greenway system, and for strong land-use policies to prevent sprawl and provide a mixture of housing and commercial facilities in each neighborhood [5]. In 2005 the city was the first in the United States to receive "platinum" bicycle friendly community status from the League of American Bicyclists. Davis residents, including many UC Davis administrators and staff, were thus inclined to be favorably disposed toward forward-looking green building projects.

Equally important, the University of California is not subject to municipal land-use planning approval for university-related growth, reducing the impact of community opposition that can often emerge as an insurmountable obstacle to development. University staff did seek to coordinate as much as possible with local citizens, elected leaders, and municipal staff. Campus planners conducted an extensive set of public workshops in the early 2000s and downsized the project substantially in response to neighbor concerns, also establishing a buffer of open space between West Village and the nearest residential neighborhoods in West Davis. These efforts pleased some city constituencies, but did not stop other neighborhood groups from contesting the adequacy of the EIR through litigation. Although the university prevailed in the lawsuit, this process set the project back about a year. If city permits had been required for development, similar neighborhood opposition would certainly have arisen, so the university's special status quite likely saved the project from a lengthy and highly contentious permitting process.

3 ENERGY EFFICIENCY STRATEGIES

The cornerstone of ZNE strategies at UC Davis West Village has been energy efficiency. Project planners realized that only by greatly lowering West Village energy demand below California's already-strict Title 24 standards through an array of energy efficiency strategies would it then be possible to consider meeting ZNE goals with renewable technologies. Energy efficiency measures were modeled as a package for different types of structures; the proposed measures produced a 65 percent saving for the single-family homes, a 58 percent savings for the multifamily structures, a 50 percent savings on common area lighting, and lower levels of savings for commercial and institutional building space [6].

Energy strategies at West Village evolved considerably during the decade between project conception and the opening of Phase 1, and it was only through flexibility and creative action by the university, the developer, energy-related institutes at the campus, outside consultants, and granting agencies that the project could consider attaining ZNE status. Passive solar design was part of the strategy from the beginning, but a major rethinking of goals occurred in 2006, when campus planners commissioned the UC Davis Energy Efficiency Center and faculty from the Graduate School of Management to develop an initial roadmap to reducing energy demand through energy efficiency. This strategic thinking enabled the university to apply for and obtain $7.5 million in grants from the previously mentioned sources to plan energy and conservation systems. This funding, much of it only available to research institutions, enabled intensive studies on energy demand, efficiency strategies, energy infrastructure design, and financing, and proved essential in developing ZNE concepts. Consultants and industrial partners such as SunPower, PG&E, Chevron Energy Solutions, Energy & Environmental Engineering, and the Davis Energy Group provided additional assistance in the following years, for example, by providing models to help calculate energy savings or production from various proposed features.

Energy efficiency performance in the West Village structures is achieved primarily through added insulation, radiant barrier roof sheathing, solar reflective roofing, high efficiency lighting fixtures, high efficiency Energy Star appliances, and high efficiency HVAC units. Exterior walls use $2'' \times 6''$ framing to give them added thickness to accommodate the additional insulation. Floors include an additional ½″ of gypcrete to increase thermal mass. Roofs benefited from both R-49 blown insulation and radiant barrier roof sheathing. Extensive hard-wired fluorescent or LED lighting in units, with vacancy sensors, reduces the need for occupants to add their own less-efficient lighting fixtures. The UC Davis Lighting Technology Center and the Davis Energy Group advised the architect of the mixed-use housing, San Diego-based Studio E Architects, on energy control systems such as occupancy sensors and dimming controls to manage lighting demand in exterior spaces. Student housing architect MVE Institutional, based in Oakland, included similar features in its portion of the development, and emphasized strategies to provide real-time data to residents on their unit's energy consumption.

Passive solar design is one of the main energy efficiency strategies at West Village. South-side windows let in low-angle winter sunlight, while strategically placed roof overhangs and sunshades above window frames keep out high-angle summer rays that could potentially overheat units (Figure 12.4). Buildings around the central square were a particular challenge for passive solar design, since long east- and west-facing frontages were unavoidable. Studio E Architects sought to avoid overheating of these buildings from late-afternoon summer sun by placing moveable wooden

FIGURE 12.4   Passive Solar Design. Passive solar design features such as this large roof over-hang on the south side of the Activity Center building help reduce summer heating of interior spacing and associated cooling needs. *Source: Stephen M. Wheeler.*

louvers on rails outside windows. In the same way that people living in Mediterranean countries have covered their windows with shutters for centuries, West Village residents can shade their rooms with these louvers. The architects also oriented large windows and patio doors in each unit to take advantage of cross breezes and to increase natural daylighting. Meanwhile, vertical corrugated metal on south and west sides of the buildings helps create a thermal shield for building walls and ventilate the facades [7].

A related emphasis of West Village is to promote energy-conserving behavior among residents through a variety of strategies. For much of the housing, web and smartphone applications are under development to provide residents with real-time information on energy consumption. These systems are designed to provide residents with access to programmable controls for lighting, appliances, and electrical outlets, although the contractor responsible for installation of these feedback systems did not meet initial performance goals, and how they will perform in practice remains to be seen. However, such informational systems when fully in place should help lead to reduced energy demand by giving residents tools to show how behaviors such as turning off lights and computers or adjusting cooling levels can modify real-time energy consumption.

Research funded by the California Public Utilities Commission grant continues into other strategies that may further reduce energy use at West Village [8]. These strategies include optimizing installed West Village storage battery capacity and use, evaluating other storage options for future use, evaluating the possibility of adding solar thermal arrays to the existing rooftop PV, revised financial models for energy systems on as-yet-unbuilt portions of the project, and developing battery-coupled solar charging stations for plug-in vehicles associated with the neighborhood's single-family housing. A number of energy systems are likely to be modified in response to actual performance. For example, initial results show that the photovoltaic arrays are producing more than 100 percent of electricity consumed by the apartment buildings. For future buildings it therefore may be possible to reduce the amount of PV in favor of combined PV and solar thermal rooftop systems.

4 ON-SITE ENERGY GENERATION

In addition to the energy efficiency measures discussed above, West Village actively generates its own electricity through extensive use of photovoltaic panels. This generation is intended to offset both electricity use and natural gas use on the site. Originally planners envisioned a large, separate solar array on university-owned land to produce electricity for the new neighborhood. However, detailed studies proved this option unrealistic due to the cost of extensive conduit and inverter infrastructure that would have been required and the lack of available financial incentives to justify this strategy. At this point, the design approach shifted to placing all PV within the footprint of the community, on building rooftops and parking lot canopies. Studio E Architects designed central buildings with a saw-toothed roof configuration that maximizes south-facing surfaces for PVs; south-, east-, and west-facing roofs of the student apartment buildings were also used for solar panels (Figures 12.5 and 12.6). SunPower Corporation, based in San Jose, then installed four megawatts of PV capacity in Phase 1 to power the student apartments and mixed-use buildings. To maximize rooftop PV, project designers refrained from adding solar thermal units for hot water, although this is still a possibility on later phases of the project. As installation of each phase of PVs is completed, annual performance data will be used to inform future phases of project design.

The 2006 West Village energy strategy also planned for construction of a biodigester in which bacteria and enzymes would break down campus dining hall waste, landscape clippings, and manure from campus barns to produce biogases such as methane and hydrogen that could be burned to generate electricity. Both food processing water and animal wastewater can potentially be used as feedstocks as well. The technology is relatively new: although past biodigester systems have used anaerobic decomposition of

FIGURE 12.5   View of the West Village Square. West Village's central square. Building roofs slope in a saw-tooth configuration to maximize southern exposure for photovoltaic panels. Strategically placed shades prevent the sun from overheating buildings. *Credit: Stephen M. Wheeler.*

FIGURE 12.6   Aerial image of West Village under construction. This view from early 2011 shows the extensive PV on south-facing roof surfaces and parking awnings. *Credit: U.C. Davis Office of Campus Planning and Community Resources.*

liquid wastes to produce energy, substantial problems with materials handling, speed of digestion, and economics have prevented commercial application of systems designed to use mixed wet and dry ingredients. In 2006 UC Davis professor Ruihong Zhang in the Biological and Agricultural Engineering Department formed a UC Davis Biogas Energy Project to address these challenges. Assisted by some of the grant funds mentioned above, her group has extensively tested potential feedstocks and has constructed and tested a model Anaerobic Phased Solids Digester on the UC Davis campus that can turn eight tons of waste per day into enough electricity to power 80 households. Zhang's previous work led to the unveiling in 2012 of the first commercially available, high-solid anaerobic digestion system in the United States, marketed under license by Clean World Partners, a Sacramento-based startup company that is also planning to build the UC Davis system [9]. This system produces energy in about half the time of previous systems, and generates a greater variety of gases that can be burned to produce energy. The same technology is planned for a larger facility on the UC Davis campus, which in the initial energy modeling was seen as necessary to meet ZNE goals (Figure 12.7).

As siting studies and energy analyses have progressed, the campus has realized that it stands to gain greater efficiencies by placing the biodigester facility at a location that may not directly serve West Village, a site near the campus landfill about one mile away. If electricity from this facility does not serve West Village directly, the West Village project may purchase biogas

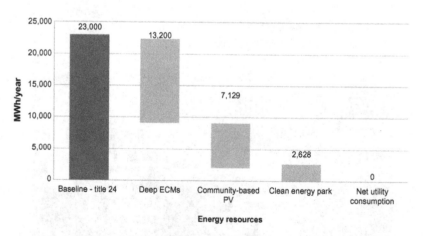

FIGURE 12.7   Modeled Path to Net-Zero Energy. Energy modeling in the late 2000s for West Village showed that ZNE status could be met through a combination of deep Energy Conservation Measures (ECMs), photovoltaic generation of electricity, and other technologies (principally a biodigester) in a Clean Energy Park. Initial results indicate that the PV alone may be sufficient to offset energy use. *Source: [6].*

offsets through PG&E in lieu of utilizing output from the biodigester itself. Since the biodigester was seen as producing about one quarter of the electricity used by Phase One of West Village, to achieve ZNE status this amount will need to be made up through the purchase of such offsets, additional PVs, and/or increased efficiencies and incentives for residents to reduce demand. However, by helping catalyze development of biodigester technology, West Village has performed a major service for more sustainable neighborhood-scale energy systems in the future.

Alternative energy features in the single-family West Village housing presented a slightly different challenge than for the apartment buildings. Architects worked to optimize the roof square footage of these smaller structures for PVs. Rooftop capacity on the free-standing alley-loaded garage structures will be kept open, with the aim of being able to accommodate additional rooftop solar if residents wanted extra capacity for electric vehicles.

Indeed, a key benefit of West Village overall will be ongoing evaluation of energy efficiency and renewable energy components to learn from the community's experience. Since the initial units only opened in late 2011, little data is available as of this writing. It is to be expected that actual performance will be somewhat different than modeled predictions, and that refinements may be needed to achieve ZNE status, such as the addition of more photovoltaic modules, the improvement of informational strategies to change behavior, or other incentive programs for residents to reduce energy consumption. Whatever the case, in the years ahead West Village is certain to become a laboratory for energy systems testing, helping to inform other projects in the future that aim for ZNE status.

5 OTHER SUSTAINABILITY ELEMENTS

Although it is likely to get the most attention for its ZNE strategy, West Village also seeks to meet sustainability goals in many areas besides energy use. Transportation, stormwater design, climate appropriate landscape design, and affordability measures (in terms of the for-sale faculty and staff housing) are other main areas of sustainable practice.

One of the main motivations for West Village from the start was to encourage as many students, faculty, and staff as possible to live within walking or bicycling distance of campus. This project characteristic should very substantially contribute to the positive environmental effects of the community, since most occupants will not be using fossil-fuel-powered vehicles for their daily commutes. West Village is located approximately 1.5 miles from UC Davis' central quad, and is even closer to a number of other facilities including the rapidly expanding veterinary medicine complex (Figure 12.8). A below-grade freeway separates the neighborhood from the central campus; however, a pre-existing bicycle and pedestrian bridge

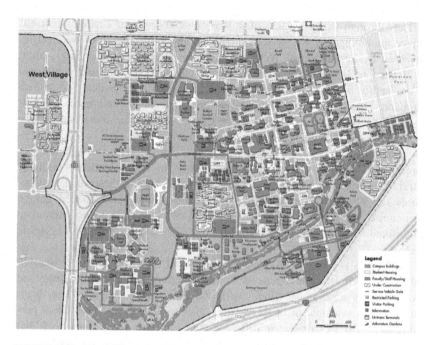

FIGURE 12.8   West Village in Relation to the UC Davis Campus and Downtown. West Village (at upper left) is located within a mile of most campus locations, and is connected by primarily off-street bicycle and pedestrian paths as well as frequent bus service. Residents are not allowed central campus parking permits, further decreasing transportation energy use.

provides a convenient connection across this obstacle. City and campus bike route systems then offer primarily off-street paths to reach many destinations throughout town. West Village provides ample bike parking near all units as well as on and off street bike paths through the neighborhood. It also provides a bus transit stop within a five-minute walking distance of all residences, with frequent service to central campus. The transportation system is not part of the ZNE calculations, although energy use for transportation should be much lower than for conventional neighborhoods elsewhere in Davis.

Many West Village residents do own cars, but must park these in peripheral parking lots that discourage daily use. To save space some of the parking is in the form of tandem spaces (one vehicle immediately behind another), a feature that also discourages frequent usage. These peripheral parking lots are ideal for photovoltaic canopies that also provide shade for the parking lots and mitigate some of the heat island effect of large paved surfaces. If these parking spaces were more closely located to each apartment building, the shade cast by buildings would reduce the effectiveness of these lots for solar energy. In large part due to neighbor preferences, campus planners omitted any road connection to Russell Boulevard, the main arterial street to the north of West Village connecting to the city's downtown. To drive into town residents must follow a more circuitous route to the south, further discouraging motor vehicle use. The Davis downtown is about a 2-mile bicycle ride from West Village. As of fall 2012, neighborhood residents will not be allowed to purchase campus monthly parking permits unless they have specific access needs, so regular commuting to the central campus by car is eliminated. ZipCar currently operates car-sharing sites on campus, and in the future the university plans to open a car-sharing pod at West Village, and to explore the possibility of using all-electric vehicles for this purpose.

Water is another area of sustainability emphasis at West Village. Apartment fixtures go considerably beyond code in terms of water conservation; toilets use only 1.28 gallons per flush, and shower faucets dispense only 1.5 gallons per minute, 40 percent below code. The project's design attempts to keep all stormwater onsite, in part through swales, landscaping, and green street features that encourage rain to infiltrate into the ground where it falls, and in part by channeling overflow to a series of seasonal ponds in the greenway to the north of the neighborhood.

West Village landscape design emphasizes native trees such as California live oaks, valley oaks, and California sycamores; native shrubs such as toyon, manzanita, and ceanothus; native grasses such as Muhlenbergia rigens and California fescue; and other native plants such as rushes and sedges. All of these species are drought-tolerant, and much irrigation utilizes water-saving drip systems or micro-sprayers. Turf areas are concentrated in locations with active outdoor uses. Landscapes such as tree strips between roadside curbs and sidewalks are planted with low water-using plants.

From the beginning campus planners intended to promote economic and social sustainability goals by providing faculty and staff housing for sale at below-market prices. In order to keep these homes affordable for future generations, the agreement between each homeowner and the university includes provisions to cap the rate of annual appreciation for the for-sale housing, a strategy that the university previously employed on its much smaller Aggie Village development built in the 1990s near the Davis downtown. West Village homebuyers will only be able to sell at fixed rates of appreciation, and the university has right of first refusal on all sales. This mechanism provides a relatively small and predictable increase in home value to the owner and avoids large positive or negative swings that may occur in the open market. The 2008 housing crash has lowered real estate prices in Davis by about 20 percent, and prices in nearby cities such as Sacramento even more, greatly increasing the difficulty of providing below-market housing. However, housing in Davis is still the most expensive in the region, and as the market recovers the need for affordable housing for campus employees is likely to grow once again. The build-out of for-sale housing at West Village is anticipated to be slower than originally expected due to market conditions, but the underlying model is not likely to change.

Other sustainability features inside the buildings include recycled quartz countertops and 50 percent recycled flooring in the student apartments, ceiling fans in all rooms, and low volatile organic compounds (VOC) finishes throughout. Much of the lighting in units was hard-wired in place, a strategy that increases the likelihood that residents will use these highly efficient fixtures rather than bringing in their own halogen or incandescent lamps. Somewhat detracting from the eco-image but certainly appreciated by students on hot days, a large swimming pool is located immediately behind the welcome center; unfortunately in the effort to maximize PV rooftop space solar thermal for the pool was not included. To ensure that units would be attractive to potential residents, the developer has given a decidedly up-market character to West Village. Every bedroom of student housing in the initial phase, for example, has its own bathroom and a walk-in closet. The recreation center and pool add to the amenities. Not surprisingly given this level of appointment, rents are at the upper end of the Davis market. However, the sheer number of new apartments offered at West Village should help moderate rental prices in the local Davis market and relieve very low vacancy rates that lead to higher rental pricing.

Perhaps the most important contribution of West Village to sustainable community planning is the demonstration that ZNE is being achieved with no greater cost to the West Village resident. The project shows that the same dollars used to pay a resident's typical energy bill can be redirected to pay for energy efficiency and green power when planned in a holistic way, resulting in a large-scale ZNE community. Overall, in terms of financially viable ZNE performance, transportation demand reduction, water use, and landscape treatment, West Village appears at the cutting edge of global eco-district development.

6 CONCLUSIONS

As the first large-scale ZNE neighborhood in the United States, West Village provides a powerful model of how society might move towards a ZNE status for primarily residential mixed-use communities. The project is somewhat unique in that it benefited from a powerful institutional sponsor, the University of California, Davis, which was willing to experiment, had access to a wide range of technical expertise, owned the land, and did not have to subject development plans to local land-use planning approval processes. It also benefited from a political climate in which interest in lowering greenhouse gas emissions was rising rapidly in the state, and from a favorable climate and geographic context for keeping energy use to moderate levels. Those factors certainly helped enormously in attaining ZNE goals. However, West Village is largely a private-sector financed and managed project, and there is no reason why many of West Village's features could not also be incorporated into private sector development projects or neighborhoods in other parts of the world. The passive solar design features and the 'reduce, then produce' approach in particular involved relatively simple strategies that should be replicable in most places.

Although detailed studies of actual energy use, renewable power generation, and resident behavior must await the availability of data, a few energy-related lessons can already be gleaned from West Village. One basic lesson is the importance of combining three main strategies – passive solar design, energy efficiency measures, and active renewable energy – within the effort to achieve ZNE status. None of these strategies alone could have met West Village's energy goals. Without halving electric usage through efficiency strategies, for example, PV would never have come close to meeting the neighborhood's needs on the available rooftop and parking lot space. Another main lesson is the importance of designing ZNE strategies into a project from the start. For example, because the first West Village rooftops were not originally envisioned to carry solar panels, they contained vents and protrusions that made later PV installation more difficult. Once the ZNE model was embraced those elements were consolidated on the north slopes of rooftops to avoid conflicts with energy systems.

However, West Village's development overall has illustrated a fortunate synergy between a willing and creative institutional sponsor, many public and private sector partners with technical and development expertise, and a supportive political climate. With these elements in place, planners are optimistic that the neighborhood will in fact meet its ZNE goal at no greater cost to the resident than a "business-as-usual" scenario. By embracing a community-wide approach that stretches from street layout to light fixtures, it also seems likely that similar projects elsewhere could reduce energy demand to the point where energy needs can be supplied by on-site renewable power.

REFERENCES

[1]   California Public Utilities Commission (CPUC). Zero net energy action plan. Sacramento. Available online at <http://www.cpuc.ca.gov/NR/rdonlyres/6C2310FE-AFE0-48E4-AF03-530A99D28FCE/0/ZNEActionPlanFINAL83110.pdf>; 2010 [accessed 08.05.12].

[2]   Royal Institute of British Architects. (RIBA). RIBA Journal sustainability award BedZED, Web resource available at <http://www.architectsjournal.co.uk>; 2003.

[3]   Dakin B, Hoeschele M, Petouhoff M, Zail N. Zero energy communities: UC davis' west village community. Working paper presented at the ACEEE summer study 2010.

[4]   Corbett J, Corbett M. Designing sustainable communities: learning from village homes. Washington, D.C: Island Press; 1997.

[5]   City of Davis. Davis sustainability, Web resource at <http://cityofdavis.org/cdd/sustainability/>; 2012 [accessed 08.05.12].

[6]   Finkelor B, et al. West village: a process & business model for achieving zero-net energy at the community-scale. Davis: UC Davis Energy Efficiency Center; 2010.

[7] Vinnitskaya I. U.C. davis west village/studio e architects. Archdaily. march 17, 2012. Web resource at <http://www.archdaily.com/215764/uc-davis-west-village-studio-e-architects/>; 2012 [accessed 03.05.12].

[8] Braun G, Hayakawa M.G. 2011 Annual report, west village energy initiative: CSI RD&D project. Davis: U.C. Davis; 2012.

[9] U.C. Davis News and Information Service. Researcher's waste-to-energy technology moves from lab to marketplace. Web resource available at <http://www.news.ucdavis.edu/search/news_detail.lasso?id=10202>; 2012 [accessed 04.05.12].

Chapter 13

Crouching Demand, Hidden Peaks: What's Driving Electricity Consumption in Sydney?

Robert Smith
Ausgrid, Sydney, Australia

1 INTRODUCTION

After over a century of persistent growth, Sydenysiders'[1] demand for electricity has suddenly waned. This apparent end to demand growth has come at an extraordinary time of erratic weather, unstable economic conditions, soaring electricity prices, heightened environmental concerns, and multiple new regulations. Falling consumption has been matched by slow but erratic growth in electricity peak demand. If these conditions persist it is a positive step in greenhouse gas reduction but creates enormous challenges for the management of the electricity grid and increasing uncertainty about the future direction of the industry. The principle difficulty in forecasting the future is untangling the myriad different forces impacting growth and determining if we are seeing a once-off pause or a permanent step change heralding the end of growth. Is demand crouching and peaks hiding, waiting to re-emerge when current conditions pass?

This chapter consists of four sections in addition to the Introduction. Section 2 looks at where all the kWh have gone, the history of residential electricity consumption and demand with particular emphasis on the largely unforeseen changes that have taken place since 2005–2006. Section 3 examines the "usual suspects," possible causes put forward for recent changes and narrows the number down to a list of the major culprits, most notably price rises. Section 4 discusses the size of the major factors and their interrelationships to separate the persistent from the temporary driver of change.

---

1. Ausgrid is the largest electricity distribution network in Australia operating in Australia's largest city Sydney as well as Newcastle, Upper Hunter and the Central Coast. Parts of Sydney and elsewhere in NSW are not covered by the network. Ausgrid's customers are the focus of this chapter because of the author's access to data and the prominence of Sydney in leading changes in Australian electricity use.

Energy Efficiency. DOI: http://dx.doi.org/10.1016/B978-0-12-397879-0.00013-X

The chapter's conclusions are summarized in Section 5, which foresees an eventual end of large price increases, testing the resilience of recent energy savings and how, without corresponding falls in peak demand, it will be a challenge to translate the benefits of energy savings into lower customer bills.

2 WHERE HAVE ALL THE kWh GONE?

Ausgrid and its antecedents have been supplying electricity in Sydney for over 100 years. Throughout the periods' world wars, depressions, energy crises, company mergers, and restructures, new technology developments, social and demographic changes, demand for electricity has followed a continuous upward path. In the past interruptions in growth have been brief and were clearly linked to individual causes, most notably economic downturns. However, this appears to have changed around 2005–2006 as can be seen in Figure 13.1, which shows long-term total energy consumption of customers in Ausgrid's area[2].

This experience has been reflected across Australian markets with the recent history of consumption in the National Electricity Market (NEM) indicating a similar break in the long-run growth trend. While these NEM numbers are not weather corrected and erratic weather has played a part in dampening demand, there is a clear trend of falling consumption in the years after 2007–2008 in the NEM as a whole as well as declines in all states except South Australia where demand has been flat as shown in Table 13.1.[3]

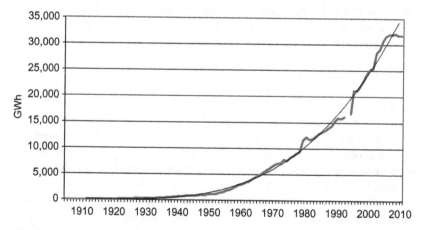

FIGURE 13.1   Ausgrid and its antecedents 2005–2011.

2. The figure includes the impact of company amalgamations and mergers, the last being in 1997. More details on history can found in Wilkenfield and Spearritt [45] (2010) and Anderson (1955) [46].

3. AER, State of the Energy Market 2011 [30].

TABLE 13.1 Electricity Supply to Regions of the National Electricity Market (terawatt hours)

	QLD	NSW	VIC	SA	TAS[1]	SNOWY[2]	NATIONAL
2010–11	51.5	77.6	50.9	13.5	10.2		203.7
2009–10	53.2	78.1	51.2	13.3	10.0		206.0
2008–09	52.6	79.5	52.0	13.4	10.1		207.9
2007–08	51.5	78.8	52.3	13.3	10.3	1.6	208.0
2006–07	51.4	78.6	51.5	13.4	10.2	1.3	206.4
2005–06	51.3	77.3	50.8	12.9	10.0	0.5	202.8
2004–05	50.3	74.8	49.8	12.9		0.6	189.7
2003–04	48.9	74.0	49.4	13.0		0.7	185.3
2002–03	46.3	71.6	48.2	13.0		0.2	179.3
2001–02	45.2	70.2	46.8	12.5		0.3	175.0
2000–01	43.0	69.4	46.9	13.0		0.3	172.5
1999–2000	41.0	67.6	45.8	12.4		0.2	167.1

[1]*Tasmania entered the market on 29 May 2005.*
[2]*The Snowy region was abolished on 1 July 2008. The New South Wales and Victorian data subsequently reflect electricity consumption formerly attributed to Snowy.*
Note: Estimates based on generation required to meet energy requirement within a region–calculated as regional generation plus net flows into the region across interconnectors.
Source: AEMO; AER.

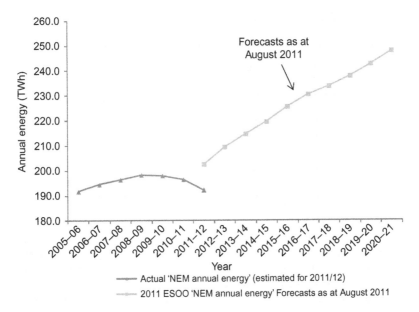

FIGURE 13.2   National Electricity Market annual energy, actual, estimates and forecasts.

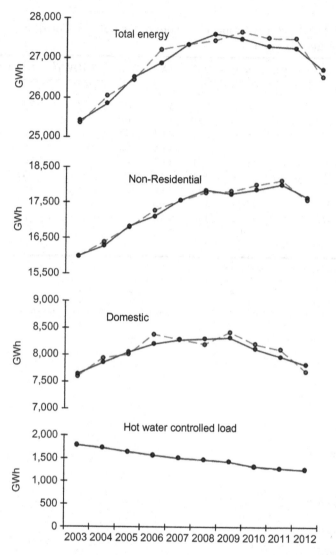

FIGURE 13.3   Ausgrid customers' energy consumption total and by major segments.

Despite these recent trends, expectations of a continual growth in electricity consumption in line with the long-term pattern persisted, with downturns apparently seen as aberrations rather portents of the future. The official growth forecasts for the NEM from the market operator the AEMC[4] shown in Figure 13.2, incorporates data of falling consumption in the first half of 2011−2012 to

4. AMEO, [1,25] Electricity Statement of Opportunities for the National Electricity Market, Update March 2012.

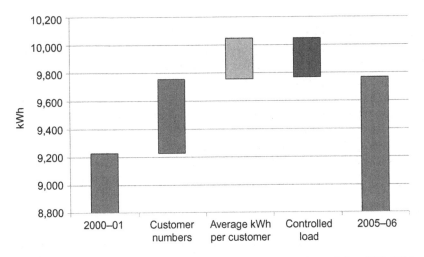

FIGURE 13.4 Composition of residential consumption growth in the 5 years before 2005–2006.

downgrade growth in that year but has not adjusted longer-term growth forecasts made in 2011. The AEMO's long-term forecast showed a traditional steady upward growth trend, which seems incongruous with recent experience. The long-term growth projection appeared to have been drawn with a ruler when recent history has been shaped by a protractor. Forecasts of electricity consumption and peak demand drive billion dollar investments, set electricity prices, and shape government policy. The clash between recent forecast and current experience suggests there is value in understanding more about the underpinnings of recent changes in demand for energy.[5]

Recent national market trends are also reflected in consumption data for Ausgrid's customers. The slowdown in consumption experienced by Ausgrid has occurred across all areas but is concentrated in the residential sector. Normally the business sector is expected to be more volatile and subject to more volatile swings in consumption and demand in line with economic activity. This has not been the case with recent changes, largely reflecting the relative strength of the Australian economy compared to other OECD nations after the global financial crisis. Declining energy intensity due to improved production practices and an increase in service sectors role in both economic activity[6] and household consumption[7] is consistent with trends across developed nations. It is a contributor to lower electricity consumption in the long-term but not recent changes. Growth in the non-residential sector, which accounts for around two-thirds of total electricity sales, has remained flat, as shown in Figure 13.3, while total consumption for all Ausgrid customers has fallen. Trends in the figure are shown with actual consumption as green dashed lines and weather corrected consumption in blue.

---

5. Newer AEMO [2], National Electricity Forecasting Report forecast have changed but do not envisage an end to growth.
6. Sandu S, Petchey P, [12].
7. Bureau of Resource and Energy Economics [31].

The overall fall in energy consumption has been driven by residential customers where consumption has experienced a pronounced downturn both in general domestic use and in the separately metered electric storage hot water segment. The decline in hot water load control is attributable to a range of government policy initiatives aimed at reducing greenhouse gas emissions, saving water, and increasing the uptake of gas, heat pump, and solar hot water systems. This is a reversal of earlier policies, before greenhouse emissions were recognized as an issue, which actively promoted large off-peak hot water systems to shift residential winter evening load and reduce system peaks. Controlled load tariffs for residential customers still support shifting of hot water load with substantial discounted electricity prices but greenhouse gas and water-saving policies are driving electric hot water load on a long path of decline.

The stark change in the composition of weather-adjusted residential customer's consumption over the last decade can be seen in Figures 13.3 and 13.4. Taking 2005–2006 as a reference point in residential electricity demand growth, in the 5 years prior to 2005–2006 residential electricity consumption grew 5.8 percent, while in the 5 years following consumption fell 5.3 percent, leaving the total just above what it was 10 years earlier. As can be seen in the composition figures, part of this is explained by the continued falls in controlled load electric hot water use and part by the slower growth in customer numbers. However, average kWh per customer excluding controlled load increased by 3.7 percent in the 5 years to 2005–2006 but fell back 6.4 percent in the subsequent 5 years. This change represents a significant turnaround as it is not only the absolute fall in consumption that needs to be accounted for but the gap between current

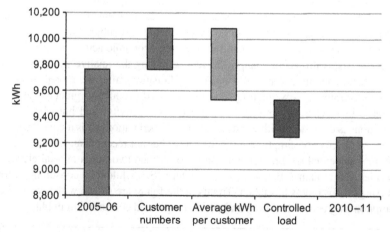

FIGURE 13.5 Composition of the residential consumption decline in the 5 years after 2005–2006.

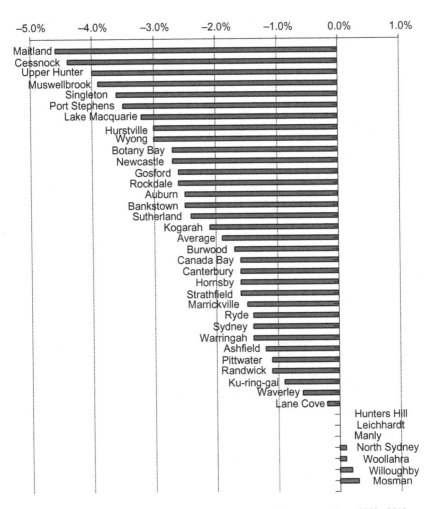

FIGURE 13.6 Change in kWh per household by Local Government Area 2009–2010 to 2010–2011.

consumption and the underlying long-term growth trend. If growth had continued on trend average kWh domestic consumption would be 13 percent higher in 2010–2011. While some of the fall in kWh per customer is due to customers' hot water use, which is not separately managed and metered as a controlled load, overall consumption must be falling in a range of areas to produce such a large decline.

The downturn in residential consumption has proven to be widespread throughout Sydney. Across 44 local government areas covered by Ausgrid's distribution network per capita household consumptions in 2010–2011 declined in 33 areas and showed no growth in a further two.[8] As Figure 13.6

8. Updated data is available at http://www.ausgrid.com.au.

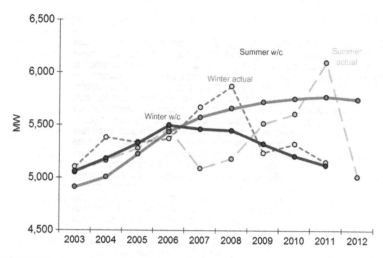

FIGURE 13.7   Ausgrid's weather-adjusted Winter and Summer peak demand.

shows consumption changes ranged from -4.5 to $+2$ percent and, although rising consumption appears to be confined to wealthier suburbs, there is no simple correlation between location, housing type, average household incomes, or any other identifiable factors to explain the different size of declines. More recent numbers show further falls in per capita consumption in all areas but have been affected by exceptionally mild seasonal temperatures and are not fully representative of longer-term trends.

2.1 Weather Effects in Crouching Demand and Hiding Peaks

Erratic weather patterns across Australia have added to the difficulty of determining if recent downturns represent an end in demand. The effect has been particularly pronounced in peak demand where across Australia a shift from a winter-peaking to a summer-peaking network has been accompanied by greater swings in temperature-sensitive demand. The summer of 2011–2012 in particular was unprecedentedly mild.[9] As a result Ausgrid's raw un-weather-corrected network system peak reverted back to winter, a result that runs against the trend of declining winter peak demand and growing summer peaks. Overall, new network investments and system capacity are based on peak demand forecasts while revenues are based on energy consumption. Networks build for peaks and bill on consumption, so when underlying peak demand growth is hidden by erratic weather patterns forecasting becomes more uncertain. With weather effects removed it is possible that overall

9. For a discussion of the weather impacts see *Recapping NEM-Wide demand (or the lack thereof) over summer*, http://www.wattclarity.com.au/2012/04/recapping-nem-wide-demand-over-summer/.

system peak demand has plateaued, as shown in Figure 13.7. If this were true it would still represent a challenge for electricity networks as growth in individual areas within the network or shifts of activity between areas can still require investment at a time when revenues from consumption are flat or falling. Decoupling consumption and revenues may become necessary if the gap between growth in peak demand and consumption continues and widens.[10]

3 THE USUAL SUSPECTS

Reasons for the recent downturns in residential electricity consumption and peak demand growth center around a set of "usual suspects" − popularly discussed possible causes whose negative impacts on consumption are widely recognized but not quantified and hence cannot be ranked or compared. The list is large because detailed appliance load research has not been done in Australia for almost 20 years[11] and so it is not possible to reliably determine where falls in residential consumption are occurring. Changes in residential customers' usage and the causes behind the changes can only be inferred indirectly. However, looking at the trends in individual factors it is possible to remove some high-profile but low-impact suspects from the list and at least create a short list of where major change is occurring.

The suspects shown in Figure 13.8 are all potential reasons for changes in consumption and can be grouped into four areas: price impacts (in blue), macrofactors particularly demographics and economics (in yellow), climate change concerns (dark green), and energy efficiency measures (light green).

While these broad areas provide a useful classification of causes they do not provide much insight into the nature of recent changes. It is more useful to separate individual influences on energy consumption and peak demand into one-off changes and persistent changes. Isolating the effects of minor and one-off changes allows more focus on the remaining factors that are ultimately likely to signal an end to demand.

Strong once-off impacts with a clear influence on domestic customers include:

- Incandescent Lighting Ban and CFL Giveaways: a major reduction in lighting consumption but with little impact on summer afternoon peak demand because of longer summer daylight hours.
- Home Insulation Program: a expected reduction in winter heating and summer cooling requirements after 2010, but not as yet reflected in consumption due to recent mild weather.

---

10. Simshauser and Nelson [3] raise the spectre of a "Death spiral" for prices and demand but seem to have overlooked the dampening effect that electricity's low price elasticity of demand would have on a spiral scenario. See also Ausgrid [4].

11. BRANZ, [5], *Energy Used in Australian Appliances − Analysis of the 1993/94 RES Appliance Energy Use Data Research. Report no. UC0170/2*, this report remains the best source of appliance end use data in Australia despite giving prominence to water beds and predating the appearance of computers and a large range of new technologies.

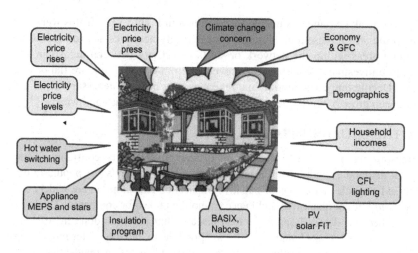

FIGURE 13.8   Factors driving falls in energy and demand.

- Rooftop Solar Feed-In Tariffs (FIT): a significant impact at a system level but not reflected in customers' metering consumption because of the gross metering of FIT customers and with a low coincidence with system peak demand.
- Water-savings Initiatives: a significant reinforcement of the decline in hot water consumption through promotion and regulation of low-flow showerheads, tap aerators, and water pricing but with minimal peak impact due to the large number of off-peak and gas systems.
- Price Rises: recent major increases in residential prices that are expected to end with a 20 percent rise in 2012−2013, which incorporates a 10 percent carbon tax.

Once-off impacts with an uncertain influence on domestic customers include:

- Price Press Coverage: sustained media coverage of price rises, particularly in newspapers, as a possible trigger of behavioral change.
- Climate Change Concern: the subject of major political debate and substantial media coverage in print, radio, and TV, particularly on the introduction of a carbon tax, issues of climate change may influence behavior.
- Global Financial Crisis: the dampening effect on confidence and behavior around an expected recession may have affected consumption even though a downturn is not reflected in Australian economic data.

Strong ongoing impacts with a clear influence on domestic customers include:

- Demographics: growth in new household formation and customer numbers has not fallen and remains positive but has slowed. This appears to be the new path but could experience a rebound if recent restraint has led to pent-up demand for housing.

- Household Incomes: continue to increase, keeping electricity's share of average household expenditures in check.
- Price Levels: when the once-off impact to a price rise has passed, higher price levels result in continuing higher electricity bills, a permanent increase in the benefits of energy efficiency actions and quicker paybacks for efficient appliance purchases.
- Residential Building Regulations: the gradually increasing efficiency of housing stock as new and renovated houses are required to meet NSW BASIX building standards creates sustained long-term downward pressure of average household usage.
- Hot Water Efficiencies and Fuel Switching: continuing government programs combine to produce a steady long-term decline in electric hot water consumption.
- MEPS and Star Ratings: changes in minimum energy performance standards (MEPS) and appliance Energy Star rating programs have a sustained long-term effects as stock turnover gradually increases the share of efficient models and as new standards are progressively implemented.[12]
- Appliance Penetration and Ownership: increasing appliance ownership is leading to a steady rise in consumption demand, notably by air conditioners, computers, and entertainment equipment,[13] which can undermine other efficiency improvements.

At an individual level, a household's consumption depends on the characteristics (the overall socio-economic factors, climate, and dwelling type) and the particular energy uses of each household. Modeling of cross-sectional data that looks at household characteristics can explain about 40 percent of the variation in household electricity consumption and bills, while modeling that concentrates on "energy uses" can explain 60 percent of the variation. As would be expected using gas and living in a coastal area reduced bills while living in a detached house, the number of bedrooms, number of occupants, and income level increased consumption.[14] While this modeling from cross-sectional data provides insight into possible reasons for differences between customers at a point in time it does not explain where recent falls in energy use have come from over time. Of the major factors identified as influencing individual demand no major changes from trend levels have been experienced in household composition, household size, pool ownership, or household income. This suggests that changes in use rather than household characteristics are driving the recent falls in consumption.

Particular energy-saving actions can also be significant for individual customers but not a major factor in recent trends. For example, modeled estimates

---

12. See the stock turnover model in DEWHA [32] and also see NERA [40,41] and PWC [42].
13. Based on Ausgrid estimates from ABS surveys, internal sources and appliance sales data. Television data is based on Energy Efficient Strategies reports and discussions with the reports' authors.
14. IPART [6].

of the impact of second fridge use, which is correlated with detached houses with larger numbers of occupants, show they have a significant impact on bills. Getting rid of an old second fridge is a standard energy efficiency tip, can reduce a customer's bill by an average of $265 each year, and government-supported fridge buyback programs in Sydney have retired over 36,000 fridges. However, while the savings to an individual customer are significant the number of buybacks is likely to have been too small to have a material impact on overall residential electricity consumption when considered in conjunction with other changes in fridge stocks and performance.

The impact of other trends in appliance use is also difficult to estimate with certainty. The rapid rise in air conditioning ownership across Australia, from just over 50 percent in 2006 to just over 70 percent in 2011,[15] is part of a long-term trend that is expected to continue until ownership rates approach saturation at about 80 percent in around 2016. If, as shown in Figure 13.6, saturation is not reached and recent trends continue, penetration could reach the South Australian and Northern Territory at levels over 90 percent. Nationwide, residential air conditioner use is believed to be increasing the temperature sensitivity of summer peak loads disproportionably more than it is a contribution to revenues from energy consumption. However, the lack of load research data on air conditioner use and recent mild summer weather make estimating the size of these impacts guesswork. Measuring future impacts of air conditioning has become even more complicated by the recent increase in ceiling insulation and the impact of MEPS, which are expected to lower the future usage and peak impact of individual of air conditioners.

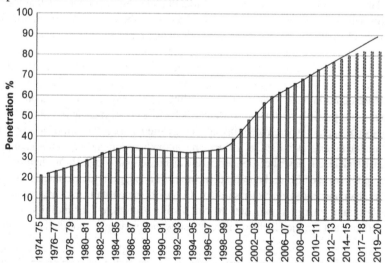

FIGURE 13.9   Australian household air conditioner ownership trend and forecast.

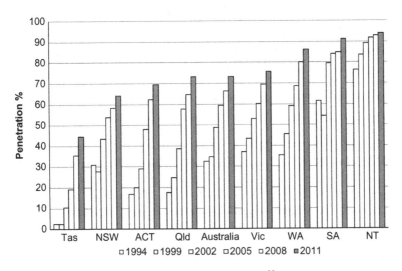

FIGURE 13.10   Household air conditioner ownership by State.[16]

Televisions are another area where the difficulties of predicted changes in appliance consumption can be seen. Televisions were predicted to produce a major increase in residential appliance consumption as penetration reached a limit at 98 percent of households but rises occurred in the number of TVs per household, hours of viewing, screen sizes, and overall energy consumption.[17] However, while accurately predicting these trends and the move away from cathode ray tube (CRT) TVs, the forecasts underestimated the speed of technology change that rapidly not only made CRTs obsolete but also saw plasma screens replaced with more efficient LCD and then LED technologies. While still an area of growing demand, televisions are likely to have a much smaller impact than was previously projected.

Of all the "usual suspects" examined price changes are the most problematic. Real residential electricity price rises in Sydney and across Australia have been large and prolonged but, without timely and detailed appliance load research on residential customers, untangling and estimating the impacts of pricing, behavior, and government policies and their inter-relationships cannot be done with confidence. Proxy measure and estimates provide some guidance to separating pure price impacts from other causes and the persistent from the temporary driver of changes.

---

16. Topp and Kulys [7] based on ABS statistics.
17. DEWHA [32].

4 PRICES AND THEIR ACCOMPLICES

Price changes are the most obvious culprit for explaining recent falls in residential electricity consumption. The Australian Bureau of Statistics (ABS) electricity price index for Sydney produced as part of the Consumer Price Index (CPI) measure of inflation[18] is shown below along with the Sydney CPI index.

Household electricity prices in Sydney have historically moved broadly in line with the CPI and were even flat and in Figure 13.11 the CPI long-run trend until round 2002. After 2001–2002 electricity price rises outstripped the average inflation rate, rising 134 percent in nominal terms in the decade to 2010–2011 or 82 percent in real terms (after adjusting for inflation). The sharpest rise in Sydney households' electricity prices, 58 percent in real terms, occurred in the 5-year period after 2006–2007. This increase was the largest price rise of any product or service measured in the CPI over this time and exceptional by international standards.[19] The most recently announced price rise, not yet reflected in official figures, will lift prices by a further 20 percent, 10 percent of which is due to the implementation of a carbon tax.

In a decade of rising energy prices recent electricity prices increases have stood out. Although other residential energy prices have seen similar overall price rises in the last decade the recent short sharp upturn in electricity prices contrasts with the steady rises in gas prices and the large gyrations around an

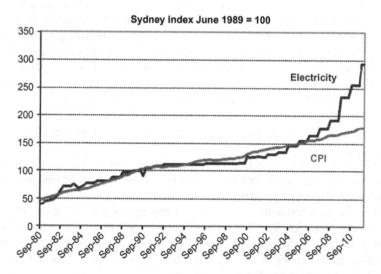

FIGURE 13.11   Consumer Price Index and electricity price indices for Sydney.

---

18. ABS [27].
19. See Mountain [38,39] and IEA and OECD for price comparisons [35], see EnergyAustralia [34] for price list.

upward trend in the case of petrol. All three fuel price indices have, for differing reasons, increased well above the rate of the CPI but electricity prices, by rising the quickest in a shortest period, have received the most attention.

So do electricity price rises explain the fall in residential demand? Isolating the impact of price changes requires an estimate of the price elasticity of electricity demand. One measure by the Reserve Bank of Australia[20] using household expenditure data between 2003−2004 and 2009−2010 estimates a price elasticity of demand for electricity of −0.26, suggesting that for every 10 percent real increase in price demand drops by 2.6 percent. This is the lowest price elasticity of any good or service measured other than tobacco but still represents a substantial price response for an essential service traditionally seen as price inelastic. Internal Ausgrid estimates suggest that shorter-term price elasticity, after adjusting for underling economic growth, could be as high as −0.35. The problem with this simple approach, however, is that it bundles both price and non-factors together into the measured price effect. Big price rises and big demand falls have coincided but correlation does not necessarily imply causation or at least not a full explanation. Independent government policy or consumer actions may be operating irrespective of price changes.

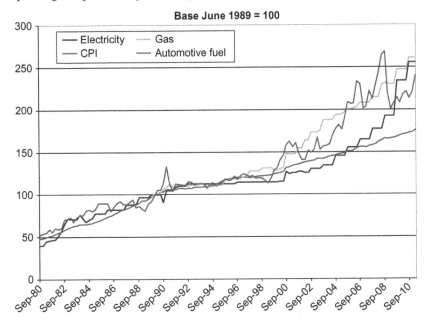

FIGURE 13.12 Consumer Price Index and fuel price comparisons for Sydney.

---

20. Jaaskela and Callan [8] while AEMO's 2012 forecasting report estimates −0.29 for the non-large industrial consumption.

The normal textbook consumer response to increasing prices for a product is to reduce the quantity demanded, but electricity is not a normal product. Electricity cannot be stored, has limited short-term substitutes, and is an input into customers' overall household production and not an end product in itself, and therefore customers' responses to price rises appear in a variety of ways. Price changes need to trigger actions by consumers that lead to a reduction in consumption. Customers can switch fuels, increase the efficiency of their usage, identify and eliminate waste, or reduce their usage through a reduction in their amenity. In all these instances, including reducing use by reducing amenity, customers' actions can be influenced by significant non-price factors that are changing at the same time as prices are rising. Recent price rises have had a large range of accomplices in driving down demand.

The mechanism by which this price change trigger operates is also important. If it is the publicity about prices or the change in price levels driving reductions then changes will be once-off. If it is the new level of prices that is prompting action demand reductions will be ongoing. There are indications that, while an integrated part of the price change process, publicity about price changes does not on its own drive changes in behavior.[21]

A nationwide survey of key factors affecting community appetite for energy efficiency by the Clean Energy Council[22] found that rising electricity prices were the primary reason influencing consumers to change their electricity consumption. This survey found that:

- 95 percent of people were concerned or very concerned about rising energy costs
- 89 percent were willing to take action to use less energy
- 73 percent wanted more information on how they could save energy
- 57 percent knew little or nothing about government programs available
- 50 percent knew little or nothing about key aspects of their energy use

The level of concern over raising energy costs is consistent with the size and publicity about recent price rises. However, a gap between action and intention remains, and customers' reported desire for information but lack of knowledge suggests that more energy efficient action by customers is possible and that independent action by customers may not have been a driving factor in recent changes.

Determining the sustained impact of higher price levels compared to the short-term impact of price changes (as measured by price elasticises) is difficult. Sustained but stable higher price levels are expected to improve the benefit cost ratios and shorten the payback periods for all energy efficiency actions including buying high Star-rated appliances, insulating a home, and

---

21. Unpublished research thesis.
22. Auspoll [9].

fuel switching. However, if it is bill levels rather than price levels that customers see and respond to electricity remains a relatively small proportion of household income for the majority of customers even after recent price rises.[23] This view is supported by the examination of peoples' everyday practices, which shows a strong resistance to change even in the face of strong economic incentives.[24]

If price rises cannot be held fully responsible for explaining recent changes in electricity consumption how much can be attributed to non-price factors?

4.1 Macro factors, Demographics, and Economics

Over the long-term macrofactors have traditionally and will continue to increase overall residential electricity consumption. Continued growth in population, household formation, household incomes, and overall economic activity consistently increase electricity consumption both for households and businesses. A change in these macrofactors may explain slightly slowed growth in 2006−2011 compared to the first half of the decade, but overall, macrofactors have not produced major downward pressure on residential consumption levels[25] as:

- The number of persons per household in NSW has remained constant at around 2.68 throughout the period.
- Population growth in NSW, and in Ausgrid's area, slowed in 2006−2011 compared to the first half of the decade but remained positive.
- Growth in customer numbers slowed but remained positive.
- Real household weekly incomes continued to rise throughout the period,
- Australia avoided the worst of the global finance crisis and subsequent financial problems and maintained strong economic growth rates throughout.

As these major macrofactors have shown a continuing consistent upward, albeit slower, growth trend they cannot explain recent downturns in residential electricity consumption.

Overall, national energy use changes can be divided into:

- An activity effect − changes in the level of activity.
- A structural effect − changes in the composition of activity.
- An energy efficiency effect − improvements in use of energy.

The longer-term trend in Australian energy use has been falling due to structural and energy efficiency effects creating lower energy intensity of the economy as a whole, but then this being offset by increases in activity.[26]

23. IPART [6,36,37] also see Australian Industry Group [28] and AEMC [29].
24. Strengers [10].
25. Ernest and Young [11] and internal Ausgrid analysis.
26. Sandu and Petchey [12].

However, for electricity the balance appears to have shifted and the activity effect from growing incomes and population is being overtaken by structural and efficiency effects.

4.2 Impact of Climate Change Concern

Climate change has been a major political issue over the last decade and one possible explanation for the fall in residential electricity consumption is individual action to address households' climate change concerns. This would result in reduced usage that is either independent of or reinforces both price changes and government greenhouse gas policies.

An increase in climate change concern in Australia appears to have coincided with the initial fall in residential demand and so could have delivered an independent and sustained shift in consumers' preferences about electricity consumption. In particular, 2006 and 2007 saw the release of the Stern report on the economics of Climate Change, the Al Gore documentary "An Inconvenient Truth," and the Intergovernmental Panel on Climate Change (IPCC) Fourth Assessment report. Evidence from the Google Insights tool show a clear increase in Australians' Internet searches about climate change, peaking at the end of 2009 but falling away since then. This fall occurred during a period when Australian parliaments and media were engaged in fierce debate about the introduction of a carbon tax that kept climate change issues prominent. Hence, while awareness of climate change issues may have lifted in the recent past it is difficult to link this to a sustained effect on energy efficiency action and reduced electricity consumption.

Further evidence of a fall-off of interest in climate change issues and limited likelihood of independent energy efficiency action being based on climate change concerns can be seen in "What matters to Australians."[27] In a ranking of 103 areas listed by Australians in order of their relative importance, climate change (part of the environmental sustainability category) fell from being ranked 12th in 2007 to be 51st in 2011. Over the same period energy prices (part of the societal economic well-being category) rose in importance from a ranking of 58th in 2007 to be 25th in 2011. Both climate change and energy prices continue to be outside the top 10 concerns for Australians. This reinforces earlier "Who Cares About the Environment?"[28] surveys that showed an increase in the proportion of people in NSW who said they often reduced their energy consumption (up 9 percent from 73 percent in 2006 to 82 percent in 2009) but predominantly for economics/incentives/cost-savings reasons (45 percent) rather than environmental concerns (18 percent). These surveys also found that 9 out of 10 people believe that others (the federal government [93 percent], industry [92 percent], other countries [92 percent], and state

---

27. Devinney, Auger and DeSailly [13].
28. Department of Environment, Climate Change and Water [14].

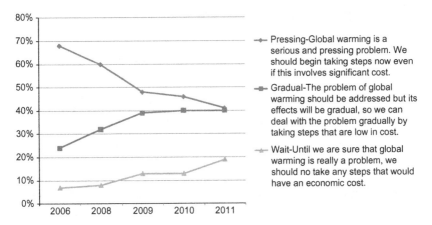

FIGURE 13.13   Lowy Institute poll – dealing with climate change.[30]

government [88 percent]), should have major responsibilities for preventing climate change or keeping it from getting worse.[29] Only two-thirds (63 percent) thought individuals should be responsible for addressing climate change. Finally, the Lowey Institute poll shown in Figure 13.13 supports other surveys in tracking a fall in climate change concern. There appears to be no survey results or research findings that indicating a major shift in behavior to save electricity based on climate change concerns alone.

The falling importance of climate change as an issue, the expectation that others are responsible for addressing the problem, and the introduction of a carbon tax suggest climate change concern, while a major topic of political and media interest is a supporting but not a significant driving factor for explaining falling residential consumption.

4.3 The Impact of CFLs and the Lighting Phase-Out

In late 2008 Australia became the first developed nation to effectively ban incandescent bulbs for general-purpose use. The ban followed a period of intensive promotion of compact fluorescent lamps (CFLs) through giveaways funded by state government greenhouse gas abatement schemes. This dramatic and abrupt change in household lighting cannot be measured directly[31] but can be seen in light bulb import statistics (as there are no local manufacturers, import statistics provide a close proxy for total light globe

29. Department of Environment, Climate Change and Water [15].
30. Hanson [16].
31. Recent moves have been made to measure impacts but base level data points of historical comparison are missing (Energy Efficient Strategies, lighting audit and REMP reports) [33].

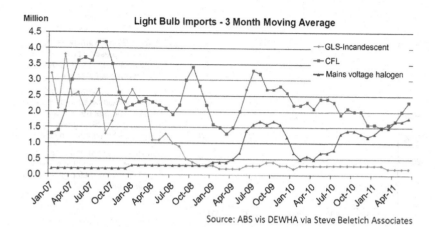

Source: ABS vis DEWHA via Steve Beletich Associates

FIGURE 13.14   Australian mains voltage general-purpose light imports by type.

sales). The progress of the inefficient lighting phase-out is shown in Figure 13.14 by:

- An increase in CFL imports from around 2004 largely for giveaway schemes.[32]
- An effective end to mass incandescent bulb sales with the import ban in 2008 and sales ban in 2009.
- The emergence of main voltage halogen (MVH) bulbs with similar shape and performance characteristics to old incandescent bulbs but energy savings of only 30 percent compared to 80 percent savings from CFLs.

These changes occurred independently of price rises and despite the strong cost savings that existed for CFLs prior to the ban.[33] Particularly interesting is the recent strength of sales of MVH lamps, which have some advantages in their characteristics compared to CFLs but deliver only a fraction of the energy savings. If the cost of electricity rather than the impact of regulation under the ban on incandescent bulbs was driving behavior there should be fewer sales of these less-efficient MVH lamps.

Figure 13.15 estimates the impact of the 26.9 million CFLs given away under the NSW greenhouse gas abatement scheme in million years of light output. Years of light output provides a better proxy for the overall share of

---

32. The NSW GGAS scheme, one of the first tradable permit emissions reduction schemes in the world, supported giveaways of around 26 million CFLs to a state population of 3 million households.

33. Productivity Commission [17,18] quotes a payback period for a CFL prior to price rises as 4 months while the expected life of a CFL is 5 to 7 years. Also see Waide and Tanishmia [44].

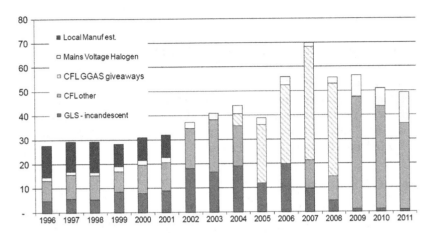

FIGURE 13.15   Estimated CFL giveaways and sales in NSW by million years of use.

CFLs than imports as they have a longer life and so should have a greater energy savings impact per light bulb sold than less efficient bulbs.

There is no official ex-post implementation assessment of the impact of the Australian governments' phase out of inefficient lighting but a broad estimate of energy savings can be gained based on available information. The benefits of an individual CFL are clear, saving 80 percent of the energy used in a traditional incandescent bulb. The impact of the overall program is harder to assess because there is no measure of how many of the CFLs given away or sold are currently installed, what lights they replaced, and how many hours they are used for each night.

Surveys of appliance usage[34] suggest that CFLs have replaced incandescent bulbs in many household applications but that an offsetting preference for halogen bulbs, particularly low-voltage halogen bulbs used as downlights for general-purpose lighting, may have limited the overall savings from the CFL ban. Informal information from researchers suggests there are around 14 CFLs in an average Australian household, or 30 percent of all lights, and that these CFLs are more likely to be located in frequent long-use situations. A bottom-up calculation of installing four CFLs as a replacement for 60-watt incandescents in high-usage areas that are on for 4 hours a day most of the year could represent a reduction of 5 percent in an average home's energy consumption. Therefore, it is possible that the incandescent lighting ban alone could account for the majority of recent falls in kWh consumption per customer, independent of price or other factors, but the actual impact remains uncertain. Without knowing what lights the CFLs replaced and how

---

34. 2010 Residential lighting survey informal background [33].

long they are switched on for the savings calculations are still broad estimations.

The future of lighting is clearly with LEDs. LED sales are moving from niche markets and speciality uses that supported high prices and limited performance into mainstream residential usage and general-purpose lighting. Performance issues are now addressed, and while product quality remains an issue, price improvements are the next step in making LEDs ubiquitous in households. This will take time but provides a technology pathway for further lighting efficiency improvements.

4.4 MEPS and Labeling

Australia's Equipment Energy Efficiency (E3) program involves a combination of pure Minimum Energy Performance Standards (MEPS), responsible for about 80 percent of savings, and labeling or labeling combined with MEPS, which delivers the other 20 percent of savings. Savings are projected to come mostly from the phase-out of electric-resistant water heaters, refrigerators, and freezers, but also from TVs and set-top boxes, standby, lighting, and air conditioners.[35]

Overall, the E3 program is projected to lead to a reduction in household electricity use per capita of 0.8 percent per annum compared to a projected increase under business as usual of 1.0 percent per capita. If achieved, a reduction of this size by the E3 program could account for 70 percent of the reduction in household per capita electricity use experienced over the last 5 years and more than offset the impact on total residential energy use of growth in customer numbers. While MEP programs include the high-profile ban of incandescent light bulbs, in general they have operated unseen and unheralded, creating a steadily building wave of energy savings.

While hot water and lighting are discussed separately, measuring the actual impact of MEPS and labeling programs across all products has not been attempted because of the lack of household appliance usage data. Nevertheless, significant MEPS improvements have clearly occurred, independently of price impacts, and existing MEPS and labeling programs are expected to underpin reduced appliance energy consumption over the next 10 years as existing appliance stocks are replaced with newer more efficient alternatives.

4.5 Impact of Hot Water Changes

Because electric hot water systems on controlled load tariffs are separately metered a much clearer picture of changes in their consumption can be seen than for other uses. As off-peak hot water comprises an average of 30 percent

---

35. Discussion of MEPS is based largely on Wilkenfeld [19].

of the cost of electricity to an all-electric household, declines in hot water use can make a substantial contribution to the "end of demand."

Controlled load hot water usage has been in steady decline for over a decade and is expected to continue to fall in the future (see Figures 13.4 and 13.5). Falling hot water usage has been driven by an ongoing series of government regulations, residential building standards, and incentives aimed at reducing greenhouse gas emissions, chiefly from fuel switching to the alternatives of gas, solar, and heat pumps but also by water-saving policies including mandating low-flow showerheads. Overall, falls in hot water use are likely to be even larger than measured by controlled load meters because electric instantaneous hot water systems and small storage systems are not controlled or separately metered and falls in their use are bundled together with other appliance consumption. These systems are generally smaller and used in apartments and are therefore not as amenable to replacement by solar hot water or gas as large controlled load systems, but it is likely that water restrictions and some systems conversions have reduced their consumption.

The phase-out of greenhouse intensive hot water heaters, which was due to come into force in 2012, would effectively ban the installation of new electric storage systems and would see the recent falls in electricity hot water use continue into the next decade.

4.6 Impact of Insulation

In response to the expected economic impact of the global financial crisis in 2008 the Australian government instituted a range of economic stimulus programs including a home insulation program (HIP) as part of an energy efficient homes package. Under this $3.9 billion package to improve the energy rating of Australian homes free ceiling insulation worth up to $1,200 was offered to all Australian homeowner-occupiers with limited or no ceiling insulation. The program ran for a year from February 2009 to February 2010 before being closed due to safety, implementation, and administration problems. Over that year ceiling installation was installed in over 1.25 million Australian households including over a quarter of a million households, or 1 in 6 households, in the Sydney area.[36] A subsequent inquiry examined the safety and the administration issues of the program but did not provide an estimate of the outcomes in energy savings.

An ABS survey of household energy use and conservation was conducted just over a year after the HIP program ended. It found that the proportion of households in NSW reporting they had insulation rose to 63.4 percent up from 53.4 percent three years earlier,[37] and 12.9 percent of NSW households with insulation listed rebates as the main reason for installing insulation.

---

36. Data provided by the Department of Climate Change and Energy Efficiency.
37. ABS [26] and Energy Efficient Strategies [20].

While there are some inconsistencies in the volume of these reported changes the government's HIP program clearly created a step change increase in the amount of insulation installed compared to the slow long-term increases experienced in the past. Estimating the energy savings and peak demand impact of this step change, however, is difficult to calculate without underlying information of customers' appliance ownership, energy use, and behaviors.

The mild climate in Sydney and across Australia has historically resulted in lower levels of insulation and lower benefits from installation than in the Northern Hemisphere. Partly because of mild climates, modeled energy savings tend to overstate the electricity saved from household insulation. Because a large share of the benefits of insulation are taken as increased comfort (rather than reduced electricity consumption) the actual change in heating and cooling use of households is not known. Evidence from New Zealand suggests customers may have a greater tolerance for temperature variation than is assumed in thermals comfort modeling.[38]

Furthermore, estimating the benefits of the program has been made more difficult because of the unusually mild and erratic weather patterns experienced since the program ended in 2010. Therefore, while clearly a potential major factor in bring about a longer-term end of demand, the full impact of increased insulation on customers' usage remains unquantified until reliable data is available.

4.8 IMPACT OF BUILDING REGULATION – BASIX

Introduced in 2004 the NSW government's BASIX residential building regulations requires all new homes to be designed to produce 40 percent less greenhouse emissions than a benchmark of the average NSW home. The estimated impact of these changes has been less dramatic as, after adjustment for household characteristics, the delivered savings are around a 10 to 20 percent reduction compared to benchmarks of an average household set in 2001–2002.[39] Therefore, while delivering real energy savings and having a positive net benefit for customers over the life of a house, BASIX changes are not believed to have created a noticeable sharp short-term impact on overall residential consumption. In the longer-term, however, BASIX regulations are acting to restrain future growth and prevent increases in inefficient housing stock.

38. Branz [21] Appliance load research in New Zealand found most of the benefits of insulation program come for improved health outcomes not energy savings, largely from under heating of housing in winter, and comfort is the reason most people (63 percent) install insulation.
39. Energy Australia [22].

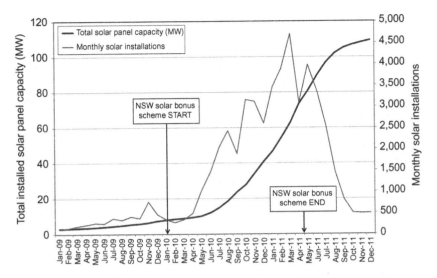

FIGURE 13.16 Ausgrid customers solar panel capacity and monthly installs 2009–2011.

4.9 Impact of PV

One of the most unexpected features in the NSW electricity market in recent times has been the rapid take-up of residential rooftop solar systems driven by the generous state government gross 60.0 c/kWh Feed-in-Tariff (FIT), which was offered from January 2010 to April 2011. The program saw new installations in Ausgrid's area peak at over 4,500 systems a month compared to a long-term average of around 500, and total installed PV capacity exceeded 100 MW.[40]

The FIT-driven boom in residential rooftop solar systems could explain a 1.5 percent reduction in residential demand over the period but, because over two-thirds of the systems in NSW are separately metered to support a gross FIT, this is seen at the transmission and generation levels but is not reflected in customers' metered energy consumption (as shown in Figure 13.5). With the end of large FIT subsidies uptake installation of residential solar is continuing but growth is back to long-term levels and is expected to remain steady but unspectacular unless new subsidies appear or the price of installed PV falls, see Figure 13.16. The boom in residential rooftop solar was a one-off phenomenon that cannot explain the recent falls in domestic consumption and is not expected to be a contributor to the "end of demand" in the immediate future.[41]

---

40. Simpson [43].
41. If customers choose to stop gross metering of their PV production when the FIT ends in around 2018–2019 this will appear as a reduction in metered energy consumption without producing an underlying change.

To the extent that customers who already had rooftop solar panels changed their metering from net to gross to take advantage of the FIT there may even have been a slight apparent increase in metered consumption for these customers, while conversely the PV output of customers who did not take up the FIT offer appear as reductions in metered consumption.

As funding for the FIT came from a reallocation of money from existing state government energy efficiency programs, mostly commercial, the impact of the FIT also needs to consider the net savings that would have accrued had the money been spent on its original purpose.

The NSW FIT has been part of an ongoing "boom and bust," "fits and starts" pattern of assistance to PV. In 2012 residential PV was promoted as reaching "grid parity" – as the cost of PV systems continued to fall and electricity prices continued to rise – the point where the c/kWh cost of electricity over the life of a system is predicted to be the same as for buying electricity from the grid. However, "grid parity' implies a 25-year payback for the large up-front investment in solar panels so that, without a FIT, this is not yet supporting a sustained change in adoption driven by the underlying economics of residential solar. Without a FIT, adoption of PV has slowed dramatically and while steady is not expected to accelerate again in the short-term.

5 CONCLUSIONS

With an end to the unprecedented increases in residential electricity prices expected in 2013–2014, the future of Australian electricity consumption and peak demand remains highly uncertain. The usual suspects have been rounded up, interrogated, and the list of likely culprits narrowed. In the end, no one factor is likely to have been the dominant cause of the dramatic change in consumption and demand.

Many of the potential suspects for explaining the fall in residential demand can be dismissed. Macro trends in population growth and household formation, household income, and overall economic activity are not consistent with a sharp fall in consumption. Customers' climate change attitudes and behaviors have not been reported to have shifted enough to independently affect their electricity consumption. Several high-profile policy changes, especially the solar FIT and Household Insulation Program, had big one-off impacts but cannot explain the recent past and are not expected to be repeated or to dramatically influence future growth. But behind the headline price and policy changes there is a strong underlying current of appliance energy efficiency, most clearly seen in hot water use but also in MEPS for other appliances, which suggest that the future will no longer be like the past and will remain as uncertain as it has been for any time in the last 100 years.

The explanation for the recent "end of demand" and the shape of the future is likely to be found somewhere in the interrelationship of price response, energy efficiency policies, and behavioral changes.[42] While exceptional price increases have continued it has not been possible to gauge the extent to which sustained structural changes due to energy efficiency policies, technological change, or customer behavior are underpinning the recent slowdown.

Unless structural change has occurred, demand could be crouching, waiting to spring back when customers habituate to higher prices and peaks hiding, waiting to remerge from behind erratic weather patterns.

So we are left with a cliff-hanger. If a simple "economic textbook" demand response to rising prices is not the culprit who is? Consumption, rather than taking a knockout blow from one big policy hit or set of price rises, appears to be being pummeled by a flurry of energy efficiency measures. Under this barrage of continuing blows growth could well have lost the ability to spring back and be down for the count. A "no growth" future for electricity consumption, which would have seemed inconceivable 10 years ago, is now a real possibility.

ACKNOWLEDGMENTS

This chapter has benefited from the input of colleagues at Ausgrid, in particular Neil Gordon, Paul Myors, Robert Simpson, and Craig Tupper. All errors and emissions, however, are the author's and do not reflect the views of Ausgrid or others.

REFERENCES

[1] AEMO, Electricity statement of opportunities for the national electricity market, update March 2012; 2011.

[2] AEMO, National electricity forecasting report (NEFR) <http://www.aemo.com.au/en/Electricity/Forecasting/2012-National-Electricity-Forecasting-Report>; 2012.

[3] Simshauser P, Nelson T. The energy market death spiral – rethinking customer hardship, AGL applied economic and policy research, working paper no.31– death spiral; 2012.

[4] Ausgrid, Directions paper on AER/EUC rule change proposals Ausgrid Submission, <http://www.aemc.gov.au/Media/docs/Ausgrid>; 2012 [16.04.12].

[5] BRANZ, Energy used in Australian appliances – analysis of the 1993/94 RES appliance energy use data research. Report no. UC0170/2; 2000.

[6] IPART, Determinants of residential energy and water consumption in Sydney and surrounds, Regression analysis of 2008 and 2010 IPART household survey data – research report; December 2011.

[7] Topp V, Kulys T. Productivity in electricity gas and water: measurement and interpretation, Australian government productivity commission, staff working paper; 2012.

[8] Jaaskela J, Callan W. Insights from the household expenditure survey. Reserve Bank of Australia Bulletin; December Quarter 2011.

---

42. Cross-sectional analysis on Sydney Water supports this contention see Abrams, Kumaradevan, Sarafidis and Spaninks [23], and AEMO [2] National Electricity Forecasting Report (NEFR) low scenario.

[9]   Auspoll, Energy efficiency - a study of community attitudes- survey for the clean energy council, key slide pack for media; 2012.

[10] Strengers Y. Designing eco-feedback systems for everyday life, CHI 2011; May 2011.

[11] Ernest and Young. AEMC power of choice, rationale and drivers for DSP in the electricity market – demand and supply of electricity; 2011.

[12] Sandu S, Petchey R. End use energy intensity in the Australian economy, Australian bureau of agricultural and resource economics (ABARE) research report 09.17, Canberra; November 2009.

[13] Devinney T, Auger P, DeSailly R. What matters to Australians: our social, political and economic values, A report from the anatomy of civil societies research project, report of australian research council (ARC) in conjunction with the university of technology, Sydney and Melbourne Business School; 2012.

[14] Department of Environment, Climate Change and Water, NSW, Who cares about the environment 2009? A survey of NSW people's environmental knowledge, attitudes and behaviours; 2010.

[15] Department of Environment, Climate Change and Water, NSW, Who cares about water and climate change in 2007? A survey of NSW people's environmental knowledge, attitudes and behaviours; 2007.

[16] Hanson F. The Lowy Institute poll 2011, Australia and the world, public opinion and foreign policy; 2012.

[17] Productivity Commission, Energy efficiency; draft report, Melbourne; 2005a.

[18] Productivity Commission, The private cost effectiveness of improving energy efficiency, report no. 36, Canberra; 2005b.

[19] Wilkenfeld G. George Wilkenfeld and Associates: prevention is cheaper that cure - avoiding carbon emissions through energy efficiency; projected impacts of the equipment energy efficiency program to 2020; January 2009.

[20] Energy Efficient Strategies, The value of ceiling insulation: impacts of retrofitting ceiling insulation to residential dwellings in Australia, September 2011 Report for ICANZ Version 4.0 (Final); 2011.

[21] BRANZ, Energy used in New Zealand households – report on the 10 year analysis of household energy end-use project (HEEP) Report no. SR155(2006); 2006.

[22] EnergyAustralia, BASIX water and energy monitoring project – electricity consumption 2007−08 and 2008−09; 2010.

[23] Abrams B, Kumaradevan S, Sarafidis V, Spaninks F. An econometric assessment of pricing Sydney's residential water use. Econ Rec 2012;88(280):89−105.

[24] ABARE. Energy in Australia 2011, <http://www.ret.gov.au/energy/Documents/facts-stats-pubs/Energy-in-Australia-2011.pdf>; 2011.

[25] AEMO, Electricity statement of opportunities for the national electricity market, <http://www.aemo.com.au/planning/esoo2011.html>; 2011.

[26] Australian Bureau of Statistics, Environmental issues: energy use and conservation Australia. Cat. no. 4602.0.55.001. Available from <www.abs.gov.au>; 2008 & 2011.

[27] Australian Bureau of Statistics, Consumer price index Australia Cat 6401.0.xls. Available from <www.abs.gov.au>; 2012.

[28] Australian Industry Group, Energy shock: pressure mounts for efficiency action, Sydney; 2012.

[29] Australian Energy Markets Commission, Possible future retail electricity price movements: 1 July 2011 to 30 June 2014: Final Report; 2011.

[30] Australian Energy Regulator, State of the energy market 2011, ACCC, Canberra; 2011.

[31] Bureau of Resource and Energy Economics, Energy in Australia; 2012.
[32] Department of Environment, Water, Heritage and the Arts, Commonwealth of Australia, Energy use in the Australian residential sector, 1986–2020 prepared by Energy Efficient Strategies; 2008.
[33] Energy Efficient Strategies. Residential Lighting Audit Report: prepared for the E3 committee draft unpublished; 2010.
[34] EnergyAustralia, Residential customer price list - 2011 and previous years. Accessed at <http://www.energyaustralia.com.au/nsw/residential/products_and_services/electricity/pricing>; 2011.
[35] International Energy Agency, Energy prices and statistics: quarterly statistics, fourth quarter; 2011.
[36] IPART, A review of regulated retail tariffs and charges for electricity 2010–2013, electricity – final report; March 2010.
[37] IPART, Changes in regulates electricity retail prices from 1 July 2012, electricity – draft report; 2012.
[38] Mountain B, Littlechild S. Comparing electricity distribution network revenues and costs in New South Wales, Great Britain and Victoria. Energy Policy 2010;38.
[39] Mountain B. Electricity prices in Australia: an international comparison, a report to the energy users association of Australia; March 2012.
[40] NERA, Analysis of key drivers of network price changes, a report prepared for the ENA; 2012a.
[41] NERA, Rising electricity prices and network productivity: a critique, a report for the ENA; 2012b.
[42] Price Waterhouse Coopers, Investigation of the efficient operation of price signals in the NEM, report for the AEMC; December 2011.
[43] Simpson R. Solar photovoltaics – what effect is this having on the grid? Ausgrid research paper available at <http://www.ausgrid.com.au/>; 2012.
[44] Waide P, Tanishima S. Light's labour's Lost, Policies for Energy-efficient lighting, In support of the G8 Plan of Action, International Energy Agency; 2006.
[45] Wilkenfeld G, Spearritt P. Electrifying Sydney: 100 years of EnergyAustralia, EnergyAustralia; 2004.
[46] Anderson G.F. Fifty Years of Electricity Supply, Sydney County Council; 1955.

Chapter 14

From Consumer to Prosumer: Netherland's PowerMatching City Shows The Way

Stefanie Kesting and Frits Bliek
DNV KEMA Energy & Sustainability, the Netherlands

1 INTRODUCTION

1.1 Global Energy Trends

The United Nations estimates that the world population will grow to between 10 and 14 billion people by 2100, with 57 percent of this growth expected to come from the growing Asian economies. As a consequence, global energy demand will almost double by 2050 [1].

How will the world meet its energy needs? The latest annual energy outlook by the International Energy Agency (IEA) indicated, for example, that large reserves of natural (shale) gas will come [2] to the market in the coming decades, keeping fossil fuel prices at an acceptable level. However, the fossil fuel reserves most likely won't be able to meet the growing demand for energy, especially when we look ahead toward 2100. Alternative sources of (renewable) energy are required to meet our energy demand.

In addition, a fuel shift to electricity can be observed, driven by electric heat pumps and the electrification of the transport system. As a result, demand for network capacity and power production will increase, requiring extensive investment in the transmission and distribution networks if these networks continue to be extended in a "traditional way." As described in this chapter, smart energy systems will allow control of the peak load on networks, reducing the need for network capacity extension. The social cost benefit analysis recently carried out by CE Delft and DNV KEMA Energy & Sustainability indicated that, independent of the future energy scenario considered, significant value can be created by making the energy system "smarter" [3].

Today's fossil-fueled large-scale power production plants are mostly at the top of their S-curves, operating almost at maximum achievable

Energy Efficiency. DOI: http://dx.doi.org/10.1016/B978-0-12-397879-0.00014-1

efficiency. These power plants are typically built far from load centers for several reasons, among them to prevent air quality issues, provide easy access to transport routes to ensure undisturbed delivery of fuel, and to allow easy access to resources of cooling water.

Small-scale combined heat and power (µCHP) units provide an attractive technology that is available today at cost levels competitive with large-scale production units. The waste heat they produce can be consumed locally, so the system operates at much higher total energy efficiency. The scale of these cogeneration units continues to decrease, which opens new markets where such technologies can be applied. Micro-cogeneration units, µCHPs, with an electric output power of approximately 1 kW, have entered the market, allowing even residential end-users to produce electricity at home. Finally, photovoltaic (PV) solar and wind power are becoming more mature technologies. Their price per kWh is still dropping and will probably end up below the cost level of large power plants.

Today, µCHP units and PV solar power are becoming more and more cost-effective. This changes the end-users' position in the energy market dramatically, as they will start to produce their own energy at home at competitive price levels. They become energy-producing consumers: prosumers.

These trends have a significant effect on how energy supply will have to be organized in the future to keep it reliable and ensure sustainability.

1.2 The Energy Prosumer

While decentralized and small-scale power production has become a common component of the energy systems in many countries, we suggest that the process of energy consumers becoming prosumers[1] could have more potential than often currently assumed. One example of drastic and unanticipated change is the impact of the Internet. The Internet changed our world in less than a decade, when end-users were starting to deliver content to the Internet. Almost no communication channel or information source was unaffected.

Similarly, prosumers could become energy up- and down-loaders, delivering energy back to the grid during times that their energy production exceeds their energy consumption. When their energy production drops below their consumption level, they start to import energy from the grid. This has multi-ple effects: it will not only make them (to a certain extent) independent from the traditional energy markets, but it allows them to

---

1. The meaning of the word prosumer as used here is similar to that of the original version as defined by Alvin Toffler in his book "The Third Wave" (1980). Toffler describes how the boundaries between the roles of producers and consumers will start to blur. However, in "The Third Wave," the process of mass customization was the underlying driver, because a high degree of customization required consumers to be involved in the design process.

choose their own sustainable energy source. At the same time, the prosumers gain detailed insight into their energy production and consumption with smart meters and energy management displays that are installed in their homes, on tablets or even smart phones. This allows them to relate their own energy consumption to their energy production. It will take significant effort to produce sufficient energy to become self-supporting. Initially, both technical expertise and investments are required, but if such technologies become more common, the initial cost can be expected to drop over time.

These developments will not only affect individuals, entire regions or suburbs could become energy-generating "virtual power plants." It is also possible that new energy communities will arise, perhaps starting with collective orders of PV panels and developing into communities that exchange energy with each other. Energy service companies might come to the market as well, as commercial aggregators or other entities providing numerous energy services we cannot envision today.

Section 2 describes the PowerMatching City project. Section 3 covers the role of end-user. Section 4 summarizes the chapter's conclusions.

2 POWERMATCHING CITY

2.1 Overview of the Pilot Project

PowerMatching City is a live smart grid project currently comprising 25 households in the Hoogkerk district in the northern Netherlands [5]. It enables residents to share electricity without sacrificing comfort. The homes are connected with each other as part of the trial project and equipped with micro-combined heat and power systems, hybrid heat pumps, smart meters, photovoltaic panels, electric vehicle (EV) charging stations, and smart home appliances. The pilot was launched in 2007 as one of the INTEGRAL project demonstrations (EU FP6-038576), and is coordinated by DNV KEMA along with its partners distribution system operator Enexis, energy company Essent, gas infrastructure company Gasunie, system integrator ICT and automatization and knowledge institute TNO. Other partners include three technology universities.

As a result of its initial success,[2] the project will expand from 25 to about 75 households and will include consumer involvement through an interactive interface and other features.

In the project, energy supply is focused on a sustainable energy solution and sourced from various distributed and renewable energy sources. Households are equipped with a micro-CHP (μCHP) or a hybrid heat pump

---

2. The project was recognized as a solution and listed in the Sustainia 100, a guide to creating sustainable societies using readily available solutions. Sustainia is an advocacy organization promoting the use of available technologies for a sustainable global future. The listing was also announced at the UN Conference on Sustainable Development held in Brazil in June 2012.

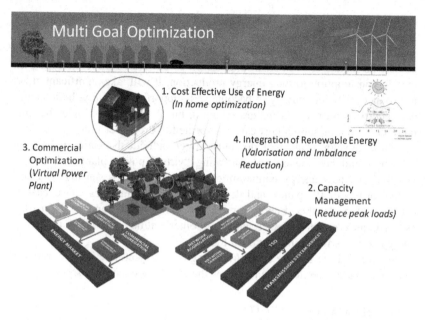

FIGURE 14.1 Optimization on different levels in the PowerMatching City project.

system, PV solar panels, and smart appliances like dishwashers or washing machines. Electric vehicles and scooters provide transportation. Additional power is produced by a wind farm and a gas-fired turbine. Smart meters are installed to measure the power production and consumption of the individual smart appliances installed. Finally, an energy management portal provides insight into the energy consumption and production of the individual end-users as well as the average of all end-users participating in this experiment.

The project centers on the development of a decentralized energy market. It aims for intelligent network operation under simulated market conditions, which allows for a simultaneous multi-goal optimization:

- In-home optimization for the prosumer
- Capacity management for the distribution system operator
- Commercial coordination for the balancing responsible party or supplier
- Effective integration of renewable energy sources Figure 14.1

2.2 In-home Optimization

In-home optimization focuses on the trade-off of energy-producing end-users to either produce or consume energy. Prosumers who have invested in their own power production facilities are looking for the optimal economic benefits of their investments. From a household perspective, the network can be regarded as a very large battery. The economic benefits for a prosumer can be maximized by continuously seeking the highest profits for energy export toward the grid and minimizing the costs for import from the grid. This provides the flexible reactive power for a smart energy system. The real-time price is used as a balancing mechanism to express the scarcity or surplus of energy in the system. With the introduction of a transport tariff or an energy tax, the in-home market becomes partly decoupled from the local electricity market. As long as the in-home energy price, $p^{in\text{-}home}$, remains smaller than the sum of the market price, p^{market}, the transport price, $p^{transport}$, and the energy taxes, p^{taxes}, it remains less expensive to consume the in-home-produced energy than buying it from the market, and vice versa for selling the energy on the local energy market.

$$p^{in\text{-}home} < p^{market} + p^{transport} + p^{taxes} +$$

This introduces a preference to consume the in-home-produced energy.

2.3 Capacity Management and Integration of Renewable Energy Resources

The experiment of PowerMatching City is set up in such a way that the entire local energy demand can be supplied with distributed energy sources in the form of CHP units and connected wind turbines and PV solar panels (both renewable energy sources). The average demand of each household is approximately 1 kW resulting in an average total energy demand of around 25 kW. Each form of power generation is in principle capable of completely covering the energy demand on its own. Thus, during ideal weather and/or market conditions, PowerMatching City is an energy-supplying area and actually delivers energy back to the power grid.

Half of the households are equipped with μCHP units that provide highly efficient local power generation. Each of these units has an electric output power of 1 kW$_e$ and a thermal output power of 6 kW$_{th}$. The maximum simultaneous output power of all these households equals 12 kW$_e$. A 210-liter hot water buffer has been applied to decouple the production of heat and electricity, allowing these units to generate electricity whenever market prices are attractive. Therefore, the price is higher than the marginal production costs. In an office building, a mini gas turbine with an output power of 30 kW$_e$ and 60 kW$_{th}$ can be connected next to the μCHP units to provide additional power and flexibility [6]. The output power of this unit can be altered between 7 kW and 30 kW.

Besides these μCHP units, households can generate power with PV solar panels connected to their home installation. The PV panels provide approximately 1590 Wp of power and annually deliver approximately 880 kWh of energy per year. The maximum output power of the complete set of solar panels in PowerMatching City is 40 kW.

Additional power for PowerMatching City is provided by a wind turbine with a nominal power output of 2.5 MW. The output is scaled down to match the cluster size of 25 households. A wind prediction model is used to accurately predict the wind power production.

In general, the large-scale introduction of distributed, renewable energy sources brings a new challenge to the energy system, due to the fact that it will introduce two-way energy flows and is an intermittent energy source. In addition, a significant part of the power production becomes dependent on weather conditions. The fluctuating nature of wind and solar power requires flexibility in the power system to ensure a continuous balance between production and demand. Demand response, interconnection capacity, storage, and flexible (e.g., gas-fired) production will allow seamless integration of these renewable energy sources. Smart energy systems can enable continuously balanced demand and production of energy.

As mentioned before, wind and solar energy are intermittent sources of energy. Ideally, their production profiles could be predicted with higher accuracy by allowing large-scale gas-fired power plants to provide the required backup power and maintain the power balance in the energy system. However, the predictability of wind and solar power is limited and power that can be brought online quickly is required. With an increasing share of wind and solar power (reflected by the upper area in Figure 14.2) the need for flexibility increases. The must-run capacity and base load power plants do not have the flexibility to adapt their output power fast enough. As a result, the system will no longer be able to maintain the essential balance between supply and demand.

Load curtailment of wind and solar power provides a technical solution, but it will hamper the business case for large-scale deployment significantly. Smart energy systems can help to overcome this problem by introducing flexibility, seamlessly matching supply and demand from various sources.

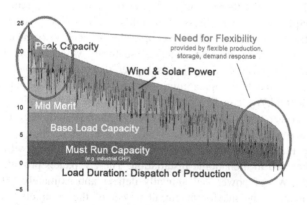

FIGURE 14.2   Load Duration: Dispatch of Production.

Capacity management in the project focuses on the need to operate the energy transportation system in a more decentralized and renewable world. Distribution system operators (DSOs) will be confronted with changing energy demands and load profiles. Moreover, electrification of the energy system will lead to increased network loads, which will strain existing networks. Augmenting existing transport capacity is very expensive, especially in cities, and labor-intensive, with a high level of impact on the built environment. Therefore, the development of advanced distribution automation systems (explained in more detail in [9]) is highly relevant to managing future load profiles, congestion, and peak loads in local grids and at distribution stations. Within PowerMatching City, the DSO can influence the load profile on the transformer by adjusting the local transport price, $p^{transport}$, in real time. Since the transport prices are defined locally, price levels become regionally differentiated at the moment transport prices are introduced. Similar to in-home optimization, the local market has a tendency to consume the locally produced power as long as:

$$p^{local\_market} < p^{central\_market} + p^{transport},$$

where p^{local_market} is the price level in that specific area, $p^{central\_market}$ is the central market price, and $p^{transport}$ is the local transport and distribution costs for congestion. In this way it can actively limit the import or export of energy.

It should be noted that the applied price signal is a soft control signal that stimulates certain behavior but does not enforce it. This means that the local control agents decide by themselves (automatically) how to react to certain price levels. Besides the price level, the user-defined settings and comfort levels are parameters weighted by the agent to decide either to turn on or off the appliance that it controls. Hence, a proper reaction during emergency situations, such as when peak load on the grid almost exceeds the maximum level, cannot be guaranteed. This could partly be overcome by programming the control agents in such a way that they force appliances to switch off at extreme market prices. However, the price settlement in the PowerMatcher algorithm operates in the minute domain and therefore is unable to react to fast power peaks. Additional control mechanisms should be introduced to deal with emergency operation of the grid that operate in the millisecond domain to provide load curtailment services.

2.4 Coordination and Virtual Power Plant

The project also includes a commercial optimization component on the supply side. The interconnection of μCHPs with a virtual power plant (VPP) is a familiar concept today. It can be used for power imbalance reduction and for trading portfolio optimization. In PowerMatching City, the whole cluster is treated as a VPP, directly controlled from the trading room of the energy

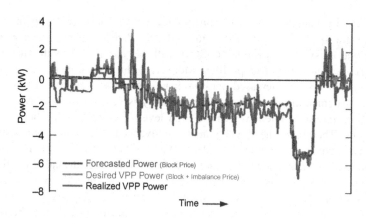

FIGURE 14.3   Example of VPP operation of the PowerMatching City smart grids.

supplier involved in this experiment. The trading room continuously dispatches all the power plants under contract with the supplier trying to maximize the supplier's profit by selling energy to their own end-users but also by trading the generated power on the various energy markets (see also explanations around market optimization below). The VPP of PowerMatching City is incorporated in the trading processes of the supplier as a physical power plant. Of course, the specific characteristics are taken into account during dispatch. The total power production of the current experiment is relatively small in comparison to the total production capacity. Nevertheless, it provides a rich source of information on how to organize the internal processes and model the VPP characteristics transparently.

In the VPP, the balance between energy production and demand can be altered by shifting the price level at the top-level market agent or auctioneer. The resulting power production or demand of the cluster can be influenced and exploited to smooth peak power demands and prevent dispatch of costly spinning reserves. Figure 14.3 shows how the VPP is operated. Power is supplied to the imbalance market.

2.5 Introduction of a Local Energy Market

Generally, a market is a place where goods and services are exchanged. The parameters of how these goods and services are exchanged are defined by controls. This implies three main activities: setting standards, measuring actual performance, and taking corrective action. More specifically, control can be defined as any process that directs the activities of individuals toward the achievement of system goals [7]. Market control involves the use of pricing incentives to regulate activities in systems as though they were economic transactions. In a typical application of market-based coordination, there are several entities producing and/or consuming a certain commodity or good [8].

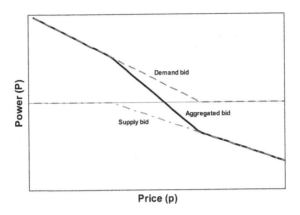

FIGURE 14.4 Bid price curves demand and supply and resulting equilibrium.

In the PowerMatcher game, each entity is represented by a local control software agent that controls an appliance in the smart energy system. This control agent communicates with a concentrator agent that aggregates all the bids of the connected agents and sends the aggregated bid either to a concentrator agent or the top-level market agent or auctioneer. This reduces the communication overhead significantly and allows the system to be scaled-up to millions of appliances that can simultaneously be optimized. During each market round, the control agents create their market bids, dependent on their state history, and send them to the market agent. These bids are ordinary, or Walrasian, demand functions d(p), stating the amount of the commodity the agent wishes to consume (or produce) at a price of p. The demand function is negative in the case of production. After collecting all bids, the market agent searches for the equilibrium price, p^e, i.e., the price that clears the market:

$$\sum_{a=1}^{N} d_a(p^e) = 0,$$

where N is the number of participating agents and $d_a(p^e)$ the demand function of agent a. The price is broadcast to all agents. Individual agents can determine their allocated production or consumption from this price and their own bid.

Figure 14.4 shows the bid-price curve for an energy-consuming device (dashed line) and an energy-producing device (dashed dotted line) and the resulting aggregated bid-price curve (solid line). The point where the line crosses the axis, where P = 0 is the equilibrium point where the demand matches the supply.

The market-based coordination mechanism, provided by the agent-based PowerMatcher technology has been extended in PowerMatching

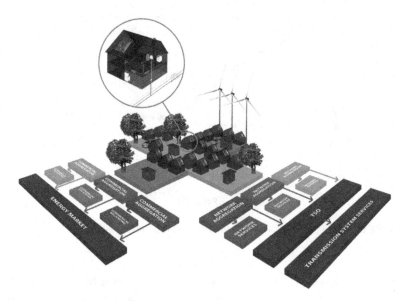

FIGURE 14.5   Setup of (local) markets in PowerMatching City.

City to allow these simultaneous optimizations. The overall purpose is to enable the participation of all entities in the local market, as shown in Figure 14.4 [9,10].

Demonstrating that it is entirely possible to control the use of energy sources and appliances on the basis of a continuously changing market rate does not, of course, reflect the current energy market situation. One of the key questions to be answered by PowerMatching City is whether a total energy system can be optimized on the basis of a market mechanism via dynamic pricing, whereby individual appliances in households will join the price "negotiation" process in this market. The control mechanism is driven by rate incentives according to the principle of peak load pricing.

The (integrated) electricity rate in the project constantly changes because during peak hours the rate for using the network increases as demand exceeds available capacity. When there is a surplus of energy (e.g., on a windy day) the sun is shining and demand is relatively low (such as on a typical Sunday afternoon), the rates will fall. It will be attractive at that moment to consume electricity to do the weekly laundry, for example. These mechanisms can also conflict. When there is little supply, the electricity rate rises, making it opportune for consumers to produce additional power. If this threatens to jam the local grid, the transport rate rises, which in turn makes it less attractive to produce electricity.

To optimally profit from rate fluctuations on the energy market, it is important to create maximum flexibility. In this context, all parties benefit over time from creating a single market for generators and consumers in which the energy rate as well as the transport costs are determined. The total costs of the energy system can thus be minimized.

Due to variances in transport capacity, it will be possible for local rates to vary, as will additional transport costs. Using the market model developed by PowerMatching City, network operators can provide insight into the costs associated with active capacity management and find a balance between investing in additional network capacity or recurring congestion costs. As a result, it is possible to limit the expansion of the power network. This can help avoid costly investments, which provides an argument for investing in the necessary intelligence with a portion of the money that can be saved. Commercial parties can successfully operate a smart energy system as a virtual power plant (VPP). Business value can be created by optimizing the production profile of the VPP and selling the energy produced on energy markets like the intra-day market where market parties can buy or sell energy during the delivery day to compensate for imperfections in their energy demand forecasts. Alternatively, the power could be sold on the imbalance market operated by the transmission system operator to compensate for market and technical imperfections, as part of the controls that ensure the power system remains in balance at all times.

3 THE ROLE OF END-USERS

3.1 Turning End-Users to Prosumers

Where residential end-users start producing their own energy on a larger scale, a lot of new power-producing units enter the existing energy system. These new small-scale energy sources are spread over existing energy networks and induce two-way energy flows. Of course, this situation is completely different from how our power system operates today, where power is produced centrally at a very limited number of large-scale power plants that seamlessly follow the demand of all the end-users. Today, the very large industrial end-users usually have access to the energy markets and are often able to adjust their energy demand in a flexible way to react to variations in energy prices and support the balancing of the power system. Some groups of medium-sized power producers, like greenhouse farmers with CHP units, have negotiated access to the power markets in Europe by aggregating their production portfolios, which offers them a very flexible way to access the power markets.

In order to allow the system to find its economic optimum, all the prosumers need to react to actual market prices by controlling the local generation as well as their energy demand. Due to the electrification of the energy system, congestion in distribution and transmission networks can be expected in the future. By altering the price levels for transport and distribution of electricity, local generation and/or demand can either be stimulated or discouraged. Therefore, nodal prices are needed to reflect the actual price at each moment for the commodity as well as the available capacity. Due to limitations in available capacity, nodal prices will vary. It becomes essential that end-users have access to local markets where prices can be formed based on the marginal costs for production.

The concept of nodal pricing allows that prices are formed on a local level. In an ideal world, when all users have transparent access to the market, all assets in the system would be dispatched in an optimal way. Such an optimization could be characterized as the next level of liberalization of the energy market, where prosumers can deliver power to the network where and to whom they wish.

To keep things manageable for households, local intelligence needs be introduced to decide on behalf of the end-users to optimize their assets, based on their status, the end-users' demands, and the current market price. Such market-based systems could be extended by other price components allowing these components to be taken into account on an economic level as well. Inclusion of demand response in a similar way is a straightforward method of allowing flexible power generation to be combined with shifting energy demand over time.

Once such markets are in place, numerous new business models can start developing, ranging from peer-to-peer delivery to aggregation services for local communities. Mass customization of energy products and services is within reach and will allow continuous profit growth, although cost and benefits might be shared in a different way from today's model.

To summarize, the following conditions must be met to allow end-users to become active prosumers as in the PowerMatching City pilot:

- A local market with dynamic prices also on end-user level, ideally with nodal pricing;
- Local intelligence systems to automate in-home optimization of energy consumption versus generation;
- The technical integration of local generation units into the distribution system; and
- The integration of all units (consumption and generation) into the smart system

In addition, the introduction of smart household appliances, smart meters, and appropriate data communication tools are major preconditions as well, which we explain in the following sub-sections.

3.2 End-Use Appliances

The demand side of PowerMatching City can be split into two groups of devices: regular household appliances that cannot communicate with the smart grid and smart appliances that can. The first group cannot adjust its demand profile to fluctuations in price levels. The second group, the smart appliances, can shift demand toward moments in time when energy prices are relatively low, allowing minimized costs for energy consumption and reacting to the availability of renewable energy available in the system. Half of the households in PowerMatching City are equipped with smart hybrid heat pumps (SHHPs) instead of μCHP units. The SHHPs combine high-efficiency air-to-water heat pumps with a condensing boiler with an electric-ity consumption of about $0.9 \, kW_e$ and an output power of $4.5 \, kW_{th}$.

The SHHPs are used for base-load heating throughout the season. During peak loads (for tap water or during cold winters when the efficiency of the heat pump drops, for example), a condensing boiler of $20 \, kW_{th}$ is used as an additional more efficient heat source. A 210-liter hot water buffer has been applied to decouple in-home heat demand from the electricity demand of the heat SHHP itself. This is not only a very cost-effective heating solution, but it also allows for switching between electricity and natural gas in response to changes in the price ratio between gas and electricity. Additionally, μCHP units and SHHPs can balance each other. In this way, the solution becomes seasonal-independent of available wind and solar power and can be self-supporting throughout the year.

Smart washing machines and smart dishwashers provide more flexibility. The load shift is achieved by programming on the control panel the desired lat-est time when the washing process should be finished. Without user interaction, the local intelligence in the devices takes into consideration the lowest expected costs to determine the optimal period to complete the washing program. A number of other smart devices are tested in this experiment as well, to demon-strate the potential and the fact that the solution is application-independent. For this reason, a smart freezer with a temperature range of $-18°C$ to $-25°C$, an in-home battery system of $4 \, kW$, and a number of fully electrical vehicles with batteries of $40 \, kWh$ are part of the experiment, too.

Of course, smart meters and effective data communication play an impor-tant role in connecting demand to the smart energy system, as outlined in the following section.

3.3 Smart Meters and Data Communication

It's important to note that smart meters are not common yet in all European countries. The introduction of smart meters is one of the most basic precon-ditions in a smart energy system. In PowerMatching City, (bi-directional) smart meters are applied to measure a) total electricity and gas consump-tion and b) energy production of each household. Each smart device in a household is individually equipped with a smart meter as well, to mea-sure the specific electricity/gas consumption. This data is fed into a home energy management system that provides detailed insight into the energy consumption and production profiles of the end-users. Several chapters in this volume notably King and Stromback and Laskey and Sayler examine the role of information and communication in influencing consumer behavior.

End-users in the PowerMatching City pilot can access their consumption and production data on their smart phones, computers, or in-home wall display. Aggregated data and detailed near real-time signals of individual devices can be monitored and compared with the average of the whole group to entice end-users to reduce their own energy consumption.

Next to the end-user online portal, an operator portal is used for remote monitoring and maintenance purposes. The operator portal provides configuration and monitoring functionalities that allow operators to detect faults and take corrective measures, even before end-users experience them. To analyze the results of the experiment, a data analysis portal is available to generate automated reports as well as individually configurable reports for data mining in all the data collected in the experiment. This allows detailed analysis by the scientists involved in the project.

Communication between the households and the central servers for coordination and control of the cluster and data collection is provided by a secure VPN network. Connections with the households are based on public broadband infrastructure based on ADSL connections. The electric vehicles use a 3G modem to communicate wirelessly with the central server system. Low-power PCs provide the platform for the in-home energy service gateways. Currently, all the in-home device agents and the in-home market/concentrator that matches the demand and supply in each home run on this platform. The local device agents exploit the flexibility provided by the smart appliances and smart local generators to economically optimize the energy demand and supply of each household. The surplus of energy is traded on a local market in the suburb, which in turn is connected to a central market aggregator where the market of PowerMatching City is attached to a trading agent. This trading agent operates the whole system as a virtual power plant and trades the surplus or demand of local power on the day-ahead market, the intra-day, and imbalance market.

The load on the distribution station is measured to determine the local network load and an objective agent connected to this distribution station influences the local market price in such a way that the network load remains within the technical boundary limits of the transformer.

3.4 Behavioral Aspects of Prosumers

The PowerMatching City pilot community is still relatively small, and behavioral aspects of end-users have been only a side topic so far. The focus has been on making the complex interactions on the different levels of optimization work, as explained in section 2. Nevertheless, some of the typical end-user feedback, perceptions and behavior observed in the course of the project so far are highlighted below.

The energy management portal plays a key role for end-users to understand their energy behavior. Today, this portal is most suitable for end-users to monitor their day-to-day behavior, more than allowing for a detailed and overall assessment of, for example, investments in new energy devices at home. End-users can see from the portal when their smart devices have been switched on and off, depending on the local market (price) situation. Consumption from household devices that are not "smart" (i.e., not connected to the portal) is visible on an aggregated level. Insight in consumption patterns is an important first step in behavioral changes. In the course of the project, we noted that participants often asked for personalized advice to check and validate their data and to also gain more insight into how to reduce their own energy consumption further. Future advanced energy management portals should provide more personalized advice based both on behavior and on the performance and maintenance conditions of technical equipment. This would ideally also allow for comparisons between peer groups with similar equipment, family composition, and home construction.

However, end-user questionnaires revealed that the participants feel that the portal provides relevant information, and almost half of the population of PowerMatching City uses it at least once a week. The other half of the group has used the portal either intensively in the first weeks after the portal has been put online, directly after an update, or after new appliances was installed in their homes. This indicates that end-user participants both verify and validate their investment and also adapt their behavior based on a better understanding of their consumption patterns. End-users usually state that they take actions to reduce their own energy demand, based on the data they can see from the system. On the other hand, once they are satisfied and understand the new situation the information becomes less relevant to them, especially when the possibilities to reduce their energy consumption further without losing comfort have reached a limit.

As already described, the in-home technologies used in PowerMatching City are fully programmed and optimize the energy portfolio automatically. Participants can start off with a basic understanding of the system and are informed via end-user meetings and interactive sessions that capture their feedback and requirements. This feedback is essential in order to further develop data systems that are comprehensible and acceptable. For example, we asked the participants how they value the various characteristics of their smart energy systems. Essential for most participants is their own comfort level and the functionality of the system, followed by cost reduction and the overall sustainability of the system.

An important change in end-users' mindsets could be regularly observed when end-users realized that their meter was running backwards (i.e., when they start to deliver energy back to the grid). Immediately, participants started looking for opportunities to get an acceptable profit out of it and often developed a competitive attitude. As they understand more about the potential effects of a local market on their own energy bills, many start developing their own ideas about how to exchange energy with their neighbors. This feedback confirms that the introduction of local markets and real-time pricing *can* contribute to increased awareness and concern about the availability of energy in general, and energy use in each specific household.

Within the timeframe of the PowerMatching City project, now in its third year, it has clearly changed the perception, understanding, and behavior of the participants. However, the speed of this learning process depends on various factors such as age, education, involvement, adaptability, etc. Since the group is relatively small, certainly more observations are required to quantify these changes and to draw more conclusions on the behavioral side of participants.

Nevertheless, the extensive interactive sessions with participating end-users have provided almost all of them with sufficient insight into the basics of smart grids and the way their appliances operate. Those who we call "highly involved early adapters" immediately started to validate the performance of the system themselves and a number of end-users also bought additional software to gain insight to the uncontrolled loads in their households so that they could complete their picture of their own energy consumption. On the other hand, a number of participants still have problems understanding the concept of real-time prices and the implications for their energy bill.

Most of the participants reported that they enjoy the fact that they are capable of producing their own energy. Until recently, most of the technology has been provided by the project team, but many participants have started to undertake investment decisions themselves. In particular, the households with heat pump systems show interest in increasing their own power production with additional PV solar panels to compensate for the power consumption of the heat pumps by themselves.

As a new communication feature, a weblog has recently been introduced. Participants are starting to communicate more frequently with other participants as they look for new ways to jointly improve their energy sustainability, jointly resolve technical issues, and exchange insight gained into their individual systems as well as the total smart energy system. It is interesting to see that jointly optimizing energy has led to broader sustainability discussions among the pilot participants, such as discussion about sustainable food supply. As we observe in many other projects, participants develop joint initiatives. Jointly investing in solar panels to reduce the investment costs is a good example of a joint initiative. Projects that are initiated by end-users themselves show a stronger social coherence, although in PowerMatching City, where an external project team started the pilot, that coherence occurred as well. In this respect, common goals and interaction certainly accelerated the processes in PowerMatching City.

Taking the feedback and observations into account, we think that a dialogue between end-users can add significant value to improving knowledge and sharing best practices. In general, households should be able to understand the consequences of their energy-related behavior. An ideal energy management system would to a certain degree be able to answer end-users' questions automatically and also enable interaction, and would help the user define realistic goals to save energy and provide support in achieving them.

4 CONCLUSIONS

PowerMatching City, a live total-concept smart grid project involving 25, and being expanded to 75 households, in the Netherlands, has demonstrated that it is possible to create a smart grid with a corresponding market model, using existing technologies. The system enables consumers to freely exchange electricity and keeps the comfort level up to par. The trial demonstrates that it is technically feasible to allow demand to track supply rather than the other way around, as is currently the case. Measurements from the μCHPs, the hybrid pumps, and electric vehicle charging all indicate that the system responds quickly to fluctuating demand and maintains comfort levels for the end-user over the long-term. This bodes well for the smooth integration of wind and solar energy, sources whose energy supply cannot be controlled. With smart appliances, end-users in the project can decide for themselves whether to switch on or off depending on the current electricity rate, for example when the rate falls because the supply from renewable sources is high.

The PowerMatching City pilot also has a behavioral component that shows the potential and effectiveness of educating end-users and making them aware of their energy consumption patterns and its effects on their energy bill. Although in PowerMatching City the focus has been on the prosumer role of end-users, it also created a higher sensitivity for energy efficiency at home. The project confirms again that people do have the willingness and ability to change their mindset with regard to their own energy consumption when they get the required information and instruments to act in a more energy-efficient way.

In Europe, energy policy currently tends to focus strongly on power generation and the shift from fossil fuels toward renewable energy sources. As a consequence, Europe faces the problem of extensive network extension that is required in the common future energy scenario. With PowerMatching City, we want to show that there might be alternatives or additional concepts to consider in solving the future energy challenge. In a decentralized energy system, the high investment cost in additional energy infrastructure can be mitigated significantly. However, a market such as introduced in PowerMatching City assumes a total market change by involving dynamic pricing on end-user level and a structural role change of consumers into pro-sumers. It goes without saying that the current energy supply chain is not set up in this way, and it would not be easy to change it in reality, for political, legal and industrial reasons.

Section 1 summarized some major trends in energy supply that increase the urgency to think about future options to solve the "energy challenge." Smart(er) grids can help solve some of the issues, mainly on the generation and capacity side. Total-concept smart energy systems such as PowerMatching City go a significant step further through the integration of end-users in a local energy market with dynamic pricing. If such a market becomes possible on a larger scale, this certainly changes the energy system dramatically as described in this chapter, but that is far from reality at this moment.

Even in the current setting of today's energy supply, there are some lessons that should be taken into account in the discussion about future energy systems:

- End-users have started to play an increasing role in generating their own energy and this trend can be expected to increase.
- Including more areas from the demand side in optimizing the energy system still offers great potential that should be used in solving future energy problems.
- Some countries are still facing ongoing discussions about the introduction of smart meters, appropriate technical standards and the allocation of related investment cost.
- Agreements in these countries should be found sooner rather than later, as smart meters are one of the preconditions to integrate demand aspects into the energy system.
- Smart household devices can help shift demand to optimize capacity on the system level and to reduce the energy bill on household level.
- Saving energy or shifting demand does not necessarily go together with compromising on comfort.
- End-users do have an interest in saving energy and contributing to a more sustainable future. A comprehensive assessment of what measures can still largely save energy as well as the introduction of further appropriate energy efficiency programs should be done in each country, considering the specific conditions and possibilities as well as technologies available.
- In other parts of the world – e.g., the United States, Australia, and a few exceptions in Europe such as Denmark – energy efficiency obligations and structural funding models for energy efficiency programs, mostly implemented by the utilities, are in place. Next to the concepts that are currently developed in the context of the energy transition towards a carbon-neutral future, the demand side and the potential to increase energy efficient end use should not be forgotten.

REFERENCES

[1] Mccall, A, FAO, UN. UN population growth estimates 1800 to 2100.

[2] World Energy Outlook. IEA, 2011.

[3] Blom M, Bles M, Leguijt C, Rooijers F, van Gerwen R, van Hameren D, et al. The social costs and benefits of smart grids, Publicatienummer: 12 3435 10, CE Delft & DNV KEMA.

[4] Toffler A. The third wave. Antwerpen 1980.

[5] Bliek FW, van den Noort A, Roossien B, Kamphuis R, de Wit J, van der Velde J, et al. PowerMatching City, a living lab smart grid demonstration. IEEE ISGT Goteborg 2010;10−13.

[6] Kester JCP, Heskes PJM, Kaandorp JJ, Cobben JFG, Schoonenberg G, Malyna DV, et al. A smart MV/LV-station that improves power quality, reliability and substation load profile. CIRED 2009.

[7] Bateman TS, Scott A. Management: leading and collaborating in the competitive world. 8th ed. New York: McGraw-Hill/Irwin; 2009.

[8] Dash RK, Parkes DC, Jennings NR. Computational mechanism design: a call to arms. IEEE Intelligent Systems 2003;18(6):40−7.

[9] Kok JK, Warmer CJ, Kamphuis. PowerMatcher: Multi-agent control in the electricity infrastructure. In agents in the industry the best from the AAMAS 2005 industry track. IEEE Intelligent Systems March/April 2006.

[10] Hommelberg MPF, Warmer CJ, Kamphuis IG, Kok JK, Schaeffer GJ. Distributed control concepts using multi-agent technology and automatic markets. IEEE PES 2007.

Chapter 15

Back to Basics: Enhancing Efficiency in the Generation and Delivery of Electricity

Clark W. Gellings

Fellow, Electric Power Research Institute

1 INTRODUCTION

This book is mostly about reducing electricity consumption once electricity is delivered to end-users. However, this chapter is about reducing electricity consumption *during* the process of production and delivery of electricity. Power producers and transmission and distribution utilities use electricity to produce and deliver electricity, and there are significant opportunities to reduce this electricity use. In power plants, these uses include power for plant auxiliaries or "parasitic loads" such as electric motors to power pumps and fans, power for environmental controls, power for building uses or "house uses" such as lighting and air conditioning, and others. On the power delivery system, primary uses of electricity include losses in cables and conductors, losses in transformers, energy use in substations and auxiliaries, and others (see Figure 15.1). Collectively, these uses represent consumption of electricity to which approaches to increase energy efficiency can be applied. This is similar to the way that approaches to increase efficiency can be applied in the transportation, industrial, and buildings sectors.

To understand the potential for energy efficiency improvements in the use of electricity in the production and delivery of electricity, estimates of current electricity use are needed. However, no known comprehensive studies have been published to document these uses so as to encourage debate on what can be done to mitigate them. This chapter discusses these issues. Recently, as elucidated in other chapters of this book, there has been substantial attention paid to increasing end-use energy efficiency (see also [2]), but little or none to increasing electric efficiency in electric technologies used for power production in generation and in electric delivery in transmission and distribution systems.

Energy Efficiency. DOI: http://dx.doi.org/10.1016/B978-0-12-397879-0.00015-3

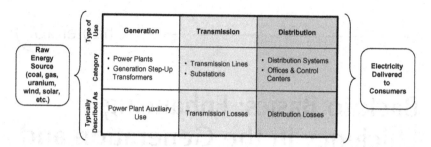

FIGURE 15.1 Electricity Use by the Electric Sector. *(Source: [1]).*

This chapter does not consider heat losses during the conversion of fuels to electricity (except to the extent that reducing parasitic losses improves conversion efficiency). For example, depending on the type and class of power plant, heat losses during fossil fuel conversion can range from approximately one-half to as much as two-thirds (for older generations of power plants) of the original fuel energy. Improving fuel conversion efficiency (i.e., improving power plant heat rate) is the subject of other extensive research and is not the subject of this chapter.

Hence, the uses of electricity discussed in this chapter do not equal the amount of energy that is "lost" between the fuel and the power outlet. This larger amount of energy lost includes the heat losses during conversion of fuels to electricity. The uses of energy discussed in this chapter are only a portion of the amount of energy lost between the fuel and the power outlet. However, these energy uses present an opportunity for energy savings – an opportunity that has not yet been examined in detail.

There are other uses of electricity related to the production and delivery of electricity. Electricity is used in coal mining, natural gas drilling, uranium enrichment, natural gas production and compression, fuel transportation, as well as the manufacture and construction of the power production and power delivery facilities themselves. These uses are *not* addressed in this chapter.

This chapter is organized into five sections. After this introductory section 1, section 2 offers an overview of this topic. Section 3 discusses electricity uses in power plants, and section 4 discusses uses in transmission, distribution systems and buildings. Section 5 highlights the chapter's main conclusions and suggests next steps.

2 OVERVIEW OF ELECTRICITY USE IN PRODUCTION AND DELIVERY OF ELECTRICITY

The results of a recent analysis prepared by the author conclude that approximately 11 percent of electricity produced in the United States is consumed in the production and delivery of electricity itself. That use is segmented as depicted in Table 15.1.

TABLE 15.1 Use of Electricity in Producing and Delivering Electricity

Electricity Use Category	Approx. Percent
Power Production	4.6
Transmission	2.9
Distribution	3.5
Total	11.0

(Source: [1])

TABLE 15.2 Largest Industry Users of Electricity

	Consumption (billion kwh/yr)	Percent of Total
Manufacturing	898	58
Agriculture	40	3
Mining	76	5
Construction	82	5
Electric Industry	451	29
Total	1547	100

(Source: [3])

Based on 2010 estimates of electricity generation, this represents approximately 450.7 billion kilowatt hours of U.S. electricity generated, making the electric sector the second largest electric consuming industry after manufacturing (Table 15.2).

Estimating electricity use for generation and delivery *globally* is more difficult, primarily because useful data is not available on which to base estimates or analysis. However, the author's opinion is that on average, this electricity use globally is similar to that in the United States. Such electricity use depends on the local and regional generation mix, as well as the configuration of the power delivery system. For example, dense load centers in Europe and Singapore, for example, would yield lower power delivery losses, whereas far-flung power delivery systems in Asia or Africa would yield higher power delivery losses.

Based on expert judgment, technologies mentioned in this chapter have the technical and economic potential to reduce electricity use in electric utilities in the United States by 10 to 15 percent. Even a 10 percent reduction is enough electricity to power 3.9 million homes.[1]

Approximately $5 billion is spent annually in the United States on end-use energy efficiency. While these efforts are critical to achieving sustainability, expenditures of a fraction of that amount on electricity use in generation and delivery will have a much broader impact on managing energy than energy end-use efficiency programs.

3 ELECTRICITY USE IN POWER PLANTS

This section covers electricity use in power plants. After an overview of this subject, this section provides important background information on how power plant auxiliaries affect plant heat rate (section 3.1). The section then describes an EPRI analysis of auxiliary loads in U.S. coal-fired power plants (section 3.2). Electricity use in emission controls (section 3.3), nuclear power plants (section 3.4), natural gas-fired power plants (section 3.5), and renewable power production (section 3.6) is covered. Section 3.7 summarizes the results of power plant electricity use studies around the world. Section 3.8 then describes ways to apply energy efficiency improvements to auxiliary power consumption. This subsection covers adjustable-speed drives in detail. Section 3.9 summarizes the key points of this section.

Electricity is used in power plants to power a variety of types of electric equipment, typically called power plant auxiliaries or "parasitic loads." In thermal power plants, these devices include:

- Electric motors used to power:
  - Pumps
  - Gas booster compressors
  - Fans
  - Air compressors
  - Material handling (conveyors, coal mills, crushers, limestone slurry feed, etc.)
  - Gas turbine starters
  - Soft starters for hydro turbines and synchronous condensers
  - Electric pre-heaters
  - Environmental controls
    Building uses or "house uses" such as:
  - Lighting
  - Air conditioning
  - Food service

---

1. Based on Energy Information Administration (EIA) data of 3,950,331 MWh × 10^3 delivered (2009). A 10% reduction or 1.09% overall reduction is equivalent to 4.3 × 10^{10} kWh.

- Domestic water heating
- Information technology, including computers, control centers, monitors, supervisory control and data acquisition systems (SCADA), etc.

Figure 15.2 illustrates the components of a typical thermal/steam-driven electric power plant. These components or auxiliaries are designed based on maximum economic performance and environmental compliance, not on minimum use of in-house electricity consumption.

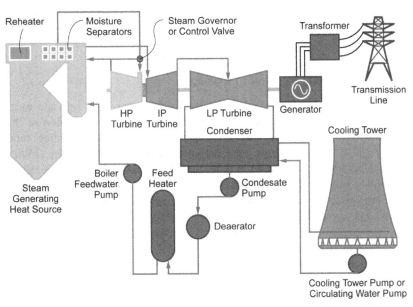

SMALL	MEDIUM	LARGE
• Cooling water pump or circulating water pump	• Cooling tower	• Steam generating heat source
• Step-up transformer	• Low-pressure turbine	• High-pressure turbine
• Electric generator	• Intermediate-pressure turbine	
• Condensate pump		
• Boiler feedwater pump		
• Condenser		
• Steam governor or control valve		
• Deaerator		
• Feed heater		
• Reheater section (if any)		
• Moisture separators		
• Transmission line		

FIGURE 15.2   Steam-Driven Electric Power Plant. (*Source: Based on Wikimedia.org, 2012*).

The configuration of electric devices applied in a power plant directly impacts the electricity usage in generation. These configurations are the result of a comprehensive optimization. When power plants are designed, there are many design trade-offs between efficiency and cost. After the plant is built, however, fuel and electricity prices often deviate from initial expectations, and energy use technology may create opportunities for fresh re-optimization.

Auxiliaries are typically oversized by 5 to 20 percent in order to ensure they meet design requirements. Many of these auxiliaries operate constantly at full output. Some pumps and fans are modulated mechanically using valves or dampers, and a few are modulated electronically using adjustable-speed-driven (ASDs) mechanisms. Modulating by mechanical means is much less efficient than modulating using ASDs.

3.1 Plant Auxiliaries Affect Heat Rate

Auxiliary uses are built into the "heat rate" calculation. Heat rate is a measure of the energy input to a power production facility as a function of the electricity output. Overall, unit heat rate is calculated by dividing total energy (BTU or kj) input by total net generation in kilowatt hours (kWh). Because net generation is used in this calculation instead of gross generation, the energy impact of plant auxiliaries is masked even though the electrical auxiliaries used to operate the plant can adversely affect the heat rate significantly. On average, electrical auxiliaries degrade utility heat rates in coal-fired power plants by 86 BTU/kWh (90.8 kJ/kWh) [4].

Sub-optimal operation of power plant auxiliaries unduly increases heat rate,[2] resulting in inefficiencies. Auxiliaries, such as pulverizers, condensate booster pumps, and hot well pumps, are needed for a specific size of power plant. However, there are many opportunities to reduce auxiliary power consumption, often during start-up and low load. For example, running the proper number of pulverizer mills for a given plant operation can help reduce auxiliary electricity use. In addition, cooling tower fans or circulating pumps depend on unit load and ambient conditions. An analysis compared condenser and auxiliary effects can determine the optimum cooling requirements. The following steps can be taken at plants when circumstances permit to reduce auxiliary power use:

- Operate equipment such as service water pumps and air compressors only as needed.
- Properly maintain equipment that uses more power as its performance deteriorates, such as motors, pulverizers, and pumps.

---

2. Higher heat rates reflect more energy needed to produce electricity.

- Maintain boiler ducts and expansion joints to prevent air leakage and hence conserve fan power.
- Install variable speed drives for fans, replacing dampers for air-flow control.
- Install outdoor lighting controlled by time clocks and/or sensors.
- Set heating and air conditioning controls.
- Turn off information technology when not in use, especially overnight.

Optimizing steam power plant air systems can also have a substantial impact on auxiliary power consumption. For example, STORM®, a company that specializes in combustion and power [5], has identified a range of heat rate variables, one of which is "auxiliary power consumption/optimization," which is defined as fan clearances, duct leakage, primary air system optimization, etc. Boiler air in-leakage contributes to wasted fan power and capacity. Figure 15.3 illustrates several of these variables.

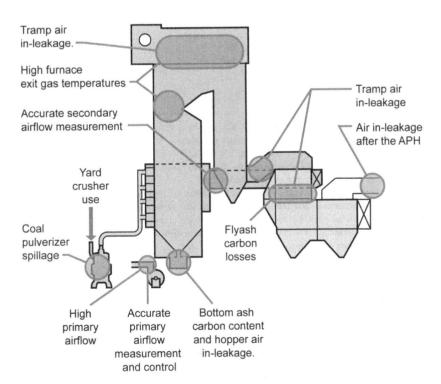

FIGURE 15.3 Stealth Heat Rate Penalties that are Controllable by Boiler Combustion and Performance Optimization. *(Source: [5]).*

3.2 Analysis of Electricity Use in Power Plants

3.2.1 Overview

The Electric Power Research Institute (EPRI) conducted an evidence-based analysis of auxiliary or parasitic loads (internal plant usage of power) in the U.S. fossil and nuclear generation fleet to better understand such power usage. According to conventional wisdom, internal power needs are approximately 5 to 10 percent of total generation, and this usage can vary by fuel type. Power needs are known to vary across such parameters as unit age, capacity, heat rate, capacity factor, number of starts, ambient operating temperature, and cooling water temperature.

Using the commercially available Energy Velocity database,[3] the EPRI team gathered data on power generation across the U.S. fleet, for coal plants, nuclear plants, and natural gas plants. The team analyzed internal power usage (as a percentage of gross generation) across each fuel-specific fleet, and statistically (i.e., through regression analysis) related usage to key characteristics across the fleet, including unit capacity, age, heat rate, and usage (based on information such as capacity factor and number of annual starts). The team used regression techniques to help parse the internal power requirement (on average) to each key contributing characteristic.

---

3. Energy Velocity: Ventyx's Velocity Suite, also known as Energy Velocity or EV, is a popular commercial source for energy data. It consists of database modules - EV Power with New Entrants, EV Market-Ops, EV Energy Map, EV Fuels, Power Transactions, EV Weather, and Grid Map - run inside the Velocity Suite and share its common interface and data tools. EPRI currently subscribes to EV Power and EV Market-Ops.

EV Power combines all the data on the electric industry with complete coverage on investor-owned utilities, generation and transmission cooperatives, distribution cooperatives, municipal utilities, non-regulated market participants, and generating assets, and updates it daily with the latest available information. EV Power with New Entrants includes:
- Existing and future generating capacity
- Power station production costs
- Operating statistics for companies and plants
- Market price data
- Regulated financial data
- SEC financial data
- Retail and wholesale power volumes and dollars

EV Market-Ops' primary components are hourly data for generation, heat rates, emissions (EPA CEMS), loads and prices (FERC and ISO). Unit costs are linked to generation and locational marginal price data to enable EV Market-Ops users to review hourly operating revenue and profit streams for nearly 2,300 of the nation's largest generating units. EV Market-Ops also compiles and formats extensive supply and demand data for use as inputs to most market models. The data is updated daily.

3.2.2 Data Sources and Quality and Statistical Approach: Power Plant Electricity Usage

Gross generation measured prior to subtracting auxiliary power for internal usage, as well as parameters such as unit age, capacity, heat rate, and capacity factor, are available by generation unit (EPA). Net generation[4] measured essentially at the busbar, which is the amount that is supplied to the grid, is available only at the aggregate plant level (EIA). The team analyzed data for five recent years − 2005 through 2009.

Analysts first matched up all unit information to its corresponding plant so that a consistent gross generation versus net generation composite database could be derived. Data sanitation led to some culling of entries[5] due to inconsistencies and imperfect plant-to-unit correspondence. Inconsistencies include data not reported for both data sets or net generation larger than gross generation. Once a sanitized database was assembled, the team ran a variety of regressions to test which explanatory variables were most critical in explaining the range of internal usage variation.

3.2.3 Overview of Electricity Usage Analysis Results: Power Plants

Across the U.S. generation fleet, there are variations in internal power usage. These may be explained in part by variations in parameters that are related to duty-cycle or overall heat rate, including unit capacity, age, heat rate, and number of starts. They may also be explained by configuration-related parameters, such as the presence/absence of particular types of pollution control equipment (e.g., scrubbers and electrostatic precipitators).

The analysis did not clearly explain variations through the detailed differences in pollution control equipment. Overall, the variations across plants and units were simply too narrow, and they were mostly superseded by macro-level indicators such as age and duty cycle. However, macro-level indicators were useful for the coal and nuclear fleets, and these indicators largely conformed to conventional wisdom as suggested above.

3.2.4 Electricity Usage in Coal Power Plants

Approximately 350 plants were in the data sample after sanitizing. These data were first separated by individual coal type, but that parsing yielded no significant difference in results from the aggregate analysis. The average power usage of internal plant auxiliaries across the sample was 7.6 percent, with a standard deviation of 2.9 percent.

---

4. Gross generation is larger than net generation, because gross generation includes the energy consumed by power plant auxiliaries.
5. A small percentage of data was discarded for the coal and nuclear generators, and a much larger percentage of data was discarded for the natural gas set (see more discussion about the natural gas fleet below).

There is considerable scatter in these results. In examining the full data set of roughly 1750 data points, the internal power fraction ranges from as low as 4 percent to as high as 12 to 13 percent. Of the key driving variables tested, plant heat rate was the most sensitive one, but extreme variations in heat rate only seem to capture about 40 percent of that range. The rest of the variation seems to be obscure, asystematic characteristics of individual plants, data inconsistencies, and the like.

In summary, the study data indicate that auxiliary power use is not as large as once thought. Larger plants tend to run more consistently, and start and stop less frequently; therefore, they tend to use less electrical energy internally. The age of the plant does not appear as relevant as once thought based on this investigation. Although newer plants are designed to be more efficient, they are required to be fully outfitted with emission controls and often with mechanically driven cooling towers. As such, newer plants are not more efficient in their use of auxiliary power.

3.3 Impact of Emission Controls on Electricity Usage

Various emission control options are driving added auxiliary power requirements for coal-fired power plants. Figure 15.4 illustrates the various options for emission controls and their typical location in a coal-fired power plant. To highlight the range of impact of these processes on auxiliary power requirements, EPRI studied a variety of flue gas desulfurization (FGD)

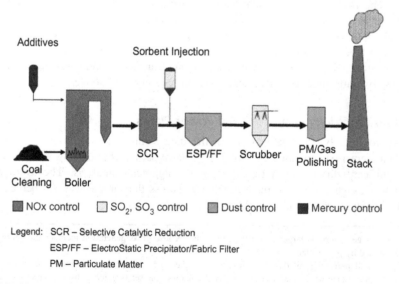

FIGURE 15.4 Emission Control Options. *(Source: [6]).*

TABLE 15.3 Auxiliary Power for Various Flue Gas Desulfurization (FGD) Processes

	Limestone Forced Oxidation (LSFO)	Chiyoda	Ammonia	Wet Mag-Lime
Auxiliary Power at Full Load (kW)				
10 Reagent Feed System =	240	240	610	135
20 SO_2 Removal System =	4,859	1,144	4,185	1,966
30 Flue Gas System =	2,610	3,636	1,657	1,657
40 Regeneration =	0	0	0	0
50 Byproduct Handling =	0	0	1,106	0
60 Solids Handling =	38	48	76	118
70 General Support Equipment =	49	50	51	47
90 Baghouse/ESP				
Subtotal	7,797	5,118	7,687	3,923
Auxiliary Power as % of Gross (500 MW)	1.56	1.02	1.54	0.78

(Source: [6])

technologies. Refer to the blue SO_2, SO_3 control elements in Figure 15.4. Tables 15.3 and 15.4 list the effects of applying several alternative FGD techniques on the connected load of a 500 MW coal-fired power plant.

These data suggest that applying FGD to a coal-fired power plant alone can increase auxiliary power usage from 0.91 to 1.56 percent.

The share of total plant auxiliary electrical power in the fleet of fossil-fuel steam plants has been increasing due to several factors:

- The addition of anti-pollution devices such as precipitators and sulfur dioxide (SO_2) scrubbers restrict stack flow and require an increase in in-plant electric drive power. About 40 percent of the cost of building a new coal plant is spent on pollution controls, and they consume about 5 percent of power generated [7].
- Additional cooling water pumping demands are needed to satisfy environmental thermal discharge requirements.
- Electric motors are increasingly used as the prime mover for in-plant auxiliary pump and fan drives instead of mechanical systems such as condensing steam turbines drives.

TABLE 15.4 Auxiliary Power Consumption of Lime Spray Dryer (LSD) and Circulating Dry Scrubber (CDS) Flue Gas Desulfurization (FDG) on a 500 MP Plant

	LSD	CDS
Auxiliary Power at Full Load (kW)		
10 Reagent Feed System =	134	134
20 SO_2 Removal System =	711	522
30 Flue Gas System =	2,840	3,404
40 Regeneration =	0	0
50 Byproduct Handling =	0	0
60 Solids Handling =	101	61
70 General Support Equipment =	40	41
90 Fabric filter/ESP	711	714
Subtotal	4,537	4,875
Auxiliary Power as % of Gross (500 MW)	0.91	0.98

(Source: [6])

3.4 Electricity Usage in Nuclear Power Plants

In the EPRI analysis discussed earlier for coal plants, 60 nuclear power plants were also represented after the data was sanitized. Approximately 280 data points were used across the five-year analysis period. The resulting analysis revealed that the average internal power usage across the sample was 4.1 percent, with a narrow standard deviation of 1.3 percent. Among nuclear plants, the variation in electrical energy usage is narrower and lower (2−6 percent) than for the coal plants. This is not surprising because operational and other variables are more similar across the U.S. nuclear fleet than across the U.S. coal fleet.

3.5 Electricity Usage in Natural Gas-Fired Power Plants

In this EPRI study, researchers found that data available on natural gas generation was problematic for determining electricity usage. There is a wide variety of duty cycle and internal station use reported. Many units that operate conventionally, sending most of their generation onto the grid, report station usage in the expected 1 to 10 percent range. Some units operate in spinning reserve mode[6] and report essentially all generation as internal usage. There are also units that seem to report no internal usage, which could be due to data error or misreporting. EPRI was able to find consistent data only among the natural gas merchant fleet, as all of this information came from the same source.

In understanding plant auxiliary use in the natural gas fleet, a key issue is data collection and reporting. Small (less than 25 MW) units are not required to report in the continuous emission monitoring system (CEMS) and may or may not be included in the plant-level data. The plant data reported in the EIA Form 923 is for "utility" power plants. Thus, merchant plants may or may not be reported, and this may change across years as the state of regulation changes in particular jurisdictions. Another likely problem could be great heterogeneity in the fleet. The natural gas-fired fleet encompasses older gas-fired thermal steam plants; combined-cycle plants that are used in baseload duty, cycling duty, two-shifting, and even strict seasonal usage; and gas turbine peaking units.

Due to these data complexities, industry estimates for auxiliary power consumption in combustion turbines (the turbines used in natural gas-fired power plants) are based on engineering estimates. These estimates conclude that the auxiliary power loads in a simple-cycle plant are about 0.5 to 0.8 percent of net power after the generator terminals (including step-up trans-former losses). This could increase if a fuel gas compressor is required.

The auxiliary power loads in a gas-fired combined-cycle plant can vary from about 1.3 to 2 percent. One factor that drives this variation is the heat rejection system design used. Direct cooling with cold water uses the least electricity, and air-cooled condensers use the most electricity. Fuel gas compressors can add another 0.5 percent. For typical combined-cycle combustion turbines with cooling towers, an estimate of about 1.6 percent auxiliary load is used, which translates to about a 1 percent decrease in overall efficiency from gross to net after generator terminals (including step-up transformer losses).

Many of these conventionally used units do report internal usage in the 1 to 10 percent range, which is consistent with the range seen in both coal and nuclear fleets. To test this, the researchers further culled the merchant fleet data to retain only those units that reported capacity factor in the range of 10 to 100 percent in order to eliminate majority spinning reserve units and other anomalies. This resulted in 3,500 observations, exhibiting a mean internal usage of 3.45 percent with a standard deviation of 2.8 percent.

Gas plants can be expected to use less internal power than coal plants because they have less internal equipment (i.e., cooling and environmental equipment and fuel handling/transport in fossil units). However, as a group, they are pressed into cycling duty much more than coal or nuclear plants, which consumes more energy relative to output. The aggregate effect is

---

6. Spinning reserve is maintaining units online and electrically synchronized, but not producing electrical energy so they are ready to immediately balance the grid against sudden increases in load or against the loss of other generation or transmission.

indeterminate. Beyond this simple observation of internal usage, the researchers were not able to further disaggregate usage by size, capacity factor, or related parameters.

3.6 Electricity Usage in Renewable Power Production

The percentage of U.S. hydroelectric power generation, based on 2010 data from EIA, is approximately 4.08 percent. Electric energy use in hydroelectric power stations is principally for excitation, with some uses for lighting, house loads, and transformer losses. There is no data available documenting this use. One reference to an Indian tariff [8] suggests an average of 0.3 percent of the energy generated is consumed for excitation. For plants with rotating exciters mounted on the generator shaft, energy consumption is considered to be zero. As a result, no energy consumption was considered for hydro generation in this study.

The percentage of the remainder of renewable power production, excluding hydroelectric power generation, is approximately 4.0 percent (, 2010). In part due to the small percentage of total renewable power production and the lack of significant data, use of electricity in renewable power production is not included in this study. Electricity is used in the following ways in renewable power generation facilities:

- Wind: The primary use of electricity in wind power production is in the inverter (power electronics), which converts the variable frequency/variable voltage output of the generators to constant frequency/constant voltage.
- Solar Photovoltaics: The use of electricity in solar photovoltaic power production is principally in the inverters (power electronics), which convert the direct current (DC) output of collectors to alternating current (AC) for use in buildings or distribution via the electrical grid.
- Solar-thermal: The use of electricity in solar-thermal power generation involves energizing technological components similar to those in thermal power plants. However, to date, there are only a handful of these types of power plants in the world.
- Biomass: The production of electricity using biomass is insignificant at present.
- Wave and Kinetic: The use of electricity in wave and kinetic energy power production is not well known or understood. At the present, these applications are limited to a few select research efforts.

3.7 Previous Studies of Power Plant Electricity Use

In other studies, various power producers have estimated the auxiliary power requirements of their units.

- Study of Power Plants in India: Table 15.5 summarizes an analysis of auxiliary power consumption in India's power plants. This analysis suggests that consumption ranges from 6.33 to 8.89 percent.

TABLE 15.5 Auxiliary Power Consumption in India Power Plants

Region	Percentage Auxiliary Power Consumption	
	2005–06	2006–07
Northern	8.89	8.62
Western	8.30	8.16
Southern	8.16	8.02
Eastern	8.39	8.37
Northeastern	6.42	6.33
All India	8.44	8.29

(Source: [9])

- Evonik Energy Services Study: Evonik Energy Services conducted an actual analysis of auxiliary power consumption at a typical power plant (see Figure 15.5). In this case, the total plant auxiliary power requirement was estimated at 9.38 to 9.85 percent of the total [5].
- Bulletin on Energy Efficiency Study: The Bulletin on Energy Efficiency published one of the most thorough breakdowns of auxiliary power (see Table 15.6). This analysis concludes that motors used to power pumps, fans, compressors, and pulverizers account for more than 80 percent of auxiliary power consumption [10].
- GE and ABB Studies: According to GE Electric Utility Engineering, for pulverized coal power plants, the auxiliary power requirements are now in the range of 7 to 15 percent of a generating unit's gross power output. Older PC plants with mechanical drives and fewer anti-pollution devices had auxiliary power requirements of 5 to 10 percent [11]. The feedwater pump power required to reach the much higher boiler pressure is approximately 50 percent greater than in drum boiler designs [12].

3.8 Applying Energy Efficiency to Auxiliary Power Consumption

Most of the discussion in this chapter addresses issues related to power plant output and heat rate. However, the bottom line in terms of revenue and operations and maintenance (O&M) costs is net unit output, defined as generator output less the unit electrical loads. Many of the unit's electrical loads are required to operate the plant, including coolant pumps, circulating water pumps, condensate pumps, and motor-driven feedwater pumps. Therefore, only a fraction of the total unit electrical load can be a candidate for consideration in improving net unit output. Despite this limitation, plant

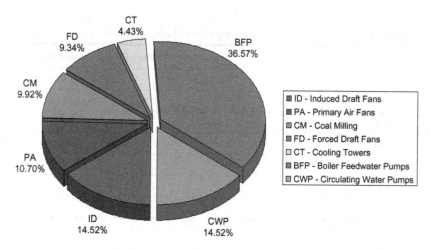

FIGURE 15.5   Auxiliary Power Consumption — Typical Coal-Fired Power Plant. *(Source: [5])*.

TABLE 15.6 Breakdown of Auxiliary Power Consumption

Subsystems	Contribution to Auxiliary Power
Draft System (forced draft (FD) fans, primary air (PA) fans and induced draft (ID) fans)	~30%
Feed Water System (Condensate extraction pumps (CEPs), LP heaters, Deaerator, Boiler feed water pumps (BFPs), HP heaters and Economisers)	25–35%
Milling system (Mills or pulverizers)	6–7%
Circulating Water (CW) System (cooling water pumps and cooling towers)	9–17%
Coal Handling Plant (CHP)	1.5–2.5%
Ash Handling System (ash water pumps and ash slurry series pumps)	1.5–2%
Compressed Air System (instrument air compressors (IAC) and process air compressors (PAC) and air drying units)	1–1.5%
Air Conditioning System	0.5–1%
Lighting System	0.8–1%

(Source: [10])

performance engineers can develop a clear list of the unit's electrical loads and considerations for reduction of power consumption and power factor improvement. Methods of power reduction include use of more efficient motors, securing some pumps during extended part-power operation, and using variable-speed motors. Of course, the effects on operating limits and component performance, such as the condenser pressure with reduced cooling flow, must be evaluated.

3.8.1 Electric Motors in Power Plants

This subsection covers the single most important technology that can be applied to reduce auxiliary power consumption in thermal power plants: adjustable-speed-drive mechanisms for electric motors. These mechanisms allow the speed of motors to be varied to match the mechanical load, resulting in reduced electricity consumption.

Induction motors are the electrical workhorses of the power production industry. Electric motors convert electrical energy into mechanical energy for operating machinery. They are used in practically every industry for pumps, fans, compressors, conveyors, and numerous other applications.

Induction motors are typically connected directly to the power system. For some "line-connected" motor applications, mechanical devices are used to control motor speed and torque − such as brakes and gearboxes. For systems that incorporate these types of mechanical devices, the flow rate of the material through the process is directly proportional to the speed of motors. Other mechanical controls − such as valves, vanes, and dampers − are used to directly control the flow of material through the process. For systems that incorporate these types of mechanical devices, the motor speed remains constant while the mechanical devices restrict the flow of material.

Mechanical control devices may provide an effective means for controlling processes, but are inefficient and somewhat limiting. Although a process motor may operate at a relatively high efficiency, the motor system − which includes the motor, the flow-control devices, and the rest of the mechanical system − may operate at a significantly reduced efficiency.

The speed of an induction motor is proportional to the frequency of its power supply and the number of poles in the machine. The vast majority of induction motors in use today are fixed-pole machines. This means that their speed is constrained to the electrical frequency of the electric power system. There are some loading effects on speed, but the typical general-purpose induction motor operates over a very narrow speed range. For example, the speed range of a typical four-pole induction motor when connected to a 60-Hz power system is approximately 50 RPM.

The Institute of Electrical and Electronics Engineers (IEEE) 112 Standard defines the efficiency of a polyphase induction motor as "the ratio of output power to input power." The efficiency of an induction motor is not consistent. As the load or output power decreases, the motor's efficiency decreases.

Motor losses can be separated into five main categories: friction and windage, core, stray load, stator, and rotor losses. All are considered electrical losses except for friction and windage, which are mechanical losses. The electrical and mechanical losses do not decrease in the same proportion as a decrease in load power. Thus, operating the motor at reduced load conditions causes the motor to operate less efficiently. It is a commonly accepted practice to oversize a motor for a particular application. Oversizing the motor for an application, which leads to reduced efficiency of the motor system, is one of the specific efficiency-design issues that make adjustable speed drives (ASDs) an attractive energy-savings alternate.

With the advent of today's microprocessor and power electronics technologies, there are alternatives to inefficient mechanical process controls. The adjustable-speed drive or ASD (also called a variable-speed drive or motor drive) is a highly efficient process-control device.

Because pumps and fans often run at partial load, energy savings can be achieved by controlling their speed with variable-speed drives. A small reduction in speed can achieve a large reduction in energy consumption. For example, a pump or a fan running at half speed consumes as little as one-eighth of the energy compared to one running at full speed (see Figure 15.6).

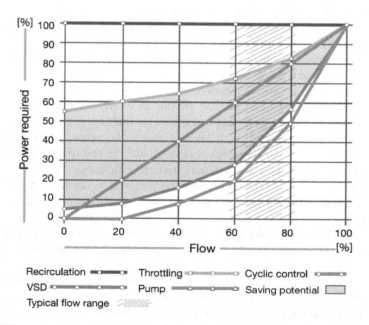

FIGURE 15.6   Energy Savings Potential of Variable-Speed Drives (VSD). *(Source: [13]).*

As a result, ASDs are replacing mechanical-control devices in industrial applications. ASDs provide a flexible and efficient motor-control scheme. Motor speed and torque can be controlled with much higher precision and accuracy because ASDs provide faster response times than mechanical controls. Motors and motor systems can be operated at significantly reduced load levels without the need for inefficient mechanical controls. Thus, ASDs can decrease system losses and increase system efficiencies when processes operate at reduced rates.

3.8.2 Adjustable-Speed Drives: Types, Analysis, and Benefits

ASDs can be broken down into two main categories: DC drives and AC drives. AC drives controlling squirrel-cage induction motors are the most common types of ASD-motor systems found in power plants. As shown in Figure 15.7, AC motor drives are composed of three basic functional blocks: an AC-to-DC converter (rectifier), a DC link, and a DC-to-AC converter (inverter). The rectifier converts AC voltages and currents into DC voltages and currents. The DC link filters the output of the rectifier and provides a DC source for the inverter. The inverter is a set of semiconductor switches that creates AC voltage and current to operate the motor.

AC drives can also be divided into two categories: voltage-source inverters (VSIs) and current-source inverters (CSIs). VSIs dominate in low-voltage industrial installations. Of the VSI drives, models incorporating the pulse-width-modulated (PWM) inversion technique are the most common. PWM inverters produce a variable RMS voltage and frequency for operating the motor at the desired speed. Figure 15.8 is a high-level schematic of a typical PWM-VSI VSD.

The speed of an AC motor is proportional to the number of poles in the machine and the electrical frequency of its power supply. VSDs allow users to vary motor speed electrically by varying the inverter frequency rather than mechanically through gearboxes and brakes. The use of ASDs allows direct connection of motors to the loads. This control technique can significantly improve the motor-system efficiency.

In addition, the need to control process parameters by mechanical means, such as vanes, bypass valves, throttling vales, and dampers, is eliminated.

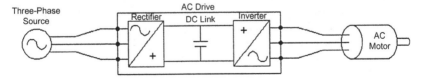

FIGURE 15.7   AC Drive Topology. *(Source: [14]).*

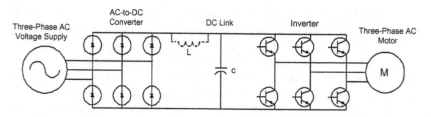

FIGURE 15.8   Typical PWM-VSI VSD. *(Source: [14]).*

Process parameters are controlled directly by varying the motor's speed and torque.

Today's ASDs provide a flexible and efficient motor-control scheme, where motor speed and torque can be controlled with high precision and accuracy. Modern digital ASDs offer much quicker and tighter control of process parameters.

There are critical issues that can spell the difference between the successful application of an ASD and one that experiences difficulties. Inexpensive issues that are not handled at the beginning of a project can be costly to remedy after the ASD has been installed. Knowing which issues to address is as important as knowing what to do about them. The following is a description of the four main issues that should be addressed with every ASD application:

- ASD Analysis: Does it make sense to use an ASD? Is it necessary or at least beneficial to your system? The answer could depend on some involved calculations to determine energy savings. Additionally, the cost of the equipment needed to control the motor depends on the goals and many other facets of the system. The goals of the analysis are to determine the economics of the project and, if economically justified, to determine what equipment is needed.

- ASD Purchase Specifications: Determining what equipment is needed is perhaps the single most difficult task facing the end-user. The modern ASD can be specified with dozens of options, all of which affect the capabilities and robustness of the motor system.

- ASD Installation Construction Documentation: For an ASD to be properly installed, a certain degree of engineering must be performed. The engineering may be accomplished in an informed fashion or in a formal way through installation drawings and installation specifications. The degree of engineering that is required depends on the size and complexity of the ASD application. To try to install an ASD with no planning or engineering invites problems.

- Start-up and Training: To have a successful start-up of a new ASD-controlled motor system depends in part on the degree of checkout

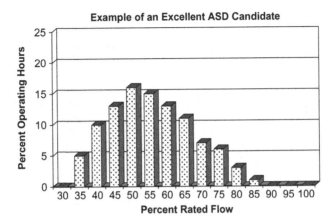

FIGURE 15.9   Load Duty Cycle: Example of Excellent Adjustable Speed Drive Candidate. *(Source: [12]).*

that is done for the total system prior to commissioning. The controls must be thoroughly checked and calibrated. All electrical and mechanical connections in the system should be checked and verified. For complex systems, training of operation and maintenance personnel is essential.

3.8.3 Adjustable-Speed Drive Applications

Examination of hundreds of ASD applications has revealed some interesting trends. Those applications that are financially justified based on energy savings alone have certain characteristics in common. The same can be said for applications that are not financially justifiable. Examination of certain system traits can help one predict whether or not an application will save enough money to justify the cost of an ASD.

There are three main types of loads that can benefit from ASDs: constant-torque, constant-horsepower, and variable-torque. Of these, only the variable-torque loads are considered probable candidates for substantial energy savings because the load power decreases significantly with speed. All centrifugal loads, which include impeller pumps and centrifugal fans, are examples of variable-torque loads that show a marked decrease in load power as speed is reduced. Axial-flow fans, turbine pumps, and compressors have a less substantial decrease in load power as speed is reduced, but there is still a significant reduction.

Figure 15.9 illustrates a motor-drive system that is an excellent candidate for an ASD mechanism. Figure 15.10 illustrates a comparison of a pump system where mechanical control and ASD control are applied. Systems loaded

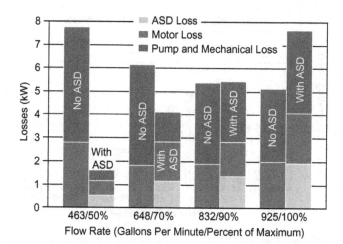

FIGURE 15.10 Example Losses in System Elements With Mechanical Control Versus ASD Control at Four Load Levels. *(Source: [12]).*

TABLE 15.7 Applications of Variable Speed Drives in Power Generation

Pumps	Fans	Other
Boiler feed-water pump	Primary air fan	Conveyor
Condensate extraction pump	Secondary air fan	Coal mill
Cooling water pump	ID fan	Oxidation air compressor
District heating circulation pump	ID booster fan	Gas turbine starter
Limestone slurry feed and absorbent		Fuel gas booster compressor
Circulation pump		

(Source: [1])

up to 90 percent can benefit from the application of ASDs. In addition, there are other benefits of variable speed control including:

- Soft starting of motor and pump
- Reduction in mechanical flow regulator wear
- Reduction in motor and pump wear
- Reduction in short circuit duty on auxiliary bus
- High power factor operation

There are a number of candidate applications for variable-speed dries in power generation. These include those listed in Table 15.7.

Even two-speed motors can offer significant improvements over simple on/off operation, particularly for air-cooled condensers and forced draft cooling towers.

In nuclear power plants, some of the largest auxiliary power uses — especially those than can sometimes be shed and/or reduced in cold weather — are mechanical draft cooling tower fans and circulating water pumps. In addition, some nuclear plants use steam-driven feed pumps that offer reduced auxiliary power consumption compared to electric motor-driven feed pumps. Some boiling water reactors (BWRs) have installed solid-state variable-speed control, which is much more efficient. Table 15.8 lists house loads for a sample of nuclear power plants.

TABLE 15.8 EPRI Plant Support Engineering House Loads Reduction Survey

Plant	# Units	Ave. Gross Generation (MWe)	Net Generation (MWe)	House Loads (MWe)
Robinson	1	752.5	718	34.5
Davis Besse	1	925	881	44
Byron	2	1175	1120	50/unit
Braidwood	2	1175	1142	33/unit
Palisades	1	821	781	40
Waterford 3	1	1147	1100.5	46.5
Grand Gulf	1	1289	1240	49
Oyster Creek	1	660	640	20
TMI-1	1	858	810	48
Clinton	1	970	929	39 winter 43 summer
Peach Bottom	2	1159	1119 winter 1093 Summer	30 winter 60 summer
Susquehanna	2	1140	1100	40
Ginna	1	492	468	24
San Onofre	2	2280	2170	113
Kewaunee	1	545	518	27
Wolf Creek	1	1226	1176	50
Average % of Gross				

3.9 Summary: Power Plant Electricity Use

Key takeaway points in the area of electricity use in power plants include the following:

- Electricity use in power plants consists of plant auxiliary equipment (parasitic loads) and building energy uses.
- Electricity use in power plants varies widely according to fuel type, unit heat rate, plant capacity, and operating duty.
- Electricity use in nuclear plants is less variable and smaller than use in coal plants; use in natural gas-fired plants is difficult to quantify.
- Environmental control systems have a significant impact on electricity usage in fossil power plants.
- Installation of adjustable speed drives on electric motors in power plants is the single most important energy efficiency technology.

4 ELECTRICITY USE IN ELECTRIC TRANSMISSION AND DISTRIBUTION SYSTEMS

Electric transmission and distribution uses of electricity occur throughout the power delivery system. The primary uses include losses in cables and conductors, losses in transformers, energy use in energizing substations and their auxiliaries, energy used in power electronics (e.g., flexible AC transmission systems, or FACTS), and losses in powering and extracting energy from pumped hydro and other storage. Storage uses include pumping and generation in pumped storage and compressed air energy storage, as well as in converting to DC for battery storage and reconversion to AC upon discharge.

Substation, control center, and office auxiliary uses include electricity to power fans, air conditioning, space heating, lighting, and information technology. In addition, in substations, uses include cooling for FACTS devices, superconducting cables, and use in pressurizing oil-fired cables.

This section begins by summarizing historical transmission and distribution use (section 4.1) and then characterizes distribution losses (section 4.2). Ways of measuring and modeling distribution electricity use are then described (section 4.3), followed by ways of increasing distribution electricity energy efficiency (section 4.4). This section then summarizes transmission electricity use (section 4.5), substation electricity use (section 4.6), and ways to improve transmission electricity energy efficiency (section 4.7). Section 4.8 then describes energy efficiency in utility offices and control centers. Section 4.9 highlights key points in electricity use in transmission and distribution systems.

4.1 Historical Transmission and Distribution Use

Figure 15.11 depicts the Energy Information Administration's (EIA) estimate of 2009 transmission and distribution (T&D) energy use (losses). It includes losses that occur between the point of generation and delivery to the customers as well as collection from differences and non-sampling error.

According to the EIA, U.S. electric power transmission and distribution (T&D) system electricity use has averaged 6.3 percent and ranged from 6 to 7 percent between 2002 and 2009 (see Figure 15.12). These uses are inherent and necessary in responding to the physics of conductance and transformation of electricity. However, there are opportunities to reduce these uses.

4.2 Characterizing Distribution Losses

EPRI performed an extensive study of U.S. electric distribution system use in what is called the Green Circuits Project. EPRI initiated this project after a series of industry workshops held in late 2007 to 2008, in which more than 30 electric utilities explored issues with distribution system efficiency. Workshop participants formed these project objectives:

- Develop and demonstrate a consistent method to quantify the use of electricity in the distribution system.
- Compile credible data to quantify the costs, benefits, and risks of using energy efficiency as a part of planning.
- Demonstrate examples in which options for efficiency improvement have been implemented, and validate realized efficiency gains.

In this project, EPRI characterized 65 circuits across 33 states and four countries to identify existing circuit losses, and prioritized potential options to improve efficiency. This effort resulted in a comprehensive database that improves understanding of the technical, economic, and implementation issues with various distribution-system efficiency measures. The results point to the following:

- Annual Energy Losses: Total distribution feeder annual energy consumption, excluding substation transformers, averaged 3.5 percent of total consumption for the feeder and ranged from approximately 1.5 to 8.6 percent.
- Primary Line Losses: Line losses averaged just less than 1.5 percent of total consumption. The study also revealed that circuit length is a reasonably good predictor of percentage of line losses.
- Transformer No-load Losses: These losses averaged about 1.4 percent of total energy consumption and ranged from approximately 0.5 to 3.25 percent. These results were the most consistent across circuits, depending on transformer age and utilization.

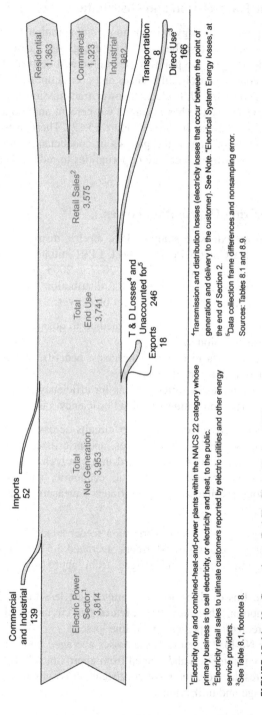

Commercial and Industrial 139

Imports 52

Electric Power Sector[1] 3,814

Total Net Generation 3,953

Total End Use 3,741

Retail Sales[2] 3,575

T & D Losses[4] and Unaccounted for[5] 246

Exports 18

Residential 1,363

Commercial 1,323

Industrial 882

Transportation 8

Direct Use[3] 166

[1]Electricity only and combined-heat-and-power plants within the NAICS 22 category whose primary business is to sell electricity, or electricity and heat, to the public.

[2]Electricity retail sales to ultimate customers reported by electric utilities and other energy service providers.

[3]See Table 8.1, footnote 8.

[4]Transmission and distribution losses (electricity losses that occur between the point of generation and delivery to the customer). See Note. "Electrical System Energy losses," at the end of Section 2.

[5]Data collection frame differences and nonsampling error.

Sources: Tables 8.1 and 8.9.

FIGURE 15.11 Net-Generation-to-End-Use Flow. (*Source: [15]*).

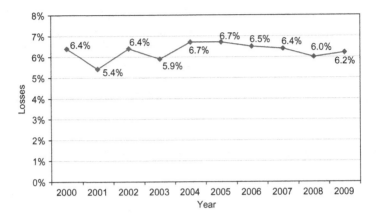

FIGURE 15.12   Transmission and Distribution Losses. *(Source: [16]).*

- Secondary Line Losses: These losses were low, averaging 0.3 percent of consumption, with a maximum of only 0.8 percent.
- Peak Demand Losses: At peak load, losses average 4.2 percent of peak demand and range from approximately 1.5 to 13.5 percent.

Historically, power delivery electricity energy efficiency initiatives, especially distribution system projects, have often been a secondary priority because of uncertainties in quantifying improvements and the difficulty in obtaining sufficient return on investment for projects undertaken. Recently, an increased industry and regulatory focus on climate change and energy efficiency has led to a renewed evaluation of power distribution efficiency initiatives.

A clear understanding of distribution electricity use is the first step in improving system efficiency. Several recent advancements have made it possible to more readily identify options for reducing distribution loss and improving overall system efficiency, including:

- Improved metering provides data on end-use patterns and diversity factors.
- Improved communication and control capabilities allow more precise voltage and reactive power (VAR) control.
- An overall improvement in modeling capabilities improves loss estimation, targeting of solutions, and ways to test and identify improvements.

While specific utility and circuit characteristics often dictate achievable efficiency actions, the wide variation in distribution losses reported from one utility to another suggests that some utilities or some circuits particularly have significant opportunity for more efficient operation.

In addition to reducing electricity use, efficiency can be increased through management of end-use customer consumption. For example, voltage control can be used to reduce energy consumption and peak demand. There is still significant work needed to quantify the potential gains through voltage reduction across regions and distribution circuits. The effectiveness of voltage control is a function of the makeup of end use loads. These loads are changing; customers are gradually using less purely resistive loads and pure motor loads and more electronic ballasted fluorescent lights, ASDs, and power electronics.

4.3 Measuring and Modeling Distribution Electricity Use

Measuring distribution system electricity use is not a straightforward process. Losses are not a quantity that can be explicitly measured at any given point in the system. Measurement of system losses requires netting the energy flowing into the system against the energy flowing out of the system at any point in time. Significant advances are presently being made in the extent and capabilities of metering on distribution systems. However, most existing systems do not have sufficient metering to directly measure electricity use. Distribution system electricity uses generally have to be calculated.[7]

Calculating the total electricity use for a distribution system is not a simple process. Electricity uses in a distribution system primarily consist of I^2R or heating losses in the distribution lines and the heating losses and core losses (hysteresis) in transformers. As I^2R implies, the heating losses vary as a function of the square of the current flowing through the line or transformer. An additional complication is that transformer no-load losses vary as a function of the square of the excitation voltage. In order to exactly calculate the total energy losses for a distribution system or circuit, planners need to represent all of the system components that contribute to losses and the varying currents and voltages through the system.

Models of a distribution system are typically used to analyze peak-demand power flows to ensure that there is sufficient power-delivery capacity to meet the peak load demand. These models typically include only the components of the primary distribution system (i.e., the medium-voltage, or MV, system) up to the service transformer and occasionally only the feeder three-phase mains. Some utilities have begun to include service transformers and low-voltage service conductors in their models. Inclusion of the full primary and secondary systems, as well as analysis of more than just the peak period, provide for a more thorough evaluation of electricity uses in a distribution system and the associated potential efficiency improvements. Representing the system in more detail is more time-consuming both from a model-preparation and analysis-computation standpoint.

---

7. Losses can also include theft of electricity. On average, theft is a relatively small percentage of the difference between energy generated and energy delivered. However, there are exceptions; theft is more prominent in certain inner cities where it is often called "energy diversion."

For the Green Circuits collaborative project, high-fidelity models of each distribution feeder were developed. The models include a representation of all of the electrical components that contribute to losses. As such, each Green Circuits feeder model includes the following:

- Substation power transformer(s)
- Primary lines (three-phase mains and single-phase laterals)
- All distribution service transformers
- Secondaries/services (not included on some feeders)
- Voltage-regulation controls (load tap-changing transformers, regulators, capacitors)

In addition to representing the full extent of the physical system, temporal variation in the load served from the circuit throughout a full calendar year was also analyzed. This is accomplished by the following:

- Individual customer loads are either assigned based on data provided by the host utility or allocated to each customer point based on the peak demand value at the head of the circuit.
- Each individual customer load is assigned an hourly-resolution annual load shape that represents the manner in which that load varies throughout a "typical" year.

The general process of developing the base case model for a given circuit is shown in Figure 15.13. The bulk of the electrical connectivity of a given circuit is obtained by converting a pre-existing model of the circuit either from the utility's own commercial analysis package format or from GIS format. The base network is then augmented with additional circuit data that is typically not available. This information typically includes the circuit voltage-control parameters such as load tap-changing transformers (LTC), regulators, and switched capacitor parameters. Characteristics of line transformer loss and secondary lines are also typically not included in base models but are added in the base-model-development process. Annual load shapes are defined from historical data and are attached to individual loads in the model.

Once the base-case model is developed, long-term dynamic simulations of the full electrical model serving all circuit loads through an annual hourly-resolution load cycle were conducted. Electrical outputs for the year

FIGURE 15.13 General Process for Developing Green Circuits Base-Case Model. *Source: [1].*

are collected from the simulation and compared with historical measured data to validate the model. Parameters such as active and reactive power flows and voltage at available locations on the circuit are very useful in validating that the modeled circuit is representative of the actual circuit operation.

Once a base-case model is validated, the base case annual simulation for each circuit is used to determine the "base case" losses that are incurred. The base-case losses are broken down as to the specific sources of the losses (primary versus secondary, load versus no-load, etc.). Losses are normalized to either the total annual energy consumption (energy losses) or the peak demand (peak losses).

4.4 Increasing Distribution Electricity Energy Efficiency

To examine distribution electricity energy efficiency, an adoption rate for each of various different technologies is needed, based on the cost of implementation and the benefits. The efficiency gains are significant and worthy of inclusion in any cost/benefit analysis of energy-efficiency initiatives.

To do this, EPRI released its first Prism analysis in 2007 [17], providing a technically and economically feasible roadmap for the electricity sector to reduce its greenhouse gas emissions. The Prism analysis provided a comprehensive assessment of potential CO_2 reductions by employing advanced technologies in the electricity sector. In 2009, EPRI updated the analysis to reflect new technologies and analysis features.

The analysis evaluates reductions in electric energy use in two main categories: reductions in end-use consumption and reductions of distribution loss. Distribution losses are composed of line losses and transformer losses, which are estimated to be 4 percent of the total energy generated in the electricity sector. While this percentage may appear relative low, the total amount of energy involved is considerable. The percentages equates to about 118 million MWh lost each year, based on a total U.S. annual generation of 3,940 billion KWh in 2008 [5].

Therefore, the top line of the efficiency Prism slice shown in Figure 15.14 is based on 4 percent of the estimated U.S. EIA's 2009 Annual Energy Outlook [18] base case for CO_2 emissions from the U.S. electricity sector. Each color represents the additional reduction in emissions based on the assumption of technically feasible levels of other technology performance and deployment. The analysis illustrates the overall reductions achievable using technologies assumed to be available.

The technical measures were stratified into two parallel efforts. The first measure reduces end-use energy consumption by employing conservation voltage reduction (CVR). Lowering voltages to end-use devices reduces

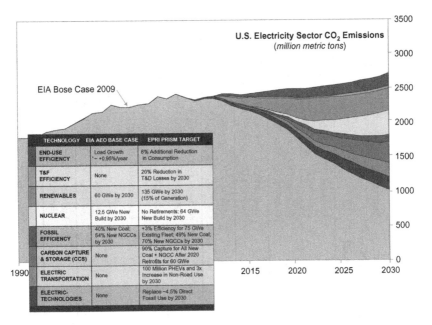

FIGURE 15.14   U.S. Electricity Sector's Potential to Reduce CO_2 Emissions Based on Deploying a Portfolio of Advanced Technologies. *Source: [17].*

consumption.[8] For every 1 percent reduction in voltage, end-use loads use approximately 0.7 percent less energy. The second measure captures approaches that will directly reduce distribution system losses. This includes approaches to reduce line losses, including phase balancing, management of reactive power needs, better application of transformers, use of more efficient transformers and conductors, and better system configurations.

4.5 Estimating Transmission Electricity Use

To estimate transmission electricity use, the author reviewed 20 U.S. studies conducted by various organizations on specific transmission corridors. These studies are summarized in Table 15.9. Using demand losses as a surrogate for annual electricity losses, the average demand loss (with respect to load) for these 20 studies is 2.97 percent, with a range of 1.7 to 6.6 percent.

---

8. In CVR, the assumption is that rather than operating circuits so as to deliver American National Standards Institute (ANSI) delivery voltage standard of 120 V ± 5%, or 114−126 volts, the circuit is operating at 108−120 volts.

TABLE 15.9 Transmission Energy Use

System/Utility/ Region	Peak Load (MW)	Demand Losses (MW)	Demand Losses (%w. r:t load)	Net Inter- change (MW)	Net Load (load + inter- change)	Demand Losses (% w.r. t NET load)	Source
Alliant Energy West (ALTW)	4.792	107	2.2%	−15	4.777	2.2%	Midwest ISO Transmission Expansion Planning 07 – for 2013 scenarios
Xcel Energy North (XEL)	12.964	380	2.9%	−2.072	10.892	3.5%	Midwest ISO Transmission Expansion Planning 07 – for 2013 scenarios
Great River Energy (GRE)	1.971	100	5.1%	555	2.526	4.0%	Midwest ISO Transmission Expansion Planning 07 – for 2013 scenarios
Hoosier Energy (HE)	855	42	4.9%	788	1.643	2.6%	Midwest ISO Transmission Expansion Planning 07 – for 2013 scenarios
Vectren Energy Delivery of Indiana (Vectren)	2.197	37	1.7%	−633	1.564	2.4%	Midwest ISO Transmission Expansion Planning 07 – for 2013 scenarios
Indianapolis Power & Light Co. (IP&L)	3.593	81	2.3%	−437	3.156	2.6%	Midwest ISO Transmission Expansion Planning 07 – for 2013 scenarios
Ameren MO	9.879	181	1.8%	−1824	8.055	2.2%	Midwest ISO Transmission Expansion Planning 07 – for 2013 scenarios
Ameren IL	11.127	268	2.4%	2025	13.152	2.0%	Midwest ISO Transmission Expansion Planning 07 – for 2013 scenarios
FirstEnergy	16.203	434	2.7%	−1.464	14.739	2.9%	Midwest ISO Transmission Expansion Planning 07 – for 2013 scenarios

Indiana Public Service Co. (NIPSCO)	3.935	66	1.7%	−208	3.727	1.8%	Midwest ISO Transmission Expansion Planning 07 – for 2013 scenarios
ITC Transmission (ITC)	12.737	295	2.3%	−448	12.289	2.4%	Midwest ISO Transmission Expansion Planning 07 – for 2013 scenarios
Michigan Electric Transmission Co. (METC)	11.522	466	4.0%	1.528	13.050	3.6%	Midwest ISO Transmission Expansion Planning 07 – for 2013 scenarios
British Columbia Transmission Co. (BCTC)	9.806	645	6.6%*	1.010	10.816	5.1%	BCTC – Loss Calculation for BCTC Transmission System – Feb. 2004 – values for 2008
A Midwestern Utility	3.385	65.09	1.9%*			5.50%	MidWest Utility Loss Analysis – Feb. 2006
Midwest ISO	134.667	4390	3.3%	−4570	130.097	3.4	JCSP Study – Eastern Interconnection – Base case reliability scenario corresponding to Summer 2018
New York ISO	36.852	977	2.7%	−2738	34.114	2.9%	JCSP Study – Eastern Interconnection – Base case reliability scenario corresponding to Summer 2018
PJM	156.542	4428	2.8%	5832	162.374	2.7%	JCSP Study – Eastern Interconnection – Base case reliability scenario corresponding to Summer 2018
SPP/CTE RC	84.839	2261	2.7%	360	85.199	2.7%	

(Continued)

TABLE 15.9 Transmission Energy Use –(cont.)

System/Utility/ Region	Peak Load (MW)	Demand Losses (MW)	Demand Losses (%w. r:t load)	Net Inter- change (MW)	Net Load (load + inter- change)	Demand Losses (% w.r. t NET load)	Source
							JCSP Study – Eastern Interconnection – Base case reliability scenario corresponding to Summer 2018
TVA RC	59.903	1519	2.5%	733	60.636	2.5	JCSP Study – Eastern Interconnection – Base case reliability scenario corresponding to Summer 2018
New York ISO	32.432	979.4	3.0%*			2.50%	NYISO – Benefits of Reducing Electric System Losses – H. Chao and J. Adams – April 2009

*Actual Annual Loss.
(Source: [1])

4.6 Estimating Substation Electricity Use

The author found insufficient data to conduct a thorough analysis of substation electricity use. Most substations have not been designed to enable internal usage to be metered separately. However, engineering estimates indicate that there appears to be substantial opportunities for reduction in substation electrical use. A preliminary study conducted by EPRI and Consolidated Edison Company of New York (ConEd) indicates that, based on 100 substation facilities, 683 megawatt hours per year are consumed to power substation auxiliary loads in these 100 substations [19].

While ConEd substations may not be the same as all other utilities, they contain many typical substation auxiliary loads. Table 15.10 lists the equipment and nameplate power ratings found at ConEd substations. All of the substation auxiliary electricity usage is embedded in transmission electricity use.

4.7 Improving Transmission Electricity Energy Efficiency

There are a number of methods by which transmission electricity use can be reduced. These include extra high voltage (EHV) overlays or upgrades, improving substation/transformer efficiency and improving transmission line efficiency, and system loss reduction.

TABLE 15.10 Typical Substation Auxiliary Loads

Auxiliary Equipment	Name Plat Power Ratings
Transformer Cooling Fans	1/6 hp—1/2 hp (per fan)
Transformer Cooling Oil Pumps	3 hp—7 hp (per pump)
Battery Chargers	10—20 KVA
Auxiliary Transformers	50—3000 KVA
Lights	35—400 Watts (per fixture)
Anti-condensation Resistive Heaters	20—200 Watts
Ventilation Fans	1/6—1/2 hp (per fan)
Space Heating	750—10 KW
Air Conditioners	Several tons BTU
Other Motor and Pumps	5 hp—350 hp

(Source: [19])

4.7.1 EHV Overlay/Voltage Upgrade (12.4 Percent Reduction in Transmission Losses)

The single greatest method to reduce transmission losses is to increase the voltage of the transmission corridor. Doubling the voltage of a line halves the required current to deliver a unit of power (because power delivered equals the current times the voltage), and cuts losses by three-quarters (because the losses are a function of the square of the current, as well as inversely proportional to the resistance). Only 23 percent of today's U.S. transmission system operates at 345 kV and above. Through aggressive industry application, by 2030, 75 percent of new lines could be installed at 345 kV and above, and 15 percent of existing low-voltage lines could either be upgraded to higher voltage (e.g., 115 kV to 230 kV) or decommissioned altogether by 2030. Voltage rationalization can also be beneficial — that is, reducing the variations in "standard" voltages. This can reduce the need for additional transformation at interconnections, thus reducing transformation losses.

4.7.2 Substation/Transformer Efficiency (1.4 Percent Reduction in Transmission Losses)

Auxiliary power. Many approaches exist to substantively reduce consumption in substation control rooms including uses of optimal heating, ventilation, and air conditioning (HVAC) units; higher efficiency fans and pumps; and automated control of components in the substation yard. While managing usage through efficiency implementations is not a new concept, many utilities have not viewed system electrical usage in the electrical substation as a high priority. Capturing and standardizing best practices through industry collaboration can yield dramatic savings. Preliminary analysis suggests potential savings of 30 percent in auxiliary loss reduction by implementing efficiency measures. Through aggressive industry application, 50 percent implementation of existing substations and 80 percent of new substations can be reached by 2030.

Transformer efficiency. Many distribution utilities have migrated over time to a lowest-initial-cost approach to procuring transformers. More efficient transformers may cost more initially but can deliver lower life-cycle cost and improve the efficiency of the transmission system. Analysis suggests an efficient transformer can reduce both the load and no-load losses by about 20 percent. While it would not be cost-effective or prudent to replace a healthy in-service transformer with a more efficient unit, electricity providers are evaluating high-efficiency transformers in new installations or when replacing a failed unit. Approximately 1 to 2 percent of transformers are replaced each year. Through aggressive industry application, approximately 20 percent of the existing transformer fleet can be changed to efficient units and 80 percent of new transformers can be efficient by 2030.

4.7.3 Transmission Line Efficiency (4.2 percent Reduction in Transmission Losses)

Use of lower-loss conductors (trapezoidal wire). Structural design is driven more by wind and/or ice-loading criteria than by conductor weight. Therefore, a traditional aluminum conductor with steel reinforcement (ACSR) can be replaced with trapezoidal-stranded conductor (TW) without significantly changing the structure design or cost. Because a trapezoidal wire has more aluminum cross-section, it has approximately 25 percent lower resistance, with the same or slightly higher diameter as the standard conductor (and a requisite 25 percent reduction in transmission losses for a given load), and provides additional transmission capacity. Through aggressive industry application, 10 percent of existing lines could be reconductored to TW, and 80 percent of new lines could be installed as TW by 2030.

Shield wire. Shield-wire losses can be reduced by approximately 50 percent by transposition or segmentation. Through aggressive industry application, 20 percent of existing lines could be retrofitted by 2030, and 80 percent of all new lines could employ transposition or segmentation by 2030.

4.7.4 System Loss Reduction (2.1 Percent Reduction in Transmission Losses)

Technologies to reduce system losses through the deployment of smart grid systems include Volt/VAR control optimization, smart transmission control of power flow controllers, and economic dispatch with loss optimization. These technologies can reduce transmission system losses by about 3 percent if 95 percent of the transmission system is adapted with some kind of smart control by 2030.

4.8 Offices and Control Center Energy Efficiency

Electricity used by the electricity sector in offices and control centers is predominantly in office building technologies. Primary uses include lighting, heating, ventilation, and air conditioning as well as uninterruptible power systems and information technology (IT). Electricity sector offices and control centers typically have a larger use of IT use than conventional office buildings.

Electric utilities have facilities that can be considered offices or "retail and service buildings" (using EIA's definitions). They typically consist of the following types:

- Office buildings including service storefront (customer service), billing, and administration.
- Distribution centers including headquarters for regional engineering staff, storehouse for hardware, service fleet housing, and maintenance.
- Control centers housing operations centers, SCADA, EMS, and other IT equipment.

To prepare an estimate of usage, the author segmented utilities into type and size and estimated the number of buildings per utilities of each size. According to the EIA, retail and service buildings use an average of 0.8 billion BTU per building per year or 234,700 kWh per year and have an energy intensity of 76.4 thousand BTU per square foot. Fifty-three percent of use is estimated to be for cooling, office equipment, lighting, and other uses. For the purpose of this analysis, it is assumed that all of these uses are electric. This is a conservative estimate, since water heating and space heating in some of these buildings is likely to be based on natural gas or fuel oil [20]. Hence, the analysis detailed in Table 15.11 was conducted.

Utility offices and control centers are not unlike other commercial building, and hence, offer significant opportunities for electricity use reduction. In some states, utilities charge themselves commercial rates for electricity use, implying a potential incentive to reduce this energy use. However, this is a function of regulatory treatment and incentives. There are numerous impediments to upgrading electricity efficiency in these buildings. For example, most utility building owners and operators have difficulty prioritizing electricity efficiency measures due to a lack of information. Another limitation is that many existing building and equipment are not yet fully amortized, complicating justification for capital improvements. Despite these limitations, there is increasing awareness of the value of increasing electricity efficiency in these buildings.

4.9 Summary: Transmission and Distribution Electricity Use

Key takeaway points in the area of electricity use in transmission and distribution systems include the following:

- Primary electric transmission and distribution uses of electricity include losses in cables, conductors, and transformers; energy use in energizing substations and their auxiliaries; energy used in power electronics; and energy for lighting, IT, and HVAC in substations, control centers, and offices.
- U.S. electric power T&D electricity use ranges from 6 to 7 percent. Of this total, distribution contributes about 3.5 percent. One study showed that this is made up of approximately 1.5 percent is primary line losses, 1.4 percent in transformer losses, and 0.3 percent in secondary line losses.
- Improvements in distribution metering, communication and control, and modeling enable improved identification of options for improving distribution energy efficiency.
- Methods of improving distribution energy efficiency include conservation voltage reduction, approaches to reduce line losses (e.g., phase balancing), management of reactive power needs, better application of

TABLE 15.11 Electricity Used in Electric Sector Buildings

Type of Utility	Total #	Small				Medium				Large			
		%	#	Bldgs/ Utility	Total Bldgs	%	#	Bldgs/ Utility	Total Bldgs	%	#	Bldgs/ Utility	Total Bldgs
IOU	342	0	—	—	—	25	86	3	258	75	256	6	1536
Co-op	893	50	446	1	446	40	357	1	357	10	90	3	270
Muni	2118	50	1059	1	1059	45	953	1	953	5	106	2	212
Total	3353		1505		1505		1396		1568		452		2018

Total buildings = 5091
Electricity use/building = 234,700 kWh/year
Total electricity use = 1.195×10^9 kWh/year
Total electricity produced in 2009 = 3953×10^9 kWh
Percent used in utility buildings = .0003%

(Source: [1])

transformers, use of more efficient transformers and conductors, and better system configurations.

- A review of 20 U.S. studies of transmission energy losses revealed an average loss of about 3 percent.
- Methods of reducing transmission electricity use include extra high voltage overlays or upgrades (potential to reduce total losses by 12 percent); improving substation/transformer efficiency (potential to reduce total losses by 1.4 percent); improving transmission line efficiency (potential to reduce total losses by 4.2 percent); and system loss reduction (potential to reduce total losses by 2.1 percent).

5 CONCLUSIONS

Table 15.12 summarizes the estimates of total uses of electricity as discussed in this chapter. As depicted in the table, approximately 11 percent of electricity produced is consumed in the production and delivery of electricity itself. This means that the electricity industry is the second largest electricity-consuming industry in the United States. The use of electrical energy in the production of electricity as well as the uses or losses in power delivery (transmission and distribution) contribute to this total.

TABLE 15.12 Total Use of Electricity

Segment	% Generation	% Use			Net % Avg
		Low	Avg	High	
Generation					
Coal	44.9	4.7	7.6	10.5	4.6
Natural Gas	23.8	1.3	1.65	2.0	
Nuclear	19.6	2.8	4.1	5.4	
Hydro	6.2		0		
Renewables	4.0		0		
Other	1.5	—	—	—	—
Transmission	100		2.9		2.9
Distribution	100	1.5	3.5	8.6	3.50
Total Use				11.0	

Based on 2011 Energy Information Agency (Electric Power Monthly Table 1.1. Net Generation by Source, http://www.eia.doe.gov)

(Source: [1])

In power production, duty-cycle or capacity factor is the key driver that influences internal power usage relative to unit output. In coal-fired power plants, the average internal power use across the sample used in this analysis was 7.6 percent. In nuclear power plants, the average was 4.1 percent. There are opportunities to reduce electricity use in power production. These opportunities may include advances in control systems for auxiliary power devices and the use of adjustable-speed drive mechanisms (ASD). Not all ASD installations are economical unless externalities such as CO_2 emissions are considered.

Electricity use in power delivery totals approximately 6.4 percent. In the distribution system, the use of efficient transformers, improved voltage control, phase balancing, and balancing of reactive power needs could substantially reduce electricity use. In the transmission system, opportunities include extra high-voltage overlays, transformer and line efficiency.

Technologies mentioned in this report and elucidated in other references have the potential to reduce electricity use in electric utilities by 10 to 15 percent. Even a 10 percent reduction is enough electricity to power 3.9 million homes. Based on recent reports [21], $5 billion is being spent annually in the U.S. on end-use energy efficiency. While these efforts are critical to achieving sustainability, expenditures of even a fraction of that amount on electricity use in generation and delivery will have a much broader impact on managing energy that energy efficiency programs.

In the power industry, there are considerable opportunities to achieve higher efficiencies end to end, from generation through power delivery and end use. To recognize these opportunities, it is important to look at each link within the electricity value chain individually and compare the effort for achieving higher efficiency within a holistic framework.

To understand the opportunity to enhance electric energy efficiency across the value chain, utilities need a detailed framework they can use that serves as a decision-making tool to optimize the impact of the utility's energy efficiency efforts. The framework should consider the various alternative technologies when identifying opportunities to improve efficiencies and enable utilities to assess, compare, and evaluate these technologies and measures. This should include a cost-benefit analysis. The framework could include the following steps:

1. Develop a comprehensive list of energy efficiency measure for each stage in the value chain.
2. Quantify the potential for efficiency for generation, transmission, distribution, and end-use programs based on estimates available from existing studies.
3. Propose metrics and methods to quantify the energy and carbon reduction benefits and costs of prospective energy efficiency measures.

4. Provide a basis to compare and prioritize capital projects across generation, transmission, distribution, and end-use functions using energy efficiency and carbon impact as the overarching criteria.

Once the analysis is complete, the utility may wish to develop an awareness campaign to inform key stakeholders, including employees, customers, shareholders, regulators, intervener groups, and the public.

This assessment is most robust if a cross-disciplinary project team of experts in power plant operations, T&D operations, end-use efficiency, and environmental impacts are used to develop the framework. This framework development process should facilitate interaction and cooperation between groups in the utility that might otherwise have limited or no opportunity for dialogue on issues of energy efficiency or loss reduction.

REFERENCES

[1] Electric Power Research Institute (EPRI). Program on technology innovation: electricity use in the electric sector, EPRI, Palo Alto, CA; 2011. 1024651.

[2] Electric Power Research Institute (EPRI). Assessment of achievable potential from energy efficiency and demand response program in the U.S.: (2010–2030), EPRI, Palo Alto, CA; 2010. 1016987.

[3] Electric Power Research Institute (EPRI). Electrotechnology reference guide, EPRI, Palo Alto, CA; 2011a. 1022334.

[4] Electric Power Research Institute (EPRI). Heat rate improvement manual, EPRI, Palo Alto, CA; 1998. TR-109546.

[5] Electric Power Research Institute (EPRI). Heat rate improvement conference, EPRI, Palo Alto, CA; 2009a. TR-1017546.

[6] Electric Power Research Institute (EPRI). Survey of impacts of environmental controls on plant heat rate. EPRI, Palo Alto, CA.; 2009a. 1019003.

[7] Masters, Renewable and efficient electric power systems, G.M. Masters, 2004.

[8] GOI, GOI Tariff Notification, India, November 6, 1995.

[9] Performance, Performance review of thermal power stations 2006–07 Section 11, Central Electricity Authority, India, <www.cea.nic.in.>; 2007.

[10] Bulletin best practices for auxiliary power reduction in thermal power stations, The bulletin on energy efficiency, ISSN 0972-3102, August–October 2006, Vol. 7, Issue 1–2, India. 2006.

[11] General Electric (GE). Electric Utility Systems and Practices. In: Rustebakke HM, editor. GE utility division. 4th ed. Wiley; 1983.

[12] ABB, The smart grid begins with smart generation[tm]. Energy efficient design of auxiliary systems in fossil-fuel power plants, ABB Inc., in collaboration with Rocky Mountain Institute, 2009.

[13] ABB, Medium voltage AC drives for power generation industry, ABB, <www.abb.com/product/ap>; 2011.

[14] Electric Power Research Institute (EPRI). Guide to the industrial application of motors and variable-speed drives, EPRI, Palo Alto, CA; 2001. 1005983.

[15] U.S. Energy Information Administration (EIA). Energy Information Administration's (EIA) estimate of 2009 transmission and distribution (T&D) energy use (losses). 2009b.

[16] U.S. Energy Information Administration (EIA). Energy Information Agency (EIA) Electricity Overview 1948–2009.

[17] Electric Power Research Institute (EPRI). The power to reduce CO_2 emissions, The Full Portfolio, Discussion Paper, EPRI, 2007. 1005461.

[18] U.S. Energy Information Administration (EIA). Annual energy outlook early release overview – December 14, 2009a.

[19] Bose. Substation auxiliary light and power efficiency initiative, consolidated edison of New York and EPRI, S. Bose presented to EPRI international PDU council meeting, London, England, June 9, 2011.

[20] U.S. Energy Information Administration (EIA). Commercial buildings energy consumption survey, Energy Information Administration (EIA). 1995.

[21] Consortium for Energy Efficiency (CEE). State of the efficiency program industry: 2009 expenditures, impacts and 2010 budgets. Consortium for Energy Efficiency (CEE), Boston, MA, December 10, 2010.

[22] ConEd, Substation auxiliary light and power efficiency initiative, Consolidated Edison of New York and EPRI, S. Bose presented to EPRI International PDU Council Meeting, London, England, June 9, 2011.

[23] U.S. Energy Information Administration (EIA). Assembled from a database section called monthly plant generation and consumption, that is assembled from Federal Form EIA-923. U.S. Energy Information Administration.

[24] U.S. Energy Information Administration (EIA). Electric power monthly data from EIA for 2010.

[25] U.S. Environmental Protection Agency (EPA). Assembled from a database section called Unit Generation and Emissions – Annual, which is assembled from federal form USW EPA CEMS (fossil units). U.S. Environmental Protection Agency.

Chapter 16

Smarter Demand Response in RTO Markets: The Evolution Toward Price Responsive Demand in PJM

Stuart Bresler[1], Paul Centolella[2], Susan Covino[3] and Paul Sotkiewicz[4]

[1]*Vice President, Market Operations, PJM Interconnection LLC,* [2]*Vice President, Analysis Group, Inc.,* [3]*Senior Consultant, Emerging Markets, PJM Interconnection LLC,* [4]*Chief Economist, PJM Interconnection LLC*

1 INTRODUCTION

Microeconomic theory teaches that in any market, the buyers of a good or service make consumption decisions based upon the price of the product and the value they place on consumption of the product, and their available budget for all goods and services they consume. For example, ask any driver their consumption of gasoline and they will likely recite back to you the prices they pay, how many times they will fill up their tanks in a week or month to commute to work, and how their driving habits change with the change in gasoline prices.

However, if the same question were asked regarding electricity prices and consumption, it is unlikely we would receive similar responses regarding price and consumption. The price of electricity can change on a 5-minute basis, and in PJM's wholesale power markets settlements are based on hourly averages of the 5-minute prices. In general, consumers of electricity do not know what the market price of electricity is at any point in time, let alone how much they would like to consume at various prices, or how their consumption patterns may change with changing market prices.

There are two main reasons for electricity consumption decisions being made blind to market prices: 1) Until recently, there has not been a cost-effective technology to both measure consumption at 5-minute or hourly

Energy Efficiency. DOI: http://dx.doi.org/10.1016/B978-0-12-397879-0.00016-5

intervals and to transmit or receive such price signals that can be acted on in an automated fashion to allow meaningful responses to market prices; and 2) To date most electricity consumers have not been exposed to market prices but have paid one flat rate for electricity regardless of the current market price in large measure due to technological limitations previously cited.

Price responsive demand (PRD) leverages the availability of cost-effective smart grid technologies such as interval meters with two-way communication and automated response to prices through appliance specific or building-wide energy management systems that can receive price information and respond in an automated fashion to allow electricity consumers to see and respond to market prices for power. The technology has become cost-effective enough to roll out to residential customers who will now have the ability learn about and manage their energy usage. Moreover, PRD relies on the fact that many state public utility commissions have commenced pilots and widespread adoption of dynamic retail rates for consumers that charge higher prices, or offer rebates, to consumers during system peaks and times of system stress as a means of translating wholesale market prices and conditions down to the level of the end-use customer.

Simply put, PRD enables consumers to respond to the market price of electricity based on the value they place on consumption and their budgets to manage or reduce their monthly bills. Stated another way, PRD is the recognition that technology has finally caught up with microeconomic theory in electricity markets. And while the focus of this compendium of chapters is focused on energy efficiency, PRD provides a gateway to enable and incentivize energy efficiency through consumer responses to prices.

This chapter lays out how PRD will integrate the wholesale and retail electricity markets and describes the confluence of events that led to the development of PRD and how PRD will enhance energy and capacity market price formation, improve reliability and operational control, and increase market efficiency while reducing transaction costs.

The chapter is organized as follows. First, the concept of PRD is defined and the way in which PRD ties together dynamic retail rates and wholesale price formation is explained. Next, the chapter describes how PRD works in energy market operations and manages reliability during system peak, which leads into the section that describes how PRD is integrated in the capacity market to ensure resource adequacy reliability. The chapter then discusses the technological advancements and regulatory drivers that have led to PRD followed by a section discussing the administrative and transactions costs that are avoided by PRD but incurred by the current form of demand response. Observations regarding the direction of PRD going forward conclude the chapter.

2 DEFINING PRICE RESPONSIVE DEMAND

Fundamentally, PRD is a predictable change (a reduction as prices rise) in electricity usage in response to changing wholesale energy market prices. End-users communicate their willingness to pay for energy consumption through energy demand curves submitted to PJM market operations by their suppliers on a daily basis. The demand curves for energy can be very simple expressing a single price at which consumption changes from a "normal" level to a "reduced" level.

PRD can also be translated to indicate an end-user's willingness to pay for capacity by linking the energy price that is likely to prevail during system peaks or emergencies to reduced energy consumption during those times. The reduced consumption during system peak and emergencies defines a cap on the quantity of capacity the customer is obligated to buy that can then be represented in PJM's Reliability Pricing Model (RPM) Capacity Market.

3 LINKING WHOLESALE PRICES TO RETAIL PRICES

PRD provides a direct linkage between the locational marginal price (LMP) in the wholesale market and the dynamic retail rates.[1] Generically, dynamic retail rates are retail rates that change with system prices or conditions and can include rate designs such as critical peak pricing, critical peak rebate (also known as peak-time rebate), or directly charging retail customers the wholesale price in real-time.

PJM's PRD manual provides that PRD eligible load must be:

Served under a dynamic retail rate structure with an LSE or subject to a contractual arrangement with a PRD provider where such rate or compensation arrangement can change on an hourly basis, is linked to or based on a PJM real-time LMP trigger at a substation location within a transmission zone as electrically close as practical to the applicable load, and results in predictable response to varying wholesale electricity prices.[2]

The requirement of linking the retail rate to the LMP at the wholesale level integrates the formation of prices in each market with those of the other.

For example, a critical peak pricing structure, such as that shown in Figure 16.1 charges the critical peak price of $1.25/kWh when wholesale prices reach a pre-defined level (e.g., $900/MWh), or when system conditions reach a pre-defined emergency level as announced by PJM. Outside of those system conditions, the retail customer may face a flat rate shown in blue or a time of use rate shown in red. It is worth noting that a time-of-use

---

1. Section 1.71F Price Responsive Demand, Reliability Assurance Agreement Among Load Serving Entities in the PJM Region, PJM Interconnection, L.L.C., Rate Schedule FERC No. 44.
2. PJM PRD Manual, Section 3 Eligibility of Price Responsive Demand, p. 3.

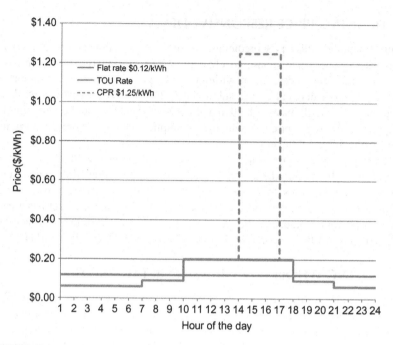

FIGURE 16.1   Sample Critical Peak Pricing Structure.

rate is not a dynamic retail rate in that the prices and hours are set in advance and are independent of system conditions. A critical peak rebate or peak time rebate structure may also be used as shown in Figure 16.2. The critical peak rebate structure in Figure 16.2 is the mirror image of the critical peak price structure shown in Figure 16.1, except when wholesale prices reach the level that triggers the rebate (e.g., $900/MWh), or pre-defined emergency condition, the customer earns a rebate for reducing consumption below a baseline, otherwise it will still be charged the prevailing flat rate or time-of-use rate.

The manner in which the link between dynamic retail rates and wholesale market response is shown is through the energy demand bid in PJM's wholesale market. Using the examples discussed above, the simplest representation would be an energy demand bid that indicates that at wholesale prices below $900/MWh, a customer would use 6,000 kW of energy because it faces a flat retail rate of only $0.12/kWh, as shown in Figures 16.1 and 16.2. At prices of $900/MWh and above that trigger the higher dynamic retail rate the customer uses only 4,000 kW of energy because it faces the higher critical peak price of $1.25/kWh, as shown in Figure 16.1, or could receive a rebate of $0.85/kWh reduced, as shown in Figure 16.2. In either case, the dynamic retail rate can be easily translated into a change in wholesale market demand

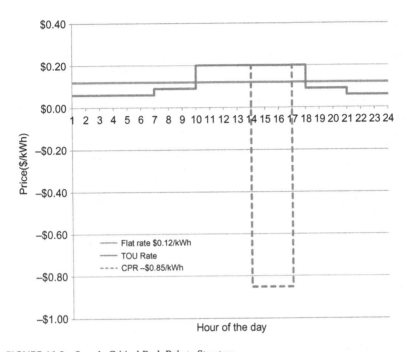

FIGURE 16.2   Sample Critical Peak Rebate Structure.

because of the linkage between the dynamic retail rate and wholesale market prices.

4 TRANSLATING PRD INTO REDUCED CAPACITY OBLIGATIONS

Because high wholesale energy market prices are coincident with system peaks upon which the need for capacity to ensure resource adequacy reliability is planned, PRD can easily be translated into reduced individual and system requirements for capacity resources to maintain resource adequacy reliability. It is easiest to view this through the example of a single, large customer building introduced in the energy market examples discussed above.

Consider a large end-user that normally has a peak hourly usage during the summer months of 6 MW (6,000 kw) based on a flat retail rate of $0.12/kWh, as shown in the examples in Figures 16.1 and 16.2. But with the implementation of the dynamic retail rate such as the critical peak rate shown in Figure 16.1, the end-user knows that it will only use 4 MW (4,000 kw) at that retail price, and also knows that critical peak price will only be in place during system peaks. The end-user can indicate in advance to its load serving entity (LSE) that it will limit its hourly usage to 4 MW

(4,000 kw) when a critical peak price is in effect thereby reducing the need for the LSE to purchase capacity to cover a peak consumption the end-user and LSE know will not take place at peak.[3] The LSE agrees that the critical peak price will only be in effect when the LMP equals or exceeds $900/MWh or during system emergencies.

To operationalize this arrangement at the wholesale level, the LSE submits an energy demand curve into the wholesale energy market to show that consumption will decrease to 4 MW or 4,000 kW when the energy price reaches $900/MWh. In the event of system emergencies that would require PJM to ensure the peak consumption was down to 4 MW, PJM can then also use the energy demand bid to set energy prices to at least $900 to ensure the end-user reduces consumption in real-time down to the agreed upon level.

5 PRD PLACES DEMAND ON THE "DEMAND-SIDE" OF THE MARKET

PRD builds on a 10-year effort to integrate demand response into the energy market. Spurred by the inability of end-users paying flat fixed rates for electricity to see, respond to, and benefit from LMPs, PJM and its stakeholders developed rules, subsequently approved by the Federal Energy Regulatory Commission (FERC), that have authorized CSPs (curtailment service providers) to aggregate and provide load reduction capability to the energy market in competition with generation.

In linking consumption decisions, as they are represented through the energy demand bid at the wholesale market level, through the LMP at the wholesale level, and dynamic retail rates at the retail level, PRD is being represented as demand for energy or capacity. This stands in contrast to the manner in which demand response (DR) has been represented in wholesale power markets to date.

In the wholesale energy market, DR is treated as a supply-side resource with the amount of supply being determined as the difference between the assumed baseline consumption absent the DR and the actual consumption in response to the wholesale energy price. There is no direct or explicit linkage between wholesale prices and retail rates. And rather than directly reducing expenditures, DR receives a direct payment for the reduction of its consumption as an offset to expenditures.

In the RPM Capacity Market, while PRD directly reduces the demand for capacity to achieve resource adequacy reliability, DR effectively requires

---

3. The CCP rate of $1.25/kWh is more than 10 times the non-critical peak price of $.12/kWh charged in all other hours in Figure 1. Numerous retail pilots have shown the price differentials needed to incent usage reduction by end-users during peak periods when wholesale prices are typically highest. See "Dynamic Pricing and Its Discontents," Regulation, Fall, 2011, by Ahmad Faruqui and Jennifer Palmer, p. 18.

two separate transactions. First, the end-use customer through its LSE buys capacity as if it were not reducing consumption in response to prices in the energy market at system peak. The second transaction is effectively a sale on the supply side of the market of DR capacity that the end-use customer will not use back to the market, where the amount of supply is measured against a baseline that is not directly related to linkages between wholesale and retail prices.

6 OPERATING VISIBILITY AND RELIABILITY WITH PRD IN ENERGY MARKET OPERATIONS

As noted in the section defining PRD, the main idea is that LSEs or PRD providers can provide to PJM an energy demand curve that provides the consumption levels of end-use customers at various prices. From PJM operational perspective, PRD allows PJM to dispatch resources in its Security Constrained Economic Dispatch (SCED) software more efficiently if there is visibility through energy demand curves into how much will be consumed at various prices. In this sense PRD is dispatchable in that PJM can use PRD to maintain the balance between supply and demand as FERC has mandated for supply-side DR under Order No. 745.[4]

In addition to the visibility provided by the demand curve for energy, the requirement that end-user's PRD be automated will provide both PJM's market and system operators with confidence that PRD will reliably not be there as prices rise including price increases associated with emergency conditions and/or operating reserve shortages.[5]

Figure 16.3 shows the dispatch of resources throughout the day without PRD. As load increases PJM would dispatch increasingly more expensive resources to meet the load without PRD.

However, explicitly incorporating and having visibility into PRD as shown in Figure 16.4, PJM can avoid dispatching more expensive resources and set prices that will ensure PRD responds. In Figure 16.4 the energy market price would be set by PRD, which has provided a price of $88/MWh for a level of consumption below what it would have consumed absent PRD. This not only reduces energy market prices, but also ensures that more expensive resources are not called and then backed down after being called but with the possibility of still having to pay the more expensive resource once it is committed to provide energy.

---

4. Demand Response Compensation in Organized Markets, 134 FERC Paragraph 61,187 (March 15, 2011).
5. The convenience of "set it and forget it" for end-users does not mean that decisions about usage in response to price are permanent. While PJM's market rules allow LSEs to alter PRD demand curves on a daily basis, retail supply contracts may be less flexible depending on how avoided capacity costs and energy price risk are treated. Likewise, PJM market rules allow LSEs to revise PRD quantities participating in the capacity market on an annual basis.

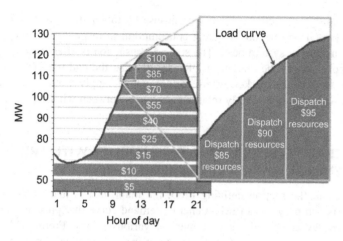

FIGURE 16.3   Dispatch of Resources without PRD.

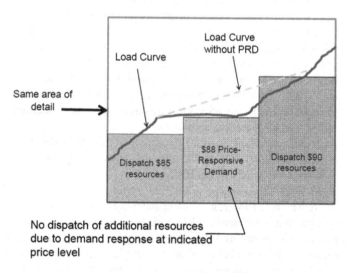

FIGURE 16.4   Dispatch Including PRD.

During peak conditions where there may be emergency procedures invoked or an operating reserve shortage, the visibility of PRD to PJM operators and the automated response that is required can have the effect of avoiding the need to curtail firm load, reduce capacity needs, and mitigate energy prices in real-time operations. Figure 16.5 shows the impact in aggregate of PRD in the energy market relative to there being no PRD available.

Without PRD in Figure 16.5 as shown by the vertical, red demand and in part overlapped by the green price responsive demand curve, there would be

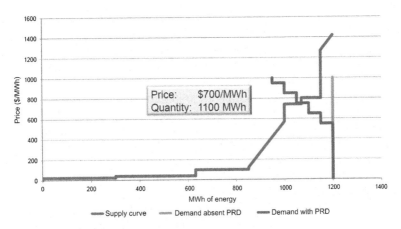

FIGURE 16.5 Role of PRD during System Peaks.

no other resources available to maintain energy balance if demand would increase by one more MW and the price would increase to approximately $1400/MWh. However, with PRD as shown in green, load is lower, and the price is lower as well. Knowing PRD is available and is automated allows more flexibility for system operators like PJM to manage peak conditions and avoid calling on emergency procedures, such as shedding load, unless absolutely necessary. Moreover, the price that would prevail would be consistent with a mandate of FERC Order No. 719 that "the market price for energy accurately reflects the value of such energy during shortage periods (i.e., an operating reserve shortage)."[6]

7 PRD IN THE RELIABILITY PRICING MODEL CAPACITY MARKET

7.1 From Energy Demand Curve to Reduced Capacity Obligation

As was shown in the previous section, PRD has the ability to reduce demand in real-time operations during system peak conditions to avoid implementing emergency procedure altogether, or to avoid going deeper into emergency procedures. From a resource adequacy reliability perspective, system operators such as PJM want to ensure there are sufficient resources to meet the forecast peak load plus a desired installed reserve margin.

---

6. See Wholesale Competition in Regions with Organized Electric Markets, Order No. 719, FERC Stats. & Regs. paragraph 31,281 (2008), paragraph 166. See also paragraph 16 for the FERC's view that "the cost of producing electricity and the value to customers of electric power varies over time and from place to place" and that "[a] well-functioning competitive wholesale electric energy market should reflect current supply and demand conditions."

The question becomes how to ensure PRD in the energy market can be translated to reduced capacity obligations in the capacity market? The answer is to determine a certain energy price threshold at which the energy consumption provides a firm service level and compare that to the forecast demand from the end-use customer on an individual level or an LSE or PRD provider on an aggregated level. Figure 16.6 shows this translation.

The PJM stakeholders have determined an energy price of $1,000/MWh as the threshold price at which the maximum emergency service level (MESL) is determined. For the PRD energy demand in Figure 16.6, this is 900 MW, while the forecast peak load absent PRD is shown as 1,000 MW, which results in a PRD capacity value of 100 MW and can be subtracted from the demand for capacity.

From a real-time energy market operations standpoint, if emergency procedures that call for maximum emergency generation are called, load that has committed to reducing consumption down to its MESL must do so, and the energy price automatically rises to $1,000/MWh to signal all load with commitment to use PRD to reduce their capacity obligations.

PRD participating in the capacity market is subject to strict performance requirements to ensure PRD reduces its consumption down to the MESL to which it has committed. If more load shows up than planned for because end-use customers, PRD providers, and LSEs have miscalculated the amount of PRD, then system reliability could be undermined. Again, PRD responses are required to be automated, and if providing commitment to reduce capacity obligations, are also subject to supervisory control and penalties imposed on LSEs for failing to register PRD MW equal to the approved PRD plan or because end-users exceed agreed to capacity caps or MESLs during an

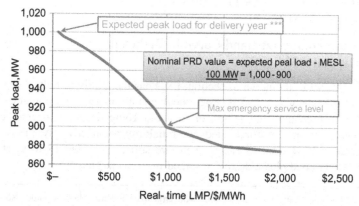

***determined consistent with the 50/50 load forecast that is the input to RPM auction

FIGURE 16.6 Using a PRD Energy Demand Curve to Determine Reduction in Capacity Obligations.

emergency or test event.[7] This ability is required to ensure that PRD does not unfairly lean on the capacity resources procured for other end-users willing to pay a higher price for capacity.

The penalties imposed on LSEs that register insufficient MW of PRD or whose PRD exceeds the MESL are intended to mirror the penalties imposed on CSPs for insufficient MW of registered LM or LM that fails to reduce in response to an emergency or test event. This treatment aims to maintain a level playing for load reduction capability participating as PRD or DR. Furthermore, while CSPs can procure capacity in an incremental auction to cure a LM deficiency, LSEs cannot substitute capacity to make up a PRD deficiency but must instead arrange or acquire more PRD.[8] Given the automation and supervisory control requirements imposed on PRD as well as the penalties, the greater rigor imposed on PRD participating in the capacity market has been characterized as "belt and suspenders."[9]

8 EFFECT OF PRD ON MARKET DEMAND FOR CAPACITY TO MAINTAIN RESOURCE ADEQUACY RELIABILITY

On a market-wide basis, the individual values can be aggregated to see the effects on overall market demand for capacity. The reduction in capacity obligation in aggregate shifts the demand curve downward and to the left as shown in Figure 16.7. Assuming no change in the supply the result is the demand and supply curves intersect at a lower capacity price shown in Figure 16.7.

In addition to the mechanics of how PRD can reduce capacity obligations through the ability to reduce usage at higher LMPs, PRD can be represented more directly to show a willingness to pay for capacity that has similar effects on the demand for capacity as shown in Figure 16.7. PRD market rules accommodate the end-user's willingness to pay for capacity by means of the PRD reservation price. When end-users' LSE specifies the a PRD reservation price, $175/MW-day for example, it effectively instructs PJM to include the forecast load absent PRD in the demand curve at capacity prices equal to or below $175/MW-day, but to exclude the PRD load from the demand curve at capacity prices greater than $175/MW-day. The PRD reservation price allows the end-user "to express its demand for capacity, by effectively stating that, above a certain price for capacity, it is not willing to pay for an assurance" that its forecasted peak load will be served.[10]

---

7. PRD subject to automation that cannot be overridden may not require supervisory control that gives the LSE the ability to limit usage at the end-use site to the MESL during an emergency event. "Price Responsive Demand," PJM Staff Whitepaper, March 3, 2011, p. 11.

8. Section 12.2 PRD Maximum Emergency Event Compliance Penalty, PRD Manual, p. 18.

9. Technical Conference Response of PJM Vice President Stuart Bresler, Docket No. ER11-4628-000, February 14, 2012.

10. Id at p. 27.

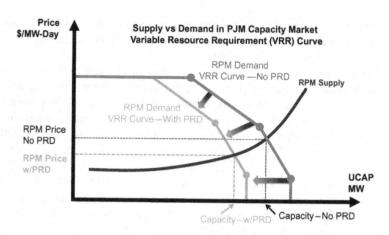

FIGURE 16.7 PRD Can Reduce the Demand for Capacity to Maintauin Resource Adequacy Reliability.

For example, an LSE like that represented in Figure 16.6 can indicate a willingness to limit hourly usage to 900 MW (from its forecast peak obligation of 1,000 MW) during an emergency event but only if the relevant locational capacity clearing price exceeds $175/MW-day. If the capacity clearing price is equal to or less than $175/MW-day, then the LSE will purchase capacity to cover its forecast peak of 1,000 MW and not want to limit hourly usage to 900 MW during an emergency event. Conversely, if the capacity clearing price exceeds $175/MW-day, the LSE will only purchase 900 MW of capacity and obligate itself to only using 900 MW of energy during emergencies.[11]

9 TECHNOLOGY AND REGULATORY DEVELOPMENTS AS NECESSARY CONDITIONS FOR PRD

The technological advancement necessary for the implementation of PRD goes by many names: smart grid, grid modernization, or digitalization of the grid. What is driving this transformation is the application of modern computing and telecommunication technologies to the electricity industry. This is most evident in the evolution of metering infrastructure and in the applications by which end-users of electricity can learn about and manage their consumption.

9.1 Advanced Metering Infrastructure (AMI)

First, it is necessary to understand the distinctive functionality of advanced meters in order to appreciate how they enable PRD. The distinct meter

---

11. Capacity charges contribute approximately 15 percent of the wholesale cost of electricity on average.

functionalities utilized by MADRI's 2005 Installed Meter Survey are clear and succinct:

- Standard Watt-Hour Meter: Electromechanical or solid state meters measuring aggregated kWh manually retrieved over monthly billing cycles; may also include the maximum rate of energy usage over a specified period of time.
- Automated Meter Reading: Meters that collect data and transmit that data one way using drive-by vans with short-distance remote reading capabilities or communication over a fixed network.
- Advanced Metering: Meters that measure and record usage data at a minimum, in hourly intervals, and which provide usage data to both consumers and energy companies on at least a daily basis; may also include built-in two-way communications capable of recording and transmitting instantaneous data.[12]

AMI's granular usage intervals, frequent reporting, and two-way communications are the critical functionalities for enabling end-use market participation in markets such as PRD. The deployment of AMI has been accelerated by the American Recovery and Reinvestment Act (ARRA) enacted in 2009 by Congress, which provided funding for "shovel ready" AMI deployments, smart grid research and development, and smart grid standards under the auspices of the National Institute of Standards and Technology (NIST). The deployment of AMI under ARRA was justified based on the development of business cases for AMI by electric distribution companies (EDCs based in part on the results of pricing pilots and the development led by the Lawrence Berkeley National Laboratory (LBL) of an open framework for automating load controls using price or other signals reflecting system conditions.[13] The Public Review Draft circulated for review and comment articulated the goal of the research:

The initial goal of the research was to explore the feasibility of developing a low cost communications infrastructure to improve the reliability, repeatability, robustness, and cost effectiveness of demand response in commercial buildings. One key research question was: could today's technology be used to automate the response of commercial buildings to standardized electricity price signals.[14]

The value of AMI has been seen at the state level. Former Commissioner Paul Centolella of the Public Utilities Commission of Ohio once contrasted

---

12. "Installed Meter Survey," by Marketing CrossRoads, Inc. for the Mid-Atlantic Distributed Resources Initiative (MADRI), April 27, 2005, p. 5, available at http://sites.energetics.com/MADRI/toolbox/pdfs/survey/survey.pdf.
13. "Open Automated Demand Response Communication Standards," by the Demand Response Research Center, Public Review Draft 2008-R-1, May 2008.
14. Id at p. v.

traditional and advanced metering functionalities by analogizing the electricity customer "stuck with a metering and billing system" to a customer in a grocery store that gets charged the same for a pack of gum and a tin of caviar and doesn't get an itemized bill at the cash register but instead gets the bill a month later.[15]

The importance of AMI, however, is the analogue of cash register in the grocery store analogy. AMI enables an end-user to receive dynamic price signals from wholesale market or dynamic retail rates, which are the best indicators of the condition of the grid, verify actual usage in response to dynamic price signals and see and understand how usage changes over time. AMI is a fundamental tool for enabling end-users and the automated devices in their homes and businesses to intelligently respond to wholesale or retail prices in order to manage electricity usage and expenditures.

9.2 Evolution of Information and Automation to Enable Response to Prices

AMI allows for applications that show users how and when they are consuming energy and makes it worthwhile to devise methods or algorithms by which response to prices can be programmed and automated. Innovative services for end-users such as the "Green Button" will allow customers to download data regarding their usage in a standard format that can be read be a variety of applications so they can better manage consumption.[16]

The other critical tool for enabling end-users to see and respond to prices is automation. The California Energy Commission supported the effort to

FIGURE 16.8   Green Button for Managing Consumption.

---

15. Remarks of Ohio Commissioner Paul Centolella, "Investing in Our Energy Future," Grid 20/20: Focus on Markets, November 30, 2011. See video at http://mediastream.pjm.com/Grid2020_Final/Grid2020_Final.html

16. The "Green Button" was Aneesh Chopra's challenge in the fall of 2011 to the nation's EDCs to enable end-users to download 13 months of hourly usage data in XML format from EDC websites. It took EDCs in California fewer than 5 months to meet the challenge. Mr. Chopra, the nation's Chief Technology Officer, borrowed the Green Button from the Veteran Administration's "Blue Button," which enables veterans to download their medical records.

automate load response by funding research and development by the Demand Response Research Center (DRRC), which is operated by the Lawrence Berkeley National Laboratory (LBL). In 2009 the DRRC published Open ADR, a framework for the underlying communication protocols needed to automate load response.[17] Subsequently, the DRRC contributed version 1.0 of the OpenADR standard to the Organization for the Advancement of Structured Information Standards and the Utilities Communications Architecture International Users Group. Work is ongoing to produce OpenADR version 2.0.[18] Open ADR has contributed to the development standard demand response signals.

In late 2009, the National Institute of Standards and Technology created a public – private partnership, the Smart Grid Interoperability Panel (SGIP), to harmonize and accelerate the development of standards for the smart grid. SGIP currently includes more than seven-hundred-seventy member organizations representing twenty-two stakeholder categories. Through its standing committees on architecture, cyber-security, testing and certification, and implementation methods and various Priority Action Plan (PAP) working groups the SGIP is developing an authoritative Catalog of Standards useful in smart grid implementations. SGIP is an "ongoing organization and consensus process to support the evolution of Smart Grid interoperability standards,[19]" that partners with a variety of Standards Development Organizations.[20]

Green Button represents the realization of standards developed through the SGIP process. Open ADR is currently being harmonized with other standards through the work of SGIP PAP 9. There is also a new SGIP Priority Action Plan (PAP) 19 that is comparing the OpenADR profile to the common business process and standards requirements for the integration of smart grid implementation developed by the Independent System Operators/ Regional Transmission Organizations Council's (IRC's) IRC SmartGrid project, which is based on the common information model (CIM).[21]

Coordinating the communication protocols, standards, and technology to allow for the interoperability of so-called Smart Grid technologies is also necessary for enabling automation of response to prices.[22] Other work within the SGIP is starting to consider standard approaches for communicating dynamic prices to end use devices and automation systems. [23]

17. OpenADR, Lawrence Berkeley National Laboratory, 2009.

18. "The OpenADR Primer," white paper by the OpenADR Alliance, p. 3.

19. Id. at p. 3.

20. Follow this work at http://www.sgipweb.org and http://www.naesb.org, respectively.

21. See the IRC SmartGrid Project materials at http://www.isorto.org/site/c.jhKQIZPBImE/ b.6368657/k.CCDF/Smart_Grid_Project_Standards.htm

22. The National Institutes for Standards and Technology (NIST) is responsible for coordinating these efforts under the Energy Independence and Security Act of 2007. See SGIP 2.0 Business Sustainment Plan (v1.0), 4/13/12, p. 2.

23. See: http://collaborate.nist.gov/twiki-sggrid/bin/view/SmartGrid/BnPSystemsandDevicesSubGroup

10 REGULATORY PUSH FOR VALUE FROM AMI THROUGH DYNAMIC RETAIL RATES

While state commissions must grapple with the costs and risks associated with new technologies, mitigated in some cases by "stimulus" funds, and uncertainty about end-user engagement, they also recognize the costs of the alternative, building new generation needed to meet forecast peak demand. State commission evaluation is further complicated in some cases by renewable resource portfolio standards as well as load and/or peak reduction mandates. State commissions have also commenced with a roll out of dynamic retail rate structures and of automation of response by end-use customers that leverage the value of AMI deployments.

PJM's filing before the Federal Energy Regulatory Commission (FERC) for approval of PRD included a number of state commission expressions of support. The long quote below from the filing includes the positions of state commissions and the other necessary condition for PRD implementation, retail dynamic pricing.

Notably, a number of retail regulators in the PJM region have been among the strongest advocates of changes to PJM's market rules to recognize and accommodate PRD. In a letter to the Chair of PJM's Members Committee last November, when the stakeholder body was considering an earlier version of PRD tariff changes, the Public Utilities Commission of Ohio (PUCO) pointed out that progress on wholesale rules directly impacts the business case for smart grid investments. The PUCO warned that "PJM's current market rules do not directly take into account Price Responsive Demand from new smart grid deployments when determining a utility's capacity obligations." Similarly, the District of Columbia Public Service Commission (DCPSC) wrote to the PJM members to explain that it is "exploring options for dynamic pricing, to take advantage of new smart metering technology which is currently being deployed throughout the District" and to impress on them the importance of implementing "wholesale pricing mechanisms" in order to "fully compensate retail dynamic pricing." Likewise, the Pennsylvania Public Utility Commission (PaPUC) wrote PJM's Members Committee to stress that "[a] successful implementation of PRD is critical to the effective implementation of our state's policy goals." To the same effect, citing "new smart grid technology" being deployed or considered for deployment in Michigan, the Michigan Public Service Commission (MichPSC) urged that "it is vitally important that wholesale pricing mechanisms be designed to fully compensate retail dynamic pricing for the value it provides in reducing utilities' capacity obligations."[24]

---

24. PJM Interconnection LLC, Docket No. ER11-4628-000, September 23, 2011, pp. 9–10. The positions expressed by these state commissions are consistent with those documented by Ohio Commissioner Paul Centolella and PJM Sr. Vice President Andy Ott in their March 2009 whitepaper, "The Integration of Price Responsive Demand into PJM Wholesale Power Markets and System Operations."

11 PRD IS SIMPLER AND REDUCES ADMINISTRATIVE AND TRANSACTION COSTS THAT ARE ASSOCIATED WITH DEMAND RESPONSE

End-use customers and their LSEs under PRD simply pay for metered usage at the prevailing price at the wholesale and retail levels. LSEs pay the wholesale price for hourly wholesale energy market activity with PJM. End-use customers pay the prevailing dynamic retail rate with the understanding that usage during critical peaks is more expensive. This is intuitive and uncomplicated. PRD does not require the estimation of what the end-user would have used if it had not reduced or an add back of the estimated load reduction for future capacity market participation or for system planning purposes as supply-side DR would require. PRD does not require what effectively become multiple transactions to arrive at the same result as DR. The following is a discussion of the types of administrative burdens and transaction costs DR entails and PRD avoids.

11.1 Costs and Controversy of Measuring Baseline Consumption is Avoided by PRD

For PRD verifying performance for settlement purposes is as simple as reading the metered level of consumption that is agreed upon or actually undertaken in the energy and capacity markets. For DR, this becomes more complicated as measurement and verification rely on measuring a baseline consumption that attempts to estimate consumption absent the DR activity.

Fundamentally, aggregators of DR known as curtailment service providers (CSPs) must demonstrate, on behalf of participating end-users, reductions in usage compared to normal usage patterns in response to LMPs. Estimating what the end-use site would have consumed but for the load reduction, known as the customer baseline load (CBL), is required in order to determine the reduction MWh entitled to be paid in the PJM energy market, and the MW of peak load to be paid for providing capacity and then added back to calculate the system's unrestricted peak load at the system peak. Estimated CBL values are for obvious reasons regarded differently than the meter readings that record generator production.

The time and money associated with evaluating various CBL methodologies in the PJM stakeholder process is instructive. First, stakeholders in late 2006 expressed concerns about paying for "happenstance" DR (i.e., load reductions indicated not by changed usage in response to price but by running meter data through the "high 5 of 10"), CBL calculation that averaged the five highest hours corresponding to the load reduction event from the 10 most recent qualifying days. A comprehensive evaluation of a number of

CBL methodologies and a lengthy stakeholder process followed by a FERC order in mid-2008 addressed the concerns.[25]

The high 4 of 5 replaced the high 5 of 10 as the default CBL and the look back for qualified days was limited to 45 days. PJM subsequently flagged end-use sites with submitted settlements of less than $5 or for more than 70 percent of available days on a rolling basis.

The issues surrounding computing a CBL for the purposes of verifying capacity commitments in the RPM Capacity Market also become critical. Currently, penalties are imposed on DR when committed MW are not registered or fail to perform during an event or a test. To date performance of DR has been good, but has declined over time:

- 2009/2010 Delivery Year: 118 percent aggregate test score/no events;
- 2010/2011 Delivery Year: 100 percent aggregate score over all events/ 111 percent aggregate test score;
- 2011/2012 Delivery Year: 91 percent aggregate score for single event/107 percent aggregate test score;[26]
- 2012/2013 Delivery Year: 104 percent aggregate score for two events/116 percent aggregate test score.

The Brattle Group, however, raised concerns about the 1 hour test requirement for Load Management in the 2011 "Second Performance Assessment of PJM's Reliability Price Model." The assessment recommends *"that PJM consider adding random PJM-initiated tests to the current testing procedures, and limit CSPs' ability to selectively pick the test results"* and extend *"the duration of the tests to a multi-hour period."*(Emphasis in the original)

Ultimately, PJM engaged KEMA to perform a comprehensive study of the accuracy, bias, and variability of 11 CBL methodologies including those used by the California, New York, and New England ISOs as well the Electric Reliability Council of Texas (ERCOT). The study results provided support for two subsequent filings affecting DR participation in the capacity and the energy markets.

Incidentally, while the California ISO's CBL methodology provided accurate, stable results it required twice as much meter data according to the CBL study by KEMA. And yet, the California Public Utility Commission (CPUC) has found it necessary to authorize the expenditure of $29 million over the 2012 to 2014 time period to further study DR measurement and verification, which only highlights further the cost and controversy surrounding the need to measure baseline consumption.

---

25. PJM Interconnection, L.L.C., 123 FERC ¶ 61,257, (2008).

26. "Emergency Demand Response (Load Management) Performance Report 2012/2013", December 2012, p. 5. The URL needs to be updated to http://www.pjm.com/ ~ /media/markets-ops/dsr/emergency-dr-load-management-performance-report-2012-2013.ashx

11.2 DR as a Supply-side Resource Creates the Need for Additional Administrative Infrastructure

End-users that participate as DR through CPSs typically buy power from a second entity, the load serving entity (LSE), and often receive electricity deliveries through transmission and distribution wires owned and managed by a third entity, the EDC. End-users in some cases use one CSP to participate in the energy market and another CSP to participate in the capacity market. The need to deal with three or four different entities to secure the full range of electricity delivery services available in the market is inefficient and inconvenient for the end-user.

The significant quantity of load reduction capability developed and aggregated as DR in large part by independent CSPs forms a strong foundation for PRD. PRD, however, presents a challenge to the business models that independent CSPs have evolved over the last decade as they have competed to build DR portfolios.

At the same time, as noted above, PRD's restoration of the end-user's load reduction capability to its natural place on the buyer's side of the market:

- Reduces accounting and administrative costs to the wholesale market including but not limited to qualifying and calculating CBLs.
- Reduces redundant account acquisition and administration costs in the retail market including but not limited to the headaches of getting metering in place.
- Reduces the inconvenience and cost to end-users of simultaneously transacting with as many as four different entities supplying electricity related services.

While a number of independent CSPs have been acquired by competitive electric suppliers and building control companies and while many LSEs and EDCs have retained the services of independent CSPs, PRD creates a challenge for the remaining independent CSPs. This challenge may be especially acute for CSPs unable or unwilling to evolve a business model for PRD in which they either become a service based LSE or their customers are LSEs rather than end-users.

Treating DR as a supply-side resource means that PJM must not only administer the wholesale supply of electricity through LSEs but also the registration, aggregation, dispatch, measurement, and verification, including calculation of the CBL used for settlement, and settlement of load reduction capability offered by CSPs, which adds both complexity and cost-to-market operations. PJM uses two electronic tools, eMarket and eLRS, to administer DR market participation. Figure 16.9 shows the data and process workflow handled by eLRS.

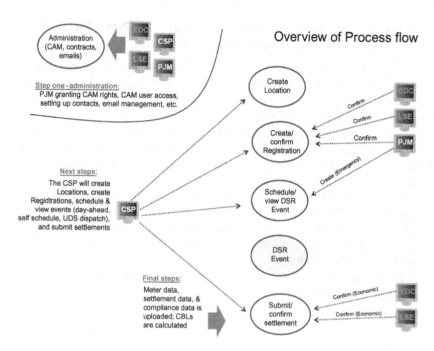

FIGURE 16.9   Administrative Infrastructure for Managing DR in PJM's Markets.

The process of accepting and approving registrations and settlements submitted by CSPs, as shown in Figure 16.9, includes an opportunity for review by the LSE and the EDC serving the end-use site, which further complicates the administration of DR and by extension the design and operation of eLRS. PJM as the wholesale market operator has no independent means to verify the existence of retail accounts, metering installed at the end-use site or other load reduction commitments made by the end-user.

12 CONCLUSIONS

The FERC's PRD order authorized an effective date of May 15, 2012, which means that PRD will be implemented in the energy market before the capacity market. The May, 2013 RPM Capacity Market auction for the delivery year beginning June 1, 2016 will be the first opportunity for PRD to be reflected in the demand curve for capacity used to the clear the auction. This lag should provide LSEs with experience in the energy market that will inform the PRD quantities included in the plans submitted for PJM review and approval prior to the applicable capacity market auction. This should also provide EDCs and LSEs working with state commissions time to fully deploy AMI, develop dynamic retail rates, and provide end-users with information about how to use these new tools to manage usage and expenditures.

PJM anticipates that PRD, like other products developed for its wholesale markets, will undergo revisions as PJM and market participants gain experience and in response to the needs of the retail electricity market and technological innovation. We also believe that the rewards of PRD implementation will prove even greater than those we have imagined. For example, the accumulated usage decisions of end-users in response to price should lead to higher capacity factors for supply assets and reduce the need to build more generation to meet "spikier" demand. These efficiencies have not escaped the notice of several state commissions in the PJM region who have pushed for PRD implementation.

The most significant challenges facing PRD including an outdated retail regulatory paradigm, end-user engagement, and an unlevel wholesale market playing field, are described below with the chapter's conclusions.

12.1 The Retail Regulatory Paradigm

Given the important retail regulatory imperatives that have necessitated PRD development, it is ironic that aspects of the retail regulatory structure will pose serious challenges to broad PRD implementation by LSEs. The current retail regulatory structure moves slowly. State regulatory proceedings necessarily take time to build records so reasoned decisions can be made. This process contrasts sharply with the pace of companies in California's Silicon Valley and New York City's Silicon Alley competing to develop PRD applications and products that entice end-users with value like a simple iPhone app that allows a residential end-user to lower the thermostat by several degrees as he is leaving work on a hot summer evening.

The statutes and precedents guiding state regulation of monopoly electric utilities were intended to prevent the exercise of monopoly power and to keep rates as low as possible based on utility provided evidence about costs incurred to deliver safe, reliable service. While this paradigm served well the last century's aid of universal service, it does not encourage the risk taking or admittance of new dynamic retail rate designs that are essential to unlock the value inherent in innovative products and services being introduced to end-use customers.

12.2 Engaging End-users – Particularly Residential End-users

Inertia and lack of information are the big challenges that LSEs implementing PRD will face. End-users, particularly residential end-users, know little to nothing about the electricity infrastructure and market that keep the lights on, or how they can interact with the electricity market. This dearth of information is understandable given the flat fixed rates they have been charged for many decades.

It is also true that the electricity bill currently represents a relatively small proportion of the household budget particularly for better-off residential end-users. Falling fuel prices and the impact of the "Great Recession" have reduced wholesale electricity prices well below the pre-recession levels last seen in the summer of 2008. Given the explosion of Marcellus shale drilling and related natural gas pipeline development in the PJM region, it seems likely that downward pressure on fuel prices will continue until new capacity is required.

Electricity markets are for better and for worse dynamic. As the region recovers from recession and old, coal-fired generation retires in response to stricter environmental regulations, the direction of prices is likely to change. Shortage pricing rules, recently approved by FERC and implemented on October 1, 2012, will also play a role in communicating operating reserve shortages to the marketplace. The resulting price increases will get the attention of many more end-users, enabling LSEs and other market participants to better explain wholesale price variability to end-users and develop and market test new products and services that will not only overcome a century of flat fixed rate inertia but also engage and delight end-users.

12.3 PRD or the Current DR Paradigm?

Just as the inertia of flat fixed retail rates pose a challenge to PRD implementation, so do inertia, the cash flows, and familiarity of market participants with the current form of DR that participates in the energy and capacity markets. Since 2002 PJM has developed and refined rules that have integrated DR into its energy, capacity, day-ahead scheduling reserve, synchronized reserve, and regulation markets. These efforts have produced over almost 16,000 MW of DR commitments and nearly 20,000 MW of DR participation through PJM's RPM Capacity Market, as shown in Figure 16.10, and forged relationships between CSPs and commercial and industrial end-users.

The Order 745 full LMP revenue stream for economic load response will promote and strengthen DR ties between CSPs and the end-users they have under contract. While DR must now be dispatchable in order to participate in the energy market, it is not required to automate load reduction in response to price. Nor is supervisory control required for participation as DR as it is for PRD.

It is difficult without any PRD experience to gauge how end-users will decide to deploy load reduction capability in the energy and capacity markets as either DR or PRD. The proportionately higher customer acquisition costs for residential and small commercial end-users may give LSEs an advantage in aggregating PRD.

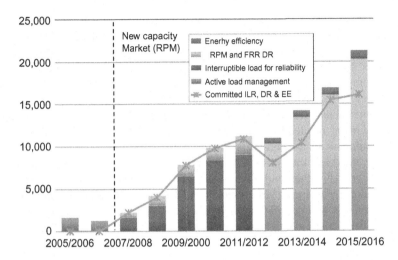

FIGURE 16.10   Offered and Committed Demand Response and Energy Efficiency in the RPM Capacity Market.

12.4 Promise of PRD

It is envisioned that PRD will integrate the wholesale and retail electricity markets and replace the vertical demand curve long a given in the wholesale energy market with an actual demand curve built and documented by LSEs using the price, quantity, and decisions of millions of end-users. PRD, when implemented, will also ensure more efficient use of grid resources, respond to the policy initiatives of state and federal regulators, and support the integration of intermittent renewable resources.

PRD will provide end-users with an understanding of the electricity market that allows them to make better decisions about when and how to use electricity. This end-user sophistication will be essential as electric vehicles (EVs) are added to the transportation fleet. Terry Boston, PJM's President and CEO, frequently reminds EV enthusiasts that 25 million EVs could be sold in the PJM region and no new generation would need to be built so long as price signals incented EV owners to charge their car's batteries during the overnight hours, the "valley" of the grid day.

Congress directed the FERC in Section 529 (a) of the EISA (Energy Independence and Security Act of 2007) to conduct a national assessment of the nation's demand response potential and report back its findings. FERC staff assessed and compared demand response potential under four different scenarios: Business-as-Usual, Expanded Business-as-Usual, Achievable Participation, and Full Participation. "The Full Participation scenario is an estimate of how much cost-effective demand response would take place if advanced metering infrastructure were universally deployed and if dynamic pricing were made the default tariff and offered with proven enabling technologies. This scenario sounds very close to a description of PRD. The report goes on to say that, [t]he reduction in peak demand under this scenario is 188 GW by 2019, representing a 20 percent reduction in peak demand for 2019 compared to a scenario with no demand response programs.

Finally, PRD will have a significant role in managing grid operations as more intermittent renewable resources are added. When the wind suddenly stops blowing or cloud cover reduces production of solar resources, prices will communicate to end-users the need to reduce usage that is flexible and/or valued less than price. Correspondingly, lower prices during the overnight hours when the wind often blows will incent end-users to shift flexible usage like EV charging. End-users with PRD will already have automated responses that will provide the flexibility needed for reliable grid operations. PRD will be a key component of the 21st century grid – "a grid that works both ways."

Part IV

Opportunities and Remaining Obstacles

Shifting Demand: From the Economic Imperative of Energy Efficiency to Business Models that Engage and Empower Consumers

John A. "Skip" Laitner[1], Matthew T. McDonnell[1] and Heidi M. Keller[2]

[1]Economic and Human Dimensions Research Associates, [2]Johns Hopkins University

1 INTRODUCTION

In 1950, the United States was an emerging economic powerhouse with a population of 152 million people – just one-half of the people who reside here today. The size of the economy, as measured by its Gross Domestic Product (GDP), was just 15 percent the size of today's number one market in the world; and, in 1950, the U.S. used about 35 quadrillion Btus of purchased energy, including electricity, coal, natural gas, oil, and wood. If the many different forms of energy were converted to a corresponding amount of petroleum, the energy needed to power the American economy in 1950 was the equivalent of about 16.4 million barrels of oil per day. That is about one-third of today's current energy use, now just over 46 million barrels of oil per day equivalent. Perhaps most surprising is that in 1950 the United States wasted about 92 percent of those various energy resources in maintaining its economy. In other words, only eight percent of all the energy used within our economy was necessary to transform matter into goods and services [1].

The good news is that the United States has improved its overall rate of converting energy into useful work by an average 0.9 percent per year from 1950 through 2010 – even as the larger productivity of the economy grew by about two percent annually over that same time horizon. The extraordinary versatility of electricity has driven a large part of that productivity improvement. Despite the many annual efficiency improvements

Energy Efficiency. DOI: http://dx.doi.org/10.1016/B978-0-12-397879-0.00017-7

445

since 1950, surprisingly, the overall energy efficiency of the U.S. economy remains only 14 percent.[1] In short, the United States continues to waste about 86 percent of the energy that is consumed in the production of goods and services.

Admittedly, not all of that energy can be recovered and put to more productive uses in a cost-effective manner. Yet, a large number of studies suggest that more can be done through better choices, smarter investments, and the development of new business models to enable those choices. Recent assessments by the American Council for an Energy-Efficient Economy (see, [2]), and the University of Toronto's L. Danny Harvey [3], among others, suggest that the technology and wherewithal exists to greatly improve the current level of energy efficiency.

On the basis of pure economics, the United States, and others, might quadruple the current level of efficiency by the year 2050 – should we choose to actively develop those prospects [2,20]. The more complete set of energy efficiency opportunities are likely to be enabled by what author Jeremy Rifkin calls a transition to Third Industrial Revolution [4]. On the other hand, if the world continues along the business-as-usual path, evidence suggests the robustness of the global economy may be substantially weakened. At the same time, however, the market is evolving and unfolding in whole new ways that may either constrain or enable future opportunities for energy productivity improvements – depending on the social and institutional arrangements enabled, and the energy or economic policies enacted.

This chapter advances three critical points:

- The first is that advances in energy efficiency rather than in energy supply have been a primary driver of economic productivity, especially since the spread of the industrial revolution, beginning in the mid-18th century. The availability of high-quality yet affordable electricity has been a significant contributor to past productivity improvements.
- Second, in order to maintain a robust economy over the next several decades, a simultaneous decrease in the cost of energy services is required as well as a paradoxical increase in "useful energy" consumption. The key to resolving this paradox is reducing the waste associated with the larger demand for raw energy.
- Finally, and looking specifically within the electricity market, the development of new business models are essential to improve system efficiencies in ways that maintain the robustness of the future economy.

---

1. Some may be surprised. The estimated efficiency of converting energy into useful work in 1950 was only 7.99 percent. The calculation to estimate the current level of efficiency over the last 60 years at an average 0.93 percent rate of improvement is 7.99 percent $* 0.93^{60} = 13.96$ percent efficient in the year 2010.

Section 2 of this chapter examines the broader aspects of converting primary energy into useful work (energy services) as well as the critical need for greater energy efficiency gains in order to maintain a robust economy. Section 3 builds on this discussion and evaluates energy efficiency as a resource and how large a role it could play in an American or global energy portfolio. Section 4 looks at the need for new electric utility business models in order to actualize large energy efficiency gains and to close the "energy efficiency gap." Using several case studies, Section 5 examines the shift in the electric utility industry from a linear, commodity-based approach (delivery of kilowatt-hours, or kWh) to that of a dynamic energy service provider. Finally, Section 6 offers concluding thoughts, namely, that in order to accommodate the greater imperative for building a more energy-efficient economy, and to leverage the full energy efficiency resource opportunity that enables a more robust economy, the global economy requires development of entirely new business models that engage and empower consumers to ensure such a transition.

2 ENERGY EFFICIENCY AND ECONOMIC PRODUCTIVITY

In the United States, economists and policy analysts have formulated many of their insights based on data collected by the Energy Information Administration (EIA). As it turns out, energy data collected by the EIA contains only part of the story on how energy moves the economy forward. The EIA routinely gathers annual data based largely on the sale of physical units of energy as tons of coal, cords of wood, gallons of gasoline, therms of natural gas, or kWh of electricity. All of these different energy forms have an equivalent heat value that allows one to compare a gallon of gasoline with, say, one kWh of electricity. The majority of other countries in the world convert these energy forms using a standard energy unit called the joule. In the United States, heat equivalent is the British thermal unit (Btu). One Btu is roughly the amount of heat given off by the burning of a wooden kitchen match.[2] According to the EIA, there are about 124,238 Btus of heat equivalent in a gallon of gasoline − or the heat energy that might be provided by the burning of 124,238 wooden kitchen matches. Similarly, there are approximately 3,412 Btus of energy for every kWh of electricity delivered to the home or office building. When comparing electricity and gasoline, one might say that a gallon of gasoline is the heat equivalent of 36.4 kWh delivered to the end-user.

When including the energy wasted in the generation, transmission, and distribution of electricity (see also chapter by Gellings on this topic), the

---

2. More formally, a British thermal unit is the amount of heat required to raise one pound of water one degree Fahrenheit at one atmosphere pressure. There are approximately 1,055.056 Joules of heat in one Btu.

production of electricity required roughly 10,697 Btus per kWh in 2010 [5].[3] In other words, the 3,412 Btus bundled in a single kWh available for use in the home is only 31.9 percent of the total energy needed to create and distribute that electricity. Similarly, one gallon of gasoline is the heat equivalent of 11.6 kWh when compared to the energy that is necessary at the generation source.[4]

While energy is typically quantified by a heat metric, a more useful metric quantifies the ability of energy to do work. In the more simple explanation, work might be defined as lifting a weight against the force of gravity or overcoming sliding friction. From an economic perspective, work is defined as the energy actually used to transform matter into goods and services. For all practical purposes useful work can be divided into three primary categories: (1) "muscle work," (2) mechanical and electrical power, and (3) a combination of low or high temperature heat, delivered to the point of actual use ([1], as updated by Laitner, 2012).

"Muscle work" is work carried out by people and animals. In 1900, for example, there was an estimated 43 million horses, mules, oxen, and other (non-milk) cattle on American farms [6]. A large number of these were draft animals providing work while another share of animals provided transportation. This did not include urban and work animals, which also provided substantial labor and transportation services. All of this "muscle work," however, helped create economic value.

While very few animals provide work or transportation services today, there has been a substantial increase in the second and third categories of work. Mechanical and electric power is work done by a variety of steam and gas turbines, gasoline and diesel engines, as well as electric generators. The final work category is a combination of low or high temperature heat delivered to the point of actual use ([1], as updated by Laitner, 2012).

The EIA now tracks data for many different amounts of purchased energy [5,7,8], and preliminary estimates for 2010 suggest the United States consumed about 98 quadrillion Btus, or quads, of total energy in that year. One quad is roughly 8 billion gallons of gasoline, which, at current levels of

---

3. Many of the values discussed here are preliminary estimates for the later years as 2010 and 2012. And those values will differ in some cases, depending on the EIA document that is used as a reference. For example, the EIA's Annual Energy Review for 2010 (the latest currently available) shows total energy per kWh as 10,189 Btus, a five percent difference than reported in the text. The Annual Energy Outlook is a later publication so we choose the value reported there. However, those small differences do not change the outcomes of the larger discussion reported here.

4. Even food can be compared on a heat or energy equivalent. A quarter pound cheeseburger from McDonald's contains an estimated 510 food calories, which is the equivalent of just over 2,000 Btus. Expressed as an equivalent of electricity, the cheeseburger has an energy value of 0.593 kWh about the same as the electricity required to light a 100-watt incandescent bulb for almost 6 hours.

energy efficiency, is sufficient fuel to run some 15.4 million cars for one full year of typical driving. It is also enough energy to provide the full energy needs for about 5.2 million households in a given year. It is sufficient energy to power $135.4 billion of annual economic activity within the United States. This, however, is only part of the story.

Ayres and Warr [1] have documented a more detailed accounting of the total energy that is actually used for heat, mechanical power, electricity, light, and muscle power. Whereas EIA and other data, for example, suggest that economic activity in the United States required a total of 98 quads (rounded) in 2010, the Ayres-Warr data (as updated and summarized in Laitner, 2012) – with their more complete accounting of actual formal and informal energy needed to power the full economy – indicate we required more like 124 quads (rounded) of total energy in that year. More critically, only 17 quads of that total energy consumption were actually used and useful in producing the total basket of consumer goods and services purchased in 2010.

As Figure 17.1 highlights below, it is not the total raw energy (i.e., the 124 quads just referenced) that is especially useful to our economy. Rather, it is the energy that is converted to useful work – as shown in Figure 17.1, the roughly 4 quads of useful work in 1950 and the 17 quads in 2010 previously referenced. In effect, it is the useful work or energy that actually enables economic activity as typically measured by the nation's GDP (again, with relevant data reported in Laitner, 2012). Ayres and Warr further clarify the point by noting that the reason the raw energy "inputs do not explain economic growth is that their inefficient conversion leads to a large fraction

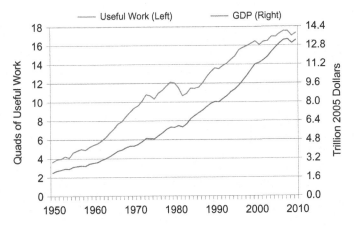

FIGURE 17.1   Useful Work (Energy) versus U.S. Gross Domestic Product (GDP). *Source: Author calculations based on EIA data (2012) and Ayres and Warr (2009) – as updated by Laitner (2012), from 2005 to 2010.*

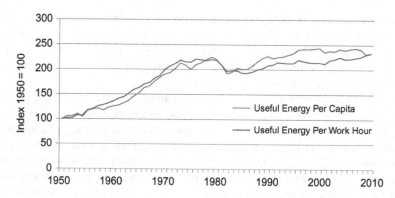

FIGURE 17.2   Useful Work Energy per Person and Work Hour (Index 1950 = 100). *Source: Author calculations based on EIA data (2012) and Ayres and Warr (2009) – as updated by Laitner (2012), from 2005 to 2010.*

of waste heat (and other wastes, like ash) that do not contribute to the economy but actually create health problems and costs of its disposal."

As a complement to the first chart, Figure 17.2 draws on the raw data published by Laitner (2012) to highlight the role of useful energy over the period 1950 through 2010. The data is shown in an index format in which the amount of useful work per person and the amount of useful work per labor hour in 1950 are shown as a value of 100. The quick observation shows a highly similar pattern for both sets of lines shown in Figure 17.2, namely, the rather steep increase in useful energy per labor hour and per capita in the 1950 to 1980 time period. By about 1980, however, the amount of useful energy flattens out for both a per capita and per work hour basis. The immediate suspicion is that the flattening consumption of useful energy per capita must be affecting other aspects of the economy as well. In short, the flattening of useful work may be constraining the larger set of economic activities. This is, indeed, reinforced by both Figure 17.3 and in Table 17.1.

Continuing, Figure 17.3 highlights the very tight and significant impact of useful energy as it drives overall U.S. economic productivity. That is, in order to amplify the ability of the U.S. population to increase its output, either on a per capita or an hourly basis, more actual work must be done. Again, looking first at 1950, we see that about 4 quads of work was necessary to generate the actual productivity recorded in 1950. The higher the productivity, the more useful work or useful energy is required.

Examining the upper-right part of Figure 17.3 we note the highest value of productivity reaches an index of 325 by 2010. Figure 17.3 also shows the largest amount of useful work, the same 17 quads of useful work described earlier. Again, this makes sense when we think of economic productivity as increasing the output per person, or output per hour of labor, by tapping into greater levels of cost-effective energy services. It actually does take energy

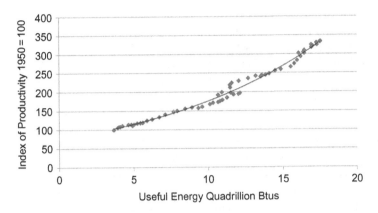

FIGURE 17.3   Useful Energy per Person (Index 1950 = 100). *Source: Author calculations based on EIA data (2012) and Ayres and Warr (2009) — as updated by Laitner (2012), from 2005 to 2010.*

TABLE 17.1 Key Historical Growth Rates in Energy and Economic Productivity

Compound Average Growth Rate	(A) Increase in U.S. Population	(B) Useful Energy Efficiency Improvement	(C) Annual Rate of Productivity Improvement	(D) Annual GDP Measured in Trillion 2005 $
(1) From 1950 to 1980	1.34%	1.45%	2.25%	3.63%
(2) From 1980 to 2010	1.03%	0.42%	1.72%	2.77%

Note: Laitner (2012) with data from 1900 taken from both Ayres and Warr [1] and updated using EIA (2012).

to move the economy forward as economic activity more broadly increases over time. In effect, by increasing the efficiency with which we convert available energy into useful work, overall costs will decrease such that we can more easily afford to integrate even more useful energy as we continue to build economic momentum.

Table 17.1 further highlights the connection between useful energy and productivity, but in a slightly different way. In this case we provide data that compares the growth rates for population (column A); improvements in the rate of converting total energy into useful energy (column B); the growth in economy wide productivity (column C); and finally, the overall growth in the nation's GDP (column D).[5] In this case, however, we are looking at two periods from 1950 to 1980 (data row 1) and from 1980 through preliminary estimates for 2010 (data row 2).

Of immediate interest is the link between the improvement in the conversion of energy into useful energy (work) reported in column B as it relates to economic productivity in column C, and finally, to the growth in the nation's economy (column D). The bad news is that the earlier period from 1950 to 1980 showed the larger improvement in columns B, C, and D. In the last 30-year period, covering the years 1980 to 2010, however, the entire economic process shows a weakening – especially as the economy-wide improvement in productivity fell from 2.25 percent to 1.72 percent. The reason for this weakening appears to be the result of a flattening rate of improvement in converting total "raw energy" into actual "useful energy" (or useful work) so that we can cost-effectively and simultaneously increase the magnitude of useful energy (or work expressed as quads of energy). Such an improvement, in turn, could drive up the nation's larger economic productivity and total GDP.[6] Presumably, then, as the rate of efficiency conversion increases, the quantity of total or raw energy required to power the economy will decline – even as the amount of useful energy continues to grow.

In effect, the data in Figures 17.1 through 17.3, as well as in Table 17.1, all reveal two important aspects of economic activity.

First, when we properly measure the links between useful energy and economic output, we find that America's overall level of energy efficiency has been stagnating at a rather anemic 13 to 14 percent for the past 20 years or so (notably over the period 1990 through 2005 with a preliminary update by Laitner, 2012). As suggested in the second data row of Table 17.1, the stagnating conversion of total energy into useful work appears to constrain economic productivity.

Second, and not immediately obvious, if we merely double our current level of efficiency (following a business-as-usual rate of improvement), we may continue to see a weaker economy over the next several decades.

---

5. Perhaps not immediately apparent, but the growth in GDP is a function of multiplying the rate of population growth times the rate of growth in productivity, the latter of which can also be expressed as a growth in GDP per capita. So, for example, if population growth in Column A of Table 17.1 averages 1.03 percent in the period 1980 to 2010, and productivity growth averages 1.71 percent, then (1 + 0.0103) times (1 + 0.0172) equals 1.0277, which results in a growth of 2.77 percent.

6. The U.S. economy provided a GDP of $5,389 billion in 1980. It more than doubled in size, reaching a total of $13,248 billion by 2010 (with both values measured in constant 2005 dollars). Perhaps not immediately apparent, but had the U.S. economy maintained a productivity improvement of 2.25 percent rather than 1.72 percent over the period 1980 to 2010, even with a smaller population grow rate, GDP would have actually grown to an estimated $15,491 billion (also in constant dollars). In other words, a very small change in productivity would have meant an economy that was about $2.2 trillion larger than we actually recorded in 2010.

Indeed, the evidence points toward a need to at least triple or quadruple the productive conversion of raw energy so that a greater magnitude of cost-effective "useful energy" is available to power a more robust economy. Improvements in the efficiency with which we both produce and use electricity will be vital in accelerating economic performance.

3 HOW BIG IS THE ENERGY EFFICIENCY RESOURCE?

Economist William Baumol and his colleagues (2009) once wrote, "for real economic miracles one must look to productivity growth." As suggested by this chapter, and by the large number of studies published by the American Council for an Energy-Efficient Economy (see, e.g., [9,10]; Gold et al. 2010; [11]), and by many other recent studies ([12], America's Energy Future 2010, and the American Physics Society [13]), by priming "the productivity pump" with enhanced or expanded energy efficiency provisions, a net positive impact for the economy will likely result. More critically, a failure to accelerate the rate of efficiency gains compared to the last 30 years or so, risks a significantly less robust economy in the years ahead.

Some might ask the question, "Just how big is energy efficiency?" Perhaps the surprising news is that the opportunity for gains in energy efficiency (energy productivity) is bigger than one might imagine, although perhaps harder to achieve than one might initially believe – and, as this chapter will suggest, new business models will be needed to achieve those larger efficiency improvements. Among the more credible estimates, is a study published by the National Renewable Energy Laboratory [14], which suggested that if all commercial buildings were rebuilt by applying a comprehensive package of energy efficiency technologies and practices, they could reduce their typical energy use by 60 percent. Adding the widespread installation of rooftop photovoltaic power systems could lead to an average 88 percent reduction in the use of conventional energy resources. Even more intriguing, many buildings could actually be producing more energy than they consume – in effect, transforming building stock into power plants (also see chapter by Anderson et al for additional insights on this topic).

Using the United States as an illustrative example, the current electricity generation and transmission system now operates at about 32 percent efficiency. That is a level of performance essentially unchanged since 1960. The United States wastes more in the production of electricity today than Japan uses to power its entire economy [2]. At the same time, a study published by the Lawrence Berkeley National Laboratory [15] suggests that a variety of waste-to-energy and recycled energy systems could pull enough waste heat from our nation's industrial facilities and buildings to meet 20 percent of current U.S. electricity consumption. And that is only the beginning of potentially large efficiency gains in power generation.

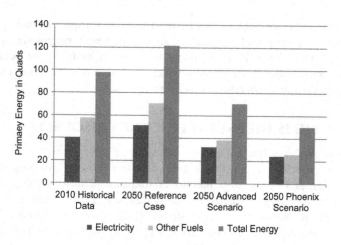

FIGURE 17.4 Electric Generation System Efficiencies. *Source: Laitner et al. (2012).*

In Figure 17.4, focusing on electricity production and consumption, Laitner et al. (2012) suggest that a combination of both device and systems efficiency improvements among all end-use applications of the economy (including the full array of residential, commercial, and industrial sector electricity needs) could deliver productivity gains that lower electricity consumption by 30 percent in what they term an Advanced (efficiency) Scenario, and 37 percent in what they call the "Phoenix" Scenario. Both scenarios compare the year 2050 impacts with what might otherwise be forecasted for the 2050 Reference Case projection (the left-most dark gray bars in each set of data in Figure 17.4). More importantly, reducing generation, transmission, and distribution losses in the production of electricity — together with lowering the consumer demand for electricity — removes a significant level of waste (shown in the middle light gray bars in Figure 17.4). These two items together suggest that total primary energy (highlighted in the medium gray bars bars) dedicated to the production of electricity might be cost-effectively reduced by as much as one-half by 2050!

Again, why the need for greater efficiencies? It bears repeating. The larger gains in energy efficiency are made necessary as a critical step toward reducing the cost of energy services, thus enabling the productive use of all resources. So without greater efficiency, and the concomitant reduction in the cost of energy services (on both the end-use and the supply side), the economy may actually become much less robust. Indeed, it may also become a little smaller. Should the economy contract by even tenths of one percent, we will have fewer economic resources available to handle things like climate change, or improvements in our nation's infrastructure.

4 THE SHIFTING UTILITY MARKET

The potential gains in energy efficiency remain large. Yet, achieving those large-scale improvements is a vital step in maintaining the robustness of our economy. At the same time, there are emerging trends in consumer behavior, regulatory policies, and new technologies that are laying the groundwork for a fresh and diversified energy market place. Perhaps, more crucially, the old market structure and technologies are creating new pressures that mandate changes in the way we generate and deliver energy services – even as the marketplace demands new business models that can deliver those badly-needed productivity improvements. This section explores both of these aspects.

4.1 Utility Pressures

Utilities are experiencing tremendous pressures to redesign the way in which customers are served and business is conducted. The electric utility's commodity-based business model is outdated and has been progressively so for the last several decades. The original model of the mid-twentieth century was based upon an ambitious "grow and build" strategy that an increasing number of analysts suggest is no longer sustainable (see also the chapter by Cavanagh in this volume).

The electric utility industry experienced broad improvements in thermal efficiency throughout the first half of the twentieth century, creating consistent declines in costs of production [16,17]. Utilities therefore encouraged energy usage and promoted the "all electric home" [18]. The increased demand for electricity continued to push large scales of production and widespread development, benefiting the consumer with low costs. According to Valocchi et al. [17], by 1967 the cost of electricity was 95 percent cheaper than in 1900. The electricity utility business model appeared to be intact. Utilities were making money and customers were paying low rates for electricity.

However, shortly thereafter in the early 1970s, the industry witnessed a plateau in the economies of scale [16,17]. Growth in the industry began to diminish and generating units reached their limit. Since the 1960s, the United States' current level of electric efficiency, or inefficiency, has remained essentially static. Despite the evolving landscape of the past 30 to 40 years, most utilities have continued to follow the same business model of "grow and build" that is no longer relevant or sustainable (again see chapter by Cavanagh).

Today, however, is different. There is now a "perfect storm" of diversified events on the horizon that the electric utility sector cannot afford to ignore [20]. If the electric sector continues on with business as usual the challenges will eventually become insurmountable and could threaten a country's security, economic vitality, and environmental quality. Lovins et al. (2011) assert that this impending cataclysm in the energy sector will be formed by:

- An Aging Electric Infrastructure: 70 percent of the electric transmission lines and transformers are at least 25 years old, while 60 percent of circuit breakers are at a minimum of 30 years old [19].
- The Electric Power Research Institute (EPRI) estimates that nearly 60 percent of the present infrastructure must be upgraded in order to ensure long-term viability and reliability [19]. Such essential upgrades could force the demand for new costly facilities.
- Environmental Constraints: Electric utilities are accountable for virtually 40 percent of today's greenhouse gas emissions [18].
- If nothing is done to change the current "Business As Usual" path, emissions are expected to intensify by 38 percent over the next 40 years [20].
- Technology Shifts: In 2011, 118 countries had implemented policies requiring the use of renewable energy. This was an increase from 55 countries in 2005 [20].
- With an emerging market for renewable energy, new customers and companies will enter the marketplace providing both an opportunity and a threat to utilities [20].
- Security Concerns: Blackouts cost the United States $80 billion each year [19].
- Businesses can lose billions of dollars if the power is out for only a few minutes. The electric grid remains not only vulnerable to electric blackouts, but also to terrorist and cyber-attacks [20].

The convergence of these highly uncertain and costly factors has forced the need for an altered way of doing business. Energy efficiency and conservation are two of the greatest tools we have at our disposal today to address the looming energy crisis and reduce our impact on the environment [21]. Yet the market is fundamentally different than just a few years ago, and we will need new business models to tap into the full resource opportunities and develop the range of energy efficiency resources throughout all regions and all sectors of the U.S. economy.

Electric utility companies are recognizing this shift and changing the way in which they conduct business. Duke Energy, headquartered in Charlotte, North Carolina, is a regulated utility providing electricity in five U.S. states. The company is changing their focus from traditional fuel source generation to renewables and energy efficiency, with plans to retire approximately 3,300 MW at eight separate coal generation sites over the next three years (Duke Energy Corporation, 2011). Following the disaster of the Fukushima nuclear plant in March 2011, Japan is considering a nuclear-free energy portfolio, leaving room for the development of renewables and a larger emphasis on energy efficiency and conservation. Becoming free of traditional energy sources, such as nuclear and coal, is possible with a revised business model. Germany is at the forefront of developing a strong sustainable renewable energy sector that is absent of nuclear and coal. The country has been studying a variety of scenarios for the past twenty years that would afford them with a 100 percent renewable electricity system. The country is poised to reduce greenhouse gas emission by 40 percent and increase their share of renewables to 35 percent by 2020 (Morgan, 2011). Their ultimate goal is to achieve 80 percent renewable energy generation and 80 percent reductions in greenhouse gas emissions by 2050 (Morgan 2011).

4.2 Understanding and Closing the Energy Efficiency Gap

With energy efficiency providing such a convincing case for delivering a low-cost energy service, why hasn't it already saturated our economy? [12]. This question has baffled the energy community of researchers, economists, policymakers, and social scientists for the past 30 years [22]. Despite developments in cost-effective energy efficient alternatives, consumers (defined as households, businesses, and governments) have routinely failed to purchase the energy efficient option and adopt the more energy efficient techniques. The phrase energy efficiency gap recognizes this failure and identifies the difference between the actual energy used for energy services and the level of energy efficiency that could be provided in a more cost-effective manner for the same services [23]. The paradox of why energy efficient technologies and practices are not more commonly utilized is traditionally explained through market failures and barriers.

While the landscape of today is vastly different from that of 20 years ago, an energy efficiency gap still remains and the full potential is far from achieved. Consumers are not rational economic decision-makers. They cannot and will not always act in a fully informed manner, despite incentives and equalized costs. Even today and as explained by Levine et al. [23], profit-centered businesses fail to control their energy use and allow thousands or even millions of dollars go to waste. Households avoid purchasing efficient light bulbs, which have tangible long-term financial payback [24]. While there are a variety of explanations to the energy efficiency gap, consumer behavior should be seen as the most prominent and influential.

Behavioral research adds a more complex component to the energy efficiency paradox, one that is more multifaceted than the traditional economic models focused on price and information. Behaviors are complex and can be influenced by a variety of habits and intangible, subconscious factors. Understanding the complexity of the many varied human behaviors can, yield significant insights on how to best influence and change public perception and action on energy efficiency [25,26].

From still another perspective, IBM's Global Utility Consumer Survey (2011) found that utilities might reap a more profound set of insights into the minds of their consumers if they were to adopt an approach based more on behavioral economics than on the more familiar "economics of technology." While there have been some efforts to educate the energy consumer through various conservation efforts, IBM also confirms there remains a broad gap in the amount of information that is accessible to consumers preventing them from making informed choices [27,28]. IBM identified in its consumer survey three behavioral factors impacting how consumers use electricity: alternative motivations, information availability, and social drivers (also see chapter by Laskey and Sayler).

Today energy consumers have alternate motivations. The health of the environment more heavily impacts choices made by the younger generations, while the prosperity of the national economy is a motivator for those over the age of 55. This is in contrast to the previously conceived notion that financial incentives were primary drivers for saving energy among all consumers. Consumer motivations are varied, and noting this variance is critical to influencing changes in behavior [27,28].

Humans are both social and competitive by nature. Developing consumer portals that allow consumers to compare their energy usage with those of others is one way of socially engaging consumers. Social competition is a proven method for persuading and engaging [27,28]. Consumers under the age of 25 are more likely follow the example of others and less likely to rely on their own preferences. "Millennials" (Generation Y) depend heavily on their social networks for news and information. Social media has now become the new driving force for all forms of decision-making. Accordingly, in order to influence behavioral change, electric utilities must carefully select their information and information channels [27,28].

4.3 Why is Today Different?

The landscape of today is vastly different than it was 20 or 30 years ago. This is true for both the electric utility industry and the economy more broadly. New approaches are required that enable the industry to more fully capture the large energy efficiency opportunity. A new world of consumer influence has evolved; a world where consumers make decisions based on social interactions and shared concerns and experiences rather than necessarily based on the purchases of a commodity such as a kilowatt-hour of electricity. In 1990 the market conditions were relatively static, while today we experience an influx of constantly changing information via multiple forms of communication – Twitter, Facebook, email, LinkedIn, tumblr, blogs, "apps," etc. [21].

Throughout the 1990s the information experience was defined by "content" [29]. Today, the framework has changed; the Internet is now focused on conversations, personal connections, and values. Social and mobile networks have eradicated the concept of simply displaying content for decision-making. "Content" has been replaced by "context," which is defined by social niches representing unique personal connections [29]. There is a social niche for everyone and the tools of the Internet allow people to develop communities based on their own interests and ideals (Jarvis, 2009). The second Internet revolution, Web 2.0, has completely altered the relationship between companies and their customers. This constant connectivity and new company-consumer relationship illustrates both a threat and opportunity for businesses, such as utilities.

Businesses that were once thought of as too big to fail, have failed. They became victims of this new age of "Digital Darwinism" and consumer transformation [29]. Borders, Blockbuster, and the Yellow Pages were all once mega companies perceived as indestructible to market conditions. Today they are defunct because of their inability to respond to the consumer and digital shift. On January 19, 2012, Kodak filed for bankruptcy following in the footsteps of Blockbuster and Borders. A company, which was an American symbol for decades, unfortunately failed to acclimate to the digital revolution. As Harvard Business School Professor Rebecca Henderson commented, "It wasn't that Kodak was oblivious to the future, but rather that it failed to execute on a strategy to confront it. By the time the company realized its mistake, it was too late" [30]. In short, Kodak focused too much on the successes of their past and failed to see the opportunities and threats of the future.

Figure 17.5 compares U.S. electricity sales with mail volume, suggesting a parallel progression and drop. The word kilowatt-hours could be substituted for "pieces of mail." If utilities fail to act and change their commodity-based business model to a service-oriented structure they could be at risk of losing their customer base and revenue stream.

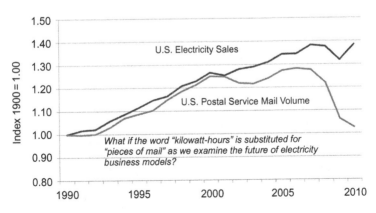

FIGURE 17.5 Comparing U.S. Electricity Sales with Mail Volume. *Source: U.S .Energy Information Administration and U.S. Postal Service.*

Consumer preferences are changing, and they want more today from their utilities. The current utility business model is geared toward delivering a commodity value, a kWh, but the new dynamic requires a shift to a service-oriented structure. Utilities must alter their revenue stream from the sale of kilowatt-hours to the sale, upgrade, and integration of energy products and services. A service-focused business model will more effectively engage the customer and strengthen the utility-consumer relationship. Such a structure is vital for utility survival in this new diversified world of economic, environmental and social change.

4.4 The New Energy Consumer

A modernized, informed public and energy consumer has materialized and is continuing to evolve. The public is no longer satisfied with the status quo and the current modes of energy production and operation. Energy consumers are ready for a change. A University of Texas Energy Poll, created by the Energy Management and Innovation Center at the McCombs School of Business, found that energy consumers have a considerable level of distrust and dissatisfaction about the current energy system [31].

Forty-three percent believe the United States is moving in the wrong direction, with only 14 percent believing we are on the right path [31].

Over the next year, 69 percent of survey participants anticipate increased household energy spending [31].

Within the next 25 years, approximately 41 percent of consumers believe the state of our energy sector will have vastly declined [31].

Such levels of distrust and angst about our current system are spurring engagement in energy efficiency and renewable sources.

In 2011, Americans declared that out of the eight actions Congress could address in 112th Congress, Americans were most in favor of an energy bill that provided incentives for using alternative energy (83 percent) [32]. In a survey of 1,016 U.S. adults taken in February 2011, *USA Today* found that "nearly three of four U.S. adults, or 71 percent, say they have replaced standard light bulbs in their home over the past few years with compact fluorescent lamps or LEDs (light emitting diodes) and 84 percent say they are "very satisfied" or "satisfied" with the alternatives" [33].

The Electric Power Research Institute conducted a consumer engagement study in 2011 on the acceptance and perceptions surrounding the Smart Grid:

Sixty-four percent of the study participants indicated that it would be very useful or extremely useful to understand how much electricity one uses, when it is used, and how much the energy usage will cost prior to receiving the their utility bill. This percentage increases to 92 percent, if "moderately useful" and "somewhat useful" are included [34].

Fifty-eight percent of respondents believe that it would be very useful to have options for owning "smart" energy-saving devices that share energy and information with the grid [34].

Customers now have more tools and an increased capacity to learn about energy efficiency and make informed decisions. Since 1990 there has been an 80 percent increase in the number of cell phone subscribers. In 2010 nearly 98 percent of the U.S. population had a cell phone (U.S Census Bureau, 2011). Cell phone subscribers now have access to social media applications on their phones and are in a constant stream of engagement. Approximately two in five social media users access social media sites from their mobile phones (NM Incite, 2011). Active social media users have tremendous influence offline with 60 percent of social media users creating reviews of products and services (NM Incite, 2011). As an example of such influence, Bank of America was forced to recall a new debit card fee following considerable consumer backlash in the fall of 2011 (Mui, 2011). The onset of a five-dollar monthly debit card fee spurred a reaction where 21,000 people were committed to closing their accounts if the fee wasn't revoked and 300,000 people signed an online petition (Mui, 2011).

In 1990 consumers waited until the end of the day to get their nightly news. Today information goes directly to the consumer, multiple times a day in the form of tweets, blogs, and status updates. The consumer expects the information to be delivered to them individually and they expect it to be personal. Business as usual has ended and companies can no longer survive on the status quo. The new information architecture has changed where consumers/people are at the center [29].

Not only do customers now have more tools to make educated decisions about energy efficiency, but they also have the capacity to take action. Between 1990 and 2008, energy service companies (ESCOs) grew at a compound annual growth rate of 19.1 percent [19]. The energy efficiency industry is expected to continue on this growth path over the next several years.

This restructuring of the company-customer dynamic provides both an opportunity and a threat to utilities. If utilities want to succeed and provide the highest quality of service, they must become adaptive and recognize the ways for improving customer satisfaction [21]. This means a shift from the delivery of commodities to providing the kinds of services and value that consumers are beginning to expect.[7] Through the digital transformation of the late 2000s, customers have become more engaged, informed, and stronger advocates for change.

5 THE NEW BUSINESS MODEL

With an increased environmental awareness and the growing availability of more productive energy technologies, consumers have more reason to be

---

7. It may not be immediately obvious but again the word consumer is more than households. The term refers to any individual, corporation or entity that consumes electricity. Yes, this includes households but it also includes businesses, industries, and governments.

connected, and potentially even become electricity suppliers themselves [35] (see also chapter by Kesting and Bliek). Energy consumers are starting to understand their potential new role in the electric power sector value chain. Volatile fuel prices, future regulatory constraints, financial infrastructural burdens, and an evolved energy consumer have created the perfect environment for electric utilities to create customer centric business models.

In order for utilities to stay competitive and not fall by the wayside, like so many big companies have done with the Web 2.0 revolution, utilities must engage and empower the energy consumer to act in ways that generate greater value-added services rather than just consume electricity. The "new utility" will find ways to leverage and understand their customers' behaviors by using segmentation and data analytics. This new evolving energy marketplace will force a fresh two-way relationship between utilities and consumers, one where the benefits and responsibilities become shared [35]. A new business model is emerging based on sharing, changing consumer attitudes, social media, and mobile capabilities [36]. This will require utilities to shift from a utility controlled decision-making model that emphasizes the sale of commodities like kilowatt-hours to a consumer focused model that builds on a two-way communication platform, emphasizing value-added services. The "grow and build" platform is no longer a viable one. In a 2010 report by IBM called "Switching perspectives: creating new business models for a changing world of energy," IBM discusses the traditional electricity value chain and how utilities need to re-invent it to include a new range of participants, an active consumer, bi-directional information flow, and distributed resources [17].

The traditional value chain was unidirectional. The new value chain of the evolving electric grid will have multiple feedback points. Power and information will be traveling in a variety of directions. Diversification will stem from new sources such as, plug-in electric vehicles, customer-owned generation, and energy storage. Customers will play a new role with their ability to provide demand response, electric power, and potentially energy storage. The consumer will no longer be in the position to simply receive value; they will be able to supply it [17].

As the value chain of the electric utility changes, it is imperative that utilities acknowledge the intersection of the new energy consumer, social media, behavioral science, innovation, and data analytics. Each of these components will be critical to developing a customer-focused business model and closing the energy efficiency gap [21]. A 2010 study by Accenture, for example, found that out of 1,000 energy consumers in the United States, 67 percent of survey participants had changed their behavior between one and five times over the course of a year because of content on a social media site [37,38]. Yet, without timely information and the capacity to do it right, those consumer changes may lead to a suboptimal or incomplete result with respect to the full efficiency and value-added opportunities. Hence, broad-scale

customer engagement will be needed to ensure improvements in energy efficiency and decreased environmental impacts, and to also assist utilities in
creating a profitable business model in the social web era of sharing and collaborating. With this confluence of concepts, utilities will be able to expand
their portfolio and continue to diversify services in ways that maintain or
appropriately enhance their profit margins.

5.1 Elements and Examples of the New Business Model

The utility business model of the twentieth century was based on increasing
electricity sales with an ever-expanding generation base – the "grow and
build model" [17]. The role of the consumer was passive; there were no costs
on carbon and the average cost of developing new generation was consistent
or diminishing [39]. The business model of today must be different because
the economic, social, and environmental landscape is different.

As suggested in Figure 17.6 below, the electricity utility industry has
been a declining growth industry since perhaps the 1950s. While the
figure above displays the U.S. annual growth rate in electricity consumption,
the overall trend is confirmed across the OECD. This circumstance alone
mandates an entirely different approach to meeting market demands. At the
same time, the consumer is no longer passive; they are active, engaged, and
aware of their economic surroundings and the potential impacts to their communities and environment. Utilities must acknowledge the financial risk
associated with carbon management and the need to account for increasing

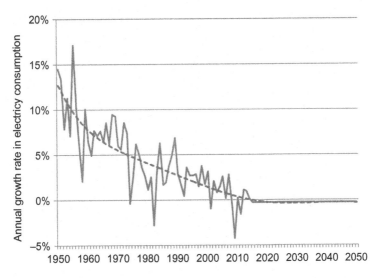

FIGURE 17.6 U.S. Historical and Future Growth Rates of Electricity Usage. *Source: EIA (2012a)
for historical data with 2010 and with estimates to 2050 adapted from Laitner et al. (2012).*

costs on carbon emissions [39]. Costs are also increasing for new capacity generation, urging utilities to look beyond their traditional service offerings and utilize energy efficiency as a true fuel source. The business model of the 20th century was simple; today it is complex, dynamic, and integrated.

In the 2011 Accenture report, "The New Energy Consumer: Strategic Perspectives on the Evolving Energy Marketplace," four separate business models are discussed as possible future strategies for success, each with their own opportunities and risks. Each business model recognizes the need for strong customer engagement. However, business models that continue to increase their value-added services will likely see the greatest increase in profit [37,38]. In order to survive it is essential that utilities continue to innovate and expand well beyond their conventional offerings. Utilities that fail to expand their array of services (as opposed to commodities only) will place themselves at risk for becoming displaced by local generation start-ups and competitor networks [40].

Additionally, in the new energy market place utilities will face further competition from companies looking to sell services that decrease electricity usage, such as Gridpoint, Johnson Controls, and AT&T ([40]; see also [11]). While Accenture lists four separate business models, becoming a full energy service provider is the only option for guaranteeing increased profits and ensuring long-term viability and success in the new dynamic energy marketplace. Figure 17.7 suggests that moving away from the standard commodities-based model of selling kilowatt-hours (see the lower-left quadrant of Figure 17.7) and toward becoming a full energy service provider

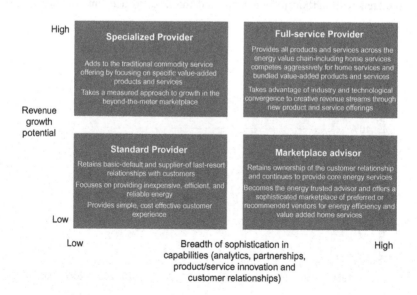

FIGURE 17.7 The Potential Emerging Electric Utility Business Models. *Source: Accenture (2011).*

(shown in the upper right of Figure 17.7) is more likely to maintain and/or enhance future revenue potential – especially in light of a declining market as suggested in Figure 17.6.

NRG Energy is a Fortune 250 wholesale power generation company headquartered in Princeton, NJ. It typifies many of the aspects of a utility that is transitioning into a full energy service provider. In effect, it is capitalizing on industrial and technological advancements to increase its bottom line. The company is expanding their offerings by creating subsidiaries in specific focus areas such as, renewable power and electric vehicle services. By developing their role in the electric vehicle systems market, NRG will acquire a new customer base. Despite expanding their services and altering their business model the company was still able to leverage existing capabilities and combine the electric vehicle services with the traditional electric customer package. Growing the value chain not only strengthens the relationship with the customer, but it solidifies NRG's positioning in the new dynamic energy marketplace [37,38].

The largest electric utility in Spain, Endesa, reinvented itself in order to prosper in the emerging multi-product marketplace. Challenged by the ability to profit in an environment with non-commodity oriented sales, the company found a way to increase their product line and customer outreach. Endesa restructured their growth on the premise that their bottom line was directly linked to customer approval and engagement. Moving beyond a traditional growth structure for a utility, Endesa added services like residential solar power, home protection services, home heating control, as well as wiring and plumbing maintenance services [37,38]. The key to Endesa's success has been integrating new product development with meeting customer needs and ensuring customer satisfaction. Endesa further complemented a new retail portfolio with an extension into online social services. The company created an online social site, www.twenergy.com, where customers can play educational games on energy efficiency and conservation [37,38].

Imaginative and innovative thinking is necessary to change the energy business model. The work behind the Pecan Street Project is one example of pushing the boundaries of the traditional electric utility structure. The Pecan Street Project is an organization through the University of Texas-Austin working to redesign the electric grid by testing new technologies and examining customer behavior with new energy management systems. In evaluating potential new business models, the Pecan Street Project released a report in 2010 that envisions Austin Energy as a fee-based service provider as opposed to a rate-charging commodity provider [41]. The report underscores the importance of transitioning to a new business model with distributed generation and increased customer engagement and involvement in energy efficiency. Under this service provider business model, customers could sign up for a plan at a fixed rate each month. The plan would engage customers as energy partners, not just receivers. For example, customers would agree to make their rooftops available for solar power, become involved in demand response programs and would acknowledge that using energy outside of their service plan would be "pay as you go" [41]. This structure would be relatively comparable to that of a cell phone bill agreement. With a cell phone bill structure, one pays for certain services (talk minutes, text messages, video downloads, etc.) with an agreed upon limit and if the limit is exceeded additional charges accrue [42]. Such a flat rate model would transition the utility's reliance from the sale of kilowatt hours to revenue generated from energy services and products.

The CEO of Duke Energy, Jim Rogers, strongly believes in developing a business model where the profits are closely aligned to the customer and their energy choices [42]. The company understands the need to go beyond the meter, foster new customer relationships, diversify the energy portfolio with clean, renewable sources, and provide a substantial platform and service for energy efficiency [42]. Utilities have the tools to advance energy efficiency, more so than any other industry with their ability to successfully raise capital. In 2007, Duke Energy proposed the "Save a Watt" energy efficiency program to the North Carolina Utilities Commission designed to allow Duke Energy to prosper in the new dynamic energy marketplace. The program would align the installation of energy efficiency measures within their basic electric service package, similar to the combined structure utilized by NRG Energy for their electric vehicle services [42]. Energy efficiency would be viewed as another fuel source and subsequently, an investment in a new generating station (Duke Energy, 2007). Duke would use the program to create a profit and also ensure that the planning and approval process was unlike the traditional, complicated, and cumbersome structure of most energy efficiency programs.

In the original model, Duke was poised to retain 90 percent of the difference between implementing energy efficiency and the cost of building an additional power plant. If a new power plant cost 6 cents/kWh and implementing energy efficiency measures costs 3 cents/kWh, then Duke could profit 90 percent of the difference, or 2.7 cents/kWh [42]. The energy efficiency program was perceived to be too profitable for Duke Energy by regulators, and the profit ratio was eventually reduced since the original proposal in 2007 [42]. Nonetheless, Duke Energy showed that energy efficiency could replace the old "grow and build" business model through customer engagement and energy efficiency measures. With considerable profits from energy efficiency investments and reduced energy cost savings, Duke Energy defined energy efficiency as a new fuel source and a replacement for investing in new capital-intensive coal plants [43]. The "Save a Watt" program has not been approved in all of the states that Duke Energy services. However, it has been successful in Ohio and additionally in North Carolina where the program was approved through 2013 (Duke Energy, 2011).

In these examples, the electric utility business model of the 21st century is all about selling services not commodities. This model has been accepted by Duke Energy's Jim Rogers, as evidenced by the "Save a Watt" program. He comments, "I would rather invest $10 billion in making my customers more productive with their use of energy than put $10 billion into a new nuclear power plant" [42].

The future of business will continue to be defined by shared experiences and customer relationships. Successful utilities must be able to create business models that incorporate customers into the company culture, process, and development [29]. The key to becoming and staying relevant today and in the future will be customer-focused engagement with an understanding of behavioral science and data analytics, added energy services that meet customer needs, and developing energy efficiency as a new fuel source. With routine and sustained engagement, utilities can spur greater adoption of energy efficiency, increase productivity, and expand profits by converting connected customers into connected energy efficiency advocates [21].

6 CONCLUSIONS

In many ways the kind of transition that is needed within the current utility business model to achieve new levels of productivity and market share might be exemplified by the post-1993 transformation of International Business Machines (IBM). In 1993, IBM was a global leader in the information technology (IT) industry – but it was in deep financial trouble. The company had reported a net loss of $8.1 billion, the largest ever in the history of that industry. Many analysts wrote off IBM as dead. Author Charles Ferguson [44] wrote that IBM's prospects for survival were very bleak. Larry Ellison, the CEO of Oracle Corporation, IBM's main competitor at that time – said, "IBM? We don't even think about those guys anymore. They're not dead, but they're irrelevant."

Faced with organizational deficiencies and the increasing commoditization of personal computers, IBM management made the bold (and at the time controversial) decision to shift its focus to a service-driven business model. In what was termed as the most remarkable turnaround of any company ever, ten years down the line in the fiscal 2003, IBM reported a net income of $7.58 billion on revenues of $89.13 billion. Today, IBM's consulting business has reinvigorated and reinvented the company, taking IBM from the brink of financial disaster to a poster child of managerial and fiscal success. In short, IBM was transformed from a company that primarily manufactured mainframes to a company that offered complete IT solutions. The company changed its focus from being product centric to being customer-centric. Put differently, IBM transformed from a commodity manufacturer with razor-thin margins, to a value-added service provider with robust profits. Today, IBM boasts a market capitalization of $232 billion dollars.

Compared to Oracle Corporation's market capitalization of $134 billion, IBM is anything but irrelevant (Bloomberg, NYSE, May 11, 2012, close; NASDAQ, May 11, 2012, close).

It is this same kind of new business model that is needed in order to move from anemic levels of (in)efficiency to one that enables a more robust economy over the next decades. As Laitner et al. [2] and other studies have documented, we have a full array of energy efficiency improvements that can deliver a 40 to 60 percent reduction in both electricity and overall energy needs by the year 2050. But market pressures are building, and consumers are changing. What was once a passive customer is quickly becoming a much more vital part of the give-and-take within the market. And, they are looking to pull more value out of purchases than ever before. All of this means that in order to accommodate the greater imperative for building a more energy-efficient economy and to leverage the full energy efficiency resource opportunity that enables a more robust economy, the development of entirely new business models that engage and empower consumers is necessary to ensure such a transition.

REFERENCES

[1] Ayres RU, Warr B. The economic growth engine: how energy and work drive material prosperity. Northampton, MA: Edward Elgar Publishing, Inc; 2009.

[2] Laitner JA "Skip", Nadel S, Elliott RN, Sachs H, Khan AS. The long-term energy efficiency potential: what the evidence suggests. Washington, DC: American Council for an Energy-Efficient Economy; 2012. Available at: ACEEE Report E121, <http://www.aceee.org/research-report/e121>.

[3] Harvey LDD. Energy demand scenarios. Energy efficiency and the demand for energy services. London, UK: Earthscan Ltd; 2010.

[4] Rifkin J. The third industrial revolution: how lateral power is transforming energy, the economy, and the world. New York, NY: Palgrave MacMillan; 2011.

[5] Energy Information Administration. [EIA]. Annual energy outlook 2012 early release. Washington, DC: U.S. Department of Energy; 2012DOE/EIA-0383er(2012). <http://www.eia.gov/forecasts/aeo/>.

[6] Bureau of Statistics. Statistical abstract of the United States: 1900. Washington, DC: U.S. Department of the Treasury; 1901.

[7] Energy Information Administration. [EIA] 2011 Annual energy review 2010. DOE/EIA-0384(2010). <http://www.eia.doe.gov/aer/>. Washington, D.C.: US Department of Energy.

[8] Energy Information Administration. [EIA]. Short-term energy outlook, April. Washington, D.C: U.S. Department of Energy; 2012<http://www.eia.doe.gov/steo/>.

[9] Laitner JA "Skip". The positive economics of climate change policies: what the historical evidence can tell us. ACEEE Report E095. Washington, DC: American Council for an Energy-Efficient Economy; 2009a.

[10] Laitner, J.A. "Skip". Climate change policy as an economic redevelopment opportunity: the role of productive investments in mitigating greenhouse gas emissions. ACEEE Report E098. Washington, DC: American Council for an Energy-Efficient Economy, 2009b.

[11] Laitner JA"Skip", McDonnell MT, Ehrhardt-Martinez K. The NPPD generations options analysis: continuing to focus on the emerging markets. A presentation given to the Nebraska Public Power District Board of Directors, February 8, 2012. Columbus, NE; 2012.

[12] [McKinsey]. Unlocking energy efficiency in the U.S. Economy. McKinsey & Company, <http://www.mckinsey.com/Client_Service/Electric_Power_and_Natural_Gas/Latest_thinking/Unlocking_energy_efficiency_in_the_US_economy>; 2009. [accessed 09.04.12].

[13] [APS] American Physical Society. Energy future: think efficiency. Washington, D.C.: American Physical Society; 2008.

[14] Griffith B, Long N, Torcellini P, Judkoff R, Crawley D, Ryan J. Assessment of the technical potential for achieving net zero-energy buildings in the commercial sector. Golden, Colo: National Renewable Energy Laboratory; 2007.

[15] Bailey O, Worrell E. Clean energy technologies: a preliminary inventory of the potential for electricity generation. Berkeley, CA: Lawrence Berkeley National Laboratory; 2005.

[16] Munson R. From Edison to Enron: the business of power and what it means for the future of electricity. Westport, CT: Praeger Publishers; 2005.

[17] Valocchi M, Juliano J, Schurr A. Switching perspectives: creating new business opportunities for a changing energy world. IBM institute for business value. IBM corporation; 2010.

[18] Gianunzio M. Changing the customer-utility paradigm. American public power association, <http://www.publicpower.org/Media/magazine/ArticleDetail.cfm?ItemNumber = 31244>; 2011.

[19] Harris Williams & Co. ESCOs – enabling energy efficiency: an introduction to energy service companies ("ESCOs"), <http://www.harriswilliams.com/pdf/HW&Co. ESCOWhitePapeFinal.pdf>; 2009. [accessed on 31.01.12].

[20] Lovins A, Rocky Mountain Institute. Reinventing fire: bold business solutions for the new energy era. Vermont: Chelsea Green Publishing; 2011.

[21] Keller HM. Why is today different? Reinventing the electric utility business model. master's thesis. Advanced academic programs, Johns Hopkins University, <https://jscholarship.library.jhu.edu/>; 2012. (forthcoming)

[22] Dietz T. Narrowing the US energy efficiency gap. Proceedings of the national academy of sciences of the United States of America 2010;107(37):16007−8.

[23] Levine M, Koomey J, Hirst E, Sanstad A, McMahon J. Energy efficiency policy and market failures. Annual review of energy and the environment 1995;20:535.

[24] Charles D. Leaping the efficiency gap. Science 2009;325(5942):804−11.

[25] Allcott H, Mullainathan S. Behavior and energy policy. Science 2010;327(5970): 1204−5.

[26] Ehrhardt-Martinez K, Laitner JA. People-centered initiatives for increasing energy savings. Washington, DC: American Council for an Energy Efficient Economy; 2010.

[27] IBM Corporation, IBM survey reveals new type of energy concern: lack of consumer understanding: behavioral economics a key factor to realizing the benefits of smarter energy. <http://www-03.ibm.com/press/us/en/pressrelease/35271.wss>; 2011a. [accessed 04.12].

[28] IBM Corporation, The social business: advent of a new age, <http://www.ibm.com/smarterplanet/global/files/us__en_us__socialbusiness__epw14008usen.pdf>; 2011b. [accessed 09.04.12].

[29] Solis B. The end of business as usual: rewire the way you work to succeed in the consumer revolution. New Jersey: Jon Wiley & Sons, Inc; 2012.

[30] Time Business, <http://money.usnews.com/money/blogs/flowchart/2010/08/19/10-great-companies-that-lost-their-edge>; 2012 January 20. [accessed on 01.02.12].

[31] Brooks S. Texas enterprise. Poll: Americans aren't optimistic about energy, <http://texasenterprise.org/article/poll-americans-arent-optimistic-about-energy>; 2011. [accessed 09.04.12].

[32] GALLOP, <http://www.gallup.com/poll/145880/Alternative-Energy-Bill-Best-Among-Eight-Proposals.aspx>; February 2, 2011. [accessed on 31.01.12].

[33] USA Today, <http://content.usatoday.com/communities/greenhouse/post/2011/02/poll-americans-ok-newer-light-bulbs/1percent29>; February 18, 2011. [accessed on 31.01.12].

[34] Electric Power Research Institute [EPRI]. Consumer engagement: facts, myths, and motivations. grid strategy 2011: consumer engagement. Palo, Alto, CA: Electric Power Research Institute, Inc; 2011.

[35] Valocchi M, Juliano J, Schurr A. Lighting the way: understanding the smart energy consumer. Somers, NY: IBM Corporation; 2009.

[36] Ganksy L. The mesh: New York. New York: Penguin Group; 2010.

[37] Accenture, Revealing the values of the new energy consumer: accenture end-consumer observatory on electricity management 2011, <http://www.accenture.com/SiteCollectionDocuments/PDF/Resources/Accenture_Revealing_Values_New_Energy_Consumer.pdf>; 2011a. [accessed 09.04.12].

[38] Accenture, The new energy consumer: strategic perspectives on the evolving energy marketplace, <http://www.accenture.com/SiteCollectionDocuments/PDF/Resources/Accenture_New_Energy_Consumer_Evolving_Marketplace.pdf>; 2011b. [accessed 09.04.12].

[39] Small F, Frantzis L. The 21st century electric utility: positioning for a low carbon future. Boston, MA: Ceres; 2010.

[40] Fox-Penner P, Wharton J. Surewest's transformation and its lessons for today's electric utilities. The Brattle Group, Inc; 2011.

[41] Duncan R. Renewable energy world. Renewable energsy and the utility: the next 20 years. <http://www.renewableenergyworld.com/rea/news/article/2010/05/renewable-energy-and-the-utility-the-next-20-years>; 2010. [accessed 22.04.12].

[42] Fox-Penner P. Smart power: climate change, the smart grid, and the future of electric utilities. Washington, DC: Island Press; 2010.

[43] Alschuler E.. 2012. Unlocking efficiency in office districts a comprehensive approach to efficiency programs. Proceedings of the 2012 ACEEE summer study on energy efficiency in buildings. Washington, DC: American Council for an Energy-Efficient Economy.

[44] Ferguson C. Computer wars: the fall of ibm and the future of global technology. New York, NY: Times Books; 1993.

Chapter 18

What Comes After the Low-Hanging Fruit?

Glenn Platt, Daniel Rowe and Josh Wall
Commonwealth Scientific & Industrial Research Organisation (CSIRO), Newcastle, NSW, Australia

1 INTRODUCTION

While the energy efficiency industry is now a relatively mature sector, there are a number of new energy efficiency approaches just over the horizon that can bring dramatic improvements, even to already-efficient buildings. In general, the current suite of commercially available energy efficiency technologies, from insulation schemes through to energy efficient lights, should be considered as "first-generation" technologies – these technologies can bring significant energy savings, and in many markets are now realizing significant uptake. At this point, an important question is "What's next?" – once the first generation of relatively straightforward technologies have saturated the market, can further energy savings be realized? This chapter will describe work from around the world working on the "second generation" of energy efficiency technologies for buildings that can not only halt demand growth, but actually reverse it. This chapter concentrates on technologies that can be cheaply and easily retrofitted to existing buildings. Importantly, such technologies have the potential to realize further savings from high-performance buildings that many would consider operating at best-possible efficiency already.

In general, many first-generation energy efficiency technologies realize their savings through relatively simple approaches – for example, compact fluorescent or LED lighting realize their gains by ensuring that instead of producing heat (as per traditional luminaires) the device energy is directed towards the production of *light*. Similarly, insulation technologies work by simply minimizing the dramatic heat flows from inside to outside in uninsulated buildings. The aim here is not to criticize such first generation technologies – these technologies can bring great savings, and their use should be encouraged. However, once these "quick wins" have been realized, something more sophisticated is needed. Consider a residential situation – once the lights are switched to low energy bulbs, the fridge seals are fixed, the second fridge is unplugged and roof and wall insulation has been installed, what's next?

Energy Efficiency. DOI: http://dx.doi.org/10.1016/B978-0-12-397879-0.00018-9
471

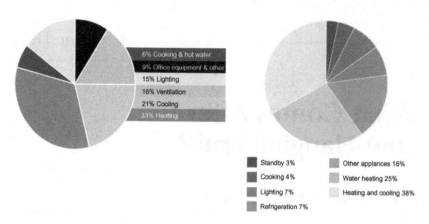

FIGURE 18.1   Energy usage in Australian buildings. *Source: http://www.yourhome.gov.au.*

This chapter describes a variety of second-generation energy efficiency technologies, targeted at realizing the next step of energy savings in our buildings. The chapter concentrates on technologies that are particularly unusual in their approach, that can be easily retrofitted to existing buildings (major structural or construction work is not needed), that are relatively low in capital costs, and that are near to market. These technologies target the major energy consumers in modern buildings — heating and cooling loads, which, as shown in Figure 18.1, make up the vast majority of energy consumption in both homes and offices. This chapter complements that by Andersen et al, where the focus is on using distributed generation as a "second generation" of energy efficiency technology. In this chapter, the focus continues to be on the consumption side of the ledger.

The following sections describe the basic principles behind each technology, as well as considering major issues that will need to be resolved before we see mass uptake of the methods described. Section 2 considers how to improve one of the highest energy consumers in most buildings — heating and ventilation, introducing a new type of air conditioning system, as well as a new way of controlling existing air conditioners. Section 3 then considers a technology-free approach to energy efficiency- how to drive energy reducing *behavior* in consumers, before the chapter concludes with investigating some of the recent social media approaches to energy efficiency.

2 IMPROVED HVAC ENERGY EFFICIENCY

Heating, ventilation, and air conditioning (HVAC) remains one of the major energy consumers of buildings worldwide — as shown in Figure 18.1, in Australia HVAC makes up around 60 percent of a typical commercial office building's overall energy consumption, and around 30 percent of a typical

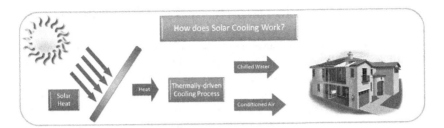

FIGURE 18.2   Solar cooling uses the sun's heat to drive a cooling process (which generates chilled water or air) to supply residential or commercial space cooling.

house's energy consumption. This section introduces some of the latest work in HVAC energy efficiency. First introduced is a completely new type of HVAC system that is becoming increasingly popular around the world, followed by a discussion of improved ways of controlling existing HVAC plant.

2.1 Solar Cooling

One of the approaches to dramatically improving the energy efficiency of HVAC equipment is to consider new approaches to chilling, or delivering chilled air. An emerging HVAC technology in Australia and around the world is solar cooling, with great potential for energy savings in the residential, commercial and industrial sectors. The technology works by converting heat collected from the sun into cooling for building air conditioning. In this way solar heat replaces electricity as the source of energy which drives the cooling process. The solar cooling process is demonstrated in Figure 18.2.

Although called a solar "cooling" system, a typical solar cooling system is also able to supply broader energy needs of a building, including space heating and hot water. Also, despite the word "solar," the technology can usually operate from a number of heat sources, including waste heat, provided certain temperature and capacity requirements are met. This may mean that residentially a solar cooling system could operate on solar heat supplemented by heat from a fuel cell, or commercially a solar cooling system could make use of waste heat from electricity generators such as gas or diesel engines.

Though solar cooling systems use some electrical power for control and moving air and water around, a well designed solar cooling system substitutes free and renewable thermal energy in place of electrical power consumption for heating and cooling. By tackling the three key thermal needs of hot water, space cooling and space heating, a solar cooling system is able to deliver useful solar energy at almost all times and, depending on design may be able to deliver all three at once or shuffle around which service is delivered at which priority level. This makes the implementation and operation of solar cooling systems very flexible, comprehensive, and powerful.

While reducing energy consumption, solar cooling systems have a broader benefit in that, by using significantly less electricity than traditional chiller systems, they can reduce the peak load on the electricity network. This peak load is often driven by conventional air conditioning systems, and as solar cooling works best when the day is hottest, the peak output of the solar cooling system is often well matched to peak consumption. In many countries like Australia a major component of electricity charges is associated with electricity network expenditure – the poles and wires needed to cope with growing electricity loads. Since the size and cost of grid infrastructure is wholly dependent on peak electricity consumption, rather than average electricity consumption, with a solar cooling system in place, grid costs can potentially be reduced significantly. Additionally, with inherent or designed storage capability, solar cooling systems can operate during cloudy weather or at night and some systems (like desiccant solar cooling) can provide partial cooling with no sun at all depending on conditions. Ultimately, the main benefits of a solar cooling system are reductions in energy, cost and greenhouse gas emissions.

2.1.1 Approaches to Solar Cooling

There are a variety of solar cooling technologies, each of which use differing components or the same components in a different configuration to turn solar thermal energy in to a cooling system. A number of variants have been technically proven, and these systems can be grouped into two general cooling approaches.

Ab/Adsorption Approaches

The most common type of solar cooling technology available are **ab**sorption-based chiller systems. These systems use a refrigerant and a sorbent (such as ammonia and water or water and lithium bromide) to transfer or "pump" heat out of a space, thus cooling it. There are minimal moving parts in the actual chiller stage of the system, with the sorption solution flowing in a closed loop. In coarse terms, evaporation of the refrigerant in this closed-loop system causes a chilling effect, the gaseous refrigerant is then absorbed into the sorbent, and finally heat (in our case, from the sun) is used to regenerate the refrigerant out of the sorbent solution ready for evaporation at the start of the loop again. Similar chilling systems are available that are based on an adsorption process, where a solid is used instead of the liquid sorption material.

Sorption-based chilling systems are a very mature technology – in fact, the first refrigerators were based on absorption-based processes. They remain a popular technology for solar or waste-heat cooling applications in large commercial buildings.

Desiccant Approaches

In general, solar cooling in the domestic and light commercial size range has suffered from a lack of sufficiently small thermally-driven cooling machines and high costs. While around a thousand solar cooling systems now exist around the world, most are based on ab- and ad- sorption technology. Unfortunately these systems are not that well suited to small scale (residential) applications because of a number of limitations. These limitations include:

- The few ab- and ad- sorption chillers available for small-scale installations are relatively expensive.
- Low-cost flat plate collectors cannot achieve high enough temperatures to supply single-effect absorption chillers, requiring the use of higher cost collector technology - contributing to high overall system cost.
- High efficiency double-effect absorption chillers require relatively high supply heat temperatures, which tends to necessitate concentrating solar collector systems. Such systems are generally considered unsuitable for residential applications, where visual amenity and maintenance become important considerations.
- Absorption chillers tend to work best with direct sunlight, and at peak sun. Often, residential HVAC systems are working hardest late in the day, when the sun is low, and householders have just returned home.

Considering these issues, a new type of solar cooling system has emerged, targeted at smaller cooling installations, and based on a *desiccant-* a material that absorbs water.

In a desiccant-based HVAC system, the actual cooling of the air is achieved in an evaporative cooling process. Evaporative coolers are a conventional cooling technology that work by evaporating water — as water evaporates, it cools the air being processed as in Figure 18.3. In hot, dry climates,

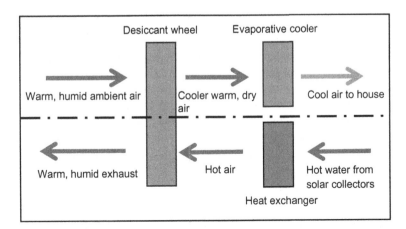

FIGURE 18.3   Basic operation of a desiccant cooling system, operating in air-cooling mode.

FIGURE 18.4 Desiccant-based solar air conditioning and hot water system.

evaporative chillers have been used for many decades to provide basic air conditioning – in a simple evaporative chiller, hot dry air is blown through a wet sponge-like material, the hot air causes water in the sponge to evaporate, thus cooling the air. While very energy efficient, evaporative coolers are relatively limited in their use, as they have a cooling range limited by ambient relative humidity and so do not work well in humid climates. For this reason desiccant evaporative coolers (DECs) use a moisture-stripping material such as silica gel to dry the air before it is cooled by an evaporative cooler. This is the approach used in the system shown in Figure 18.4 where the solar desiccant air conditioner can provide heating, cooling and hot water for residential use.

DECs are simple, robust and scalable devices using a small number of widely available components (eg. conventional fans and pumps). They are typically unable to achieve temperatures as low as ab- and ad- sorption chillers, however the simplicity of DEC systems results in significant cost-reduction potential with high-volume manufacture. Further, building integration can be as straightforward as "pre-treating" the fresh air entering a conventional air conditioner.

2.1.2 Market Challenges

Although the number of solar cooling system installations is growing around the world, they currently remain a relatively niche application. There is no fundamental reason this cooling technology cannot realize

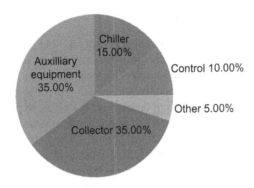

FIGURE 18.5   Breakdown of prices in a typical solar cooling system. *Source: Preisler, 2008.*

large-scale uptake – the technology does not rely on particularly exotic materials, the solar cooling systems are relatively simple to construct, and so on. Ultimately at the moment, as with other renewable energy technologies, cost is a challenge – solar cooling systems remain significantly more expensive (on initial purchase) than conventional chiller-based HVAC systems, predominantly due to a lack of sales volume. Currently, typical prices for a solar cooling system are around $3000 per kW of cooling capacity [1], although more recent anecdotal reports have suggested prices as low as $2000 per kW cooling, with these costs being split as shown in Figure 18.5. These prices are well above those for more conventional air conditioning approaches that have been mass-produced for many years.

One of the solutions to the solar cooling cost challenge may be to consider the wider financial benefits solar cooling can offer the grid. In Australia, it is estimated that a 2 kW conventional air conditioner requires approximately $7000 of investment in the electricity grid to supply energy to that air conditioner [2]. Solar cooling systems dramatically reduce this additional cost, which is spread across all users of the electricity network. Currently, this broader benefit is not passed on to the end-user who purchases the solar cooling system. Here, a broader consideration of subsidies may help the commercial proposition around solar cooling – solar cooling rarely enjoys the same subsidies or encouraging legislation given to (for example) solar photovoltaic energy systems. If there were subsidies available that reflect the wider grid benefits (and cost savings) from solar cooling, such subsidies could make solar cooling cost-competitive with conventional technologies, right now. Ultimately though, with no fundamentally expensive component to a solar cooling system, and as the benefits of solar cooling are more widely realized, the uptake of solar cooling is expected to grow dramatically in the coming years, which will drive a reduction in cost, and then even greater uptake across the community.

2.2 Intelligent HVAC Control

While new types of HVAC approaches such as solar cooling hold great promise for saving energy around the world, ultimately these are relatively capital intensive, and typically reserved for new buildings or major retrofits of existing buildings. More generally, traditional approaches to realizing savings in HVAC systems have often focused on the physical plant (e.g., installing more efficient compressors or air handling/distribution systems), or the building envelope (e.g., insulation, reducing solar gain or air exchange). Such approaches remain relatively capital-intensive, are often expensive to install, and disruptive to building occupants through this retrofit process.

This section considers how to reduce the energy consumption of HVAC systems in *existing* buildings, without requiring significant changes to the building envelope or HVAC plant. We introduce a number of new *control* strategies for HVAC systems, suited to realizing energy savings from traditional HVAC technologies, or indeed buildings using new HVAC technologies such as solar cooling. They can realize significant energy savings, while maintaining occupant comfort, and also being very easy to retrofit. The focus here is commercial buildings, which are very large energy consumers, and provide relatively easy links for deploying an improved control system.

2.2.1 Building Management Systems

In a typical large commercial building, the various HVAC componentry (compressors, fans, heaters, air louvers, etc.) is controlled by a building management system (BMS), also known as a building automation system (BAS). This building management system will consider inputs such as the indoor temperature, humidity and CO_2 levels, and then manage the HVAC equipment to achieve a particular indoor environment. Building management systems are typically a personal or embedded computer, operating dedicated BMS software, and communicating through an industrial control network to the HVAC equipment.

In general, a BMS's main HVAC goal is to maintain a particular indoor temperature, humidity and CO_2 level (or "setpoint") – 22°C is a common setpoint temperature in many office buildings. Traditionally, "energy," or "energy consumption" was not a consideration of the BMS – the BMS would use whatever energy was available to achieve its setpoint. Typically, this setpoint temperature is left constant throughout the year.

2.2.2 Building Demand Response

Traditionally, building HVAC loads have been treated as relatively constant, and HVAC/BMS systems have operated in isolation from other loads, but now this situation has begun to change. In recognition of the significant load

HVAC systems place on the broader electricity network, commercial buildings are now being asked to participate in the "demand response" programmes of electricity utilities.

Demand response programmes rely on dynamic control of electricity loads to reduce peak consumption periods on the electricity network — by better coordinating which loads are on or off at a particular time, the total load on the electricity network at a particular time can be reduced, improving system reliability and reducing the need for new electricity infrastructure. In the context of HVAC systems, a typical demand response programme may involve the electricity utility issuing a command to a participating building's BMS to reduce HVAC load at a particular time of day. The BMS will achieve this by allowing the indoor temperature to increase or decrease depending on the season, or even turning the HVAC system off, during this agreed period.

Demand response programmes are being rolled out in electricity networks around the world, and have driven a great variety of technical standardization and interoperability improvements. It is now relatively commonplace for a BMS to be able to communicate with other computers outside its own building, and there are a number of standardized "languages" for communicating with BMS's from different vendors.

Importantly, demand response programs themselves are not designed to improve overall energy consumption — in fact, in some cases a demand response event can result in greater energy consumption, as a building moves its HVAC operation out of one relatively efficient state, in to another less efficient state. As discussed later in this chapter though, the core *infrastructure* provided by demand response schemes *can* be used to realize energy savings.

2.2.3 Human Comfort

In recent years, while BMS control strategies have remained relatively constant and simple, our understanding of how people experience the environmental conditions inside a building has increased dramatically. Following seminal research from scientists such as Fanger [3], we now understand how factors such as temperature, humidity, air flow, light levels, and even color, can affect the comfort level an individual experiences in a building. Importantly, our experience of comfort is not fixed — for example, on a particularly hot day outside, individuals can find a temperature of 25°C quite comfortable, whereas on a cooler day this would be slightly warm for many. With the latest science and "human comfort factors" models, it is now relatively straightforward to predict how the majority of people will respond to a particular building environment.

2.2.4 Comfort-Based Control

With an improved understanding of human comfort factors, and the dynamic control facilitated by more interactive BMSs, we are now able to realize significant energy savings in large commercial buildings, simply by being *smarter* in how the building is controlled. In particular, by focusing on human comfort, rather than achieving a particular temperature setpoint, BMS systems can realize significant energy savings. A simple example here is that of adaptive BMS operation, where the building will operate at an increased indoor temperature (realizing significant energy savings) on particularly hot days, and a reduced indoor temperature on particularly cool days. Other examples of intelligent HVAC control include optimized start-up and shutdown operations, where the BMS will adapt the time at which it starts the HVAC system to ensure a building is comfortable before the occupants arrive to start work. After a particularly hot evening, the HVAC system may need to start well in advance of people arriving for work, but on a cooler day, the HVAC system can start much later, saving significant energy. Calculation of the optimal start-up and shutdown time involves modeling the building's thermal response for different outdoor conditions, and an understanding of how the building responds to particular HVAC actions.

The broad concepts discussed above are able to realize significant energy savings, while they are relatively simple to retrofit to an existing building. Typically, all that is required is an upgrade to the BMS software, or addition of a new controller that then instructs the BMS on what low-level setpoints (temperature, humidity, etc.) to aim for at a particular time.

2.2.5 Comfort-Based Control – Example Results

The following graphs demonstrate the energy savings possible from improved HVAC control for a particular building, over a 25-day period. The following metrics are used:

- Energy consumption (kWh or GJ) calculated using hourly averages.
- Predicted mean vote (PMV) is a continuous scale between −3 and +3 based on an ISO [4] and ASHRAE [5] standard for human comfort reporting. A PMV of −3, +3 and 0 correspond to "Cold," "Hot," and "Neutral," respectively.
- Predicted percentage dissatisfied (PPD) is calculated directly from the PMV and estimates the number of people who will be dissatisfied with their thermal conditions for a given PMV. According to the ISO standard model [4], the minimum possible PPD is 5 percent. A PPD of 5 to 20 percent is a normal target range for commercial buildings.

Like-Day Comparisons

To give a simple, yet representative comparison of energy and cost performance, two consecutive days with similar ambient conditions were chosen

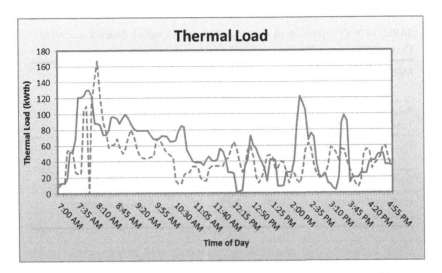

FIGURE 18.6   Average thermal load over two like-days compared, under standard BMS control (solid) and improved (comfort-based) control (dashed).

TABLE 18.1 Comparison of BMS and Comfort-Based Controller Like-Day Performance

Metric	BMS	Comfort Controller	Change
Ambient Temperature	12–21°C	14–21°C	Minor difference - lower BMS minimum temperature
Thermal Energy	2.0 GJ	1.63 GJ/day	0.37 GJ/day or 18% reduction
Predicted Percentage Dissatisfied	8.6%	8.9%	0.3% increase

for experimentation. On one day, the incumbent BMS was left in control of the HVAC system while the other allowed comfort-based control. Two particular days were chosen, where the temperature profiles match closely for most of the day, with a maximum of 1°C difference.

Figure 18.6 shows the difference in energy consumption between a traditional BMS, and a comfort-based control strategy. Table 18.1 shows that the new controller achieved an 18 percent reduction in energy use, with only a slight 0.3 percent increase in PPD, which is insignificant compared to comfort model error in predicting real occupant comfort perception. These results

TABLE 18.2 Comparison of Average BMS and Comfort-Based Controller Performance Over Winter

Metric	BMS	Comfort Controller	Change
Average Ambient Temperature	16.5°C	15.8°C	0.8°C colder under comfort-based
Thermal Energy Consumed	2.4 GJ/day	1.9 GJ/day	0.46 GJ/day or 19% reduction
Predicted Mean Vote	0.29	0.09	0.24 improvement
Predicted Percentage Dissatisfied	6.4%	5.2%	1.2% improvement

suggest that this approach, comfort-based HVAC control, can achieve energy savings in the order of 20 percent, with essentially no impact on human comfort.

Long-Term Performance

For a more complete analysis of the performance of the comfort-based control system, experiments during successive weekdays of HVAC operation were run. The experiments were undertaken in July and August, corresponding to a range of winter conditions in Australia. In short, data from 51 days of experiments during July and August 2011 showed an average of 19 percent reduction (0.46 GJ/day) in HVAC energy consumption when a comfort-based control system was running compared to standard BMS control. The average outside temperature during standard BMS operation was 16.5°C and during comfort-based operation was 15.8°C, resulting in a modest 0.7°C difference. In this case, a lower outside air temperature requires the HVAC system to consume more heating energy to achieve the same indoor thermal comfort levels. Hence, the savings figures presented below are likely understated.

The summarized energy and comfort results from the two months of experiments are presented in Table 18.2 with the expected yearly savings extrapolated from these found in Table 18.3. It can be seen that the comfort-based controller can realize significant energy savings over traditional control strategies, of the order of 20 percent, while actually *improving* human comfort.

Figure 18.7 shows the average thermal comfort PPD with the BMS controller's average PPD ranging from 5.2 percent to 12.3 percent, while the comfort-based controller's PPD average remains between 6.2 percent and

TABLE 18.3 Comparative Yearly Energy and Emissions

Estimate	BMCS	Comfort Controller	Change
Thermal Energy per year	1589 GJ	1287 GJ	302 GJ reduction
Electrical Energy per year	258 MWh	222 MWh	36 MWh reduction
CO_2 per year	254 $tCO_2 - e$	214 $tCO_2 - e$	40 $tCO_2 - e$ reduction
CO_2 Intensity per year	76 $kgCO_2 - e/m^2$	64 $kgCO_2 - e/m^2$	12 $kgCO_2 - e/m^2$

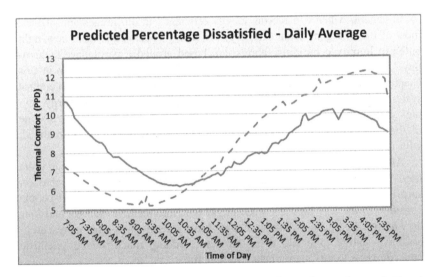

FIGURE 18.7   Average thermal comfort, in PPD, over all test days, under comfort-based (solid) and BMS control (dashed).

10.7 percent. Both ranges are considered good for commercial buildings, in which a PPD less than 20 percent it is generally considered acceptable, with less than 10 percent desirable.

2.2.6 Comfort-Based Control – Market Opportunities

The results above have shown that by controlling HVAC systems in large buildings based on a comfort setpoint, instead of temperature, energy savings of approximately 20 percent can be realized, with no negative impact on human comfort. This is a significant result, and points to a huge opportunity for these technologies. As a comfort-based control system is not based on any significant plant or building modifications, they are relatively cheap to install – typical *equipment* costs are in the order of thousands of dollars. When matched against the cost savings associated with a 20 percent reduction in energy consumption, these systems can have very short payback times. Recognizing this opportunity, there are now a number of new HVAC control strategies appearing, from traditional vendors such as Johnson Controls, as well as new companies such as BuildingIQ.

3 BEHAVIOR CHANGE

Although technologies such as solar cooling or new HVAC control show great promise in reducing building energy consumption, there is one huge opportunity for additional savings quite revolutionary compared to any of the incumbent approaches.

In general, when considering energy efficiency in our homes and offices, most of the first generation of energy efficiency endeavours, from new light bulbs to improved building fabrics, are based around a particular "technology." Indeed, the "second generation" approaches described earlier in this chapter also rely on a new technology – from HVAC chiller plant to control systems. There is an enormous opportunity to realize significant energy savings without relying on a new hardware or end-use appliance- through some of the latest *behavior change* science, significant savings can be realized, with very little new or additional technology needed.

Although governments, utilities and environmental groups have for years tried to educate consumers regarding better energy consumption behavior, surprisingly little is known about what actually works here- *why* do people consume in the way they do, *what* might motivate them to change, and *how* do we drive long-term behavior change. Most recently, a number of researchers have started to tackle these problems, adding much-needed scientific understanding to the energy behavior domain.

3.1 Understanding Energy Behavior

In order to know how to stimulate behavior change, it is helpful first to understand current behavior. For example, what do people think about when they consume energy, and why do they consume energy? Insights from sociological research (Kurz, 2007; [6]) provide some answers to these questions by explaining that:

- Energy consumption is a part of daily life and routine – it is "expected";
- Energy is essentially an "invisible" commodity needed for "non-negotiable" functions such as keeping food cool or people comfortable, and is rarely spoken about;

- Energy is considered as something "produced" rather than being a finite, precious resource (as is the case with water); and
- Feedback on energy consumption is usually delayed – for example, energy bills arrive months after the consumption has occurred.

Traditionally, energy usage has largely been driven by unconscious, automatic and habitual behaviors, and is not typically a something people think often about. This situation is changing – recent rises in energy prices mean that people are paying more attention to their energy consumption, and there is an opportunity to drive behavior change.

3.2 Energy Feedback

Having recognized the growing awareness around energy consumption, there are a number of companies now seeing this as an opportunity to drive further behavior change. The most prevalent activity here is in the area of energy "dashboards," where consumers are presented with essentially real-time information on their energy usage. Many such dashboards are based on discrete meters that consumers would install on their power supply, and which would then communicate back to a digital display in the house. More recently, with the growth in automated metering systems, dashboard products have moved to multimedia presentations that are often web based. An example dashboard is shown in Figure 18.8. This particular system was developed by VRT Technologies (http://www.vrt.com.au), but there are many vendors

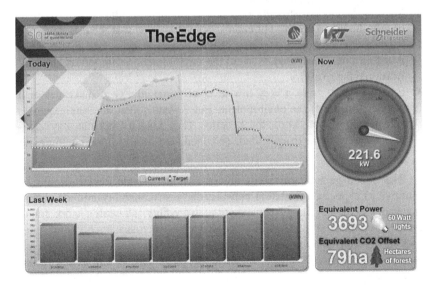

FIGURE 18.8 An example energy dashboard. *Source: http://www.vrt.com.au/products/visualisation/index.html.*

of these systems now, from traditional BMS suppliers, to new start-up companies. Some further examples of this type of work are provided in this book, in the chapter by King et al, and separately, in the chapter by Laskey and Sayler.

Dashboards are a great improvement over the heavily delayed consumption feedback traditionally provided, but unfortunately there is little evidence that energy dashboard products can realize sustained energy consumption reductions. A variety of explanations have been proposed here – from the units used (dashboards typically report in engineering (kW or kWh) or dollar units, which may not be particularly relevant to many consumers), to a more general lack of interest in checking the dashboard over time. Ultimately, despite the plethora of rich and timely consumption information, something more is needed.

Unfortunately, understanding what motivates particular energy consumption behavior is a highly complex area. Recent research by Newton et al. [7] is a classic example of the complexities here. After surveying 1200 Australian consumers, Newton's team were able to form three distinct "sustainability attitude" groupings, based on how people identified themselves with general sustainability measures, and their attitudes towards making personal changes towards sustainability:

- Committed greens indicated a willingness to pay more tax as well as higher utility charges, if it would benefit the environment. These people associate with "green choice" behaviors such as recycling, purchasing energy efficient appliances, and so on.
- Material greens are strongly opposed to paying more taxes or charges to benefit the environment, but express a general view that there is a need to balance environmental concerns against economic gain, and this balance is a fine one. They tend to view the environment as important, but not worth paying extra for.
- Enviro-sceptics have a very low level of preparedness to make personal payments towards the environment, and generally feel that any expenses here would not be worth the benefits. They tend to feel that any environmental crisis is exaggerated, and there are more important things to focus on.

The above general groupings are readily identifiable, and it would be intuitive to expect a correlation between one's sustainability attitude grouping and their level energy consumption. Incredibly, Newton's work found quite the opposite – there was *no* correlation between sustainability attitudes and energy consumption- across the group, in aggregate, committed greens consumed just as much energy as enviro-sceptics. This is a troubling result, and suggests just how hard it may be to motivate behavior change from the committed greens, let alone the enviro-sceptics.

3.3 Energy Messaging

When considering *how* to motivate behavior change, research from other disciplines such as health (in trying to motivate healthy lifestyle choices) suggests that, rather than providing additional data to consumers, the key to driving behavior change is the *messages* used. In contrast to presentation of energy consumption data, which is relatively straightforward, there are a variety of approaches to "framing" the same message. For example:

- A message may be framed in terms of a *gain*: "If you reduce your energy consumption, you will save $50 on your electricity bill," or a *loss* — "If you continue to consume like you are, your electricity bill will increase by $50."
- A message may include comparison against others in a group that you care about: "Your consumption was much more than others in your street." An example of such a campaign is shown in Figure 18.9, in this case from the company OPower, who are described in more detail in the chapter by Laskey et al.
- A message can include a variety of media forms: from dynamic or static images, through to text and audio messages.
- A message could include fact-based information, or testimonials: from (for example) a local celebrity.

As can be seen, there are many approaches to the same messaging campaign, yet surprisingly little is known about which approaches work best, particularly in an energy efficiency context. In short, although there have been many trials of marketing campaigns, dashboards, and other interventions, very few of these trials have been performed in a scientifically rigorous way — very few included a statistically rigorous sampling process, and even fewer have included a control group to allow direct comparisons. While research in this area improves, for now we turn to the health behavior domain for insight, where there has been a significant amount of research investigating which messaging approaches work best in trying to improve the prevalence of healthy living behaviors.

When considering the best message approaches to encourage healthy living behaviors, in general it has been found that loss-framed messages have been more effective in encouraging behavior change [8,9]. In the case of energy efficiency, some examples of loss-framed messages include:

- "If you leave that extra refrigerator on, your electricity bill will increase by $200."
- "Why be cold? Insulate your home."
- "Myth: Cars give you freedom. Fact: You give your freedom to cars" (referring to traffic jams — encouraging public transport use).

Though loss-framing seems to be most effective at driving behavior change, some care is needed here. For example, Obermiller [9] found that loss frames were more effective for encouraging energy conservation, but that gain frames were more effective for encouraging waste recycling. Examining things further here, Loroz [10] concluded that the type of message to use depends on *who* the message is referring to − if the message is referring to how one's behavior will affect themself, or *self*-referencing, then loss-framing is most effective. However, if the message is referring to how one's behavior will affect oneself, or 'self-referencing', then loss-framing is

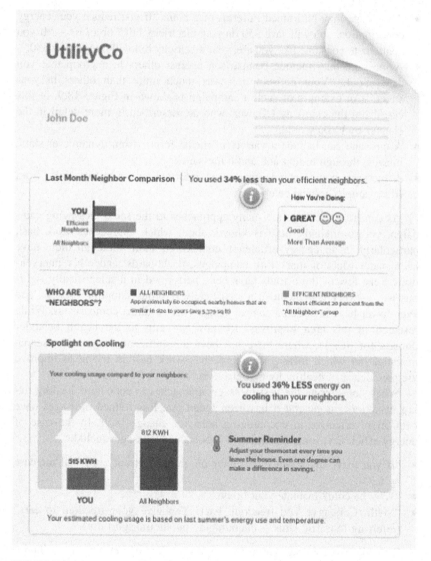

FIGURE 18.9   Example utility bill with energy consumption messaging − in this case, comparison to one's social group. *Source: http://opower.com.*

most effective. However, if the message is advising how one's behavior may affect oneself *and* others (e.g., one's children or community), or 'self-others-referencing', then gain frames appear to be more effective.

3.4 Gamification

A very new development in the broader energy efficiency space seeks to side-step the issue of how to frame messages that encourage behavior change, and instead turns energy efficiency in to a *game*. So-called *gamification* seeks to take the focus away from self-improvement or self-denial, and instead aims to communicate energy consumption information, and encourage energy-efficient behavior, in a game-structured format that targets people's affinity for play and competition.

A gamification approach to energy efficiency typically involves setting some sort of target or goal, and then having a system of rewards linked to achievements against this target. These rewards may take the form of "points" that can be exchanged for a particular good or service, or reductions off the utility bill, etc. For example, a typical energy efficiency game operated by an energy services company may offer 100 points for reducing total household energy consumption by 10 percent by the end of the month, or 200 points for reducing it by 20 percent. Alternatively, the game may offer points for the consumers that realize the greatest energy reduction amongst their community — so for example, the top three energy-reducing households all receive 200 points. These points can then be exchanged for retail items, or used to reduce one's electricity bill.

Gamification approaches to energy consumption can be used to encourage both long-term behavior change, as well as short-term behavior change. For example, a gamification campaign may reward the consumer for changes in their quarterly energy bill, as well as change realized just in the past week. Gamification can also be used to realize the demand response programs discussed earlier in this chapter- where (for example) a consumer would be rewarded for reducing their energy consumption exactly at 2.00 pm today, compared to other houses in their area.

Although based on human behavior, gamification approaches typically rely on the automated metering and control mechanisms discussed earlier in this chapter. There are a number of fundamental components to any gamification approach used with energy efficiency:

- A method to communicate the goal and related rewards to the consumer. Approaches here range from information included in a utilities bill or corporate website, through to the use of social media platforms for communications.
- A way to measure the consumer's energy consumption and validate that energy consumption was actually reduced (or the game's goal was met). These techniques typically rely on smart meters and automated metering

infrastructure for near real-time energy consumption measurement, and then a comparison against long-term averages – for example, comparing last month's total household consumption with this month's.

- For gamification that rewards short-term behavior change, a mechanism is needed to predict "what would have happened" – verifying that in the short-term window that is the focus of the game's goal, the consumer did genuinely change their behavior compared to what was typical for them. Such methods usually use an algorithm that, based on granular historical data, predicts the consumer's future energy consumption in the given game window, so that an energy saving against this typical consumption can be verified. Two examples of energy gamification that have recently entered the marketplace are the companies OPower (http://opower.com/company) and Lowfoot (http://opower.com/company). Both companies are contracting with utility customers and households, using games approaches to drive energy savings with games similar to those introduced above. OPower claims that their customers have now realized 1 TWh of energy savings – a detailed description of one particular OPower campaign is given in the chapter by Laskey and Sayler of this book. A screen shot of our own gamification system is shown in Figure 18.10. This system is currently in small-scale trials – some of the innovations here include how the system automatically forms groups of users, dynamically aggregating other participants that a particular consumer is likely to care about in to that user's current game group.

3.5 Gamification – Challenges

The gamification concept in general, and particularly applied to realizing energy savings across a wide range of consumers, is a very new idea, whose effectiveness has yet to be proven. While gamification appeals intuitively, and early results are promising, much remains before we can be confident in where this approach works well, and where it remains ineffective.

Perhaps the biggest concern with gamification is whether it can realize long-term energy consumption behavior change. While early data suggests that consumers will change their behavior as participants of an energy-savings game, it is unclear whether, over the long term, they will resort to previous behavior, and their consumption will increase again. It is also unclear what happens when their consumption savings have "saturated" – when a consumer cannot realize any further energy savings (without significant expense), will they lose interest in the game? At this point, will their consumption then start increasing again?

Research by our own organisation towards realizing long-term savings from gamification involves very careful design and automation of the game. By dynamically tailoring the rewards offered to a particular consumer, we hope to keep them engaged, and saving energy long term. For example,

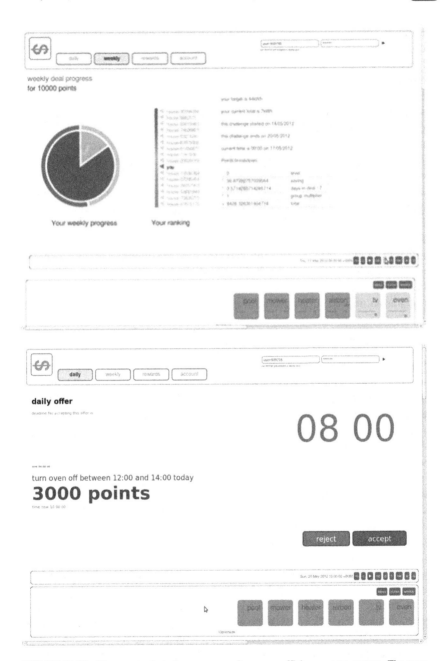

FIGURE 18.10   Two screenshots from an example energy efficiency game system. The page on top shows a consumer's behavior compared to others in their group. The page below shows the rewards available in this particular game.

initially, the energy savings required to realize a 200-point reward may be quite high — for an early participant in the game, it may be expected that there are a number of opportunities available to easily save energy. For participants who have participated in the game for a longer period, or whose energy consumption is already relatively efficient, comparatively less energy savings may be needed to realize 200 reward points. For an already-efficient participant, the effort needed to realize a comparatively small energy saving may be quite significant, and thus they should be rewarded accordingly.

Ultimately, gamification holds massive potential to realize significant energy savings across a wide range of consumption domains. By avoiding any need for significant plant, infrastructure, or equipment modifications, gamification should prove to be a very cost effective way to save energy. However, the long term benefits of such strategies remain to be proven. Backed by the psychology and social science insights introduced above, research in this area is growing rapidly, and the number of deployments increasing also. Ultimately, with the potential of dramatic energy savings, at near zero cost, and with enthusiastic customer participation, gamification remains an important prospect to watch.

4 CONCLUSIONS

Sadly, many of today's buildings are relatively inefficient, and there are massive opportunities to realize energy savings in these buildings, as detailed throughout this book. With around 60 percent of many buildings' energy consumption associated with heating and cooling, this should be the major area of focus for energy efficiency efforts. This chapter focused on the next step — once we have relatively high performance buildings, once we have insulated roofs and walls, installed efficient HVAC plant, glazed windows and so on, how do we continue to reduce their energy consumption. At this point, relatively sophisticated technologies are needed, that dramatically reconsider the fundamental approaches behind building heating and cooling.

This chapter first considered energy savings specific to heating, ventilation and air conditioning systems, one of the greatest energy consumers in modern buildings. It was found that by dramatically changing the core refrigeration source — by utilizing the heat from the sun to cool a building, significant energy savings can be realized, while also dramatically reducing the peak demand on electricity networks. The chapter then introduced a new way of controlling HVAC systems, including conventional and solar-cooling-based systems. By controlling the air conditioning based on human comfort factors, rather than aiming to achieve a particular temperature setpoint, significant amounts of energy can be saved, while actually improving occupant comfort. Such systems are suitable for retrofitting to most large commercial buildings around the world.

Lastly, this chapter considered a complete paradigm shift in energy efficiency − rather than focusing on reducing consumption through technical means, we should approach energy efficiency by focusing on energy behavior. Taking advantage of the latest psychology and social science research, the chapter investigated how to frame messages to encourage energy efficient behavior. This research supports some of the latest in energy efficient behavior development − the gamification of energy efficiency, which shows significant promise, but whose long-term benefits have yet to be proven.

As described in other chapters in this book, the first generation of energy efficiency approaches are realizing significant savings. As these technologies start to saturate, something more will be needed. This chapter shows that by reconsidering our approach to consumption − by using the sun to cool, controlling on human comfort, and aiming to change behavior, dramatic savings can be realized, in a very cost efficient manner. Ultimately, there is no floor to the opportunities available in energy efficiency − with technologies such as those introduced herein, not only can we halt demand, we can drive it backwards by fundamentally changing how we approach our consumption.

REFERENCES

[1]   Preisler A. Reduction of costs of solar cooling systems final report- publishable part, European Project No. TREN/05/FP6EN/SO7.54855/020094; June 2008.

[2]   Draft Energy White Paper 2011, Strengthening the foundations for Australia's energy future, Australian Department of Resources Energy and Tourism, 2011.

[3]   Fanger PO. Thermal comfort, Danish Technical Press, 1970 (republished by McGraw-Hill, New York); 1973.

[4]   ANSI/ASHRAE, Standard 55-2010, thermal environmental conditions for human occupancy; 2010.

[5]   International Standards Organisation, ISO 7730:2005 Ergonomics of the thermal environment − analytical determination and interpretation of thermal comfort using calculation of the PMV and PPD indices and local thermal comfort criteria; August 2005.

[6]   Kurz T, Donaghue N, Rapley M, Walker I. The ways that people talk about natural resources: discursive strategies as barriers to environmentally sustainable practices. Br J Soc Psychol 2005;44:603−20.

[7]   Newton PW, Meyer D. The determinants of urban resource consumption. Environ Behav 2012;44(1):107−35.

[8]   Davis JJ. The effects of message framing on response to environmental communications. J Mass Commun Q 1995;72:285−99.

[9]   Obermiller C. The baby is sick/the baby is well: a test of environmental communication appeals. J Advert 1995;24(2):55−70.

[10] Loroz PS. The interaction of message frames and reference points in prosocial persuasion appeals. Psychol Mark 2007;24(11):1001−23.

[11] Kurz T. The psychology of environmentally sustainable behaviour: fitting together pieces of the puzzle. Anal Soc Issues Public Policy 2002;2(1):257−78.

Chapter 19

Energy Convergence: Integrating Increased Efficiency with Increased Penetration of Renewable Generation

Ren Anderson[1], Dave Mooney[2] and Steven Hauser[3]

[1]Manager, Residential Systems R&D Group, National Renewable Energy Laboratory, [2]Director, Electricity, Resources, and Building Systems Integration Center, National Renewable Energy Laboratory, [3]New West Technologies

1 INTRODUCTION

This chapter evaluates opportunities for integrating advanced efficiency measures with the increased penetration of renewable generation by focusing on results from recent research efforts aimed at determining the most cost-effective combinations of efficiency measures and on-site renewable energy generation. These combinations technologies and measures are developed to minimize overall operating costs and energy use in new homes, leading ultimately to zero net energy homes (ZNEH) that can produce as much source energy as they use on an annual basis [1,2]. This chapter focuses on questions related to the expected near-term convergence of the cost of delivered energy savings from investments in efficiency and site renewable generation in the residential sector. Related information on zero energy policy initiatives and zero energy commercial buildings can be found in Chapters 10 and 11.

To determine the best design for a high-performance home, the net annual source energy use and net-energy-related operating costs for a series of building design options ranging from a home that meets minimum code requirements to a zero net energy home are determined by evaluating the net source energy use, utility costs, and energy-related financing costs for a series of designs on the pathway to zero energy while providing credit for any excess electric energy that is exported back to the grid via a net metering agreement with the local utility.

The total series of "least cost" building designs for each savings level between a code home and a ZNEH traces a technology pathway that defines

Energy Efficiency. DOI: http://dx.doi.org/10.1016/B978-0-12-397879-0.00019-0

the minimum cost required to achieve each energy savings level between the two extremes. Because of the current reduction in home values in many parts of the country and the slow economic recovery that will continue to limit consumer spending in the near term, this chapter will focus on energy choices that are expected to become broadly available to consumers over the next 3 to 5 years.

ZNEH have received steady interest since the mid 1990s when the U.S. Department of Energy and the National Renewable Energy Laboratory (NREL) began to evaluate opportunities for developing homes that can produce as much energy as they use on an annual basis. For an overview of previous DOE research efforts related to ZNEH and high-performance commercial buildings, see Chapter 10 by Rajkovich, Miller, and Risser.

While previous designs for ZNEH were relatively expensive compared to conventional homes, zero energy homes are currently receiving increased attention as the cost of efficient construction techniques and residential PV systems continue to decline.[1] One of the key questions we are now facing, particularly as minimum energy codes continue to improve for new homes, is how to determine which energy option to invest in when savings provided by investments in efficiency and site renewable energy generation cost about the same as purchasing power directly from a utility. This question is expected to become an even more important problem over the next 3 to 5 years as new high-performance wall systems begin to enter the market and residential photovoltaic (PV) systems begin to break the $3 / kW barrier that represents rough parity with the retail cost of residential electricity in many areas of the United States.

When the point of cost neutrality between utility power, increased efficiency, and residential PV occurs, the purchaser of a new home will have a broad range of seemingly equivalent-cost choices when making decisions about the design of their home and the impacts those choices will have on their home's operating cost. Utilities that supply power to new developments of high-performance homes will face a similar dilemma. What are the best investments in increased efficiency and site renewable generation from the perspective of a utility? What impact will future consumer and builder choices have on the design and operation of the grid?

The number of individual choices and constraints that can influence the design of new residential homes can be overwhelming to say the least. The United States does not have a national energy code. Codes, building requirements, and incentive programs vary by state and local jurisdiction making it difficult to define standard practices and costs. Homes are built with many different floor plans using many different construction techniques. Product and service supply chains are diverse and disaggregated. Home energy use is

---

1. http://www.scientificamerican.com/article.cfm?id = net-zero-energy-buildings-in-us

strongly affected by local weather, floor area, building orientation, local wind and solar shading, and occupant behavior. Some products may not be available in some locations and/or there may be insufficient demand for a new product in current markets to create the economies of the scale needed to reduce product costs. Profit Margins and research budgets are small compared to other market sectors leading to delays in development of new products. The list challenges facing builders and home buyers goes on and on.

Having acknowledged the broad range of specific issues that can affect energy improvement outcomes in individual homes, there are also a small number of big picture performance parameters that can be constructively used to describe the average behavior of residential buildings and develop an understanding of opportunities for cost-effective energy upgrades that deliver major reductions in home energy use. There are also an evolving set of easy to use residential analysis tools that use micro-analysis techniques to quickly establish the link between the engineering details of specific building materials and more global parameters such as annual energy use and annual energy cost that can directly impact choices in the marketplace.

The analysis presented in this chapter is based on the use of one such tool, the BEopt Analysis Tool developed by researchers at the National Renewable Energy Laboratory (NREL). The BEopt analysis process consists of a series of standardized, menu-driven building elements that are based on current, best available information about different classes of building products. Users can define new building elements and can also make direct modifications to the underlying simulation engine.

To facilitate a general approach to the evaluation of whole building performance, the BEopt tool uses a set of standard rules and assumptions known as the House Simulation Protocols[2] that clearly define the reference conditions and rules for the analysis including the definition of a standard reference house based on the IECC 2009 model energy code. A companion site to the BEopt tool, the National Residential Measures Database,[3] provides standard descriptions of common residential efficiency measures and associated cost ranges. The House Simulation Protocols and National Measures Database have been designed to establish a set of standard assumptions to provide a common foundation for detailed analysis of residential energy use based on the most accurate publicly available information. To further facilitate knowledge transfer, the BEopt interface has been designed to bridge the gap between researchers and practitioners and accelerate the incorporation of recent research results and emerging technologies into design practice. This is accomplished by providing a high-level menu that describes standard building assemblies while also providing more advanced users with direct access to user libraries and the core energy simulation engines. The

---

2. http://www1.eere.energy.gov/buildings/building_america/house_simulation_protocols.html
3. http://www1.eere.energy.gov/buildings/building_america/measures_costs.html

development of analysis tools with integrated capabilities provides better understanding of the potential impacts of new technologies before they are deployed, creating opportunities for better targeting of market transformation strategies and better understanding of the remaining gaps that need to be filled to achieve even higher levels of energy savings.

According to Department of Labor statistics, home energy costs are only one part of a typical household budget, mixed in with car payments, house payments, house maintenance, tuition bills, cell phone bills, car and auto maintenance costs, insurance bills, medical expenses, and entertainment. Nonetheless, home energy use is an important expense for most consumers and is roughly equivalent to annual costs for clothing or healthcare. On average, home energy use accounts for about 10 percent of the total annual housing costs. Therefore, one of the best ways to measure the impacts of improvements in building energy systems is to measure the impact of improvements on average home energy use. Torcellini and co-workers have provided a useful survey of common metrics applied to discussions of zero energy homes [3]. The global performance parameters that will be used in this chapter for evaluation of alternative approaches to delivering high-performance homes are shown in Figure 19.1.

Three simple examples of pathways leading to higher performance homes are included in Figure 19.1 to demonstrate the impacts of different technology choices on home cost and performance. Total annual energy-related costs for different building designs, including both costs for energy purchased from the utility and financing costs for energy upgrades, are plotted on the vertical axis of the figure. The annual energy savings for each investment are plotted on the horizontal axis. The combined energy savings for homes that use both electricity and natural are calculated by converting utility-delivered electricity and natural gas into source energy measured in millions of Btus using the EPA national average conversion factors for

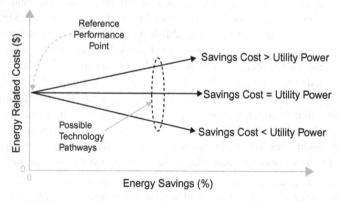

FIGURE 19.1 Primary performance parameters Used in this chapter.

electricity and natural gas. This approach assumes that a net metering agreement is in place with the local utility. Onsite renewable electric generation from a PV system is counted as contributing savings in addition to those provided by investments in increased efficiency.

The energy cost at the starting point on the left-hand side with zero energy savings and no investments in energy upgrades is equal to the utility bill for the initial house design. As energy upgrades are added to the building we move to the right on each curve and the cost of financing the upgrades becomes an additional energy-related cost that is added to the utility bill to determine the total energy related costs for the redesigned building. If the annual financing cost of the next upgrade on a pathway is larger than the corresponding reduction in the annual utility bill, then the slope of the curve depicting the next step in the technology pathway is positive and the cost of the corresponding energy savings is larger than the cost of utility supplied power. A horizontal line indicates that the cost of the savings provided by the energy upgrade is equal to the cost of utility-supplied power, and a curve with a negative slope indicates that the cost of the savings provided by the energy upgrade is less than the cost of utility supplied power.

The chapter is organized by considering an increasingly complex series of technology pathways starting with an introduction to pathway analysis in section 2. Section 3 considers the specific case of high R wall systems in a cold climate without PV and section 4 considers cases where single and multiple efficiency measures are used in combination with PV in both hot and cold climates. Section 5 reviews future opportunities and challenges for integration of high-performance homes and communities with the local grid followed by the chapter's conclusion in section 6.

2 TYPICAL CHARACTERISTICS OF RESIDENTIAL TECHNOLOGY PATHWAYS

The chapter starts with analysis of the incremental benefits of different choices for energy upgrades in new homes focusing on the limiting case of a continuous investment in the improvement of a single building component. As can be seen from inspection of Figure 19.2, when only one efficiency measure is used to reduce home energy use − for example, the additional savings achieved by adding an additional thickness of wall insulation − the diminishing marginal returns associated with each increase in efficiency quickly limit the total level of cost-effective energy savings that can be achieved.

Initially the value of energy savings from improved efficiency increases faster than the incremental cost of financing efficiency improvements. Because of the diminishing magnitude of energy savings associated with each increment of efficiency improvement, the incremental cost of the efficiency improvement is eventually greater than the incremental savings provided by the corresponding reductions in energy use. A typical technology

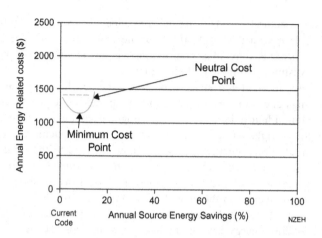

FIGURE 19.2   Technology pathway for a single efficiency measure.

pathway combines all three of the trends from the simple examples shown in Figure 19.1. The 100 percent savings point in Figure 19.2 is defined as the point where total annual energy use for the home is equal to total site renewable energy generation. In other words, a ZNEH produces as much energy as it uses on an annual basis.

The point where incremental costs equal incremental savings represents the minimum cost point for the pathway. Beyond the minimum cost point, costs increase faster than savings until total costs are equivalent to costs at the starting point on the left-hand side of Figure 19.2. The point where annual costs are equal to the cost of utility supplied energy at the starting point represents the neutral cost point. The first cost of the home at the neutral cost point is higher than the first cost at the starting point but because of the energy savings associated with increased investments in efficiency, the net annual energy related costs[4] are the same at the neutral cost point as the annual utility bills for the less efficient home at the zero savings point. It is possible to go beyond the neutral cost point and achieve larger energy savings; however, because of the increasing steepness of the cost/performance curve beyond the neutral cost point, the benefits associated with larger investments in energy efficiency are relatively small compared to the increase in energy-related costs.

The limits associated with use of a single efficiency measure can be partially mitigated by using multiple efficiency measures. An example of the technology pathway that results from the use of multiple efficiency measures is shown in Figure 19.3. When multiple efficiency strategies are used, system

4. Net annual energy related costs include the sum of the annual bill for utility-supplied energy, the incremental increase in financing costs required to build a more efficient home, and the savings resulting from the net metering agreement with the utility.

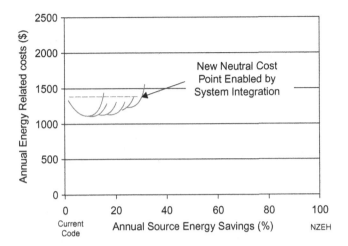

FIGURE 19.3   Technology pathway composed of multiple efficiency measures.

integration benefits lead to increases in energy savings without large increases in energy costs. The trend line in Figure 19.3 represents this by showing how a combination of curves for separate technologies results in the cumulative effect of those technologies acting together as a system.

The most cost-effective overall whole house design does not result from using just the most efficient or the least costly efficiency measures but from the least cost combination of all measures. As the efficiency of a home is improved, there are discrete transition points where the next step in efficiency improvement in one component generates a reduction in the cost of another component.

For example, in hot climates with large cooling loads, investments in insulation and better windows reduce both the size and the cost of the cooling system and the first cost savings resulting from use of a smaller air conditioning (AC) system can partially offset the cost of efficient windows and walls.

Costs associated with different efficiencies and sizes of AC systems are shown in Figure 19.4. As can be seen from inspection of Figure 19.4, if building efficiency is increased enough to reduce the required AC size from 5 tons to 1.5 tons, the cost of the AC system can be reduced by $3,000 while also nearly doubling the performance of the system from SEER 13 to SEER 24.5. The importance of improvements in enclosure efficiency are not just measured in terms of direct reductions in energy use but also in terms of additional savings in first costs due to reductions in equipment size. The impacts of improvements in enclosure efficiency can be multiplied by reinvesting savings in equipment costs in additional energy improvement options. These internal cost savings are included in all of the energy related

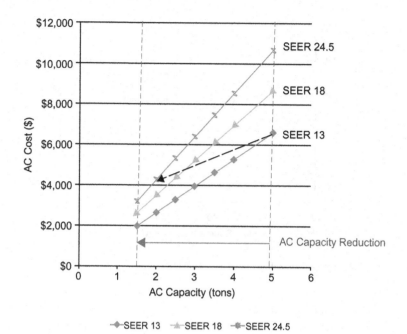

FIGURE 19.4 Reductions in air conditioning cost due to improvements in enclosure efficiency.

cost and energy savings curves shown in this chapter. The use of multiple efficiency measures produces larger overall energy savings without increasing overall costs [4]. System integration strategies that result from the use of combinations of energy measures allow the diminishing returns associated with a single efficiency measure to be offset by taking advantage of the system-level benefits provided by other efficiency measures.

When the incremental cost of energy savings associated with the next step in efficiency improvement is larger than the incremental cost of energy provided by a residential PV system, then the least cost solution is provided by investing in PV rather than making additional investments in efficiency. Because the cost of PV systems scales approximately linearly with system capacity for the multiple kW size systems used in residential applications, the least cost curve is linear after the PV start point in Figure 19.5.

The maximum energy savings are achieved by adding the system integration advantages of multiple efficiency measures shown in Figure 19.3 with the onsite power generated by the residential PV system shown in Figure 19.5. The fully developed least cost curve that develops from this combination is shown in Figure 19.6. As can be seen from the increase in energy-related costs in Figure 19.6, even though it is technically possible to

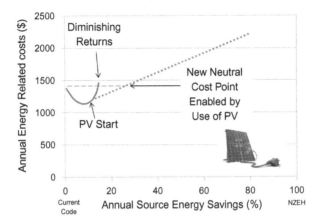

FIGURE 19.5   Technology pathway resulting from investments in efficiency and onsite renewable energy generation using a residential PV system.

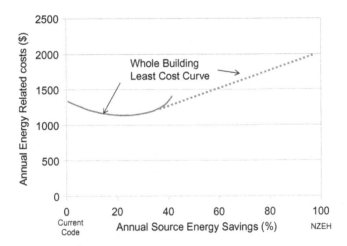

FIGURE 19.6   Fully developed ZNEH least cost curve, including the benefits of whole building system integration, multiple efficiency measures, and residential PV.

achieve a zero net energy home today in most major U.S. climates, residential PV systems at current average costs of $5.50/W generate power that is more expensive than power from the grid, leading to high-operating costs for conventional homes with savings levels larger than 30 to 40 percent.

Figure 19.6 represents a simplified view of possible pathways to more efficient homes by focusing only on a single curve. In the more realistic case shown in Figure 19.7 involving multiple technologies, there are a large number of individual design choices, all of which use less energy than the starting building design. Because of the large number of possible combinations

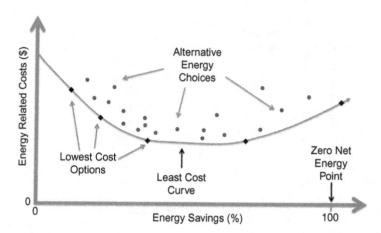

FIGURE 19.7   Determination of the least cost curve as the lower bound of possible combinations of multiple technology choices.

to be considered, the comparison of different technology choices and design approaches is simplified by focusing on the choices that provide the lowest cost energy savings. The lower bounding curve formed by the designs that deliver the lowest cost savings are used to define the least cost curve.

3 SPECIFIC TECHNOLOGY PATHWAY EXAMPLE: HIGH R WALLS

Before moving on to more specific discussions of the impacts of efficiency and renewable generation on home performance, it is necessary to define the home that will be used to make comparisons between different choices. Each residential building type has its own unique performance characteristics. Townhomes share common walls and have a lower percentage of exposed wall area than single family homes, larger homes have higher exterior wall area than smaller homes, and so on. It is not possible within the scope of this chapter to discuss all types of homes so the simple 2,500 ft^2, two-story, single-family home shown in Figure 19.8 will be used to explore options for reducing energy use in new homes.

The initial performance characteristics of a typical home design are determined by the benchmark home defined in the Building America Home Simulation Protocols and is representative of a new home built to the requirements of the 2009 International Energy Conservation Code (IECC). The front of the home faces east. When PV systems are used, they are mounted level with the roof on the back of the house facing west. The analysis assumes that PV capacity is constrained to a maximum of 8 kW by available roof area. The additional building input and operating assumptions used to complete the analysis are documented in the Building America Home Simulation

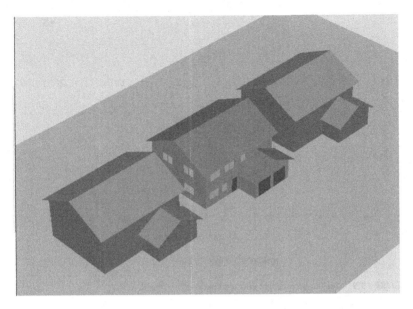

FIGURE 19.8   Image showing design details for the home used for evaluation the impacts of different energy upgrades.

Protocol [5].[5] As is the case in new homes in most major U.S. metropolitan areas, the home uses both natural gas and electricity. It is important to note that in approaching an evaluation of a specific, building performance question the inputs appropriate for that question would be used; for example, the design for a specific model home or the archetypes for a residential load profile. Armed with these definitions, one can compare the impacts of different technology choices on overall building performance.

To better understand the tradeoffs between investments in efficiency and renewable generation, we will first focus on the cost tradeoffs associated with optimizing wall insulation level or R-value in a cold climate including potential improvements in wall R-value that can occur over the next 3 to 5 years. An evaluation of the future impacts of additional energy innovations is included in the chapter by Platt, Rowe, and Wall.

One option for increasing wall R-value is to integrate varying thicknesses of insulating sheathing to the outside of a standard wood-framed wall with cavity insulation. Cost and performance assumptions for walls with 1 to 4 inches of insulating sheathing are shown in Figures 19.9 and 19.10. The base wall is a 2×6 wall with $24''$ wide cavities filled with fiberglass insulation. Future costs shown in Figure 19.10 are assumed to be reduced by about 40 percent due to economies of scale as High R walls begin to be implemented

5. http://www1.eere.energy.gov/buildings/building_america/house_simulation_protocols.html

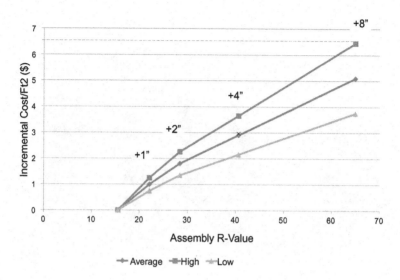

FIGURE 19.9 Current incremental cost assumptions for high R walls.

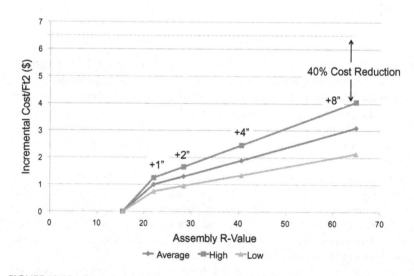

FIGURE 19.10 Future incremental cost assumptions for high R walls.

when the IECC 2012 model energy code is adopted. Each figure also shows the thickness of foam sheathing that has been added to the base wall to achieve each R-value and the high, low and average-cost estimates for current and future wall costs.

Figure 19.11 shows the results from an evaluation of a new home with the geometry shown in Figure 19.8 in Chicago with wall R-values covering

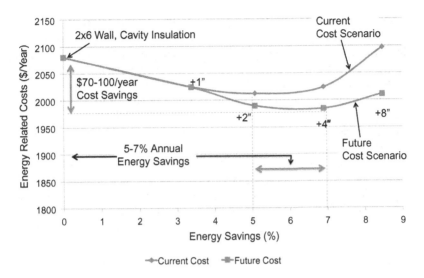

FIGURE 19.11   Estimated impacts of high R walls in a cold climate.

the range from a 2×6 wall without exterior insulating sheathing to an exterior wall with $8''$ of insulating sheathing. All calculations were based on the average cost curves in Figures 19.9 and 19.10.

The incremental costs for the wall improvements are assumed to be financed as part of the mortgage for the home, and the incremental annualized financing costs for the upgrades are added to the annual utility bills for the home to determine the total energy-related costs shown in Figure 19.11. Other approaches to financing for energy upgrades are discussed in the chapter by Hesser. It is interesting to note that even using current cost assumptions, the wall with $2''$ of insulating sheathing is cost effective in Chicago. If the first costs for insulating sheathing can be reduced by 40 percent as shown in Figure 19.10, then a wall with $4''$ of insulating sheathing also becomes cost effective in Chicago.

4 TECHNOLOGY PATHWAYS INVOLVING MULTIPLE TECHNOLOGY CHOICES

The previous discussion focused on the impact of high-performance wall systems on the cost of saved energy in a cold climate. Low-cost residential PV systems are expected to become available over the next 3 to 5 years whose cost of delivered electricity will begin to approach that of utility supplied power.[6] To evaluate the impact that lower cost PV systems will have on

---

6. DOE's SunShot initiative has the goal of developing residential PV systems with installed costs approaching $1.50/Watt. Current average installed costs without incentives are about $6/Watt: http://www1.eere.energy.gov/solar/sunshot/

energy savings in high-performance buildings, the analysis of the previous section will be extended in this section to include consideration of low-cost PV in addition to high R walls. An intermediate PV price point of $3/W was chosen to explore the expected impacts PV cost reductions over the next 3 to 5 years.

Figure 19.12 shows results for the case when the impacts of low-cost PV are added to the low-cost wall scenario shown in Figure 19.10. PV system use starts in Figure 19.12 when the next step in energy savings from efficiency costs more than the cost of energy savings provided by PV. Building designs that include the 2×6 wall (R-15.5) are the least cost competitive and PV becomes the least cost solution for this family of designs for savings levels greater than 15%. By comparison, PV system use does not start until after the 30 percent savings level for building designs that use the most cost-effective wall system with 4″ of insulating sheathing (R-40.6). The lower a curve is in Figure 19.12, the more cost effective it is compared to other solutions. Points where the curves cross indicate the transition from a lower performance wall to a higher performance wall as the least cost design approach for that energy savings level. If specific savings goals are specified as a condition for energy upgrade incentives, performance curves like those shown in Figure 19.12 can be used to determine the optimum building design to qualify for the incentive.

In addition to the evaluating impacts of building design choices on annual energy use, it is also important to consider the potential contributions of energy upgrades to grid stability during periods with critical peak demand. In many U.S. climates residential cooling is a major contributor to critical system peaks. To develop an understanding of the relative contributions of efficiency and PV to peak demand reductions in high-performance homes, it

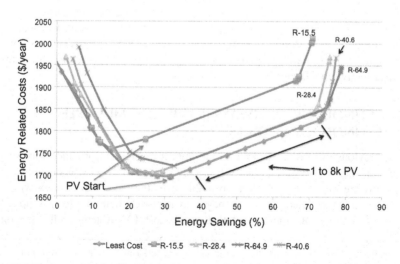

FIGURE 19.12   Impact of low-cost PV on design tradeoffs involving high R walls in Chicago.

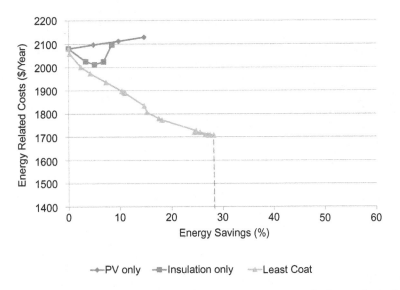

FIGURE 19.13   Savings for investment of $9,000 in three different technology pathways in Chicago.

is useful to consider the separate cases of PV only, high R wall only, and least cost energy savings pathways that utilize a complete range of energy efficiency measures during the hottest summer day based on typical weather data in Chicago and Phoenix.[7]

Assuming a fixed budget, a first cost limit of $9,000 for the total investment within each pathway was used in developing Figures 19.13 and 19.14 to determine which pathway would provide the largest reduction in peak electric demand. The impacts that a $9,000 first cost limit has on annual energy costs and annual energy savings for each pathway are shown in Figures 19.13 and 19.14. This approach to the energy upgrade decision process defines the classic issue faced in all resource allocation problems: Which choice provides the most bang for the buck?

In both climates the "insulation only" strategy provides the smallest total energy savings followed by PV, while the "least cost strategy" provides the largest energy savings by about a factor of two compared to PV only. In both climates, the maximum investment in wall insulation (8" of insulating sheathing) provides almost the same savings at the same cost as an investment in 2 kW of PV.

In both climates a $9,000 investment in PV will provide almost twice the savings as a $9,000 investment in wall insulation. Total energy savings are larger in Phoenix than they are in Chicago for the least cost pathway largely

7. http://rredc.nrel.gov/solar/old_data/nsrdb/1991-2005/tmy3/

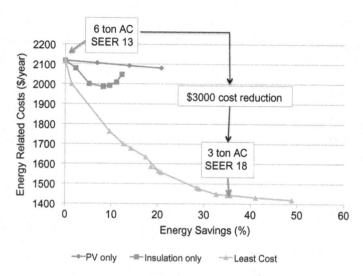

FIGURE 19.14   Savings for investment of $9,000 in three different technology pathways in Phoenix.

because of a $3,000 reduction in cooling system cost in Phoenix that occurs when the AC size is reduced from 6 tons to 3 tons as the overall efficiency of the home is improved. These results will not be as dramatic for builders who already design their homes to perform beyond the minimum code.

In addition to the many similarities between the two locations, there is one significant difference. When a $9,000 first cost limit is imposed in Chicago, the least cost solution is composed entirely of investments in efficiency. The same investment in Phoenix includes investments in both efficiency and PV.

The corresponding reduction in critical peak demand on the hottest typical day in each climate is shown in Figures 19.15 and 19.16. These figures show results for the building designs at the beginning and end of the curves shown in Figures 19.13 and 19.14. All of the strategies show large reductions in critical peak demand relative to the reference design. The least cost design with multiple technologies provides the largest reduction in peak demand as well as providing the largest annual energy savings. The least cost design reduces peak demand in Chicago by about 30 percent and reduces peak demand in Phoenix by over 50 percent — highly significant improvements.

5 FUTURE OPPORTUNITIES AND CHALLENGES FOR ZERO NET ENERGY HOMES AND COMMUNITIES

As the analysis results from the previous section demonstrate, relatively modest investments in energy upgrades in new homes can provide significant reductions in energy use and peak electric demand in both cold and hot

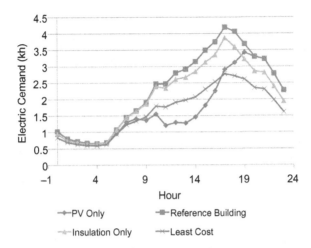

FIGURE 19.15   Tradeoffs in peak demand reduction with $9,000 first cost limit in Chicago (Typical cold climate case).

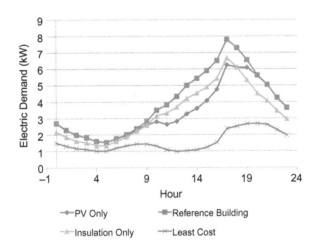

FIGURE 19.16   Tradeoffs in peak demand reduction with $9,000 first cost limit in Phoenix (Typical hot climate case).

climates. The impacts of these improvements on the local grid will first be felt in new developments with specific design requirements that will aggregate a large number of high-performance homes in a single location. Some of these impacts are discussed in more detail in section 5.

Seasonal variations in electric demand will also be an important operational challenge for utilities in hot locations like Phoenix. As has already been shown in Figure 19.16, a combination of investments in efficiency and PV provides major benefits during the cooling season. Figure 19.17 shows

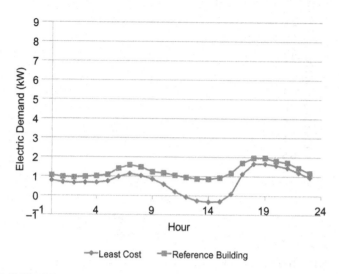

FIGURE 19.17   Winter performance of cost constrained least cost solution with 2 kW of PV in Phoenix.

the performance of the least cost design with 2 kW of PV in Phoenix on a clear day in January. In the winter in Phoenix, a 2 kW PV array has a peak electric production of a little over 1 kW and produces a negative peak during the middle of the day. Any additional kW increase in array size above 2 kW will supply an additional 0.5 kW back to the grid. For concentrated communities of efficient homes that all have PV, array sizes greater than 2 kW are a clear tipping point that will increase the complexity of integrating the community with the local grid.

The longer term ability to achieve large penetrations of high-performance homes with large PV arrays will depend on technology improvements that allow the local electric grid to reliably accept and utilize intermittent energy production from distributed generation sources like zero net energy communities, particularly during periods when energy demand is low. Additional capabilities that can improve utilization include building integrated energy storage and integration of intermittent energy supplies with the growing use of electricity by electric vehicles. A concept for a future electric grid that can adapt to communities of highly efficient homes with integrated PV systems is shown in Figure 19.18.

Future projections for cost reductions and performance improvements in residential energy systems as discussed in the chapter by Schurr and Hauser promise to make ZNEH cost competitive with conventional homes and utility supplied energy by the 2020 timeframe.[8,9] Current research efforts

8. http://www1.eere.energy.gov/solar/sunshot/
9. http://www1.eere.energy.gov/buildings/building_america/

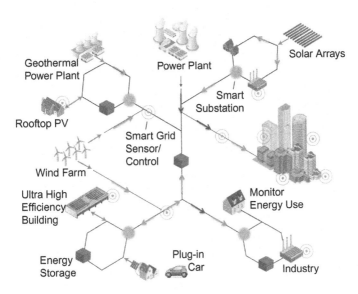

FIGURE 19.18   Schematic of future electric grid showing reliable integration of low-load buildings with a broad range of renewable generation sources.

supporting the U.S. DOE Sunshot Initiative are aimed at achieving installed residential PV system costs that make electricity from residential PV directly competitive with current residential electricity rates in most regions of the country by 2020. In regions with ample solar installation and high-utility rates, grid parity may occur at an earlier date.

Under this longer term scenario energy savings of over 75 percent are readily achievable, although the first costs of the large PV arrays that are required to achieve these savings levels remain a significant barrier in the near term as shown in Figure 19.19. Larger savings levels can be achieved in sunny locations like Phoenix that have higher solar insolation levels or Miami that has lower annual energy loads.

In addition to enabling the use of very low-energy homes, moving toward a smarter, more sophisticated electric system as depicted in Figure 19.18 also offers the opportunity to more fully optimize the entire system while incorporating cleaner, more efficient technologies. A number of new technologies and associated features could be possible in new system configurations such as that shown in the figure. These include:

- Large increases in distributed generation, including ZNEH.
- Continued increases in variable generation such as wind and PV deployed in central plants.
- Plug-in, electrified transportation.
- Bulk grid-tied storage.

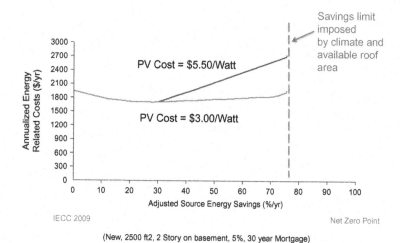

IECC 2009   Net Zero Point

(New, 2500 ft2, 2 Story on basement, 5%, 30 year Mortgage)

FIGURE 19.19   Long term projection for cost-effective savings in a two-story home in Chicago.

- Increased deployment of advanced communications and controls on the grid.
- Grid-interactive technologies incorporated into the built environment.

Additional discussion on efforts to integrate distributed generation in the Netherlands is included in the chapter by Kesting and Bliek. To reliably operate this type of system, a transformation of how the system is planned, constructed, and operated will have to take place. To exemplify the challenge that may result from a future electric system, one can imagine a distribution feeder that is loaded exclusively with zero net energy buildings. Utility revenue models aside, this scenario presents a dramatically different operating paradigm for a utility. Alternative future business models for utilities are considered in the chapter by Schurr and Hauser. Instead of dispatching conventional, controllable generation to meet demand, a utility would have to maintain a system with no net load over the course of the year but with instantaneous load that is variable and uncertain. There are potentially many new technologies and operating strategies that could be deployed to make this operating paradigm feasible. Most of these approaches remain conceptual to this point. This is a growing focus for scientists at NREL.

Building on systems modeling techniques coupled with optimization theory, NREL is beginning to look at the development of dynamic models that will allow utilities to virtually construct a system such as that depicted in Figure 19.8, look at optimization strategies for technology deployment and integration, and develop operating techniques in a virtual environment. This approach can enable system transformation with much lower risk to overall reliability.

While considerable progress must be made to fully develop these dynamic models, significant effort is underway to understand the impact on a conventional system related to high penetrations of new energy technologies. As an example, NREL has partnered with the Sacramento Municipal Utility District to understand the impacts on a specific distribution feeder of high-penetrations of PV [6].

Early findings in this type of research suggest that increasing penetrations of variable generators will definitely impact key power-quality metrics such as voltage and frequency. In a conventional system, excursions of voltage or frequency are typically addressed using either generator controls or devices such as capacitor banks to "prop up" voltage. If generation sources are variable and uncontrollable and that variability is also uncertain, conventional techniques for maintaining power quality may be more difficult and expensive to implement.

A conventional technique to manage variability and uncertainty for down ramps in variable generation, for example, is to maintain generation reserves that are "spinning" and ready to respond by putting more power out onto the grid whenever needed. This is clearly an expensive proposition because it requires the capital investment of the generator as well as the fuel to keep the generator running so it can be responsive on a moment's notice.

A less expensive and cleaner method to deal with these variable down ramps may be in considering loads that are more flexible rather than relying exclusively on flexible generation to manage system variability. Demand response has been the subject of considerable research recently and has gotten attention as a viable method to integrate variable generators [7]. Five typical demand response approaches are shown in Figure 19.20 [8]. As noted in

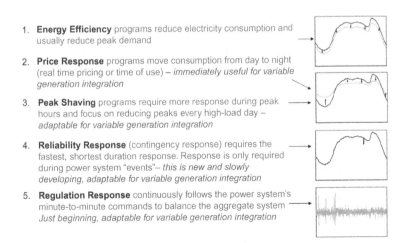

1. **Energy Efficiency** programs reduce electricity consumption and usually reduce peak demand

2. **Price Response** programs move consumption from day to night (real time pricing or time of use) – *immediately useful for variable generation integration*

3. **Peak Shaving** programs require more response during peak hours and focus on reducing peaks every high-load day – *adaptable for variable generation integration*

4. **Reliability Response** (contingency response) requires the fastest, shortest duration response. Response is only required during power system "events"– *this is new and slowly developing, adaptable for variable generation integration*

5. **Regulation Response** continuously follows the power system's minute-to-minute commands to balance the aggregate system *Just beginning, adaptable for variable generation integration*

FIGURE 19.20 Typical demand response strategies.

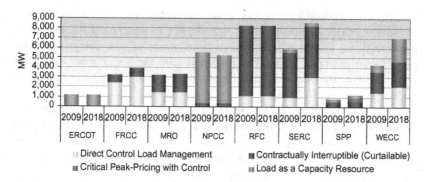

FIGURE 19.21   Potential load control by region. *Source: [9].*

Figure 19.20, four of these five can be expanded to respond to the challenges of managing a highly variable system.

Looking at these strategies in aggregate, the North American Energy Reliability Corporation (NERC) has estimated the potential of load control by region as shown in Figure 19.21 corresponding to about 10 percent of total summer peak demand in 2018 [9].

While NERC has estimated the potential to be large, realizing this potential in a manner that adds significant flexibility to the overall system while also maintaining high levels of reliability poses significant challenges. Future developments that will be needed to enable distributed and increased efficiency include:

- A smarter grid with a high degree of automation.
- New pricing structures to incentivize participation in DR.
- Advanced grid performance models.
- Distribution system models that include new energy technologies and are directly linked with load models for high-performance buildings.
- Real time optimization techniques that use a broad range of generation and efficiency technologies to make the system clean, efficient, reliable, and affordable.
- Highly dynamic control systems that lower adoption risks and advance operating techniques.

6 CONCLUSIONS

While it is intriguing to consider the possibility of cost effective zero net energy homes in the long, it is also easy to over invest in specific advanced technology pathways in the near term, leading to approaches that provide significantly lower whole building savings than solutions that are based on optimizing the cost and performance of investments in all energy upgrade opportunities. Internal cost recovery in new homes in climates with large cooling loads currently provide larger percentage savings per dollar of upgrade cost than investments in energy upgrades in cold climates. Community-scale implementation of PV arrays larger than 2 kW on high-performance homes in warm climates can lead to large negative peaks during the winter months when daytime demand for electricity is low, requiring the development of a more dynamic electric grid. New optimization and simulation techniques are being developed that link accurate predictions of the load profiles for high-performance homes with automated grid operation tools to enable broad use of distributed generation in highly efficient homes [10,11].

REFERENCES

[1] Christensen C, Barker G, Stoltenberg B. Optimization methodology for buildings on the path to zero net energy. In: Campbell-Howe R. editor. Proceedings of the solar 2003 conference including proceedings of 32nd ASES annual conference and proceedings of 28th national passive solar conference, 21−26 June 2003, Austin, Texas. Boulder, CO: American Solar Energy Society, Inc., (ASES); 2003. p. 323−328. NREL Report No. CP-550-34751.

[2] Anderson R, Christensen C, Horowitz S. Analysis of residential system strategies targeting least-cost solutions leading to net zero energy homes. ASHRAE transactions: papers presented at the 2006 annual meeting, 24−28 June 2006, vol. 112, Pt. 2. Quebec City, Canada. Atlanta, GA: American Society of Heating, Refrigerating and Air-Conditioning Engineers, Inc. (ASHRAE); 2006. p. 330−341. NREL Report No. CP-550-41004.

[3] Torcellini P, Pless S, Deru M, Crawley D. Zero energy buildings: a critical look at the definition. Preprint 2006. p. 15. NREL Report No. CP-550-39833.

[4] Werling E, Rashkin S. Top system innovations from the building America program. ACEEE summer study 2012.

[5] Hendron R, Engebrecht C. Building America house simulation protocols (Revised). 2010. p. 88. NREL Report No. TP-550-49246; DOE/GO-102010-3141.

[6] McNutt P, Hambrick J, Keesee M, Brown D. Impact of solarsmart subdivisions on SMUD's distribution system. 2009. p. 41. NREL Report No. TP-550-46093.

[7] Milligan M, Kirby B. Utilizing load response for wind and solar integration and power system reliability. 2010. p. 21. NREL Report No. CP-550-48247.

[8] Kirby B. Demand response for power system reliability: FAQ. ORNL/TM 2006/565, Oak Ridge National Laboratory, December. Available at <http://www.consultkirby.com/Publications.html>; 2006.

[9] NERC, Long-term reliability assessment 2009−2018, North American Electric Reliability Corporation, October. Available at <http://www.nerc.com/files/2009_LTRA.pdf>; 2009.

[10] Horowitz S, Christensen C, Brandemuehl M, Krarti M. Enhanced sequential search methodology for identifying cost-optimal building pathways. 2008. p.11. NREL Report No. CP-550-43238.

[11] Horowitz S, Christensen C, Anderson R. Searching for the optimal mix of solar and efficiency in zero net energy buildings: Preprint. 2008. p. 13. NREL Report No. CP-550-42956.

Chapter 20

Energy Efficiency Finance, A Silver Bullet Amid the Buckshot?

Theodore G. Hesser*

1 INTRODUCTION

A lack of project financing consistently ranks near the top of the many problems facing energy efficiency implementation in commercial, multi-, and single-family residential buildings. A global survey of decision makers responsible for energy use in the built environment conducted by Johnson Controls, the International Facilities Management Association, and the Urban Land Institute revealed that the single greatest barrier to pursuing energy efficiency opportunities was a lack of funding [1]. Creative financing solutions can solve this problem by moving debt off balance sheets for businesses, property owners, and individual homeowners. The trick is navigating the interests of existing lien holders on a property while simultaneously resolving many of the split incentives issues that currently stymie viable projects.

Public buildings have found such a balance in the form of an energy savings performance contract (ESPC). Of all the viable financing models, the ESCO industry's ESPC financing model, at $4 to 5 billion/year, is the only debt mechanism to have gained any scale. Yet this model has seen very little penetration in commercial, multi-, or single-family residential buildings. Fortunately, there are numerous debt financing models that could open up these segments, such as energy efficient mortgages (EEMs), property assessed clean energy (PACE), efficiency services agreements (ESAs), utility on-bill finance, and virtual utilities. Each possesses unique challenges and advantages and is better suited for particular building segments. Analysis by Bloomberg New Energy Finance indicates that a higher rate of building retrofits, enabled by such mechanisms, could lead to an annual market for energy efficiency expenditure in the built environment of $28 to

*This chapter is based on work by the author while he was with Bloomberg New Energy Finance and is based, in part, on proprietary research conducted by Bloomberg New Energy Finance, which is hereby acknowledged.

519

30 billion per year in the medium term. Such a surge in retrofit finance would reduce projected U.S. load growth by 20 percent — reducing from an annualized average of 0.9 percent per year to 0.7 percent between now and 2020.

For reference, total energy efficiency expenditure in the built environment is estimated to be $18 to 20 billion in the United States in 2010. This figure is still largely representative today. Out of this total, $14.4 billion can be accounted for through particular mechanisms (Figure 20.1), while an additional $3.5 to 5.5 billion is estimated to be spent directly by homeowners, landlords, small business owners, real estate companies, and corporations.

By far the largest sources of funding for energy efficiency in the United States are direct utility expenditure and ESCO financing. While the latter is financed by third-party debt, the former is direct equity investment — as is the majority of efficiency investment today. Energy efficient mortgages (EEMs), property assessed clean energy (PACE) bonds, and efficiency services agreements (ESAs) are notable exceptions, and are discussed in detail in later sections.

Having said this a small portion of utility expenditure does go towards on-bill finance programs, where energy efficiency debt is repaid through a utility bill, and the collateralization of revolving loan funds that are seeded by an equally small portion of government stimulus dollars. Analysis by Bloomberg New Energy Finance indicates that these utility and stimulus funds tally to $226 million in debt financing per year [5]. Add to this the

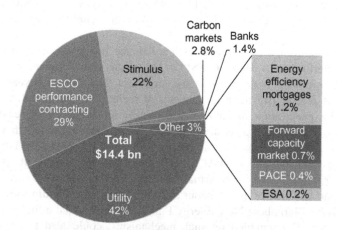

FIGURE 20.1   U.S. energy efficiency expenditure (disclosed amounts only), 2010 ($ billion). *Source: Bloomberg New Energy Finance DOE, CEE, PJM, ISO-NE, public announcement. Note: ESCO ESPC revenue is projected from 2008. "Stimulus" dollars represent strictly energy efficiency retrofit dollars from various programs. Banks represent commercial bank funds earmarked for sustainable infrastructure funds and sustainable energy utilities. PACE stands for "property assessed clean energy." ESA stands for "efficiency services agreement."*

$461 million/year from ESAs, EEMs, PACE bonds, and bank loans to sustainable energy utilities and efficiency infrastructure funds and debt financing for efficiency sums to $657 million/year (excluding the $4.2 billion in ESCO revenue) [2,3,4,6,10].

Thus, only 25 percent of the $18 to 20 billion annual investment in energy efficiency is financed via debt vehicles. Compare this with the $16 trillion U.S. housing market, which is financed by over 60 percent debt through mortgages, and it is clear that even allowing for the different sizes of these markets, energy efficiency financing is far from mature and efficiency itself is still far from an integral part of the real estate industry. Thus, the energy efficiency industry, as a facet of the wider real estate industry, is woefully underfinanced.

Aside from the numbers game of how much money comes from which source, the most salient point to draw from Figure 20.1 is that utilities sit in the driver's seat of the entire energy efficiency industry. A review of the rapid rise of state-level energy efficiency resource standards and decoupling policies (Figures 20.2 and 20.3) offers a persuasive reason as to why. Of particular interest is the spike in the adoption of such policies beginning around 2006. Total utility expenditure on energy efficiency incentives tripled from $1.4 billion to $4.6 billion over the same timeframe. Modeling by Bloomberg New Energy Finance indicates that this market size will increase to a conservative $10.7 billion by 2020 as states adopt new policies, reconcile conflicting policy measures, and ramp up reduction targets.

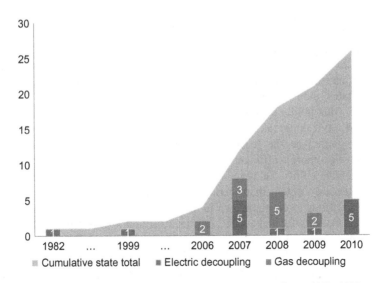

FIGURE 20.2 Number of states adopting decoupling in the United States, 1982−2010.

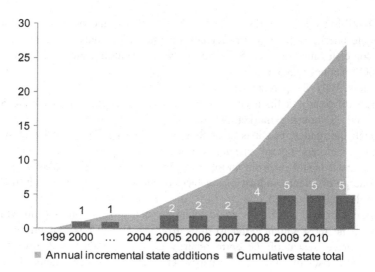

FIGURE 20.3 Number of states adopting EERS in the United States, 1999–2010. *Source: Bloomberg New Energy Finance, American Council for an Energy Efficient Economy (ACEEE), regulatory filings. Note: For the purposes of categorization, this paper considers a state "decoupled" if one or more utilities have been granted PUC approval to decouple.*

2 DECOUPLING

In a traditional utility business model, revenue and profit are tied very closely to kWh sales. Decoupling is designed to break a utility's throughput addiction by separating its ability to recover fixed costs from volume of sales. Non-decoupled utilities cannot recoup the cost of efficiency investments and find that their revenue shrinks as electricity or natural gas use declines. Such a policy structure was advantageous during the grid's expansion as it incentivized utilities to grow quickly, which in turn accelerated the United States' industrialization.

Today, this policy structure is viewed by many as economically inefficient and environmentally irresponsible. Energy efficiency measures are widely regarded as the least-cost, fastest-acting resources in a utility's portfolio. For energy efficiency to prosper, utilities must be rewarded for selling fewer kWh — a reversal of decades-old incentives. Still, some thought leaders content that decoupling is fundamentally flawed in that it relies on utilities to eat their own lunch. Such a line of reasoning contends that open markets for negawatts and negawatthours to compete toe-to-toe with generation units are the only viable way to scale the energy efficiency industry.

Successful energy efficiency programs in decoupled service areas generally lead to increased electricity rates, unless otherwise usurped by unexpected economic growth. In an industry-wide assessment of 88 gas and electric utilities, the Regulatory Assistance Project found that one-fifth of the retail rate adjustments exceeded a 2 percent increase with an average residential impact of $2.00 a month [15]. However, the successful implementation of energy efficiency programs also implies less overall consumption. The combination of these two effects results in lower energy bills if the levelized cost of energy (LCOE) of the efficiency resource in question is lower than that of any alternative supply-side resource. The LCOE of efficiency measures between 0 and 5 cents/kWh, averaging 3 cents/kWh. This is indeed cheaper than constructing a new combined-cycle natural gas facility, put at closer to 8 to 13 cents/kWh. The Institute for Energy Efficiency (IEE) calculates that the average yield of utility efficiency investments has risen from 3.6 cents/kWh to 4.1 cents/kWh over 2007 to 2009 with inflation being the primary culprit.

The economic efficiency benefits of decoupling have been generally accepted for decades; yet the breakthrough in utility decoupling did not occur until 2007 (Figure 20.2) − driven by decades of advocacy and mounting institutional acceptance − rather than one immediate catalyst. After 2007, the recession caused the growth rate of electricity retail sales to reverse from 1.6 percent annually (1996−2007) to −2.3 percent (2007−10). This reversal of 3.9 percentage points could have further incentivized utilities to pursue mechanisms for lost revenue adjustment.

The year 2010 saw another five states decouple, at least in part due to the American Recovery and Reinvestment Act's (ARRA) appropriation of $3.1 billion for state energy efficiency and conservation grants contingent upon PUCs "seeking" electric and gas utility decoupling. As of February 27, 2011 the DOE had spent $1.03 billion (33 percent) of the state energy program appropriation. State PUCs will likely be more inclined to approve decoupling bids over the next few years, and utilities will be more inclined to place those bids, as a consequence of the ARRA's efficiency grants. To place the size of these grants in perspective, they effectively double the 2009 market size of ratepayer-funded efficiency expenditures in the United States.

3 ENERGY EFFICIENCY RESOURCE STANDARDS

Decoupling removes the disincentives for utility energy efficiency investment but it does not actually incentivize efficiency investment. This requires a policy structure that sets mandatory reduction targets with teeth or allows utilities to profit from the system's increased economic efficiency. Such a policy mechanism can be found in the form of an energy efficiency resource standard (EERS).

An EERS is an energy efficiency target written into law by state legislature. It typically specifies reductions in electricity or gas use as a percentage of annual sales and in some cases as a percentage reduction in peak demand. Participants are incentivized to meet the target through penalty charges and/or performance incentives. The former are levied by the secretary of the state in question and are set at a minimum of $ 50/MWh, $ 5/MMbtu gas, of unrealized savings below a state target. A second tier of civil penalties can be levied at a minimum of $100/MWh, $10/MMbtu, if a utility fails to adequately document savings.

In many ways, a power company is exposed to the most risk when implementing an EERS. It is held responsible for the design and implementation of the efficiency program and is subject to the penalties associated with noncompliance. Hence, it is atypical for a utility to lobby for an EERS. Any funds collected through penalty fees are reinvested in additional energy efficiency program. An EERS penalty fee, or an alternate compliance payment, effectively caps the maximum obtainable price of an energy saving certificate.

EERS targets have generally evolved from mandating *spending* to mandating *savings*. While it is administratively easier to mandate spending, it is politically preferred and economically more sensible to mandate savings. Incentivizing savings − and allowing utilities to share in the system-wide benefits achieved − is regarded to be the most effective motivation to meet − and exceed − EERS targets. ACEEE reported in a survey of selected "best practice" utilities that the average utility incentive earned is approximately 11 percent of the net benefits achieved [16].

A few states have included efficiency targets within existing renewable portfolio standard (RPS) policies. Because efficiency investments are less expensive per kWh than renewable energy investments, a floor on renewable energy procurement (or a ceiling on energy efficiency contributions) is typically established. Figure 20.4 gives a geographical overview of state efficiency policies, including states with EERS carve-outs within RPS policies.

Figure 20.4 illustrates the patchwork of state approaches to electric utility efficiency policy. The light grey states that have not adopted electric efficiency policies are all stalwart Republican states, while Democratic states have shown more willingness to adopt utility efficiency policies. Energy and economic efficiency are not themselves partisan issues, but the methods by which efficiency is achieved is inherently political. If deregulated competitive markets are synonymous with Adam Smith's invisible hand, PUC regulated efficiency spending is very much its Keynesian antithesis. The political dialogue surrounding new state EERS and decoupling bids are grounded in this philosophical divide.

Of the 14 states that have adopted electric decoupling, just three (AZ, MT, ID) are historically red (Republican). Eight of the 22 states that have adopted gas decoupling are Republican (AR, NC, IN, NV, AZ, WY, UT). Of the 21 states that have *not* legislated EERS, only two are historically blue

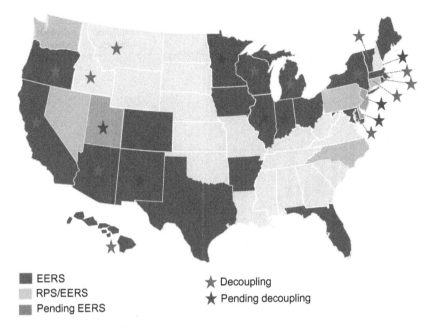

EERS
RPS/EERS
Pending EERS

⭐ Decoupling
⭐ Pending decoupling

FIGURE 20.4 Electric utility efficiency policy in the United States. *Source: NRDC, DSIRE, regulatory filings, Bloomberg New Energy Finance.*

(DE, NJ), and both of these have pending decoupling bids [17]. These political realities could limit the expansion potential of EERS and decoupling policies in the United States, as well as the continued growth of the total market size. Nevertheless, it is likely the case that even Republican dominated states will ultimately see the benefits of increased efficiency and that utility efficiency expenditure will continue to increase as states ramp up reduction targets, resolve conflicting policy structures and adopt new EERS and decoupling policies.

Of the various states with an RPS/EERS policy structure, Connecticut has been the only one to supplement it with electric utility decoupling. Most of these states are unlikely to implement decoupling because a REC-driven market approach to an RPS policy does not align with the consolidation of regulatory power through decoupling.

Similar to decoupling, EERS policy adoption has exhibited tremendous growth during the past five years (Figure 20.3). The complementary nature of the two policy structures has led efficiency advocates to push PUCs, state legislators and utilities towards the use of both mechanisms together. Figure 20.5 illustrates the increase in per capita efficiency budgets in decoupled states with EERS. The average electric utility budget for states with both mechanisms was $28.4/capita while it was only $14.1/capita for

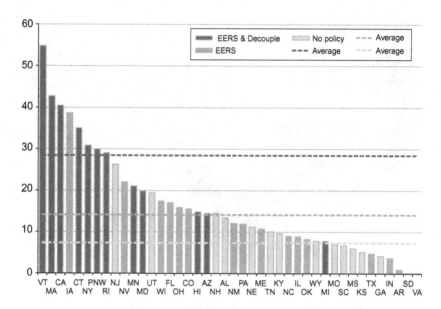

FIGURE 20.5   Electric program budgets for energy efficiency and load management by state, 2010 ($/capita). *Source: Consortium for Energy Efficiency (CEE) and Bloomberg New Energy Finance. Note: CEE's data is collected via voluntary survey and represents 73 percent of the utility service companies in the United States. The missing 27 percent represents small municipalities and a few states that did not provide information (DE, LS, WV, and AK). PNW represents the Pacific Northwest states of ID, OR, WA, and MT.*

states with either decoupling or an EERS. States with neither mechanism had an average of just $7.4/capita.

Hawaii (HI), Arizona (AZ), and Michigan (MI) are the only three with both mechanisms to sit below the state average of $16.2/capita. However, they only decoupled in 2010 (HI and AZ in the last quarter of the year), after state budgets had been predetermined. These states are expected to significantly boost their efficiency budgets and expenditure this year.

On the whole, EERS mandates have proven to be a cheap and effective way to produce efficiency savings. However the most effective framework for energy efficiency leverages both EERS and decoupling policy. Figure 20.5 illustrates that average per capita energy efficiency expenditure doubles on a statewide basis when decoupling policy is added on top of an existing EERS.

5 SUMMARY OF NEW FINANCING MECHANISMS

This section details existing financing mechanisms, strengths, weaknesses, and their current state of play.

The ESCO energy savings performance contract (ESPC) model is the only debt vehicle that has reached scale, but for the moment it is successful only in the relatively small federal, municipal, university, schools, and hospital (MUSH) building segments. These two segments are responsible for 82 percent of ESPC revenue, but only 12 percent of the total energy spent in U. S. buildings. The markets with the most potential – residential and commercial buildings – remain barely touched.

The apparent lack of activity in debt financing for energy efficiency belies a huge amount of effort behind the scenes to innovate and create new avenues for capital to flow into projects. There is recognition that on the one hand, investors are looking for ways to put money into efficiency projects and funds, while on the other hand, landlords and tenants (whether commercial or residential) need to see barriers lowered. The new financing mechanisms that are the focus of this chapter aim to solve both sides of this equation.

The barriers to energy efficiency finance differ by building segment, and so do the potential solutions. For most individuals, energy efficiency loans exist in the financing twilight zone between $3,000 and $15,000 – too big for a credit card and too small for a home equity loan. The average American holds $3,000 to $7,000 in available cash-on-hand, and would need financing to undertake any sizeable retrofit. However, roughly 60 million Americans lack bank accounts and as many as 40 million more possess "nuked" credit scores (i.e., a FICO score less than 600). Therefore, a full one-third of all U.S. citizens lack viable access to financial services. It is possible that tens of billions of dollars in energy efficiency projects are not being financed each year as a result of this financing gap.

In the multi-family residential market, low-income households have the greatest energy costs relative to income, averaging 14 to 20 percent of annual earnings compared to 4 to 7 percent for median households [6]. This low-income demographic tends to live in multi-tenant buildings that are, on average, older and less energy efficient than single-family homes. As a result, the multi-tenant market holds both the greatest need and the highest potential project returns for retrofits. Energy efficient mortgages (EEMs)[1] fit this tricky market segment well because they shift the assessment of credit worthiness from low-income tenants to landlords and housing development authorities [7]. Also, the federal government has established numerous tax credits for low-income development projects that can be accessed through mortgage financing.

In the commercial market, building owners are typically unrated limited liability entities with fully pledged mortgages; these owners are often

---

1. An energy efficient mortgage is the same as a normal mortgage, only with an energy efficiency component added to the underwriting and mortgage approval process. This is akin to a building code, only enforced by mortgage lenders (i.e., the private sector).

unable to attract additional credit and are unwilling to place additional debt onto their balance sheets. Furthermore, lease structures frequently misalign tenant-landlord incentives by having tenants reap the benefits of investments made by landlords. Property assessed clean energy (PACE) bonds; utility on-bill finance and efficiency services agreements (ESA) are well suited to the commercial market. PACE solves the credit rating issue by channeling debt through property taxes – which are paid to special entity municipalities that raise capital through credit-rated bond issuances. On-bill solves the same problem by tacking energy efficiency debt onto utility bills. This has low default rates as tenants tend to pay utility bills before rent or property insurance. Both forms of financing solve the problem of split incentives in "net" lease structures; where both tenants and landlords pay property taxes and utility bills. ESAs solve the same issues by stepping in front of utility payments with private-equity funded retrofits and siphoning savings into profit.

The residential market holds the most potential, but is also the most difficult to address. The administrative costs of underwriting and servicing small, distributed loans, plus the marketing costs of customer acquisition, often break the economics of financing residential retrofits. The additional complications of onerous consumer lending laws also tend to cause lenders to shy away from this sector. Although individual savings are only an "inch deep," aggregate potential is "miles wide." Residential PACE was a good fit, but ran aground during the mortgage crisis because property taxes hold a senior lien status to mortgage repayments. This left mortgage companies with increased exposure to foreclosure-induced losses, with no additional benefit, and they subsequently withdrew their support for the schemes. On-bill finance has since become the preferred financing method in the residential market. The threat of utility service shut-off provides lenders with the security they need to keep rates low while maintaining lien status that is junior to a mortgage. EEMs also hold significant potential, despite very little activity in the single-family market to date.

Every building is unique and retrofit finance takes many forms. Yet each mechanism outside the ESCO industry's ESPC model has achieved less than 1 to 2 percent of its potential (Figure 20.6). The overall goal of providing a low-cost conduit for money to flow from the capital markets to profitable retrofits has yet to be realized at any scale. Figure 20.6 displays the theoretical potential and current realization of the most prominent financing mechanisms. To place the large potential range into context, from $8 to 37 billion/year, it is helpful to know that home equity loans totaled $110 billion/year in 2006 and, following the financial collapse, now tally to $32 billion/year [13]. Each of these financing mechanisms is described in further detail in this chapter. Such investment could reduce energy consumption in the United States by hundreds of TWh per annum (approximately 200–500 TWh).

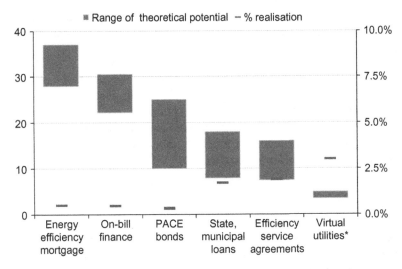

FIGURE 20.6 Realization and theoretical potential of retrofit financing mechanisms. Theoretical potential ($billion/yr) Realization (percent). *Source: Bloomberg New Energy Finance. Note: PACE stands for "property-assessed clean energy"; Virtual utilities are more akin to an energy efficiency business model than a financing mechanism, but are included nonetheless.*

5.1 Energy Efficiency Mortgages

EEMs allow a building owner to add energy efficiency debt on top of a normal mortgage. EEMs are the only financing construct capable of valuing both decreased utility bills and a green building's market premium. They achieve this by letting the mortgage provider determine interest rates, which should implicitly account for the capital market's valuation of greener buildings. As a result, the potential market for energy efficiency debt derived through EEMs is greater than any other financing mechanism and could theoretically total up to $270 billion in outstanding energy efficiency debt on top of the $13.5 trillion U.S. mortgage market.

The success of EEMs could lead to the creation of a secondary debt market where energy efficiency debt repayments are bundled, securitized, and then purchased by larger institutions with lower costs of capital. Pension funds and insurance companies could conceivably purchase large volumes of securitized efficiency debt at a low cost of capital because it matches their annuity obligation profiles, which pay out on fixed, monthly time periods over multiple decades. Furthermore, boosted efficiency investment would create job opportunities for labor unions (who in turn invest their 401(k) retirement packages with pension funds). For many in the energy efficiency space, the creation of a liquid secondary market for efficiency debt remains a "holy grail" of sorts − highly desirable, but elusive. To date insufficient loan volume has been originated to establish such a secondary market.

EEMs have historically struggled as lenders did not want to go through extra paperwork to approve a mortgage. Institutional lenders have also been doubtful that actual energy savings will match projections − performance risk remains a hurdle to acceptance. Additionally, the credit risk of 90 percent of the residential mortgage market is federally insured. There is virtually no incentive for mortgage providers to maximize portfolio returns or minimize losses beyond what is required by portfolio performance regulations − which are typically motivated by politics as opposed to profit. As a result of these barriers, energy expenses have yet to factor into the underwriting process of the residential mortgage market.

The potential for EEMs must also be placed in the context of a risk-averse mortgage market that has yet to recover from the collapse of the real estate market. From 2007−2009 average rental rates dropped 30 percent, vacancy rates went up 40 percent, unemployment increased by 115 percent and foreclosure rates quadrupled from 1 percent to 4 percent. With so much uncertainty surrounding the repayment of existing mortgages, virtually all lenders have worked to deleverage their portfolios. A few experimental CDFIs, in contrast, focused on reducing building expenses to hedge potential losses [14]. Energy bills represent nearly half of all home-incurred expenses − a ratio that is higher for renters in the low-income multi-tenant sector. The rationale behind the CDFI approach is that reducing energy expenditure through retrofits puts money in the pockets of tenants and building owners, which should decrease a mortgage portfolio's foreclosure and delinquency rates. This logic implies that EEMs would be less at risk than traditional mortgages, which further implies that they could support lower interest rates if traded on open markets.

The vehicle through which an EEM would trade could be called a green-mortgage backed security (GMBS). Fannie Mae plans to launch a GMBS initiative in 2012 that will label mortgages on multi-family residential buildings with LEED or Energy Star ratings as "green" within a conventional "vanilla" MBS structure. While this is an important first step, the most significant opportunity for GMBS lies in the commercial market. Figure 20.7 illustrates the tremendous growth of LEED and Energy Star rated commercial office space − both new build and retrofits − along with the precipitous decline of commercial MBS issuances following the collapse of the sub-prime mortgage market. The rise of LEED and Energy Star certified floor space can be attributed to a perception shift within real estate developers. To compete in the real estate market, virtually all new build is certified.

Fannie Mae is also experimenting with refinanced EEMs in the multi-family segment. The "Green Refinance Plus" program initiated in May 2011 will let low-income and affordable housing building owners add additional

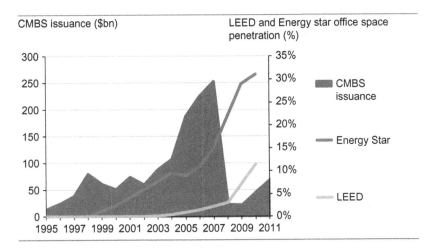

FIGURE 20.7   LEED and Energy Star U.S. office buildings and CMBS issuance, 1995–2010. *Source: Bloomberg New Energy Finance, USGBC, EPA, CBRE, University of California Berkeley. Note: office space makes up 24 percent of the outstanding CMBS market. And Bloomberg data only encapsulates 25 percent of outstanding CMBS debt.*

energy efficiency debt onto a refinanced mortgage. The Federal Housing Administration (FHA) agreed to insure losses of up to 5 percent on the additional energy efficiency debt carried on Fannie Mae's balance sheet. It is too early to tell whether these programs have worked, but they may well lead to a much larger concerted push by institutional mortgage providers in the energy efficiency space.

In addition to decreased utility bills, energy efficient buildings hold a premium market value over less efficient buildings. Occupants place a higher value on buildings that possess natural sunlight, improved air quality, and comfort through building controls, all of which are enhanced through retrofits. Table 20.1 summarizes the largest and most reputable studies on a green building's market premium. Each study uses different data sets and statistical techniques to control for differences in building age, location, and time of purchase. The sample size of each survey ranges from roughly 0.1 percent to 5 percent of total U.S. commercial office floor space — equivalent to 10,000 large commercial buildings. The general trend is clear: green buildings increase rental rates, occupancy rates, and resale values by single digit percentages.

5.2 Utility On-Bill Finance

Funding from the 2009 American Recovery and Reinvestment Act (ARRA), in combination with the demise of residential PACE, has propelled on-bill finance from a handful of programs into 31 programs in 14 states, with six

TABLE 20.1 A Summary of Statistical Studies on the Market Premium of Green Buildings

Source	Author	Year	Rental Rate Premium	Sale Price Premium	Occupancy Premium
CoStar Group	Miller	2008	16%	6%	3%
Cambridge	Fuerst, McCallister	2009	5%	31%	3%
American Economic Review	Eicholtz, Kok, Quigley	2010	3%	12%	3%
Journal of Real Estate Research	Pivo, Fisher	2010	5%	9%	1%
Journal of Real Estate Finance and Economics	Whiley, Johnson	2010	8%	–	–
Australian Property Institute	–	2011	0.4%	3.75%	5.2%
CBRE	–	2009	15%	3.5%	–
CBRE	–	2010	7.8%	3.4%	–
CBRE	–	2011	4.1%	2.6%	–

Note: Rent, occupancy and resale values are statistically controlled for differences in building age, location, and time of purchase in all studies.
Source: Bloomberg New Energy Finance, various sources listed in the table.

additional states planning new programs (Figure 20.8) [3]. All existing programs use rate-payer funds as a primary source of capital (on-bill finance) as opposed to third-party funds from the capital markets (on-bill repayment). In on-bill repayment, utilities act as a conduit between lenders and debtors, collecting fees for servicing loans, similar to Visa or PayPal. Scaling on-bill lending into the billions will require programs to break away from rate-payer coffers, and tap into outside credit from the capital markets. California is in the midst of implementing the first on-bill repayment program that is likely to take effect in 2014.

There is a lot of experimentation amongst these programs. Structures differ by targeted building segment, allocation of credit risk, financing structure and source of funding. Most on-bill programs lend to the commercial and industrial segments, use a mix of rate-payer funds and federal stimulus and implement basic financing structures with loan loss reserves or revolving loan funds. More innovative programs are beginning to use subordinate debt structures within revolving loan pools to attract a wider

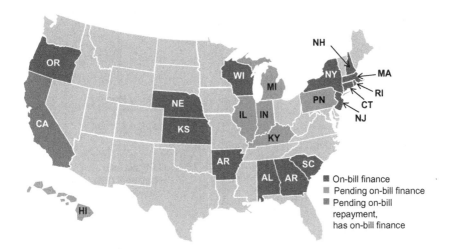

FIGURE 20.8   On-bill finance programs in U.S. states. *Source: ACEEE, DSIRE, CPUC, Bloomberg New Energy Finance.*

degree of outside capital from credit unions, CDFIs and local banks – all of which have differing risk appetites. Local lenders are cautiously optimistic about energy efficiency lending because it represents a new revenue stream and allows them to differentiate their brands from commercial banks that continue to soak up market share.[2]

The structure of on-bill repayment is schematically shown in Figure 20.10. On-bill finance functions identically, only without the bridge to the capital markets. Of particular note is the loan security provided by Uniform Commercial Code-1 (UCC1) statements. UCC1s are legal documents that allow creditors to take possession of a debtor's asset if repayments cease. In the case of on-bill repayment, UCC1 filings can be used in the court of law to pressure a utility to terminate services. A second nuance of note is the difference between on-bill loans and on-bill tariffs. On-bill loans are tied to a customer and are paid off when a property is sold whereas on-bill tariffs are tied to the meter. Most existing programs utilize on-bill loans. However, on-bill tariffs are considered to be preferable because they allow for longer financing terms, which decreases periodic debt repayments and allows for a higher pay back to the individual debtor.

5.2.1 The UK's "Green Deal" on-Bill Structure

The UK's Department of Energy and Climate Change (DECC) has made great strides to formalize a national model for avoided cost energy efficiency

---

2. In 2009 commercial bank assets totaled $13.4 trillion while credit union assets totaled only $885 billion.

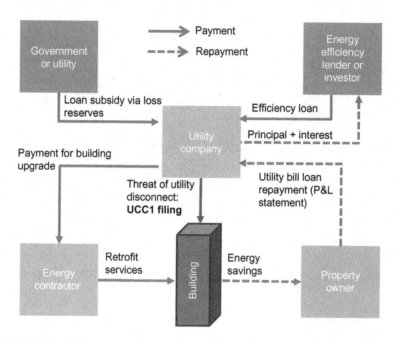

FIGURE 20.9 Utility on-bill repayment. *Source: Bloomberg New Energy Finance, WEF, GE Capital Real Estate, Bloomberg New Energy Finance [18].*

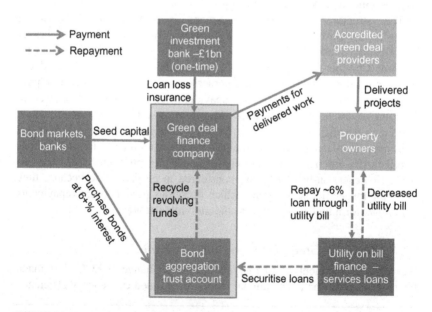

FIGURE 20.10 Schematic of UK Green Deal. *Source: Bloomberg New Energy Finance, The Green Deal Finance Company.*

financing (Figure 20.10). Due to launch this year, DECC is implementing an on-bill repayment program with a revolving loan fund supported with $1 billion in credit backstop from the UK's Green Investment Bank. Government loan loss insurance on this scale differentiates the Green Deal from the panoply of local on-bill programs in the United States. The work of administering, servicing, underwriting and provisioning loans will be centralized to a newly instituted non-profit partnership between energy providers and commercial banks called the Green Deal Finance Company. Commercial banks involved in the partnership are expected to profit from fees collected through bond issuances on the back end of securitized efficiency obligations.

DECC anticipates that this structure will support up to GBP 14 billion (USD $22 billion) in energy efficiency lending over the next 10 years. While this figure appears very high relative to the U.S. experience, it bears mention that the UK currently requires utilities to invest GBP 1.3 billion/year (USD $2 billion/year) in energy efficiency measures. Thus, if the UK's on-bill program is successful, it will shift the onus of mandatory rate-based efficiency investment towards voluntary financing from the capital markets.

5.3 Property assessed Clean Energy

Property assessed clean energy (PACE) bonds allow a building owner to repay debt from retrofits or renewable projects through a special purpose municipal property tax. The schematic structure of PACE financing is schematically illustrated in Figure 20.11. The central difference between PACE and on-bill financing is the conduit through which energy efficiency debt is repaid. On-bill utilizes a buildings utility bill, whereas PACE leverages a buildings property tax.

To date, 20 Democratic and six Republican states (party designations based on the majority of their representation in Congress) have implemented PACE-enabling legislature, yet only California, Colorado and New York have made significant progress financing energy efficiency measures through PACE. Vermont and Maine will soon enter the fold with their own programs.

PACE programs stalled in most states in 2010 after the Federal Housing Finance Agency (FHFA) advised Fannie Mae and Freddie Mac against purchasing mortgage loans with outstanding first-lien PACE obligations. On January 26, 2011, the United State District Court in Oakland, California, opened up this implicit moratorium by mandating the FHFA to comply with the outcome of a public commenting period on PACE. If the public comments prove favorable for PACE, which is expected given the large number of commenting non-governmental agencies, the Oakland case could set a precedent for numerous other cities and counties that were previously unable to stand up to FHFA.

Despite challenges in the residential market, PACE has always been a contender in the commercial market, where it continues to make progress. However, commercial buildings tied to mortgage backed security products

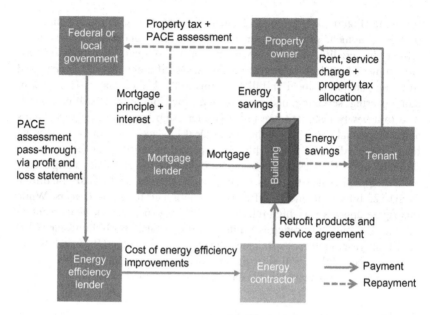

FIGURE 20.11 PACE financing model. *Source: Bloomberg New Energy Finance, WEF, GE capital real estate.*

are likely to avoid PACE due to its senior lien status. The credit-worthiness and size of aggregate PACE bonds tied to a local municipality appeals to investors who wish to securitize energy efficiency debt.

5.4 Sustainable Energy Utilities

Sustainable energy utilities (SEU) represent an attempt by multiple utilities within a geographic region to place all energy efficiency operations under one roof, are more akin to an energy efficiency business model than a financing strategy. However, there is overlap. In Delaware, for instance, Citibank issued a $57 million tax-exempt bond through the state's AA+ rated municipality to fund energy efficiency projects. Delaware's sustainable energy utility contracts with energy service companies that execute projects and which repay the SEU's debt service to bond holders through shared savings agreements. By 2020, Delaware's SEU anticipates that up to 93 percent of its revenue ($56.2 million) will be self-sustaining through shared savings agreements and REC sales. Today, REC sales and energy efficiency savings account for roughly half of the utility's revenues – estimated to be between $15 to 20 million in 2011. The remaining half derives from a utility public benefit fund, which is seeded through a utility surcharge.

The principle difference between a public benefit fund and an SEU is that an SEU is attempting to accrue revenue through shared energy savings.

Public benefit funds are essentially a mandated hand-out while SEUs represent a capitalistic approach to energy efficiency. Vermont is the only other state to have formalized an SEU.

5.5 Efficiency Service Agreements

Efficiency services agreements (ESA) have emerged as one of the more promising financing schemes in the commercial real estate market. At its core, an ESA is a hedge on utility rates for a building owner, and a structure to turn energy savings into profit for investors. Most corporations and commercial real estate companies view energy management as a cost center, not a profit center. Thus investment for the sake of project returns is rarely pursued. Instead, the focus is normally on lowering electric and gas supply prices through financial hedging, as the benefits are immediate and price certainty is highly valued by any business. ESAs allow a third-party energy efficiency company to provide building owners with the equivalent of a financial hedge through physical kWh reductions. ESAs typically finance signed contracts through a mix of private equity funds, bank loans, and special-purpose energy efficiency funds. Investors usually claim tax incentives and the accelerated depreciation of any heavy equipment installations.

The general structure of an efficiency services agreement is schematically shown in Figure 20.12. There is no public record of disclosed ESAs, yet

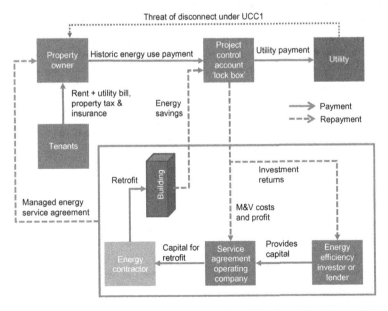

FIGURE 20.12 Efficiency services agreement. *Source: Bloomberg New Energy Finance, WEF, GE Capital Real Estate.*

analysis performed by Transcend Equity and Deutsche Bank suggests that approximately $500 m has been contracted through 100 to 125 ESA deals to date. These numbers require the caveat that much speculation exists as to the size of the ESA market. A survey conducted by Bloomberg New Energy Finance indicates that ESA deals total only $20 to $30 million/yr.

6 CONCLUSIONS

Similar to arrows in a quiver, energy efficiency financing mechanisms differ in applicability depending upon building type, landlord-tenant-lender relationships and the macro trends of the real estate industry. Energy efficiency mortgages represent the largest potential market opportunity because the mechanism fuses energy efficiency finance with traditional real estate finance. Unfortunately, additional debt for energy efficiency mortgages is currently out of the question for mortgage providers because the real estate industry is deleveraging. PACE bonds may soon be issued on heels of stimulus money that has seeded demonstration programs throughout the United States. In the past, PACE bonds have largely been held back by mortgage providers who fear the mechanism's senior lien status. On-bill finance resolves PACE's shortcomings by tagging energy efficiency debt onto a utility bill, which provisions a junior lien status relative to a mortgage. However on-bill finance is inherently tied to the cooperation of utilities many of which are slow to adopt changes and oppose aggressive efficiency measures. ESAs represent the wild card in the bunch. This mechanism solves numerous principle agent issues, and may well crack the code in the commercial market. The complexity of the mechanism, however, creates tedious and lengthy sales cycles that slow down the models adoption. The strengths and weaknesses of these mechanisms outstanding, energy efficiency finance is expected to swing into the fold in a big way over the coming years. Energy prices are expected to increase, buildings will continue to age, and grid infrastructure will cost a fortune to replace. All of these factors point to a bright future for those who innovate off-balance sheet solutions for energy efficiency projects.

REFERENCES

[1] Institute for Building Efficiency. Energy efficiency indicator, 2011 global results, <http://www.institutebe.com/Energy-Efficiency-Indicator/2011-global-results.aspx>.   Johnson Controls Inc.
[2] Benningfield Group. U.S. Multifamily energy efficiency potential by 2020. The Energy Foundation; 2009. <www.livingcities.org/related/downloads/?id = 8>.
[3] Bell C, Nadel S, Hayes S. On-bill financing for energy efficiency improvements. Washington, D.C.: American Council for an Energy-Efficient Economy; 2011Report E118. <http://aceee.org/research-report/e118>.

[4]   [CEE] Consortium of Energy Efficiency. Consortium for energy efficiency 2012 behavior program summary. Boston, Mass: Consortium for Energy Efficiency; 2012.

[5]   Goldman C, Stuart E, Hoffman I, Fuller M, Billingsley M. Interactions between energy efficiency programs funded under the recovery act and utility customer-funded energy efficiency programs. Lawrence Berkeley National Laboratory; 2011.

[6]   Bourland D. Incremental cost, measurable savings: enterprise green communities criteria. Enterprise green communities, <www.seattle.gov/dpd/cms/groups/pan/@pan/.../ dpdp018267.pdf>; 2010.

[7]   Deutsche Bank, Living Cities. The benefits of energy efficiency in multifamily affordable housing. New York, NY, <https://www.db.com/usa/img/DBLC_Recognizing_the_ Benefits_of_Energy_Efficiency_01_12.pdf>; 2012a.

[8]   Deutsche Bank, Living Cities. The benefits of energy efficiency in multifamily affordable housing; 2012b.

[9]   Kats G, Menkin A. Energy efficiency financing – models and strategies: pathways to scaling energy efficiency financing from $20 billion to $200 billion annually. Cap-E: American Council for an Energy Efficiency Economy; 2011. <http://www.cap-e.com/ Capital-E/Energy_Efficiency_Financing.html>.

[10]   Bryne J. Understanding sustainable energy utilities. Center for energy & environmental policy. University of Delaware; 2009.

[11]   Hayes S, Nadel S, Granda C, Hottel K. What have we learned from energy efficiency financing programs? Washington, D.C.: American Council for an Energy-Efficient Economy; 2011Report U115. <http://www.aceee.org/research-report/u115>.

[12]   Mark F, Baker J. United States building energy effficiency retrofit. New York, NY: Deutsche Bank Climate Change Advisors; 2011<http://www.dbcca.com/dbcca/EN/invest- ment-research/investment_research_2409.jsp>

[13]   Mortgage Bankers Association. Mortgage debt outstanding 2010. <www.mortgagebankers .org/files/.../Q210CMFDebtOutstanding.pdf>.

[14]   Opportunity Finance Network. CDFI market conditions second quarter 2011. Philadelphia, <http://www.opportunityfinance.netCDFI>; 2011.

[15]   Lesh, P., 2009, *"Rate Impacts and Key Design Elements of Gas an Electric Utility Decoupling: A comprehensive review."* Regulatory Assistance Project, *June.*

[16]   Hayes, S., Nadel, S. Kushler, M., York. D., 2010, *"Carrots for Utilities: Providing Financial Returns for Utility Investments in Energy Efficiency"*, ACEEE, *January.*

[17]   New Hampshire is considered to be historically red for this analysis.

[18]   World Economic Forum, "A profitable and resource efficient future: Catalysing retrofit finance and investing in commercial real estate". 2011. Industry stakeholder report. http:// www.weforum.org/reports/profitable-and-resource-efficient-future-catalysing-retrofit- finance-and-investing-commercia.

Chapter 21

The Holy Grail: Consumer Response to Energy Information

Chris King[1] and Jessica Stromback[2]

[1]*eMeter, a Siemens Business,* [2]*VaasaETT*

1 INTRODUCTION

Since the first championing of energy efficiency by Art Rosenfeld and others in the 1970s, consumers have suffered from a lack of energy usage data, data that can greatly increase the level of achievable improvement — and further contribute to zero demand growth. How much? A recent study, discussed in detail below, found such data could lead to average usage reductions of 8.7 percent among those who engaged with the programs;[1] at today's retail prices,[2] this would translate into savings for U.S. residential consumers of $15 billion per year.

Energy information feedback revolutionizes energy efficiency, creating "Intelligent Efficiency."[3] In general, energy efficiency is the receipt by an energy user of the same value or functionality through utilization of fewer kWh of electricity.[4] There are three main mechanisms for achieving this result:

- New versions of appliances or equipment can be manufactured using more energy efficient designs, often driven by standards such as California's Title 24 program.[5]
- Energy users can turn off equipment or appliances that provide no value, such as lighting in a vacant room.
- Consumers can make proactive decisions to upgrade their equipment, appliances, insulation, windows, or other items to more efficient devices.

---

1. VaasaETT, "Empower Demand," October 2011. Also see Footnote 18.
2. Energy Information Administration, U.S. Department of Energy, "Electric Power Monthly," April 2012.
3. American Council for an Energy Efficient Economy, "A Defining Framework for Intelligent Efficiency," June 2012.
4. U.S. Department of Energy at http://www.eia.gov/emeu/efficiency/definition.htm
5. California Energy Commission, "Building Energy Efficiency Standards," December 2008.

Energy Efficiency. DOI: http://dx.doi.org/10.1016/B978-0-12-397879-0.00021-9

Unlike the first point – standards that take effect naturally but over a long time period – both the second and third mechanisms require active decision-making based on information – hence the term, "intelligent energy efficiency." At the heart of such decisions is energy usage data.

Appliance and building standards have been the heart of policymakers' implementation strategy for energy efficiency. Now, information technologies – above all the Internet and smart meters – make it possible to achieve the holy grail of intelligent energy efficiency, because there is now cost-effective availability of and access to detailed and useful energy information.

Utility websites provide access to usage and cost data from bills but often go further. Some add in other data sources – e.g., weather and housing data – or comparison data for similar customers. Other utilities have online audit software that shows users, after filling out a form, how much of their usage is going to different end uses. Such additional data can be delivered through other means as well, including monthly utility bills and even mailed reports. At the other extreme to mailed reports of month-old data are in-home displays (IHDs) that show real-time consumption information sent directly from the meter to the display.

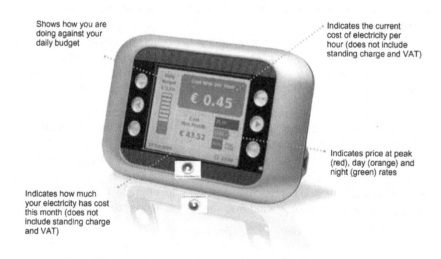

Shows how you are doing against your daily budget

Indicates the current cost of electricity per hour (does not include standing charge and VAT)

Indicates price at peak (red), day (orange) and night (green) rates

Indicates how much your electricity has cost this month (does not include standing charge and VAT)

FIGURE 21.1   Display, electricity smart metering customer behavior trial in Ireland.[6]

6. Commission for Energy Regulation, "Electricity Smart Metering Customer Behaviors Trials Findings Report, May 2011.

The most effective approach combines convenient access to the data with regulations that protect consumer privacy and security.[7] The first element of physical access is letting consumers see their data by putting usage and cost data online, where it can be accessed, normally at no cost. The second element is giving consumers their data, by enabling them to download it, import it into a spreadsheet, or authorize a third-party to receive it for various analyses. The third element is, where smart meters are installed, to enable data to be sent in real-time to local displays or devices. In all cases, up-to-date security measures are essential for consumer protection.

Utilities have conducted many pilot programs to test the effectiveness of information feedback. These programs find that consumers use less energy when provided with additional information. As described in section 5 of this chapter, the level of savings varies from a bare minimum to over 10 percent, averaging 8.7 percent.[8] Various factors influence the results. Generally, the more timely the data, the greater the efficiency gains.

Building and appliance standards have achieved impressive results for energy efficiency, first in California, then throughout the United States, and finally internationally. The addition of information feedback has significant promise for making further progress. Utilities across the United States began providing monthly data on their websites in the first half of the last decade. Those with smart meters have added interval usage data to their websites, and the trend is to add more functions, including bill comparisons and budget alerts. While utilities in Ontario have followed — or sometimes lead — the U.S. utilities closely, other countries have made less progress. There have been many experiments, especially in Europe, but far less wide-scale implementation outside North America.

Programs providing real-time data have been active in Texas and the UK, where such data is provided via specialized in-home displays to tens of thousands of customers.

As smart meters are more widely adopted and as regulators further develop data access policies for consumers, online and real-time data access are expected to grow rapidly in both availability and effectiveness.

This chapter has six sections. Section 1 is the introduction. Section 2 describes the elements of the physical infrastructure needed to deliver energy usage data to consumers most effectively. These include online access, enablement of third parties, and real-time data, when smart meters are available. Section 3 identifies the key industry standards that apply to the physical infrastructure. The focus is on access via utility websites, the interface to

---

7. "Decision Adopting Rules to Protect the Privacy and Security of the Electricity Usage Data of the Customers of Pacific Gas and Electric Company, Southern California Edison Company, and San Diego Gas & Electric Company," D.11-07-056, California Public Utilities Commission, July 29, 2011.

8. VaasaETT, *op. cit.*

deliver data to authorized third parties, and the real-time interface into the home or building. Section 4 describes essential regulatory policies to protect customer data privacy and security. Some policies are specific to energy consumption data, while others leverage existing consumer protection legislation. Section 5 presents empirical results. These data are from a variety of information feedback pilots and programs. They range from enhanced information on monthly bills to multi-functional utility websites to real-time IHDs. The chapter's conclusions are summarized in section 6.

2 ENERGY INFORMATION FEEDBACK PHYSICAL INFRASTRUCTURE

Maximizing the effectiveness of data in helping consumers achieve energy efficiency goals requires a three-part conceptual infrastructure. To begin, customers need to be able to view data already held by the utility in the way most convenient to them − online. Then, they need a means of getting the data themselves so they can use it with other software or share it with third parties that can help analyze it. Finally, if smart meters are installed, consumers should have access to real-time usage data. To minimize cost, this framework leverages the Internet.

2.1 Element One: Letting Consumers See their Cost and Usage Data

Lord Kelvin is credited with saying, "You can't manage what you don't measure." Beyond recording usage, energy measurement values need to be readily available for energy consumers to utilize the data. In today's world, "readily available" means online. Importantly, this first element of our energy information feedback infrastructure does not require the collection of any new data. It is as simple, conceptually, as taking what's already in the billing system and putting it online.

This ability to gain access to one's personal cost and usage data online is the third, and perhaps last, major step in the evolution of utility websites. The first, beginning in the late 1990s, was the initial creation of websites. These were generally what was known as "brochureware," because the web pages contained static, rarely changing information. Such sites were a source for general descriptions of utility services and perhaps posting of annual financial reports or other company-oriented information. Many small utilities still have such websites.

Some examples of functionality introduced at this stage are as follows:

- Home page
- Contact us
- About the utility

- Frequently Asked Questions, such as "How to read your meter"
- Products and services, often including tariffs
- Search
- Site map

The second evolutionary stage was to create the ability for customers to transact with the utility – using the website instead of calling up customer service representatives. At this point, energy users began to use websites to make payments, sign up for budget billing plans, or even initiate or move service. These online transactions reduce utility labor costs and increase customer satisfaction, making such websites effective and popular.

The overall concept is to provide the same services and options as "offline" customer service, but to make those services available online. Few new services have been introduced in this evolutionary stage. By now most websites of large utilities have various online transaction capabilities. These typically include the following:

- Sign up for automatic bill payment
- Start/stop/transfer service
- Online meter read submission
- Online home energy audit

The third stage of evolution is the addition of energy information feedback supporting customer energy efficiency activities. In essence, it involves making the utility website an information portal for a customer's own information. A common first feature is online presentment of customer total bill amounts – but not details such as consumption. Over time, consumers are given access to more information, including monthly usage. Where smart meters have been installed, most customers in North America have access to their interval usage data.

Some utilities are going further and adding personalized analysis. For example, California's three major investor-owned utilities – Pacific Gas & Electric, Southern California Edison, and San Diego Gas & Electric – have added bill comparisons. Customers of those utilities can click a website

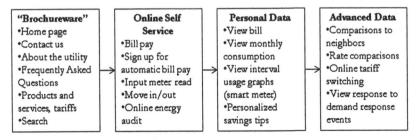

FIGURE 21.2 Evolution of utility websites from brochure-ware to interactive, personalized data viewing and analysis, including rate comparisons and tariff switching.

FIGURE 21.3   Consumer daily electricity cost detail shown on typical utility website, including projection to end of bill period. *Courtesy Westar Energy.*

button and see what their annual energy bills would total according to the different rate options available to them, generally a tiered flat rate, a time-of-use rate, and a dynamic pricing option (e.g., critical peak pricing). They can then switch tariffs online, selecting the tariff that best fits their needs.

As will be discussed in section 5, data that is more timely is more useful for managing energy consumption. The online information is only as timely as the underlying data. For example, monthly data is not available until the end of the billing month. On the other hand, for customers with smart meters, the standard practice for retrieving data to the utility data center – and which could be considered the best practice – is to retrieve meter data daily and post it within the next day or so.[9]

Next-day data provides near term feedback for energy-consuming activities conducted yesterday and still fresh in the consumer's memory. It allows

9. Daily data retrieval of smart meter data is the de facto standard in North America, the UK, Australia, and New Zealand. Most European utilities, including those in Italy and Scandinavia, typically read their smart meters only once a month.

for timely calculation of month-to-date bills and corresponding projections. It supports sending email or text message energy alerts for crossing a usage or budget threshold. And next-day data enables email or text message notification that a customer's energy price has gone up to the next price tier.[10]

Real-time data is available through the home area network (HAN) interface available in most smart meters, though delays in industry standards development have limited the number of utilities that have activated the HAN interfaces.

2.2 Element Two: Letting Consumers Download or Share their Data

The second element of our energy information feedback infrastructure is allowing consumers to download their data or authorize the sharing of their data with third-party service or application providers to use the data – at the consumer's choice and only with the consumer's authorization – to help manage consumption. Aneesh Chopra, former Chief Technology Officer of the United States, calls this "liberating the data."[11]

2.3 Element Three: Real-Time Access to Usage Data

The third element regards real-time data. Our information framework also accounts for smart meters but does not require them. The idea is that *if* there is a smart meter, then the smart meter should have a built-in communications interface to send real-time usage data to an IHD, gateway, or other device, whether in a home or in a business. Figure 21.1, above, is an example of an IHD that retrieves real-time data directly from the electricity meter and shows the results to the consumer. Similarly, *if* there is a smart meter, the emerging best practice is daily retrieval of the data and posting on the utility website for customer access within 48 hours.

3 STANDARDS

The U.S. National Institute of Standards (NIST) leads the Smart Grid Interoperability Panel (SGIP) to produce and maintain a catalog of standards. The catalog is a compendium of standards and practices considered to be relevant for the development and deployment of a robust and interoperable smart grid, including standards for exchanging energy usage information.

---

10. In many U.S. states, customers have tiered electricity prices. For these tariffs, the first block of usage each month – say 400 kWh – is at one price, while usage above that has a higher price. Some states have as many as five price tiers, though two tiers is most common.

11. Pacific Gas & Electric Company, "White House Challenge Met: PG&E's Green Button Now Live," January 2012 at http://www.pgecurrents.com/2012/01/18/white-house-challenge-met-pges-green-button-now-live/

3.1 The Role of Standards for Energy Usage Data

These standards specify how energy consumption data from meters, especially smart meters, is to be stored and shared. For example, there are standards for how data should be stored in utility databases and in electricity meters. Other standards are data models that specify how different data elements are interrelated − such as the connection between the meter, the consumer, the transformer, and the distribution circuit. These data models are used in software applications that manipulate or analyze energy usage data, including presentment of data to consumers via online websites.

The most important use of standards is in exchanging data between computer systems operated by different entities. Examples include exchanging data between one utility system (e.g., billing) and another (e.g., demand response management), between the utility and the customer (e.g., downloading energy usage data to a laptop), or between the utility and a third-party authorized by a customer (e.g., a company that provides demand response services to energy customers).

Without such standards, the exchange of data between parties requires custom development of software interfaces. This is costly in both time and money. The standards also allow competitive markets to develop. Standardized data exchange with authorized third parties allows multiple companies to develop products and services for energy customers, knowing that they will be able to obtain the energy information from the utility via the standard interface − once the interface has been adopted and implemented by the utility, of course.

The standards process and concepts can be daunting at first glance. There are literally hundreds of current and potential standards that affect smart meters and the grid. Also, standards development organizations (SDOs) operate around the globe.

There are four key points regarding standards that apply to energy usage information, as described below.

3.2 The Standards Context

The first point to consider is that standards are a journey, not a process. New standards are introduced into the market all the time. There is no "end point." Smart grid standards get adopted as they become available − and as utilities and other companies decide it is sensible to use them.

In addition, standards are an opportunity, not a requirement. Independently operating SDOs such as IEEE and IEC develop standards. Governments often recognize standards, but only rarely mandate adoption of specific ones. For example, regarding the NIST smart grid standards catalog, George Arnold, NIST's National Coordinator for Smart Grid Interoperability, explains: "Entries in the Catalog of Standards constitute the first items in what will be a useful toolkit for anyone involved in the Smart Grid — whether they are utilities that generate and distribute power, companies developing new electronic devices, or consumers who buy and use them."[12]

In another good example of standards best practice, the Public Utility Commissions of California and Texas have ordered utilities in those states to use an open standard for the HAN interface on their smart meters. In both cases, they declined to specify which standard.

3.3 Interoperability

Interoperability is the most important goal of standards. Interoperability means devices able to send data to one another — for example, computer wifi networks. Another form of interoperability is when different entities send data to one another — for example, a utility sending data to its customers via a website. Interoperability is what made the Internet possible, not to mention huge leaps and bounds in the functionality of almost every kind of electronics.

For the smart grid, interoperability spurs two important kinds of competition that ultimately benefit consumers, utilities, and the smart grid ecosystem:

- Competition in available products. When devices from different manufacturers and vendors can talk to devices from other manufacturers and vendors, those manufacturers and vendors must compete for utility or consumer purchasing dollars.
- Competition among service providers. Interoperability allows data to be exchanged between multiple parties so utilities and consumers can use data services of different parties.

In both cases, interfaces use a published, open interface — usually without paying royalties to use the standard.

3.4 Key Standards for Energy Information for Energy Users

Of the hundreds of smart grid standards currently in development and use, two are crucial for the smart grid. These concern two key interfaces (see Figure 21.4):

- Open Home Area Network (OpenHAN), allowing in-premise energy devices to communicate with each other.

---

12. http://www.nist.gov/smartgrid/sgip-072611.cfm

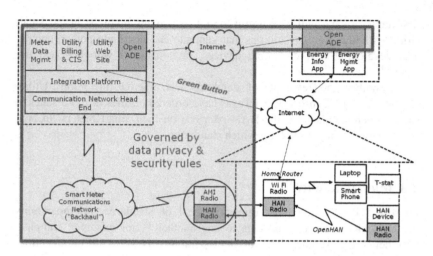

FIGURE 21.4 OpenADE, including Green Button, is a key standard for delivering data to energy users or their service providers. OpenHAN enables the provision of real-time meter data to in-premise devices, including homes and businesses.

- Open Automated Data Exchange (OpenADE), allowing utilities to exchange information with customers or other parties authorized by customers to receive data.

OpenHAN (which could be ZigBee, wifi, or something else) enables low-cost, high-function smart appliances, lighting systems, and smart thermostats − such as the Nest thermostat developed by the iPod's creator. OpenHAN applies equally to businesses.

OpenADE allows utilities to share with their own customers or with third-party service providers to help energy users understand and manage their energy usage. Importantly, parts of OpenADE (other standards) will make it easier for utilities to manage smart meter data internally − as well as exchange it between internal systems.

OpenADE is another name for the Energy Services Provider Interface (ESPI) standard adopted and administered by the North American Energy Standards Board (NAESB). ESPI is a standardized process and interface for the exchange of a retail customer's energy usage information between their designated data custodian (i.e., electric utility) and an authorized third-party service provider. It includes a consistent method for the authorization of third-party access to retail consumer's usage information and a standardized interface for the exchange of that information. The goal of ESPI is to support the development of innovative products that will allow consumers to better understand their energy usage and to make more economical decisions about their usage.

The ESPI standard provides model business practices, use cases, models and technical details that describe the mechanisms by which the orchestrated exchange of energy usage information may be enabled. NAESB works closely with NIST and SGIP.

3.4 Green Button Initiative

Green Button is a subset of the ESPI standard. It is based on the idea that electricity customers should be able to securely download their own easy-to-understand energy usage information from their utility or electricity supplier. Armed with this information, consumers can use a growing array of new web and smartphone tools to make more informed energy decisions, optimize the size and cost-effectiveness of solar panels for their home, or verify that energy-efficiency retrofit investments are performing as promised. Consumers are also seeing innovative apps that allow individuals to compete against Facebook friends to save energy and lower their carbon emissions.

Green Button is an industry-led effort that responds to a White House call-to-action: provide electricity customers with easy access to their energy usage data in a consumer-friendly standard format via a "Green Button" on electric utilities' websites.

Voluntary adoption of this consensus standard by utilities across the United States is allowing software developers and other entrepreneurs to leverage a sufficiently large market to support the creation of innovative applications that can help consumers make the most of their energy usage information. Initially launched in January 2012, by May 2012 utilities committed to provide Green Button capability to over 30 million energy customers by the end of 2012.[13] At the same time, application developers have already created over 50 different applications.[14]

4 CONSUMER DATA PRIVACY AND PROTECTION

Security of customer information in wireless applications, and how personal data characteristics (such as customer usage information) can be protected are issues often mentioned in discussions of the Smart Grid and cybersecurity. Encryption of data (with limited decryption for data checking), and aggregation of data at high levels to mask individual usership have been mentioned as ways to protect the identity of individual customers. Customer-specific data

---

13. Executive Office of the U.S. President, "New Industry Commitments to Give Over 3 Million New Households and Businesses Tools to Shrink Their Energy Bills," May 2012, at http://www.whitehouse.gov/sites/default/files/microsites/ostp/green_button_release_2012_05_02.pdf
14. GreenTech Grid, "Green Button Apps: How Innovative Are They?," May 2012, at http://www.greentechmedia.com/articles/read/green-button-apps-55-and-counting/

FIGURE 21.5 Leafully is an example of a Green Button application and the winner of the U.S. Department of Energy Apps for Energy competition.

stored in HANs, or customer-specific data communicated between the HAN and distribution utility (or load aggregator) must be secure to protect the privacy of information.[15]

The National Regulatory Research Institute recommended in a recent report[16] that public utility commissions should define the information that utilities will collect, determining with whom and for what purpose it should be shared, and assess the need for protecting the data. While allowing states to develop their own policies may be accomplished more quickly, the report also advocates a national approach to smart grid privacy issues.

4.1 California Public Utilities Commission Regulations

In 2011, the California Public Utilities Commission (CPUC) adopted the world's first comprehensive set of rules to ensure that consumers can access the detailed energy usage data gathered by their smart meter — while also protecting the privacy and security of their data. Figure 21.4 illustrates the

15. Congressional Research Service: The Smart Grid and Cybersecurity — Regulatory Policies and Issues, Richard J Campbell June 15, 2011.

16. NRRI, "Smart Grid Data: Must There Be Conflict Between Energy Management and Consumer Privacy?" December 2010.

entities and data exchange interfaces that are covered by the CPUC rules, essentially regulated utilities and third parties authorized by customers to receive their data.

The CPUC clarified that its regulatory mandate includes exercising jurisdiction over regulated utilities with respect to energy data, privacy, and security. Furthermore, to protect consumer privacy and data security, the CPUC is exercising jurisdiction over third parties who receive data (via the utility's data center) in the course of providing services to utilities, or when authorized by consumers. However, the CPUC is not exercising jurisdiction over third parties who receive energy usage data directly from a device installed at a residence or business that receives data via the HAN interface.

In this decision the CPUC relied mainly on existing privacy law, using the Fair Information Practice Principles that the U.S. Department of Homeland Security developed as its privacy framework. To clarify the application of these principles, the CPUC decision includes an appendix with details of its privacy rules.

Here is a condensed summary of the FIP principles as utilized by the CPUC:

- Transparency: Utilities and covered third parties must provide customers with meaningful, clear, accurate, specific, and comprehensive notice regarding the accessing, collection, storage, use, and disclosure of energy usage information.
- Individual Participation: Utilities and covered third parties must provide to customers convenient and secure access to their energy information in an easily readable format. They must also provide customers with convenient mechanisms for granting and revoking authorization for uses of energy information other than using it solely for the delivery of purchased electricity.
- Purpose Specification: The notice provided to energy users must include an explicit description of: 1) each category of information collected, used, stored or disclosed by and the specific purposes for which it will be collected, stored, used, or disclosed; 2) each category of information that is disclosed to third parties, and the purposes for which it is disclosed; and 3) the identities of those third parties to whom data is disclosed.
- Data Minimization: Covered entities must collect, store, use, and disclose only as much covered information as is necessary to deliver electricity or for a specific other purpose authorized by the customer. In addition, the entities must maintain data only for as long as necessary or authorized by the Commission and may not disclose to an authorized third-party more data than is necessary to fulfill authorized purposes.
- Use Limitation: Utilities and third parties must use the data only for delivering power and only those other purposes specifically authorized by customers.

- Data Quality and Integrity: Covered entities must ensure that energy data they collect, store, use, and disclose is reasonably accurate and complete or otherwise compliant with applicable rules and tariffs regarding the quality of energy usage data.
- Security: Covered entities must implement reasonable administrative, technical, and physical safeguards to protect covered information from unauthorized access, destruction, use, modification, or disclosure. Any security breaches involving 1,000 or more customers must be reported immediately, with annual reports of all security breaches.
- Accountability and Auditing: Utilities and third parties are held accountable for complying with the CPUC's requirements and must make available upon request or audit the privacy notices that they provide to customers, their internal privacy and data security policies, and the categories and identities of agents, contractors, and other third parties to which they disclose energy data. They must also provide customers with a process for access to their data, correction of inaccurate data, and addressing customer complaints about data covered by the rules.

With this ruling, the policy for California is somewhat settled. Implementation details need to be worked out for each user scenario such as providing data to third parties.

4.2 Should Other Jurisdictions Follow California's Example?

The California example is a good one as a base. As a general rule, it adopted existing data privacy and security laws and regulations put in place to protect other consumer data, such as financial data. The CPUC determined that there are no fundamental differences between private energy data and other sensitive data, such as telephone records or bank transactions. Where energy data is breaking new ground – in particular, the transfer of energy data to third parties authorized by consumers – the CPUC put in place protections adopted specifically for the electric and gas utility industry. The implementation on the details of third-party access is in progress in a current CPUC proceeding, with a ruling expected in 2013.

Each jurisdiction – state, province, country – will develop its own rules. For example, the U.K. is establishing policies as the deployment of smart meters commence. Another instance is the Netherlands, where data policy is ensconced in legislation that was needed to address data policy issues that arose following earlier law approving a smart meter rollout. Germany's regulators have issued a data protection policy finalized in late 2012.

4.3 Consumer Data Privacy Concerns

While all data can be considered sensitive in some way or the other; energy consumption data is somewhat less sensitive than other consumer data already widely shared over the web or other public/private networks. Utilities have handled detailed consumption data for large industrial customers with smart meters going back more than two decades. This data is quite sensitive, because it can reveal operating schedules, relative efficiency of production facilities, and other information that can affect a company's competitiveness. All utilities have such data and most have made it available to their large customers on line without any known issues. Utilities have used the data for their own operations, as well as shared the data with third parties with the customer's permission.

For mass-market customers, the availability of detailed usage data is new. However, the fundamental principles remain the same. There are concerns about data privacy, because, for example, the data could reveal occupancy patterns for a premise — though smart meters cannot reveal what appliance a consumer is using, contrary to some statements in the press. Occupancy can more easily be determined by observation for a person intent on learning that information. In general, the privacy considerations for consumption data are less serious than for other consumer data, particularly banking and credit card data.

Nevertheless, the energy data privacy concerns are important and should be respected, with concomitant data protections put in place — as we have seen in California.

5 RESULTS OF CUSTOMER PILOTS AND PROGRAMS

A recent meta analysis of 100 pilot programs broken down into 460 sub-pilots — with 74 focused on energy information feedback — provided extensive detail on the level and scope of consumer response to detailed energy information. The display of almost real-time energy consumption data on in-home devices (IHDs) led on average to an 8.7 percent reduction in energy consumption.[17] Lower but still significant reductions of 5 to 6 percent on average were achieved through enhanced, more informative bills and access to usage data on websites.[18]

---

17. VaasaETT, *op. cit.*
18. It is important to note that during a national or regional program rollout not all consumers will be willing to engage with feedback and therefore the total energy reductions will be dependent on the level of regional consumer uptake and engagement with the program.

5.1 Research Methodology

The aim of the research was to discover the potential and limitations of a range of feedback and dynamic pricing programs enabled through smart grid technologies, including smart meters. The findings and conclusions based on a large pool of pilots are designed to gauge repeated results and surrounding requirements for successful implementation by other utilities.

Pilot organizers usually form sub-groups within their pool of participants and try different solutions with different groups. A typical case would be to measure the response of participants when given an IHD and when given detailed informative bills. We call "samples" these sub-groups within a pilot. Impacts of trials on individual samples were not calculated. Instead they were calculated and reported by the pilot organizers in their final reports, academic papers and presentations. These were collected for impacts reported with statistically significant results at a 90 percent confidence level and above. This review took the mean of the individual impacts in order to understand what the key determinants of successful pilots were. The mean impacts were calculated by averaging the individual impacts on each sample with each sample equally weighted.

5.2 Research Samples and Data

The research involved collecting and comparing about 100 pilots. Typically, organizers divide participants in a pilot into sub-groups in order to test different solutions, for instance different feedback types, different dynamic pricing schemes, a group with home automation and one without, etc. Hence, the pilots were broken down into 460 samples. The samples were then analyzed according to 22 different variables selected to gauge internal structural pilot variables influencing success as well as outside market factors that might also affect a pilot outcome. In all, over 450,000 residential consumers were involved in the reviewed pilots. Feedback pilots are designed to help participants reduce their overall energy consumption, lowering distribution and supply costs.

A total of 74 information feedback trials were analyzed during this research. The sample comprised 290,000 residential households from five regions; Australia (3 samples), Canada (12 samples), Europe (35 samples), Japan (3 samples), and the United States (21 samples). The majority of the pilots from Europe were conducted in Great Britain. Over 60 percent of the pilots took place within the last ten years and almost half after 2005.

5.3 Feedback Methods

The pilots analyzed in the research tested three main types of energy information feedback.

IHDs are displays that hang on the wall or sit on a counter and provide near real-time information about household electricity consumption. They also provide a variety of other data. For example, the display provided in the "Electricity Smart Metering Customer Behavior Trials" (see Figure 21.1) allows people to set daily budgets for how much they want to spend, informs them of their success, what the current price of electricity is, and provides information on how much they have spent so far this month. The home screen for the dynamic display unit is the key screen that the customer sees when the device is switched on, while further information can be gained if desired by navigating to other screens.

Importantly, there is significant discussion about the value and role of specialized IHDs. Some industry researchers believe real-time energy information will show up on smart phones:

...some smart grid technology vendors are getting out of the home display/hardware business because they anticipate that people will be able to use apps on smart phones (as well as controlling in-home devices through Internet-ready televisions) to program in-home devices, set trigger prices and set-point temperatures, and change those settings remotely if necessary. Home energy apps will be sufficiently powerful and flexible, and Internet connectivity will be sufficiently ubiquitous, that they are likely to be attractive to homeowners for home energy management and automation.[19]

Given that people are relying more and more on their smart phones, this vision appears likely to materialize.

Websites offer an alternative way to provide the consumer with information about their electricity consumption. As noted earlier in this chapter, websites are complementary to real-time displays by providing other information. The websites considered in the analysis rely on smart meters to collect the necessary consumption data and therefore the granularity of data provided to consumers depends largely on how often the meters are read or how often the information is transferred from the meter to the utility (or retailer). The standard frequency for the pilots was to have the information — typically hourly or quarter-hourly intervals — sent in a packet from the meter to the utility once a day.

The third category is called "informative billing." Informative billing is an example of indirect feedback. Most residential consumers in Europe now receive estimated bills that are adjusted for the time of year and the household's average consumption. They therefore do not accurately reflect the actual usage for a given month. The difference between the estimated average consumption and the actual usage is made up at the end of the year or when a resident changes electricity supplier.

---

19. Lynne Kiesling, "Dematerialization, smart phones, and smart grid," June 2012, at http:// knowledgeproblem.com/

Informative billing follows the typical U.S. practice by issuing a monthly invoice for the actual consumption and provides either historical information comparing what the customer used this month to last month or to last year during the same period. The bill may also provide information on how much the household consumed in comparison to other dwellings of the same description.

5.4 Feedback Content

For the pilots included in the analysis, information about consumption presented in the different feedback programs typically had one or several of the following content types:

- Peer Comparison: Comparison of household energy consumption levels between participants and similar-sized households. It enables participants to see if they use more or less electricity than their peers. (See chapter by Laskey et al for further discussion of such comparisons.)
- Price of Electricity: The current price of electricity per kWh.
- Historical Comparison: The household's current electricity consumption levels in comparison to pre-pilot consumption levels. Participants can see if they reduced or increased their consumption compared to the same period last year.
- Disaggregation of Consumption: The household's electricity consumption is broken down as per household electrical appliances based on the consumer inputting information about the appliances in the home. The depth and degree of the breakdown can vary but in most cases the consumption of the oven, the fridge, the TV, and the lighting are estimated. It enables participants to see how much electricity individual appliances use and act upon it (and perhaps buy more energy efficient ones).
- Bill-to-date Consumption: The current up-to-date consumption level of the household in kWh.
- Bill-to-date Cost: The current up-to-date bill in currency units.
- Savings Compared to Previous Periods: Compares the energy savings of households to previous periods.
- Environment (CO_2 emissions): The amount of CO_2 emissions caused by the household due to electricity consumption.

5.5 Research Results

Figure 21.6 shows the energy conservation results for different types of feedback. Customers with an IHD had the highest response, with average energy savings of 8.7 percent. The remaining channels for feedback; webpage, and informative billing; produced almost equal consumption reduction levels and in some cases they were used in combination. A key benefit of IHD seems

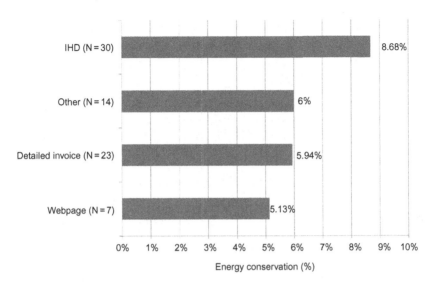

FIGURE 21.6   Energy conservation associated with different types of feedback, based on meta analysis of 74 pilot programs.[20]

to be the "almost real-time" aspect which enables participants to link their actions to their energy usage practically in real-time.

Program results are reported as a percentage reduction of kWh consumption by the household.

Program success of pilots/trials is directly dependent on consumer involvement, and research findings indicate that "more is more" at every stage of the piloting and roll out process. For example, programs using consumer segmentation to create directed marketing campaigns for a particular consumer groups, increase consumer uptake and results. In the program structure, feedback and dynamic pricing used together tend to achieve better long-term overall results than either program type alone. Education improves dynamic pricing and informative billing programs. Multiple types of information on a display or a bill (current consumption, price, historical consumption, etc.) tend to achieve higher results than a display or a bill with only one message.

5.6 Extrapolating the Results of the Meta Analysis

The findings in Figure 21.6 reflect the average reductions made by the participants in the pilots. What they do not quantify is the potential reductions if the feedback were introduced into an entire population, something that has

---

20. VaasaETT, *op. cit.*

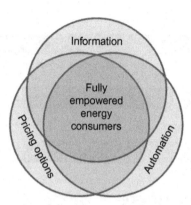

FIGURE 21.7   Energy feedback information is the first among equals in the consumer empowerment triad that allows maximization of energy saving and management benefits.[21]

happened with website feedback in California and Texas. When a utility launches an offering into a market, the level of energy consumption reductions made will depend heavily on the number of people who become engaged with the program − in other words, on the uptake rate. For example, a 10 percent electricity consumption reduction by 1 percent of the population will lead to a 0.1 percent total reduction, whereas a 10 percent electricity consumption reduction by 100 percent of the population will lead to a 10 percent total reduction. Potential program uptake can be estimated by the percentage of people who are willing to take part in the pilot. If this number is high, say around 30 percent, as it was the case in the CER pilot in Ireland, a national rollout of a similar program is also likely to gain good customer support.[22]

The fact that uptake plays such a definitive role in program success highlights the importance of marketing and consumer education. It also can mean that low technology programs, such as informative bills which reach everyone, can have similar or better results to high technology solutions, such as home automation or displays, if these interest only a small percentage of the population. In this case, a range of appropriate feedback choices for differing consumer groups is optimal.

---

21. VaasaETT, *op. cit.*

22. During rollout, success can be utility dependent. Some utilities have simply been better than others at communicating and marketing pricing programs to their customers and will succeed better with the same offering in the same market than their competitors. Salt River Program, for example, has achieved a participation rate of over a third in its time-of-use rate program; see Association for Demand Response and Smart Grid, "The Persistence of Consumer Choice," June 2012.

6 CONCLUSIONS

Energy users need three things to maximize their ability to save energy and money on their electricity bills: the consumer empowerment triad of information, pricing options, and automation (see Figure 21.7). Information — to provide understanding — is first among these equals and the starting point for intelligent efficiency. Time-based pricing provides additional financial incentive to lower peaks, and automation allows for "set-and-forget" energy management in our busy world. But information remains the key to all of these.

As we look forward, it is essential to keep in mind that while energy information is the "what," the "how" is just as important. The analysis of the pilots showed that, if customers do not find a program interesting, accessible and attractive, it will fail. This is equally true for every type of smart grid program. Customer segmentation offers utilities and technology providers the opportunity to study which customer groups are reacting best and how programs could be improved during rollout. This knowledge can then be used to create messages and material that have a direct and central impact on the number of consumers who successfully engage with a program in a given market.

Education should be included within all consumer energy initiatives. Providing detailed usage feedback supercharges dynamic pricing programs, especially those involving automation, as it helps to decrease total consumption rather than only peak consumption. By itself, usage feedback offers the opportunity to achieve billions of dollars in energy savings every year; leveraging information with dynamic pricing and automation could triple those savings, according to McKinsey and Co. — potentially to $59 billion per year in the United States.[23]

---

23. "U.S. smart grid value at stake: The $130 billion question," Summer 2010.

Chapter 22

Trading in Energy Efficiency – A Market-Based Solution to Market Failure, or Just Yet Another Market Failure?

Iain Macgill[1], Stephen Healy[2] and Rob Passey[1]

[1]*School of Electrical Engineering and Telecommunications and Centre for Energy and Environmental Markets*, [2]*School of Humanities and Centre for Energy and Environmental Markets, University of New South Wales, Sydney, Australia*

1 INTRODUCTION

1.1 The Challenge of Energy Efficiency Policy

1.1.1 A Role for Energy Efficiency

The opportunity for, and challenges with, improving end-use energy efficiency to help meet our growing energy challenges have been widely noted.

Energy efficiency is, however, only one of a range of possible "means" to the "end" of delivering desired end-use energy services most effectively. These desired services are themselves difficult to define as they not only cover "needs" but also "desires," and these vary within and across societies and change over time. Furthermore, satisfaction of these energy services sits within diverse and somewhat conflicting objectives including the affordability of an essential public good, energy security and, increasingly, environmental concerns. These objectives and constraints aren't fully or transparently represented in existing energy markets that suffer from a wide, some would argue near complete, range of market failures.

Energy efficiency itself is hard to define in a meaningful way. The technical concept of energy efficiency can be described as "the relative thrift or extravagance with which energy inputs are used to provide goods or services. Increases in energy efficiency take place when either energy inputs are reduced for a given level of service or there are increased or enhanced services for a given amount of energy inputs" [28]. However, at a societal level,

Energy Efficiency. DOI: http://dx.doi.org/10.1016/B978-0-12-397879-0.00022-0

the primary objective of energy-related decision-making should be to maximize the broader societal benefits delivered through a carefully chosen mix of energy services and the means by which they are best delivered. The "bottom-up" technical concept of energy efficiency can be difficult to separate from, and reconcile with, this broader energy services perspective. Top-down indicators such as energy intensity, on the other hand, conflate a wide range of factors and so can be very difficult to interpret, limiting their usefulness as a tool for guiding decision-making.

Despite these challenges, improvements to end-use energy efficiency are almost certainly one of our best options to address our energy accessibility, affordability, security, and environmental challenges, and the benefits are acknowledged by virtually all governments (e.g., [1]).

1.1.2 The Need for Policy

Despite the evident opportunity and need, there is a clear need for policy intervention to promote energy efficiency. Such intervention is often framed in terms of market failure – i.e., policy efforts may be appropriate when the market does not provide economically efficient outcomes for society (often with the proviso that such outcomes wouldn't be made worse by "government failure" when intervening). The energy industry certainly features every possible form of market failure including:

- Monopolies and oligopolies, due to a generally concentrated supply-side, and typically monopoly networks;
- Public goods, given energy's role as an essential public good and its contribution to economic welfare which means that private actors may not receive all the benefits of their actions, and so be less likely to take them;
- Incomplete markets, as energy infrastructure requires high levels of coordination;
- Information failures, including information asymmetry between government and industry as well as generally poorly informed energy users;
- The "business cycle," a particular issue given the capital intensive, long-lived investments for energy infrastructure; and last but not least,
- Externalities, including climate change, energy security, and social impacts.

In such a framework, impediments to energy efficiency are typically cast in more general terms of energy market failure, including:

- Many of its benefits are market positive externalities – i.e., their environmental and social "value" are public goods rather than private benefits captured by individual market participants, and;
- There is clear evidence of widespread market failure in demand-side decision-making as energy users fail to undertake even highly cost-effective

energy efficiency options – options resulting in reduced energy costs that more than cover the additional investment that may be required.

The challenge of developing policies and measures that correct for market externalities has received a great deal of attention. One particular difficulty is in appropriately "valuing" such externalities so that these costs and benefits can be "introduced" into energy markets.

The greater challenge, however, appears to be in solving existing market failures in decision-making. The reasons for such energy market failures are complex, however, a poor understanding of energy efficiency options and insufficient attention to its importance by key decision makers are certainly major factors. For many consumers of energy services, the low cost of energy and effort required to contemplate energy efficiency options means decisions are often driven by other priorities. Even where decision makers have knowledge and motivation, there are still impediments to them actually being able to take action.[1]

This broader decision context is changing. In particular, there are growing global efforts to restructure energy, particularly electricity, industries away from vertically integrated monopoly structures towards greater market-based competition in many jurisdictions. This is changing, in part, the context of energy efficiency policy development for the sector. In particular, restructuring is changing some of the key decision-making responsibilities of industry participants.

It might be expected that greater market-based, competitively driven, decision-making in the electricity sector would help address some of the existing market failures in delivering energy efficiency. Experience in the electricity industry, however, has been mixed to date. While some impediments can be overcome through restructuring, others remain, new impediments can appear, and falling electricity prices for many consumers in restructured markets have reduced the value of improving energy efficiency for them.[2]

A key issue is that the efficiency of a competitive industry model depends critically on informed decision-making by consumers. However, the

---

1. See, e.g., the Energy Savings Trust [2] which states "There is a broad consensus that the key barrier to energy efficiency is related to individuals' knowledge, motivation and ability to optimize their energy use (p. 5)."
2. For example, the EC SAVE programme [3] found that "Although there are some economic incentives inherent in the market system for energy companies to engage in end-use energy efficiency, the incentives are too weak for consistently increasing such activities to levels motivated by the potential for energy efficiency and the broader energy and climate policy objectives... In those Member States, which have combined the implementation of the EU Internal Markets for electricity and gas with a supportive policy framework, energy efficiency programmes by energy companies are continuing or even expanding in volume and scope. In Member States without such a policy, such activities have gradually reduced with the introduction of retail competition, and are carried on only by a smaller number of more innovative companies."

complexity of electricity markets makes this particularly difficult to achieve, a fact that electricity industry restructuring has generally ignored.

Note that some argue the apparent failure to undertake cost-effective energy efficiency actions doesn't reflect market failure, but is instead because of factors including different perceived quality of energy services (e.g., the "harsh" light of high efficiency lighting systems), risks and uncertainties associated with novel technologies, the heterogeneity of users, relatively low returns as a proportion of expenditure and high discount rates applied to such actions [4]. While these factors undoubtedly contribute to the suboptimal uptake of energy efficiency, as we discuss, it is unlikely they fully account for this.

One key question, that in our view has not yet received sufficient attention, is whether the "market failure" model itself is a sufficiently complete and appropriate basis for undertaking energy efficiency policy. For example, Kay [5] argues that seeing government intervention and specifically policy-making purely through the framework of market failure risks leading policy-makers to an impoverished view of politics, democracy and collective decision-making. One thing that is very clear is that energy efficiency related decision-making involves many decision makers — including end-users of course, yet also infrastructure developers, equipment manufacturers and suppliers, service providers, installers, owners, and managers — and many of these do not see themselves as primarily energy market actors. Another clear limitation of the market failure model is its disinterest in equity outcomes — a significant omission in the context of energy-related decision-making, which has often had significant equity implications and motivations.

1.2 Energy Efficiency Policy Options

Despite these concerns, energy efficiency policy efforts would seem to be increasingly designed and implemented in the context of addressing market failures. Given the broad range of such failures and the many decision makers involved, there are many reasons to believe that no single policy instrument will suffice to drive optimal levels of energy efficiency across the economy [1]. We are certainly seeing many diverse international, national, regional, and local policy measures being undertaken worldwide targeting different aspects of energy efficiency — desired energy services, end-use equipment, and infrastructure — as well as the many and varied decision makers involved.

Policy measures intended to promote energy efficiency (amongst possibly a number of objectives) can be broadly categorized into a number of different categories (e.g., see Figure 22.1, [6]):

- Suasive approaches revolving around the provision of information and support to guide decision-making such as guidelines, codes of practice, and star rating schemes;

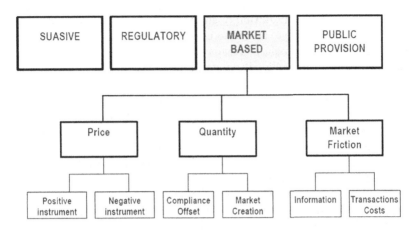

FIGURE 22.1   One possible formulation of energy efficiency policy options [6].

- Regulatory approaches generally revolving around standards and licensing such as minimum energy performance standards (MEPS) and building codes, with penalties for non-compliance;
- Market mechanisms including environmental taxes, tax credits and subsidies, and now energy efficiency certificate trading that change the effective "price" seen by decision makers for different energy options; and
- Public provision that can be used where provision of "public goods" is difficult or uneconomic to manage through the private sector.

Determining the optimal mix of such policy measures is a great challenge, and the subject of considerable ongoing work. The context within which such policy resides is also changing. Of particular note is the relatively recent interest and growth in the use of market-based mechanisms and, in particular, the development of designer markets for "energy efficiency" itself.

Such approaches, going under names including white certificates, energy efficiency portfolio standards, energy savings, energy efficiency obligations, and energy efficiency certificate trading (EECT) are receiving growing policy attention around the world including in Europe, the United States, and Australia. They are often portrayed as a market-based solution to a "market failure."

While details vary significantly across the schemes implemented to date, they all involve the measurement and commodification of some types of energy efficiency activities, the creation of demand for these savings (often through legislated targets), and a range of flexibility mechanisms such as trading of "energy efficiency" certificates between liable parties and energy efficiency providers. The flexibility and potential scope of such schemes is very large — for example, an end to energy consumption growth can effectively become a direct policy instrument objective, rather than just a possible outcome of other policy efforts with different aims, such as improved social outcomes, increased economic growth, or reduced greenhouse gas emissions.

EECT schemes, as we will now refer to this broad range of policy efforts,[3] are now operating in a number of countries around the world including the UK, Italy, France, Denmark, Australia, and Brazil. Schemes are being developed for Poland and Ireland, while the Netherlands, Portugal, Romania and Bulgaria are also interested in this policy instrument [7].

The benefits of such approaches are argued to be many. Such schemes are compatible with restructured competitive retail markets and focus cash flow through specialized participants who can assist energy users to undertake particular types of energy efficiency actions. They can also spread energy efficiency costs across a range of energy consumers and harness energy service companies (ESCOs) to drive energy efficiency in areas that are less responsive to price signals (e.g., because of split incentives and upfront costs with long payback times), such as the residential sector. In theory, they also allow the market to identify the cheapest way to deliver energy savings rather than relying on fallible government policymakers.

1.3 Structure of this Chapter

The key question addressed in this chapter is whether such schemes do, indeed, represent an effective market-based solution to the evident market failures in energy efficiency deployment. Or do they, instead, merely introduce new market failures? Also, is the "market failure" model itself a hindrance to effective action on energy efficiency due to the broader societal context that must be addressed in order to progress energy efficiency effectively?

The chapter is structured as follows. Section 2 presents a conceptual framework for trading energy efficiency, describes how some approaches might be assessed and considers their potential strengths and weaknesses by comparison with other approaches. In section 3 we describe some of the key experience to date with such schemes in jurisdictions including Europe, the United States, and Australia. Possible lessons for policymakers are then presented in the final section, which includes discussion on scheme design but also, more generally, on the governance challenges of such schemes and, most importantly, the need to move beyond framing energy efficiency as a

---

3. Not withstanding the fact that some schemes don't involve the trading of certificates (e.g., Denmark and the South Australian Residential Energy Efficiency Scheme), a considerable number of the schemes do involve trading of some form.

market failure problem and to address the broader challenges of engaging energy users.

2 A CONCEPTUAL FRAMEWORK FOR TRADING ENERGY EFFICIENCY

2.1 Market-Based Policy Options for Energy Efficiency

Market-based instruments are intended to encourage improved energy efficiency through market signals to key decision makers. Such price signals might be introduced into the energy markets within which energy efficiency decisions are taken via increases to the cost of energy itself through policies such as emissions trading schemes or carbon taxes. Such price increases might be expected to motivate energy users to choose more energy efficient equipment to reduce energy bills. Alternatively they might be placed on energy efficiency more directly such as, for example, through subsidies or tax rebates on energy efficient equipment. The focus here is on a fourth option, which is the use of "quantity"-based instruments that "create" a market for energy efficiency directly.

Market-based approaches in general have many claimed benefits over other approaches. There are clear limits to what suasive approaches can achieve alone and limits to voluntary restraint, while public provision also has limits given the important role of the private sector in most economic activity. Regulatory approaches can promote inefficiency and inhibit innovation when imposing uniform requirement on decision makers who may have very different capabilities, costs and benefits. There are also reasons to be skeptical about the ability of government to effectively and efficiently regulate decision-making directly.

By contrast, market-based approaches are intended to *encourage those who can most cost-effectively improve outcomes to do so*. They may also assist in avoiding perverse interactions between different policy measures as price "signals" just stack up.

Designer markets for energy efficiency may also take advantage of existing competitive pressures on participants, and offer considerable flexibility in how market participants respond. They offer considerable design flexibility for policymakers − both a strength and potential weakness as highlighted below. In theory, regulators can "set and forget" the desired energy efficiency outcome and then transfer the actual decision-making and risks to "better" informed parties.

However, in reality, there is a trade-off between achieving more accurate price signals and operational simplicity. Transue and Felder [30] note that "At the design level, rebate and white certificate loosely represent opposing dilemmas: the imperfect government and the imperfect market. Each approach has distinct advantages and disadvantages, often trading one set of costs and inefficiencies for another... Although rebate programs are easy for customers to understand and pursue, in practice they risk inefficiency in pricing as their administrators often lack the feedback and authority necessary to rapidly adjust incentive levels during program implementation... White certificate programs contain the feedback loops, flexibility, and design options missing from rebate approaches but regrettably sacrifice simplicity... [and] require the development and establishment of new processes and institutions that depend on all parties having full and complete information about market supply, demand, and price."

Thus, EECT shares many of the usual policy challenges of any regulatory approach while potentially adding new ones. These are still relatively novel mechanisms so mistakes will be made and learning is likely required. We have more, hard earned, experience with many other approaches. There is the inevitable complexity in attempting to match a commercial market with physical actions that reduce energy consumption. The loss of government control over decision-making might see the market undertake actions with unexpected and adverse impacts with other policy objectives. Most importantly, *these are designer markets: the greatest competitive advantage for participants may lie in gaming the market rules, and especially the rule design process, rather than through innovation in competing within the actual market itself.*

2.2 EECT Framework

EECT schemes have four main attributes (Figure 22.2):

- Energy efficiency certificates (EECs) representing a measured and verified unit of energy savings from energy efficiency (e.g., 1 saved MWh of electricity) undertaken by some party;
- Parties that are able to undertake energy efficiency actions that can be measured and verified in order to create certificates;
- A government-directed legal obligation on some group of parties that they regularly acquit some number of these certificates as part of their societal obligations (voluntary initiatives marketing energy efficiency "benefits" to concerned consumers are also possible); and
- Trading so that parties obliged to acquit certificates can choose to buy certificates from other parties as an alternative to undertaking their own energy savings (liable parties may also choose to engage third parties to undertake energy efficiency activities and thereby create certificates on their behalf).

Within this framework, however, lies an enormous number of design choices that will determine the effectiveness of the scheme, yet also involve significant compromises and uncertainties.

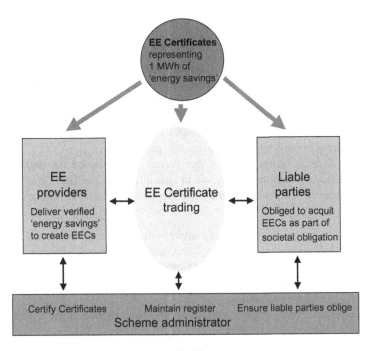

FIGURE 22.2   Schematic framework for EECT Schemes.

2.3 Design and Assessment Criteria for Policy

Frameworks for the design and assessment of policy efforts often include their:

- "Effectiveness" in actually driving greater energy efficiency;
- "Efficiency" in doing this at reasonable cost and effort compared to both the benefits of meeting policy objectives and other possible energy efficiency measures; and
- "Equity" implications in terms of how the costs and benefits of the policy measure are distributed across all relevant stakeholders.

Effectiveness lies at the heart of policymaking — is the measure likely to deliver the policy objectives? At the same time, efficiency in achieving objectives is also vital. Two measures of efficiency are relevant in the context of EECT [8] — static efficiency or the cost effectiveness of the scheme and dynamic efficiency that refers to organizational and technical change and progress that leads to longer-term market transformation. The equity implications are also vital. Beyond the reduced likelihood of the policy being implemented if it is seen to be unfair to particular key stakeholders [9], there are fundamental questions of fairness and affordability in energy-related decision-making.

2.4 EECT Design Choices and Challenges

2.4.1 Measuring and Certifying Energy Efficiency

EECT schemes are by necessity schemes where energy "savings" are created by specified activities. The scope of such activities is a design choice with a broader scope including potentially more low cost options for improving energy efficiency. However, as discussed below, energy efficiency is inherently counter-factual, creating challenges in measurement and certification, and a wider scheme scope exacerbates this problem.

EECT is built on the premise that parties are rewarded for undertaking energy efficiency actions that result in measured and verified "energy savings" compared with what would have happened otherwise. The challenge then is how to:

- Separate changes in energy consumption due to energy efficiency actions from all the other possible factors that might change consumption;
- Identify those energy efficiency actions that are specifically motivated by this energy efficiency policy, and hence additional to what would otherwise have happened; and
- Measure and verify energy savings arising from these actions so that they can be appropriately rewarded.

The usual approach is to create a baseline from a "Business As Usual" (BAU) view of future changes in energy efficiency without any EECT policy measure in place. Energy efficiency initiatives must then prove their additionality above and beyond this baseline, in order to be credited. Such mechanisms are known as "baseline and credit" schemes. The inescapable problem with proving this additionality is that it is impossible to verify what would have happened in the absence of this policy measure. Adding to this challenge is the clear evidence that energy-related decision-making is often irrational because cost-effective energy efficiency options are not always taken.

This vexed question continues to plague baseline and credit schemes in general, and certainly energy efficiency policy mechanisms relying on measurable and verifiable energy savings. Assessments of additionality involve identifying the degree to which:

- A particular energy efficiency action reduces energy consumption;
- The action doesn't represent common practice;
- There aren't other policy measures that would have resulted in the energy efficiency action being taken anyway; and
- Investment in the energy efficiency project would not have occurred without the financial incentive of certificate sales (e.g., because of technological progress, changes in primary energy availability, changes in primary energy price relativities, or changes in demand characteristics).

Unfortunately these tests still leave mechanisms open to gaming by participants or "free-riding" off BAU technological progress and other changes to the decision-making context. To add to these problems, there is also the question of how baselines should be adjusted over time given technological, common practice and policy progress. Further, being project-based, EECT schemes don't take into consideration the activities of projects not registered in the scheme. This means that one facility could reduce its output and energy use and so create certificates while a separate facility, not participating in the scheme, could increase its output and energy use – with the result that net energy use had stayed the same yet certificates had been created. The difficulties of additionality and its importance in establishing verifiable and credible energy savings from energy efficiency initiatives have been noted by proponents of EECT yet there are no obvious solutions – a point to which we will return.

2.4.2 Establishing Targets and Liable Parties

The setting of scheme targets resides within choices about the energy sectors and customer classes involved in the scheme. Setting these targets often involves additional assumptions regarding BAU energy consumption developments into the future, an issue that is growingly problematic as this book highlights. Many of the schemes to date assume continued growth in energy consumption that the scheme will only reduce from "what it otherwise would have been," rather than result in an absolute reduction in energy use compared to the start of the scheme.

The meaning of these targets hinges on the actual additionality that the scheme achieves – a major target for energy savings might mean very little for actually driving energy efficiency if additionality is compromised. It is entirely possible to design schemes that appear to deliver significant outcomes (e.g., in terms of certificates created) yet don't actually drive any real changes to decision-making or levels of energy use.

2.4.3 Establishing Trading

Trading in energy efficiency is not a requirement for EECT schemes but offers the potential to increase the economic efficiency with which an overall "energy savings" target is met by allowing a market, albeit a highly abstracted "designer" market, to determine which of the energy services, end-use technologies, and associated decision makers included in the scheme actually create certificates. There are three types of trading within a system for supplier obligations and white certificates [10]:

- Horizontal trading between obliged parties;
- Vertical trading whereby obliged parties purchase certified savings or projects from third parties; and

- Temporal trading, most notably banking, whereby in case of over-compliance participants carry over part of their savings to the next compliance period.

Policymakers can have only limited knowledge of the best available "energy efficiency" options when designing highly end-use or technology-specific programs. A market-based approach may facilitate energy-related decision makers to discover and implement attractive options that have been missed by policymakers. However, there is also considerable potential for some of these decision makers to be unaware and uninterested in the choices available to them. Certainly, some of the relevant decision-making groups are going to be better informed, organized and responsive than others − broadly targeted market-based instruments can still fail to "reach" uninterested groups even if they have excellent energy efficiency options. However, and as noted earlier, trading can also create opportunities for highly motivated and specialized ESCOs to facilitate new energy efficiency actions by such groups.

Another important issue with trading certificates is the need to measure, verify, certify, register, trade and finally acquit the certificates − potentially billions of them if the certificate unit is 1 MWh "energy savings" as typically proposed. Schemes that involve certified savings or projects have equivalent problems. Each of these stages adds another layer of transaction costs.

One challenge with trading is particularly relevant for EECT given the difficulties in verifying and certifying measurable energy savings. This is the well-known "Market for Lemons" problem, outlined by economist George Akerlof [11].

If buyers in a market are unable to verify the quality of what they are buying then sellers of poor products (lemons) are encouraged to enter the market. Unfortunately, cautious buyers then won't be prepared to pay the high prices required to fund high quality products. The result is that good products are penalized even as poor products are subsidized.

Where buyers are in a market only because of legislated obligations then they may well not be particularly interested in the "quality" of what they are buying (beyond ensuring that it meets "certification"). They will, instead, seek out the lowest available prices, and the "lemons" problem becomes even worse. For a highly abstracted commodity such as energy efficiency savings, the risk of a market for lemons emerging would seem to be significant.

2.5 The Broader Context for Scheme Design

EECT scheme design invariably sits within a broader context of policymaking and broader decision-making including:

- How well it integrates with other energy efficiency policies;
- How well it integrates with policy more generally including energy, climate and wider social policy efforts such as, for example, emissions trading;

- Broader market issues including those associated with the commodification of energy efficiency (establishing a measurable and fungible commodity that can be traded) and Financialization (the role of financial markets trading this commodity, and derivatives associated with it); and
- Broader issues not directly amenable to market logic and therefore concepts of "market failure," including energy user engagement beyond that achieved through such market-oriented framing.

3 EXPERIENCE WITH ENERGY EFFICIENCY TRADING

EEC schemes are now operating in a number of countries around the world – for example, in a number of states in the USA, the UK, France, Italy, Denmark, Poland, and China [12]. The Netherlands, Portugal, Romania, and Bulgaria are interested in this policy instrument [7], and Brazil also has a scheme [13].

In Australia, three states, Victoria, South Australia, and New South Wales (NSW) have EEC schemes, although the South Australian scheme doesn't involve trading. NSW had a form of EEC since 2003 as a component of a more general "baseline and credit" scheme called the NSW Greenhouse Gas Reduction Scheme (GGAS) [14]. The Clean Development Mechanism (CDM) also provides a framework for EECT within a broader "baseline and credit" scheme for emissions reductions.

3.1 Implementation

A detailed exposition of the different schemes implemented to date is beyond this chapter. Bertoldi [7] and Bertoldi and Rezessy [10] provide a valuable summary of the European schemes to date, and Passey et al [16] a review of the NSW GGAS scheme. The Regulatory Assistance Project [15] also provides a very detailed review of current EECT schemes within the broader context of Provider obligation schemes. Here we present a high level and necessarily incomplete description of some of the key design choices seen in scheme implementations to date. A key finding is the very significant differences between scheme designs. This highlights both the range of policy objectives targeted through such approaches to date, and the different technical, commercial, and institutional contexts within which the schemes have been implemented.

3.1.1 Measuring and Certifying Energy Efficiency

The scope of the schemes varies considerably. Some schemes consider a range of energy sectors while others include only electricity. The class of energy users covered also varies from residential only to also including commercial and industrial activities.

There is also considerable variation in the methodologies used to establish energy "savings" from energy efficiency both within and between different schemes. Most include some forms of "deemed" energy savings for particular energy efficiency activities where measurement is impossible, difficult, or expensive. Some, such as the CDM, allow project proponents to put forward their own methodologies, while others have only a select list of standardized methods. Some, such as the CDM, have very formal additionality processes, while others don't actually mention additionality anywhere in the scheme legislation and associated regulations (e.g., the NSW GGAS).

3.1.2 Establishing Targets and Liable Parties

Scheme targets vary from carbon emissions reductions to primary energy savings and final energy saving, typically calculated over some imputed lifetime for the measures implemented. The obligated parties vary from electricity suppliers and retailers, to electricity and gas distributors. For the CDM, the liable parties are national governments with targets under the Kyoto Protocol or companies within the EU emissions trading scheme or some other climate change policy framework that have emission reduction obligations.

Penalties for failing to achieve these targets vary greatly. Some are fixed, while others take into account the size of underperformance and other potentially relevant factors.

3.1.3 Trading

Some schemes only permit trading between obligated parties (e.g., UK), some don't permit trading at all (e.g., Denmark), while other scheme designs facilitate high levels of trading both over the counter (OTC) and spot markets, with associated derivative markets also emerging.

3.2 EECT Experience to Date

3.2.1 Measuring and Certifying Energy Efficiency

A particular feature of scheme performance to date has been that in almost all jurisdictions, most energy savings have come from one dominant measure. In the UK this has been insulation, in Italy, compact fluorescent light bulbs (CFLs), in France, household heating systems, and in the NSW GGAS and Victorian VEET, CFLs.

It is possible that this is an outcome of a well-functioning market identifying the least-cost approach to driving energy efficiency. As such it would reflect the different technical opportunities in these countries such as residential construction techniques and climate. No doubt it also reflects the varied scheme scopes and available "energy savings" methodologies [8] However, at least in part, it is likely to be an outcome of relatively modest targets and the rule-making processes by which different types of activities, involving different levels of time, money and effort, are effectively made fungible through inevitably complex and occasionally flawed rules. For example, in the NSW GGAS households were still being given free CFLs to create certificates after the available CFL market had been well and truly saturated [16]. Any "additionality" lapse in the rules may quickly be taken up by fast moving, highly motivated, and entrepreneurial market participants.

3.2.2 Establishing Targets and Liable Parties

Almost all the schemes have achieved their mandated certificate creation targets to date, and a number would seem to have driven energy efficiency actions well beyond them. Two schemes support energy efficiency activities within a wider range of "emissions reductions" options. For the CDM, energy efficiency represents less than 4 percent of projects to date and less than 1 percent of certified emission reductions [29]. The NSW GGAS scheme's energy efficiency (officially termed demand-side abatement) activities were initially a minor contributor to overall scheme actions but this changed markedly over several years due to programs deploying CFLs to households that saw energy efficiency become one of the most significant sources of "claimed" emissions reductions.

3.2.3 Trading

Trading within most schemes to date has been somewhat limited. For some schemes there are only limited opportunities for market participation other than the liable energy companies. In Italy, it would seem a range of factors has supported considerable trading. Interestingly, both quantities and prices of bilateral OTC deals have to be registered in order to increase the transparency of market operation. More than 80 percent of certificates have been created by energy efficiency projects implemented by market participants other than the liable parties [8]. The NSW GGAS has also seen considerable trading activity and considerable spot price variability over the duration of the scheme. At times, certificate prices were trading at near the "penalty" price for noncompliance. In later years of the scheme, certificate prices fell very markedly.

3.3 Scheme Performance

Given the novelty and high expectations placed upon EECT schemes it might be imagined that they would receive extensive assessment. In practice, however, there would seem to have been remarkably little independent assessment of the effectiveness, efficiency, and equity outcomes of existing the schemes. Instead, most assessments have been carried out by the body responsible for designing and/or implementing the scheme, or largely based on their analysis, or according to terms of reference set by government [16]. According to Mundaca and Neij [17], even the extensive and widely cited Euro WhiteCert Project has not undertaken any detailed assessment of EEC outcomes in Europe because of "the enormous complexity of such an evaluation process and the need of having a better understanding of policymaking processes and detailed data of technologies, market actors, and segments, energy efficiency policy instruments already implemented, etc., for each European country."

Standard assessments have generally concluded that the schemes are effective in achieving their certificate creation targets and relatively efficient in doing so at reasonable cost, and with those costs recovered from a large proportion of energy users (e.g., [18]). Giraudet and Finon [8] suggest that the static efficiency of the European schemes seems favorable although not necessarily optimal in terms of minimizing scheme costs — investment follows market incentives rather than the social optimum.

Equity outcomes have depended on the scheme design. Some schemes, for example, specify that a certain proportion of energy efficiency activity must occur in low-income, or in other ways vulnerable, residences. Where such measures aren't in place, the schemes generally drive activity in those types of projects and energy user classes that offer the lowest cost and risk "energy savings."

As noted above, many of these assessments are not genuinely independent and, in our view, don't appropriately define and assess the effectiveness, efficiency, and equity outcomes of the schemes. Nevertheless, the assessments to date raise considerable concerns regarding the performance of some schemes [16,15].

3.3.1 Measuring and Certifying Energy Efficiency

This has proven one of the most problematic areas for the schemes and lies at the heart of effectiveness, efficiency, and equity outcomes (i.e., who earned how many certificates and for doing what?). As noted earlier the key challenge is that energy savings have to be calculated against some counterfactual BAU baseline. The most common baselines used in the existing schemes are:

- The performance of the most commonly used (median) or average appliance, equipment, system, or process available on the "market"; or
- The performance of the median or average installed stock of such appliances, equipment, system, or process.

Note that neither of these tests ensure additionality as there will generally be some investment in such products and processes of higher performance than the average in the market, while using average stock estimates provides even less assurance of additionality.

Policy additionality is another key issue, and some schemes such as the UK require that a particular energy efficiency activity counted with an EECT scheme can't be a legislated requirement, but also shouldn't be financially rewarded by another policy. In other schemes such as that in Italy, this isn't considered and some activities may receive credit, and be counted as an outcome, of several or even numerous policies. The temptation for double counting is clear, both for scheme participants but also the governments seeking to claim credit for improved energy efficiency outcomes [7].

As noted previously, some schemes do not seek to formally address additionality — for example, the original NSW GGAS legislation and regulations didn't even use the term or concept. While it is difficult to separate the additionality of the energy efficiency activities within this scheme from other actions claiming to deliver emission reductions, these non-additional activities grew to have a significant role in meeting the target, and the overall additionality of the scheme was estimated by the federal government at around 5 percent for 2009 (i.e., 5 percent of the claimed emissions reductions of the scheme were actually driven solely by the scheme itself).[4]

More broadly, a recent review of 19 energy efficiency obligation schemes around the world notes that "None of the schemes have established robust procedures to verify whether energy savings are additional" [15].

Additionality is very formally addressed in the CDM, but there are still significant concerns. Reasons given for the relatively small role of energy efficiency activities to date include the high transaction costs and complexity of such projects by comparison with some other options, due at least in part to the process for establishing additionality.

3.3.2 Targets

Assessing scheme performance relative to mandated targets is only a meaningful measure of effectiveness to the extent that energy efficiency activities leading to certified "energy savings" are actually additional. Estimates of the impacts of policies based on hypothetical estimates of what would have happened in the absence of the policy are problematic.

For example, a UK DEFRA official appearing before a House of Lords Science and Technology Committee Energy Efficiency inquiry [19] and being questioned about this issue noted *"They are real relative savings. They are measured against the baseline that was projected... they are genuine reductions on what would otherwise have happened had these policies not been put in place"* The Committee responded in its report *"If savings are real, they cannot be relative — it is meaningless to talk of savings against what might have happened had certain policies not been in place... Levels of carbon emissions should be grounded in clear historical data, not hypothetical projections. We recommend that the Government ground its targets*

---

4. Note that the NSW GGAS scheme was in transition in 2009 given the stated intention of the Federal Government to introduce a national emissions trading scheme that would, in the view of both the Federal and NSW Governments, negate the need for GGAS.

more firmly in reality." Clearly, this isn't just an issue for energy efficiency or climate change policy more generally, but a broader policymaking challenge. However, the relatively abstracted nature of energy savings is particularly problematic.

3.3.3 Support for Dynamic Efficiency

It is worth considering the dynamic efficiency outcomes of the schemes separately to the static efficiency assessment above. This is because a primary policy goal of EECT schemes is to facilitate the processes and institutional capacities that can drive diffusion of energy efficient technologies. Although these aren't quantitative measures of energy efficiency, the ability of the schemes to drive major technology and business model transformation are useful guides. Again, the outcomes vary greatly across the schemes. Giraudet and Finon [8] argue that the UK scheme has achieved high dynamic efficiency, but the Italian and French schemes have had far more modest outcomes in this regard with an emphasis on picking "low-hanging fruit."

Other assessments have taken a different position. For example, Lees [13] also highlights that EECT objectives include the development of energy service companies (ESCOs) "...that they see themselves moving from being "suppliers of a commodity" to providers of sustainable energy solutions." In the UK, Lees argues that progress with ESCOs has been very limited in the residential sector for reasons including "...the complexity of the concept and the basic mistrust by customers that any energy company would wish to sell them less of their product!." By contrast, Italy does have significant non-liable parties playing a significant role in the marketplace. Experience to date with the NSW GGAS suggests that there has been relatively little technology innovation, however, the scheme has supported the activities of ESCOs and created some novel new business models for both the residential and commercial sectors [16,20].

4 POSSIBLE LESSONS FOR POLICYMAKERS

4.1 Market-Based Policies to Address Market Failures can also Suffer from Market Failure

The first lesson from the conceptual framework of EECT and experience to date is that these market-based energy efficiency policies intended to address energy and energy efficiency market failures can suffer from their own market failures. Examples include (building upon the more general energy market failures noted earlier):

- Potential oligopolies, for example, where the liable parties within EECT are the existing highly concentrated energy suppliers and these can control who does what sorts of activities within the scheme;

- Public goods, given energy efficiency's key role in the essential public good of energy provision and the implications of EECT on this;
- Incomplete markets, as the EECT schemes are invariably limited in extent and scope, and many energy efficiency opportunities require very high levels of coordination that seem beyond current scheme designs;
- Information failures, including generally poorly informed potential EECT participants;
- The "business cycle," a particular issue given capital intensive, long-lived investments for energy efficiency infrastructure, and the difficulty posed here by certificate market prices that have exhibited significant variability and uncertainty;
- Externalities, because while EECT offers a means to address some externalities it doesn't cover all of them, and can create new ones such as perverse equity outcomes.

4.2 There are Wider Potential Market Failures Associated with the Schemes

4.2.1 Abstraction

One key issue with "baseline and credit" schemes, and certainly EECT, is that of the significant *abstraction* required in their implementation. Concepts such as energy efficiency, energy savings and additionality have to be defined, and this requires assumptions, choices, and tradeoffs. All of these necessary abstractions, and the process of determining them:

- Add to the complexity of such schemes;
- Make it far harder to determine the real outcomes of the measure; and
- Create moral hazards for both scheme designers as well as participants.

Thus, the process by which the policy intent of improved energy efficiency is translated to actual energy efficiency activities becomes highly complex and abstracted, as shown in Figure 22.3. As discussed below, abstraction is particularly problematic when establishing a fungible commodity for trading.

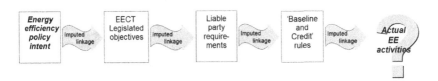

FIGURE 22.3   The many implementation steps, linkages, and hence potential abstractions with EECT schemes.

4.2.2 Additionality

As noted at length above, additionality is a vexed but essential aspect of effective and efficient EECT design. Furthermore, there are no obvious fixes. There are, of course, many possible energy efficiency projects that deliver energy reductions beyond any likely BAU baseline. Other projects might be less clear-cut, yet still be widely accepted as offering credible energy savings beyond BAU. To the extent that projects can be properly assessed at all, there is likely to be a trade-off between accuracy (and hence credibility) as well as economic efficiency, against high administration and compliance costs.

One might argue that even though proving additionality may be near impossible, EECT schemes can still play a role in energy efficiency policy. If particular "energy efficiency" targets and measurement rules are insufficient to require any real effort beyond BAU then there is little harm done. If sufficiently challenging, then the scheme will promote additional activity in order to deliver these objectives.

There are several problems with this view. One is the considerable effort by policymakers and demand-side decision makers to establish EECT schemes — effort that might prove to be wasted. Another problem is that some sort of baseline methodology has to be established and this will determine the winners — and possible losers — amongst scheme participants. Society is best served when the winners are actually those that most contribute to the underlying policy objective.

The potential implications of low additionality are highlighted in Figure 22.4. Effectiveness is likely to be low because the scheme doesn't actually deliver much towards the policy intent of additional energy savings. Efficiency is low because there are inevitable transaction costs associated with the scheme that are borne for all claimed energy savings, including those that are not additional. Finally, equity outcomes are also likely to be adverse because of the potential for very significant windfall profits to some

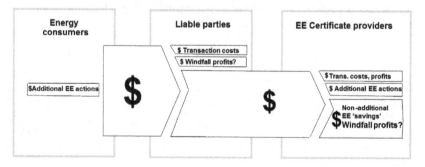

FIGURE 22.4 The potential effectiveness, efficiency, and equity implications of low additionality within an EECT.

participants — likely participants with high political influence that persuade the rule makers to establish a framework that permits these non-additional energy savings to be included.

4.2.3 Commodification and Financialization

Commodification has several common meanings. One is the process by which something that traditionally doesn't have an economic value is assigned such a value and then entered into commercial relationships. It also describes the process by which distinct goods with different attributes and values end up as simple fungible commodities within undifferentiated price competition. Financialization commonly describes the process of capturing all of the value associated with the exchanges of goods or services into financial instruments.[5]

It can be seen that EECT involves a commodification of a diverse range of potential energy efficiency activities undertaken for a potentially wide range of reasons into an "energy savings" commodity that can be traded.

It is notable that one of the first companies to look at energy services and the role of energy efficiency through the lens of standardization, commodification, and tradability was Enron [21]. Energy efficiency doesn't appear likely to be a good fit with these processes — energy efficiency opportunities are highly diverse, non-standardized, and context-specific. Standardization lies at the heart of commodification. However, such standardization in at least some aspects of energy efficiency does offer the potential for rapidly scaling up efforts. Certainly, some energy efficiency options are more straightforward than others and schemes will likely tend to focus on these. That may be acceptable, and the involvement of Enron in pioneering such commodification approaches doesn't mean they are wrong. However, the failure of Enron does highlight some of the risks with this approach, and its limits. Commodity markets have proven troubling enough — energy efficiency commodity markets are almost certainly even more troubling given the higher levels of abstraction and the political nature of market design and settings.

Some of the risks and limitations of financialization have been discussed above, such as the "market for lemons" problem. Broader risks have recently been highlighted by the Global Financial Crisis [22].

Price volatility is an inherent characteristic of EECT and emissions trading schemes more generally. Once there is a perception in the market that there are sufficient certificates to meet the target into the longer term, the certificate price will likely drop quite dramatically. There is also significant uncertainty in such markets because market participants never have perfect

---

5. Definitions of these terms vary. The definitions used here are adapted from those provided at Wikipedia.org.

access to reliable information regarding current and future energy efficiency costs and demand for certificates. Note that these challenges, and the potential for some non-additional types of activities to be permitted to earn certificates creates considerable risks for parties considering taking genuine "additional" activities that have a real financial cost [16].

Another inherent characteristic of tradeable certificate schemes is that the spot price of certificates for liable parties should theoretically settle at the marginal cost of energy efficiency actions that achieve the target despite the fact that the vast majority of certificates might actually be created at less than this cost. In strict economic terms, the extra surplus (profit) to parties with very low-cost efficiency options does not represent inefficiency at a societal level because money is not wasted, merely transferred. However, considerable "windfall" cash flow may go to parties with very low-cost activities, or even "no-cost" non-additional activities, which would have occurred regardless of the scheme. More generally, there would seem to be significant risks in permitting energy efficiency policy outcomes to be driven, at least in part, by the outcomes of financial markets with all of the attendant speculation, potential bubbles and crashes.

In conclusion, the challenges for EECT are significant. The UNCTD [22] highlights that high transparency of market participation can assist, and that "beyond this kind of "soft regulation," a number of direct commodity price stabilization measures should be considered to address potential financial market failures.

4.3 The Challenge is Designing Policy Frameworks, not Policies

EECT has been put forward by some policy researchers and makers as a way to avoid the use of other energy efficiency policies in a similar manner to the claimed advantages of emissions trading over a suite of more targeted policies for reducing greenhouse emissions. Unfortunately, it is now clear that this is not the case. No single policy instrument will drive optimal levels of energy efficiency across the economy. There is also considerable uncertainty in determining the capabilities of EECT through comparison with other policies. Given the importance of making progress on energy efficiency the most appropriate way forward is almost certainly a suite of complementary policies.

An important issue, therefore, is how EECT might work in tandem with targeted energy efficiency programs. This is the situation with existing schemes that have all been introduced within wider policy frameworks, albeit not necessarily comprehensive and coherent policy frameworks.

While the contribution of EECT versus other policies towards driving energy efficiency varies considerably, it is notable that the key energy efficiency policies at present in many jurisdictions are regulatory. Given that there will inevitably be a need for a suite of energy efficiency policies, it is important that EECT doesn't impede progress on other energy efficiency policies, or that such policies don't adversely impact on EECT. A key challenge for EECT with such mixed approaches is to actually drive measurable and verifiable "additional" change beyond all these other measures, or at least reduce the overall cost of achieving the same energy savings. Otherwise, why bother with the complexities, costs and effort of EECT.

4.4 The Governance Challenge for Policymakers

The design and implementation of government policies targeting improved energy efficiency is one of the most important and difficult challenges facing modern society. To date, most electricity is produced using fossil fuels and so energy efficiency policies have mostly impacted on incumbent power sector participants with political influence. Meanwhile, energy efficiency doesn't have an obvious or exclusive set of influential stakeholders, although some powerful stakeholders may have an interest in undertaking energy efficiency for themselves and others. Thus, there are clear imbalances between the various stakeholders that may influence the decision-making process used to develop government policy.

This is particularly problematic for EECT as they are novel "designer" markets that provide enormous flexibility to policymakers working under considerable uncertainty. As a result, effective governance is the key to successful market-based energy efficiency policies, including [23]:

- Robustness to ensure the desired energy efficiency objectives are achieved even if particular favored policies such as EECT fail. This is almost certainly going to require policy "portfolios" where EECT potentially makes a useful, adaptive contribution within a broad suite of policies;
- Robustness against inappropriate rent-seeking (often but not always by incumbents). Incumbents have important knowledge and expertise to contribute to the scheme design but are seeking competitive advantage – market competition begins during the policy design process. New entrants have a key role to play in driving dynamic efficiency but are generally poorly represented in stakeholder processes;
- Formal and highly transparent processes for ongoing review and improvement – "market and investor" certainly must not over-ride necessary fixes and review requires separation of powers between setting objectives, making the rules, implementing them, and judging their effectiveness; and
- Broader policy processes ensuring that any EECT is compatible and complementary to other key energy efficiency policy efforts.

To date, governance arrangements for energy efficiency policy efforts have often been far less formalized and assured than those for supply-side policy efforts reflecting, perhaps, their lower priority. Formal governance processes must be able to manage these current asymmetries between supply and demand-side stakeholders.

Present retail electricity markets and associated regulatory frameworks provide a largely uninformed and unresponsive interface between end-users and wider decision-making. Effective end-user advocacy arrangements can be challenging to achieve. Supply-side participants are generally large, focused almost exclusively on the electricity industry and have considerable shared interests. End-users are generally small, far more numerous and diverse, and generally have only a limited interest in electricity itself.

4.5 Moving Beyond the Market Failure Model for Energy Efficiency

As noted earlier there is a growing literature on the limitations of the "market failure" model for policy assessment. Bozeman [24] argues that the concept of market failure has its uses but also important shortcomings as a framework for understanding the public value aspects of public policy and management. He highlights, instead, the importance of understanding that government policies oriented to address "market failures" may fail to deliver public value for reasons including a failure to properly establish and incorporate underlying community values associated with particular objectives, their "privatization" of public benefits, their focus on short-time horizons, and when social and market transactions threaten fundamental human values and needs. Kay [5] has expressed concerns, more generally, that policymaking within a market failure framework may neglect fairness, processes, and equity.

Such concerns would seem particularly relevant to market-based approaches to energy efficiency with their reliance on price signals and the promise of private returns to motivate key decision makers in order to improve public welfare. Energy use, and potential energy efficiency actions, are associated with a wide range of community values — for example, some cultures place great importance on a warm and well-lit "welcoming" home while some people are motivated to undertake energy efficiency for its environmental benefits. Short-time horizons can be a significant problem for energy efficiency — for example, housing has relatively slow capital stock turnover yet its design and construction has a very significant impact on what energy efficiency can be achieved within the residential sector over the longer term. Finally, energy, and hence energy efficiency, has a vital "public good" role. Measures that seek the most economically efficient delivery of energy efficiency may fail to deliver benefits to particularly vulnerable sectors of the community.

Beyond the broader behavioral aspects noted above lies the vexed issue of energy user engagement in the electricity industry. Current conceptual frameworks for energy efficiency don't adequately address the way contemporary consumer choice has come to be predominantly conditioned by the low-cost and seemingly boundless electricity availability. Indeed our current energy-intensive lifestyles may well be as much about the systemic framing and conditioning of the choices available to consumers as they are about the exercise of individual consumer choice [25].

It seems almost certain that a complete transformation in energy-related decision-making is required to address our growing climate change challenges [1]. Reduction in energy use has a vital role to play in any such transformation and a key question for policymakers is whether we can continue to focus on energy efficiency without addressing present community desires for an ever expanding range of energy services. Energy efficiency, distributed energy generation, and a growing suite of "smart" equipment offers the potential for energy users to play a far greater role in the provision of their energy services, and potentially an opportunity to reflect upon the sustainability of these services and their real importance to their wellbeing [26]. Whether energy consumers desire such engagement is another question.

EECT schemes have interesting implications in all of these regards. While being a market-based mechanism, they do create an opportunity for knowledgeable, skilled, and motivated organizations to assist disengaged energy users to undertake energy efficiency actions that they would otherwise ignore. The engagement of these firms with energy users will almost always involve financial transfers of some form (even if it is the "gift" of CFLs) yet may involve a wider relationship. For example, a number of environmentally focused ESCOs emerged within the framework of the NSW GGAS. Schemes can be designed so that a range of stakeholders privately benefit from the private yet also public benefits of increased energy efficiency. The timeframe of energy efficiency actions is a function of the particular activities and associated rules within an EECT scheme. While it can be expected that participants will first chase the easy wins, appropriate scheme design could drive focus on longer-term energy efficiency opportunities. Equity concerns have been a key design imperative in some EECT schemes given their clear potential to drive activities in only some energy user groups. For example, the UK EET required that a given proportion of energy savings had to be achieved in "vulnerable" households given the potential for scheme participants to target, instead, high income and high consumption households.

The issue of wider energy user engagement with EECT schemes is particularly vexed. Giraudet and Finon [8] note that, "In general, consumers are barely aware of their participation to the scheme." This is, in some regards, a key strength of the approach as noted above. However, there are likely to be limitations to what can be achieved in reducing energy use without greater energy user engagement. The schemes, as currently implemented, also don't link to the broader suite of opportunities that energy users might have to engage in their energy services or address the perverse parallel incentive to increase consumption implicit in the ongoing conventional growth orientation of existing economies [27].

5 CONCLUSIONS

EECT schemes are a novel market-based policy approach to energy efficiency that are receiving growing policy attention around the world. They have some attractive characteristics including their potential role in facilitating motivated, knowledgeable and skilled ESCOs to assist often poorly motivated, unknowledgeable and unskilled energy consumers in undertaking energy efficiency actions. Experience with these schemes is growing, however, they are still relatively novel by comparison with many other well-proven policy approaches. Furthermore, the experience to date is mixed. Of particular concern is the important challenge of ensuring the additionality of energy efficiency actions that fall within the scheme. Poor additionality outcomes adversely impact the effectiveness, efficiency and equity outcomes of the schemes. More generally, EECT schemes necessarily involve significant levels of abstraction that is problematic in both establishing a meaningful market for energy savings, and governance of the scheme design, implementation and ongoing development. Furthermore, the commodification and financialization of energy savings actions raises significant questions given the diverse and context-specific nature of energy efficiency opportunities and recent poor performance of markets in abstracted financial products.

To conclude, well-designed EECT schemes may be able to play a useful role in a coherent and comprehensive energy efficiency policy framework that includes suasive and regulatory approaches as well as other market mechanisms. As such, they can represent a market-based solution to some market failures in energy-related decision-making. Poorly implemented schemes will, however, certainly just create further market failures. And, finally, there are challenges in energy efficiency, and engagement with energy users more generally on the sustainability of their energy services, that would seem to lie beyond the "market failure" framework that is currently setting the policy agenda. This would seem to be an important area for future work.

REFERENCES

[1] International Energy Agency (IEA). Energy technology perspectives. Paris; 2012.
[2] Energy Savings Trust. Putting climate change at the heart of energy policy, submission to the UK energy white paper; September 2002.
[3] European Commission (EC SAVE). Bringing energy efficiency to the liberalised electricity and gas markets, report of the EC SAVE programme, Brussels; December 2002.

[4] Jaffe AB, Newell RG, Stavins RN. Economics of energy efficiency. Encyclopedia of Energy 2004;2:79–90.

[5] Kay J. The failure of market failure, Prospect Magazine; 2007 August 1(137).

[6] BDA Group. Market-based instruments decision support tool, Report to the queensland, department of natural resources and water, Available at <http://www.marketbasedinstruments.gov.au/Portals/0/docs/DST_%20final_web.pdf>; 2008.

[7] Bertoldi P. Assessment and experience of white certificate schemes in the European union. Presentation to IEA PEPDEE Workshop, Sydney; December 2011.

[8] Giraudet L, Finon D. White certificate schemes: the static and dynamic efficiency of an adaptive policy instrument CIRED working papers series no 33–2011, <http://www.centre-cired.fr/IMG/pdf/CIREDWP-201133.pdf>; 2011.

[9] Passey R, Bailey I, MacGill I, Twomey P. The inevitability of 'flotilla policies' as complements or alternatives to flagship emissions trading schemes, Energy Policy (accepted for publication); 2012.

[10] Bertoldi P, Rezessy S. Energy saving obligations and tradable white certificates. A report prepared by the joint research centre of the European commission, Available at <http://ec.europa.eu/energy/efficiency/studies/doc/2009_12_jrc_white_certificates.pdf>; 2009.

[11] Akerlof G. The market for lemons: qualitative uncertainty and the market mechanism. Q J Econ 1970;84:488–500.

[12] Joshi B. Best practices in designing and implementing energy efficiency obligation schemes, Research report task XXII of the international energy agency demand side management programme, prepared by: The regulatory assistance project; 2012.

[13] Lees E. European and South American experience of white certificates, a WEC-ADEME case study on energy efficiency measures and policies, <http://www.ffydd.org/documents/ee_case_study__obligations.pdf>; 2010.

[14] IPART, Compliance and operation of the NSW greenhouse gas abatement scheme during 2006, Independent pricing and regulatory tribunal of NSW; July 2007.

[15] Staniaszek D, Lees E. Determining energy savings for energy efficiency obligation schemes, a report to the regulatory assistance project and the ECEEE, <http://www.raponline.org/document/download/id/4898>; April 2012.

[16] Passey R, MacGill I. Energy sales targets: an alternative to white certificate schemes. Energy Policy 2009;37(6):2310–7.

[17] Mundaca L, Neij L. Work package 5, policy recommendations for the assessment, implementation and operation of TWC schemes, task report, Euro whitecert project, European commission intelligent energy program; 2007.

[18] Bertoldi P, Rezessy S, Lees E, Baudry P, Jeandel A, Labanca N. Energy supplier obligations and white certificate schemes – comparative analysis of experiences in the European union. Energy Policy 2010;38(3):1455–69.

[19] House of Lords Science and Technology Committee. Report of the committee inquiry on energy efficiency. London; 2005.

[20] MacGill IF, The new black, some new grey or now passé presentation to the Australian alliance to save energy summer study, Sydney; February 2012.

[21] Mathew P, Kromer S, Sezgen O, Meyers S. Actuarial pricing of energy efficiency projects: lessons foul and fair. Energy Policy 2005;33(10):1319–28.

[22] UNCTD, Price formation in financialized commodity markets: the role of information, study prepared by the secretariat of the United Nations conference on trade and development, United Nations, New York and Geneva, <http://unctad.org/en/docs/gds20111_en.pdf>; June 2011.

[23] Passey R, MacGill I, Outhred H. The governance challenge for implementing effective market-based climate policies: a case study of the New South Wales Greenhouse Gas reduction scheme. Energy Policy 2008;36(8):3009−18.

[24] Bozeman B. Public-value failure: when efficient markets may not do. Public Administration Review 2002;62:145−61.

[25] Passey R, Betz R, MacGill I. An energy efficiency policy model that addresses the influences of the 'infrastructures of provision', eceee 2009 summer study, La Colle sur Loup, Côte d'Azur, France; June 2009: 1−6.

[26] Healy S, MacGill I. From smart grid to smart energy use. In: Sioshansi FP, editor. Smart grid: integrating renewable, distributed, and efficient energy. Academic Press; 2011.

[27] Jackson T. Prosperity without growth? economics for a finite planet. London: Earthscan; 2009.

[28] Energy Information Administration (EIA). Defining energy efficiency and its measurement, Available at <http://www.eia.gov/emeu/efficiency/ee_report_html.htm>; 1995.

[29] Hinostroza M, Cheng C, Zhu X, Fenhann J, Figueres C, Avendano F. Potentials and barriers for end-use energy efficiency under programmatic CDM, working paper no. 3, CD4CDM working paper series, UNEP Risø centre on energy, climate and sustainable development, ROSKILDE, <http://www.cd4cdm.org/Publications/pCDM&EE.pdf>; 2007.

[30] Transue M, Felder FA. Comparison of energy efficiency incentive programs: rebates and white certificates. Utilities Policy 2010;18(2):103−11.

Chapter 23

The Ultimate Challenge: Getting Consumers Engaged in Energy Efficiency

Alex Laskey[1] and Bruce Syler[2]

[1]*Opower*, [2]*Connexus*

1 INTRODUCTION

Customer engagement is an increasingly important topic of converstation within the energy industry. The central challenge: Most Americans don't spend a lot of time thinking about their home energy usage. A recent study from Accenture estimated that the average American interacts with their utility just 7 minutes a year.[1]

How do you help people save energy in just 7 minutes a year? One approach is to take the consumer out of the loop through increased automation. Under such a scheme, a smarter grid communicating directly with smarter home energy appliances achieves a fully automated home energy usage experience. In this case, the grid, the meter, and the appliances are operating a home's energy use and controlling peak demand in a way that is invisible to and uanffected by the consumer. An example of this is a two-way communicating and Internet-enabled thermostat connected with a smart meter. These devices can estimate when peak demand is expected to occur and pre-cool a home prior to the event so they can cycle the air conditioning system during the event without sacrificing the comfort of the inhabitant – and without the inhabitant playing an active role or even noticing any discomfort.

While this model is promising, Opower has focused on a different approach that at once complements and contrasts this energy-internet model. The most important "node" in the smart grid – and the most important "appliance" in the smart home – is the customer. Rather than work around the customers, the idea is to focus on, and empower the customer. This

---

1. See Accenture report, Actionable insights for the new energy consumer, available here, <http://www.accenture.com/us-en/pages/insight-actionable-new-energy-consumer.aspx>; 2012.

Energy Efficiency. DOI: http://dx.doi.org/10.1016/B978-0-12-397879-0.00023-2

591

notion is critical. If the customer is working against the technology, you're just not going to get very far. And if you want the customer to make big decisiuons or investments – such as buying a more efficient refrigerator or retrofitting their home – you need to engage them.

The potential for this type of engagement is significant. ACEEE estimates 39 TWh in energy saving potential by 2030 for established behavioral efficiency programs, or 2.4% of expected TWh in demand.[2] Even this enormous potential only scratches the surface. Opower conducted a survey of households in 2010 that suggests up to 21% of residential energy usage is wasted from inefficient behavior.[3] This vast potential indicates the importance of customer engagement.

This chapter discusses the innovative services in which utilities invest to empower their customers with insights and information. Connexus Energy, the largest customer-owned cooperative utility in Minnesota, has been able to consistently save customers over 2 percent on their energy bills, results that have been independently verified. The Opower-Connexus partnership is featured in this chapter to provide empirical evidence.

This chapter is organized as follows: Section 2 describes the current customer engagement deficit. Section 3 discusses Opower's solution for filling this void, namely a utility platform for customer engagement. Section 4 describes a case study implemented by Connexus Energy followed by the chapter's conclusions.

2 WHAT'S THE PROBLEM?

The problem can be explained by two numbers that describe the divide between household interest and current time spent on saving energy.

85: 85 is the percentage of people that think saving energy is a good idea.[4]

7: 7 is the number of minutes per year the average person spends interacting with their utility.[5]

Clearly, there is a gap between the proportion of people that want to do something to save money and energy, and the amount of time they spend thinking about energy use.

---

2. York et al., January 2013, "Frontiers of Energy Efficiency," ACEEE, available here: http://aceee.org/research-report/u131

3. Corcoran, Christopher, November 2010, "Behavioral Efficiency Survey Results," ACEEE BECC Conference.

4. See Gallup poll results, Green behaviors common in United States, but not increasing, available here, <http://www.gallup.com/poll/127292/green-behaviors-common-not-increasing.aspx>; 2010.

5. See Accenture report, Actionable insights for the new energy consumer, available here, <http://www.accenture.com/us-en/pages/insight-actionable-new-energy-consumer.aspx>; 2012.

Those 7 minutes largely go toward checking and paying utility bills. As a result, consumers are completely in the dark about their inefficient energy use, leading to massive amounts being wasted every year. According to a recent McKinsey & Company study, this waste amounts to an estimated $400 billion a year globally, which equals the energy needed to power more than 330 million homes.[6] McKinsey also estimates that the United States alone could reduce energy consumption by 23 percent and save families and businesses more than $200 billion on their electrical bills in the next 10 years through increased energy efficiency.

There is an immediate, cost-effective solution to reducing this waste: behavior change. Changes in behavior have both direct and indirect impacts on the environment. More efficient energy usage behavior can result in energy and bill savings for households. The same programs that drive this change in behavior can also accelerate the adoption of other impactful energy improvements – such as deriving more power from renewable sources or making structural changes to peoples' homes. Still, behavior change is often overlooked for several reasons.

First among the reasons may be that energy is relatively inexpensive. In the United States, only 1.7 percent of an average household income goes to energy bills. Most customers aren't motivated to make changes to save enough for an extra meal out once a month. Even in the environmentally progressive U.S. cities of Berkeley, California and Boulder, Colorado, a recent study found that only 0.18 percent and 0.64 percent of the population, respectively, participate in available energy efficiency programs.[7]

The second reason may be that energy data is generally uninteresting. For a subject that most people spend so little time thinking about, presenting an overwhelming amount of numbers and charts on energy usage won't inspire change.

Finally, energy is confusing. The average consumer doesn't know what kilowatt-hours or therms are. When they receive their bill, customers don't know whether using 200 kWh per month is high or low.

There are a lot of misconceptions about using energy. For example, 81 percent of people leave their heating or cooling system running when they aren't at home[8] with the false belief that it takes more energy to turn them on and off rather than leaving the systems running continuously. This simply isn't true.

6. See McKinsey report, Unlocking energy efficiency in the US economy, available here, <http://www.mckinsey.com/client_service/electric_power_and_natural_gas/latest_thinking/unlocking_energy_efficiency_in_the_us_economy>; 2009.
7. Bailey M, Johnson CB. Innovative energy efficiency financing approaches; June 1 2009.
8. See post here, <http://www.ksvc.com/blog/2010/11/16/becc-day-1-report/>.

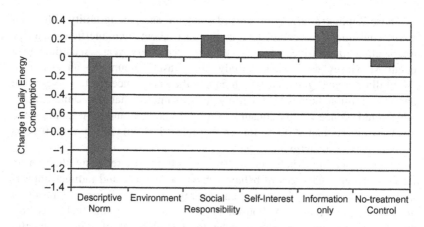

FIGURE 23.1 Energy savings across types of messages used in Dr. Cialdini's San Diego State study. *Source: Cialdini (2004).*

3 WHAT IS THE SOLUTION?

To arrive at a solution, it may be helpful to examine the scientific underpinnings of what drives the motivation and action that are essential to behavior change. In 2005, behavioral economist Robert Cialdini, the author of *Influence*, set out to answer this question. Cialdini and his students at San Diego State University ran field tests during a hot summer in California, leaving notices on door handles at homes with four different messages printed on them.[9]

The first said: "Turn off your AC [air cooling system] and turn on a fan — you can save money."

The second said: "Turn off your AC and turn on a fan — you can save the environment."

The third said: "Turn off your AC and turn on a fan — it's your civic duty."

The fourth said: "4 in 10 of your neighbors turned off their AC and turned on a fan."

After three weeks, they analyzed the homes' energy use and found there was no impact on consumption as a result of getting any of the first three messages. However, homes that received the 4th message used on average 6 percent less electricity than the control group. Figure 23.1 is a chart from the study that displays the daily kWh savings comparison across these categories, as well as a group that received information only, and the control group (the two on the right of the chart).[10]

---

9. Cialdini R, Schultz W. Understanding and motivating energy conservation via social norms, available here, <http://opower.com/uploads/library/file/2/understanding_and_motivating_energy_conservation_via_social_norms.pdf>; 2004.
10. Id., p. 6.

The discovery of the impact of *social norms* was a catalyst for the creation of Opower. Since then, Opower's work — now with more than 70 utilities, including 9 of the 10 largest in the United States, and 10 million homes across the United States and the United Kingdom — has led to a much deeper understanding of the mechanisms needed to harness the power of behavior change to achieve energy savings.

As Cialdini's study identified, normative comparisons — like the fourth example above — appear effective in motivating behavioral change, as do other tools like goal setting, usage ranking, and historical usage comparisons that tap into humans' innate competitiveness with their peers and with their neighbors. But motivation alone is insufficient. It must be linked to actionalbe insights and advice to drive behavior change.

While this is a relatively new concept in the utility industry, the basic idea is not completely foreign. Personal finance tools like Mint.com, for example, provide users with insights about their spending history and investments that are beyond the numbers. The service has evolved into personalized recommendations on services and steps that people can take to save money.[11]

Continuous engagement, like the low balance account notice you might receive from your bank or the minute overage alerts from your mobile service provider, is essential in the energy use context as well. Utilities have an opportunity to prompt action when it counts; not at the end of the billing cycle, but in real time. With new smart metering technologies being deployed, information needed to provide this type of information and service to customers is starting to come online as described in chapter by King and Stromback in this volume. This will enable utilities to alert customers if they are on track for an irregularly high charge during the current billing period, offering tips to avoid that eventuality.

Even with these technologies, consumers still can't be expected to keep their energy use as a top priority. There are too many distractions in life, and too many competing priorities. Instead, the best way to sustain changes in energy behavior is with regular and well-designed reminders. Just as speed limit signs are repetitively placed along roads to keep speeds at reasonable levels, energy insights should be repetitively delivered to drive and sustain efficient behavior.

To keep customers engaged, one needs to deliver the right message at the right time while catering to different customer segments differently. This is a complex challenge and a global problem that offers opportunities to engage energy users of all ages, interests, and demographics.

---

11. See, e.g., <https://www.mint.com/how-it-works/>.

3.1 How Can Behavioral Change Lead to Energy Savings?

Opower was founded on a simple premise: it's time to engage the millions of people who are in the dark about their energy use. As a result, a new form of behavioral energy efficiency was formed – one that acheives measurable and verifiable savings predicated on two concepts:

- First, normative comparisons tap into the innate competitive spirit that motivates customers to engage.
- Second, data analysis produces relevant actions and advice that a household can take to be more efficient.

Combining motivation with action leads to more efficient energy usage behavior.

Opower collects energy use data from the utility and merges it with third party data to create individual customer profiles. The neighbor comparisons, for example, lets households know how they are using energy relative to similarly sized homes in their neighborhood – and targeted messages delivers tips on how to save energy and money.

Opower provides utility customers with personalized reports that compare their energy usage to their neighbors through a variety of channels – online, text messages, smart phones, and phone calls. The most important channel, however, is very traditional, the mail. By delivering its targeted message through the mail, Opower is able to reach an astonishing 99 percent of customers, including low-income and senior households, and measure the impact of the specific schemes on typical usage. Figure 23.2 provides an illustrative comparison of the penetration rates of various technologies for engaging households regarding their energy consumption and energy efficiency.

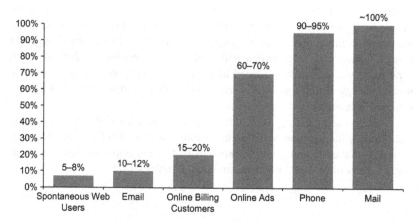

FIGURE 23.2 Illustrative example of approximate penetration of technologies for engaging households on energy. *Source: Opower (2012).*

Importantly, these mailed communications are opt-out — meaning that homes receive them proactively from Opower without taking any action of their own. This opt-out design eliminates the need for disinterested homeowners to initiate participation in the program, which, in turn, allows Opower to reach many more households than otherwise possible.

Despite the scientific evidence behind its approach and the high penetration rate in deliveing its targeted message, Opower still faces multiple challenges to achieving sustinable behavior change at scale. To succeed, the solution has to be affordable, actionable, reliable, and measurable:

• Affordable: while others were trying to convince consumers to adopt some sort of new energy monitoring device in the home, Opower focused on leveraging information that the utility already has and delivering it through channels they already use: online and through the mail.
• Actionable: most people do not spend a lot of time thinking about energy use — the 7 minutes a year dilemma — so the message has to deliver key insights that are easy to act on. Most people don't know what to do with raw data.
• Reliable: Utilities already have the data on how people use energy and an established customer relationship. The challenge is how best to partner with utilities to make home energy reporting a big part of engaging customers with better energy services.

Being mindful of these considerations results in reliable, scalable, and verifiable behavioral changes that have delivered over 1.7 TWh of savings and over $190 million in bill reductions.[12] Twelve external evaluators have verified these claimed energy savings and the measurement methodology. Among these is a study by Allcott who examined nearly 22 million utility bills from Opower's 17 longest running deployments[13] and concluded that the program generated electricity and gas savings of 1.4 to 3.3 percent for all targeted households, with an average of 2 percent, across all geographies, and that these savings persist over time. These savings are also cost effective. Allcott estimates investment in Opower cost 3.31 per kWh saved — about a third the cost of conventional energy resources.[14]

Naturally, 2 to 3 percent savings adds up across millions of households. Figure 23.3 is a scatter plot of savings from Opower deployments with different utilities over time. The consistency in these savings is represented by

12. See, <http://www.opower.com>.
13. Allcott H. Social norms and energy conservation, J Public Econ, <http://web.mit.edu/allcott/www/Allcottpercent202011percent20JPubEcpercent20-percent20Socialpercent20Normspercent20andpercent20Energypercent20Conservation.pdf>; October 2011.
14. Binz et al., April 2012, "Practicing Risk-Aware Electricity Regulation," Ceres, Figure 10, available here: http://www.rbinz.com/Binz%20Sedano%20Ceres%20Risk%20Aware%20Regulation.pdf

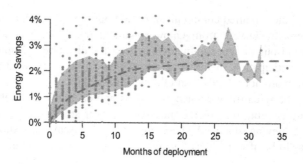

FIGURE 23.3   Average savings across all Opower deployments over time.[15] *Source: Opower (2011).*

the average savings line. And Opower has only achieved a portion of the potential. ACEEE estimates behavioral programs like Opower's could save 39 TWh by 2030.[16] To put this in context, this is greater than the amount of energy to be delivered from solar photovoltaic installations by 2030 based on estimates from the U.S. Energy Information Administration ("2013 Annual Energy Outlook," U.S. EIA, Table 17, available here: http://www .eia.gov/forecasts/aeo/er/pdf/tbla17.pdf).

The savings are important for another reason: they help utilities engage their customers.

4 DOES CUSTOMER ENGAGEMENT MATTER?

Recently there has been a growing concern about customer engagement within the utility industry. There is a general sense that customer relationships are increasingly important, and there is a range of research − mostly from other industries − that demonstrate the value of customer engagement. Higher levels of engagement lead to reduced customer attrition, increased customer acquistition, and expanded margins. These benefits are highly valuabel in the banking, telecom, and consumer products industries, but less relevant for regulated utilities that have service territory monopolies. This begs the question: is customer engagement truly important for utilities?

Historically, there wasn't much need for strong customer engagement − as long as the utility kept the power on and customers paid their bills. The typical utility-customer relationship was minimal. It therefore isn't surprising to find that most consumers haven't engaged much with their utility.

15. See, < http://opower.com/utilities/results >.
16. Estimated using the carbon emissions using http://www.carbonfund.org. from BOSto LAX flights.

Moving forward, however, customer engagement is likely to become an increasingly important ingredient for utility success, especially in areas where there is retail competition, such as in Texas. A variety of new utility goals, strategies, and technologies are highly dependent on customer acceptance and, in some cases, action. Energy efficiency and peak reduction programs, smart grid infrastructure deployments, and time-differentiated rates are examples of initiatives that can't succeed without at least a basic level of trust, awareness, and involvement from customers. Relationships based on nothing more than monthly bill payment will no longer suffice. Over time, the most successful utilities will be those that have the most customers engaged as partners to manage energy usage and cost. Katherine Hamilton, the former President of the GridWise Alliance, aptly observed: "The smart grid is not smart until the consumer is actually engaged."

4.1 What does Customer Engagement Mean for Utilities?

Customer engagement is a combination of emotion and behavior. Put another way, it is the interaction of how customers feel and what they actually do. Engaged customers have positive sentiment toward the company and are actively involved with the company's products or services. While there are many ways to conceptualize and measure engagement, there are two primary dimensions:

- Home Energy Activity – i.e., what customers do; and
- Trust in the Utility – i.e., how customers feel.

There is a spectrum of awareness and action with regards to home energy use. At one end of the spectrum, there are customers who are unaware, uninterested, and passive. These customers do not spend time thinking about their energy use or utility company until there is a specific problem such as a high bill or power outage. They are unlikely to proactively take action or make purchases to increase their energy efficiency, and they have no or limited awareness of smart grid-enabled products.

At the other end of the spectrum are customers who are very aware, interested and active. These customers regularly think about their home energy consumption and adjust their behaviors and purchases to use energy more efficiently. They are also likely to be aware of smart grid initiatives and other technologies that can impact energy usage and costs in the future. It's important to note that the underlying motivations for interest and action are varied, which has implications for developing customer communication strategies. Some people are driven primarily to reduce their utility bills, some are focused on reducing carbon emissions, and others are eager to be early adopters of new technology.

Without trust, even the most energy-aware and energy-active customers will be unlikely to engage with the utility. Customers with low trust in, or perceptions of, the utility are skeptical of their utility's intentions and/or capabilities. These customers may ignore communications from the utility, and — even if they pay attention — they may not find the information credible and thus be unlikely to follow recommendations.

Customers with a high level of trust believe their utility generally is benevolent and does a good job of serving the community's needs and of looking out for its customers' interests. They are likely to view the utility as a good partner and source of useful and credible information and advice related to home energy use. These customers are more likely to support and participate in utility-driven initiatives. Figure 23.4 provides a visual depiction of this continuum from low to high engagement.

These two dimensions create a framework for utilities to assess the readiness and willingness of customers to engage with them. Highly engaged customers have both a strong awareness and interest in home energy use, and also trust their utility. It's useful to consider customers in four different groups spanning these two dimensions:

Passive
Generally unaware or uninterested in energy issues
Don't read or respond to communications
May or may not have trust in the utility
Unlikely to engage unless a significant problem or surprise arises

Independently Active
Moderately to highly aware and active with energy issues
May be taking significant actions to manage energy usage independent of utility programs or advice
Skeptical and possibly very mistrustful of the utility
Potential for vocal opposition of utility programs

Slightly Engaged with the Utility
Care at least a little about energy usage
Have at least a moderate level of trust in the utility
May already be taking some actions to manage usage
Good prospects for utility programs and increased engagement, but unlikely to make major investments

Highly Engaged with the Utility
Small group of people who are both highly energy aware and active, and highly trusting of the utility

FIGURE 23.4 Continuum from low to high utility customer engagement. *Source: Opower (2012).*

Likely visitors to utility website and other sources of energy usage advice

Ideal audience for more aggressive utility EE and DR programs

Potential to leverage as promoters for utility programs

In an ideal world, all customers would have a high trust for the utility and also be highly energy aware and active. However, this is neither a reasonable expectation nor a necessary outcome. Even with very large investments in outreach, customer service, and program development, the majority of customers are not going to care enough about their energy usage to become highly engaged active participants. It's important to keep in mind that most utilities are starting with minimal customer engagement, starting with 7 minutes of engagement described at the outset.

The more critical objective is to move as many customers as possible into the Slightly Engaged segment from Disengaged or Independent, which means achieving at least minimal levels of trust and energy activity. It's especially important to minimize the Independent group, as even small groups of untrusting customers can create significant challenges. They can become vocal opposition to important initiatives, and they may be unwilling to engage enough to understand the underlying facts and utility's position better.

In addition to a basic level of trust, it will be important that most customers give their home energy use at least a small amount of their attention. Some level of engagement is needed to facilitate the rollout of more complicated time-based pricing plans and gain needed participation in energy efficiency and demand response programs. Slightly Engaged customers are likely to be aware of important utility programs, and are more likely to participate − or at least be supportive. This group will realistically be a significant majority of the entire population, and it provides a good base for building deeper engagement over time with the right communications and interactive tools.

Historically, utilities have primarily focused on delivering reliable power at a low cost. Utility customer relationships were transactional and based solely on billing and troubleshooting. Most utilities did not make large efforts towards building deeper customer relationships.

But new utility business goals are increasingly dependent on customer awareness and involvement. These goals include: energy efficiency, peak reduction, smart meter acceptance, and customer satisfaction. In some instances, these goals have been formalized in performance-based regulation. Under this approach, regulatory authorities assess utility performance according to both cost of service and additional metrics, such as customer service. Utilities are unlikely to be successful achieving these goals if they have large numbers of untrusting or unaware customers. Accenture's "Engaging the New Energy Consumer" report concludes that "Up to two-thirds of all smart metering and energy efficiency business case benefits depend on achieving some level of consumer change."[17] This can't be done without engaged customers.

---

17. See Accenture report, Actionable insights for the new energy consumer, available here, <http://www.accenture.com/us-en/pages/insight-actionable-new-energy-consumer.aspx>; 2012.

4.2 Energy Efficiency

Motivating customers to manage and reduce their energy consumption is increasingly important for utilities. According to data from the U.S. Department of Energy's Energy Information Administration, megawatt-hour (MWh) energy savings increased by over 50 percent between 2000 and 2009 as utilities have invested in efficiency to meet energy efficiency resource standard (EERS) goals. Looking ahead, regulators in many states have already established increasingly aggressive EERS targets for the coming years. [18] Achieving the growing energy savings targets is a major challenge, even with increased marketing and incentive budgets that are typically granted by regulators in conjunction with the higher goals.[19]

Hitting these growing EE targets will require utilities to get more of their customers to take more action to reduce their energy consumption. A particular challenge in the coming years will be the reduction in deemed savings for compact fluroescent lightbulb (CFL) programs due to new federal lighting standards as the market approaches saturation. Currently, CFL programs account for 25 to 50 percent of total energy efficiency portfolio savings.

Most utilities will need to backfill a reduction in CFL-driven savings at the same time that they're striving for larger overall savings. Hitting these ambitious efficiency goals will require significant boosts in customer awareness and action.

Historic participation rates in residential efficiency programs are very low — typically in the 1 percent range — and surveys show that even awareness of programs is limited. A recent Accenture study found that 66 percent of consumers are not aware of any energy efficiency programs offered by their utility.[20] Even at utilities that have actively promoted their programs, there are large awareness gaps. Last fall, a Commonwealth Edison survey found that over half (54 percent) of its customers were unaware of any programs.[21]

The fundamental problem is that any amount of marketing will fall short of its goals if customers are not ready and willing to engage. This is true for both big-ticket programs such as Energy Star HVAC rebates, as well as free advice such as reminders to turn off lights.

Without at least minimal levels of interest in energy management and trust in the utility, customers won't pay attention to attempts at outreach.

---

18. For more on EERS, see, Sciortino, et al., June 2011, "Energy Efficiency Resource Standards: A Progress Report on State Experience," ACEEE, available here: http://aceee.org/research-report/u112.
19. For more on utility incentives for efficiency, see Hayes, Sarah, January 2011, "Carrots for Utilities," ACEEE, available here: http://aceee.org/research-report/u112.
20. See Accenture report, Actionable insights for the new energy consumer, available here, <http://www.accenture.com/us-en/pages/insight-actionable-new-energy-consumer.aspx>; 2012.
21. Blackstone Group, Opower report satisfaction: pilot study results; 2010 July (N = 451).

One of the conclusions in Accenture's Understanding Consumer Preferences in Energy Efficiency report is, "utilities/electricity providers must enhance their customer relationships and earn sufficient consumer trust before they see broad-based adoption of energy management programs."

There is encouraging data from some recent customer engagement programs. Engaged customers can be motivated to sustainably reduce their energy consumption through informative messaging alone. These information-based EE programs are a powerful, new way to achieve energy savings on a large scale, as they can easily be deployed to hundreds of thousands of households.

Even when those programs are not directly marketed, customers placed in the Opower program are 15 to 30 percent more likely to participate than customers who aren't receiving the reports.[22] These results show the power that increased customer engagement can have to tangibly and significantly accelerate the adoption of energy efficient appliances and measures. Combined with the savings potential of the information itself, this lift in efficiency program participation can drive considerable energy savings. As behavioral efficiency migrates to more digital channels, the opportunity to interact with customers regarding efficiency will increase. The result will be significant increase in energy efficient decisions.

4.3 Peak Reduction

The challenges and opportunities for peak reduction are similar to those for energy efficiency. As with efficiency, there is a clear trend of growing expectations from peak reduction programs. From 2006 to 2011, investment in load management increased from ~$350 million to over $1.25 billion.[23]

To continue reducing peak demand at this pace, utilities will need a higher level of customer engagement with their demand response programs. Without customer action, inefficient behavior will continue to create greater load during peak hours, and the considerable costs associated with generating power during these high-usage hours. Driving increased participation will be much easier when reaching out to engaged customers instead of those who are disinterested or distrustful. Signing up for an AC cycling program, where the customer is ceding some control to the utility, requires a higher level of engagement and trust in the utility than the typical efficiency program.

Similarly, programmable thermostats can drive large peak reductions, but only if people learn to use them appropriately. The effectiveness of a programmable thermostat depends on how frequently and deeply a user sets

22. See, e.g., Opower results, <http://opower.com/utilities/results>.
23. Annual Industry Reports, Consortium for energy efficiency, <http://www.cee1.org/ee-pe/AIRindex.php3>; 2006 and 2011.

back the temperature. A user that sets their thermostat every heating degree day from the average temperature of 72 to 68 degrees is much more efficient than someone who sets their thermostat one in every seven heating degree days from 72 to 71 degrees. Programmable theremostats have historically demonstrated inconsistent results when evaluated for how deep and frequent a user set back the temperature. In the absence of monitoring, feedback, and better coaching in the initial setup, an "information gap" was created between the capabilities of the device and the user taking advantage of those capabilities.[24]

In conjunction with Honeywell, Opower has created a behavior-enabled thermostat solution that addresses this information gap.[25] Using a web-supported, 2-way-communicating programmable thermostat, Opower engages the user beginning at the initial setup, and these tactics encourage continues to engage them throughout the life of the product using constant feedback and remote control of the device, deeper and more frequent setbacks over time. The peak reduciton and energy efficiency potential of this solution is promising. These peak reductions will greatly ease the burden on the system at times when cost is highest, generating significant value for customers and utilities alike.

Looking ahead, it's likely that dynamic rates such as time-of-use (TOU), peak time rebates (PTR) and critical peak pricing (CPP) will play a significant role in peak reduction. These rates provide clear and powerful incentives to customers to reduce usage at peak times, and pilot programs have delivered median savings of 14 to 18 percent.[26] Peak time rebates in particular a promising way to drive peak reduction. Under this approach, utilities pay homes for reducing consumption during peak hours (rather than charging them more for using energy during these hours). By providing a positive incentive for participation, these rates are more attractive from a customer engagement perspective. Service providers like Opower are working with utilities to implement sufficient communications platforms for engaging households to encourage increased participation in these rate plans. Broad customer engagement is essential for these dynamic pricing programs to expand successfully to full-scale rollouts from small pilots (which are typically populated with a non-representative sample of agreeable engaged customers). Without some level of customer interest and trust, these pricing plans are unlikely to drive significant shifts or reductions in peak consumption, and it's more likely that the utility will face a backlash from upset and confused customers.

24. For more on this information gap, see: Peffer et al., March 2011, "How people use thermostats in homes: A review," Journal of Building and Environment, 46 (2011) 2529–2541.
25. See, e.g., <http://opower.com/company/news-press/press_releases/37>.
26. Sergici S, Faruqui A. Dynamic pricing: past, present, and future, The brattle group, <http://www.brattle.com/_documents/UploadLibrary/Upload956.pdf>; June 2011.

Even without new rates or devices, there are opportunities to drive peak reductions through customer engagement. Exciting early results from information-based engagement programs show that energy savings are the largest during peak periods. Detailed interval analyses of the Opower Home Energy Reports program show that the energy reductions during peak seasons and hours are often twice the average level of savings. By educating and engaging customers about energy usage and ways to save, many will take action to reduce their consumption meaningfully during peak periods.

4.4 Smart Grid

For many utilities, the most immediate and pointed need for customer engagement is to facilitate the effective deployment of smart grid infrastructure. The customer-driven opposition to smart meters has been widespread. While the opposition typically comes from a small number of customers, they have been highly effective at getting the attention of the press and regulators. This has led to significant delays, additional costs, and even legal challenges for utilities.[27]

The root cause of the problem is that most utilities have managed their smart grid rollouts primarily as technology and operational projects without focusing on customer benefits and consumer engagement.

While most utility technology projects are behind-the-scenes, and therefore don't raise questions or concerns from customers, smart meters are visible: physically installed on each property and oftentimes itemized on customer bills. Without the right outreach and interaction with customers, it's not surprising that questions and concerns arise.

The reality is that most customers don't know or care about smart meters. A 2010 survey by GE found that 79 percent of American consumers are not familiar with the term smart grid.[28] Even more interesting are the results of a 2010 Boston Consulting Gropup study, which showed smart meter awareness is under 50 percent in areas where they have been deployed – no different from areas where mart meters have not been deployed.[29]

---

27. For example, a small but vocal group successfully lobbied the California PUC to require PG&E to develop a non-wireless alternative for customers afraid of the smart meter health risks. The utility will need to give customers the option to deactivate the radio signal and have the meters read manually, which has both technical and operational complications and costs.

28. General Electric. Energy. National survey: Americans feel a smart grid will help reduce power outages, personal energy usage. GE news center. GE, 23 Mar. 2012. Web. 20 Apr. 2012. <http://www.genewscenter.com/Press-Releases/National-Survey-Americans-Feel-a-Smart-Grid-Will-Help-Reduce-Power-Outages-Personal-Energy-Usage-26c9.aspx>.

29. BCG, The smart meter opportunity, <http://www.drsgcoalition.org/resources/other/BCG_Smart%20Meter%20Awareness%20Survey_Key%20Findings_May%202010.pdf>; May 2010.

This lack of awareness can either be a negative liability or a positive opportunity. For naysayers, this void is a chance to create a negative storyline and loudly raise concerns that are heard, whether they are legitimate or not. But utilities can also use this void as an opportunity to provide households with the right information on the benefits of smart meters, while setting the record straight on their risks.

Even more important than providing information is delivering on the promise of customer benefits from smart meters. The key is to engage customers with new insights and advice — not just data at the sub-daily interval — that helps the customer to understand and manage energy use. Customers will be more supportive of smart grid investments when they have seen the tangible benefit of new, useful information.

While there will always be vocal skeptics, with the right customer engagement strategy the magnitude and impact of this opposition can be significantly reduced. In this regard, there is a disconnect among utility executives between their recognition of this reality and the actions they are actually taking. While 71 percent of utilities say that securing customer buy-in is a key step needed to drive the success of the smart grid, just 43 percent say they are educating their customers on the smart grid's value proposition.[30] This finding is echoed by an Oracle report that indicates that while the majority of utility executives believe that customer buy-in is critical to driving smart grid success and energy savings practices, less than half are preparing their customers and providing actionable information they can use to change usage patterns.[31]

JD Power recently released results from their electric utility satisfaction study that shows a measurable positive impact of actually delivering clear communication and awareness about smart meters. Their research showed that, "[C]ustomers who are aware of the smart grid and smart meters, as well as their utility's efforts to implement them, are notably more satisfied than are customers without this awareness."[32] Patricia Hoffman, who leads the Office of Electricity Delivery and Energy Reliability at the DOE sums it up well: "The success of grid modernization initiatives hinges upon the ability of electricity providers nationwide to respond to their customers' concerns and actively involve their customers in the process."[33]

---

30. Oracle, smart grid challenges & choices, part 2, <http://www.smartgridnews.com/artman/uploads/1/Oracle_Utilities_2011_C-Level_Utilities_Survey_Report_FINAL.PDF>; May 2011.

31. Id.

32. J.D. Power and Associates. 2012 electric utility residential customer satisfaction study, <http://www.jdpower.com/content/press-release/d7cFGW5/2012-electric-utility-residential-customer-satisfaction-study.htm>; July 2012.

33. Hoffman P. Redefining customer service is essential to modernizing grid, <http://energy.gov/articles/redefining-customer-service-essential-modernizing-grid>; December 2010.

4.5 Customer Satisfaction

While there are numerous drivers of utility customer satisfaction, including power reliability and price, customer engagement is another increasingly important factor in scoring high satisfaction marks. Consumers have growing expectations and needs from their utility, informed by experience with companies in other industries. Just as their banks provide flexibility to make payments and to receive statements and alerts through multiple channels, customers expect the same type of choices from their utility to ensure information is delivered promptly and conveniently. And just as their banks provide tools and services to help monitor and manage spending, customers increasingly expect their utility to provide tools and services to help monitor and manage energy usage. These higher expectations for support provide an exciting opportunity because they're based on customers' increased willingness to engage.

To assess the impact customer engagement programs can have on customer satisfaction, Opower conducted a survey in early 2011.[34] The study included 1,800 customers receiving Opower's Home Energy Reports across multiple utilities. The results showed a 10 percent average increase in the perception of the utility as a benevolent energy partner and a 5 percent improvement in overall customer satisfaction.

There are real benefits to a utility from a customer base that believes the utility is trying to help them save energy and money. Not only will these customers play a larger role in achieving energy efficiency and peak reduction goals, but they will also be more satisfied and generally supportive of utility initiatives.

Customer engagement will continue to grow in importance for most utilities as demand-side management and smart grid programs expand. It's therefore critical for utilities to have a clear framework for understanding and measuring customer engagement. One of the key points to remember is that the goal is not to turn all customers into highly engaged utility advocates and "energy geeks." This is neither realistic nor is it necessary. A sobering reminder came from an assessment of the SmartGridCity efforts in Boulder, Colorado — an environmentally-oriented town, where they found "that for a vast majority of people, it's exceedingly difficult to get them to do much of anything."[35]

A more practical and valuable goal of a customer engagement program is to build a basic level of trust and home energy awareness and activity across a utility's customer base.

---

34. See results here, <http://opower.com/utilities/results>.
35. Tuttle B. Why people aren't sold on energy efficiency in their homes, <http://moneyland.time.com/2010/02/15/why-people-arent-sold-on-energy-efficiency-in-their-homes/>; February 2010.

With this foundation in place, the utility will be well positioned to develop deeper relationships with customers who are willing to engage at another level with their energy usage. This increased level of customer engagement can then enable a range of positive business outcomes for the utility.

5 CONNEXUS CASE STUDY

Connexus Energy is Minnesota's largest customer-owned cooperative utility, serving approximately 125,000 households. Connexus and Opower have one of the longest running behavioral energy efficiency programs in the country, and results from the program have been independently verified multiple times.[36]

In 2008, Connexus opted to use Opower's Home Energy Reports program to meet Minnesota's 1.5 percent savings target and increase customer satisfaction in its service territory. In January 2009, Opower enrolled 80,000 households in its program. These households were randomly separated into statistically equivalent contorl and treament groups of 40,000 households each so experimental design could be set up for clean, accurate measurement of savings.

Opower began sending Home Energy Reports and developed an efficiency-oriented web portal to the 40,000 treatment households in February 2009. Figure 23.5 is a sample report that was sent to Connexus households. The two pages are printed double-sided on a one-page report. As is evident in this figure, the report motivates the household using a normative comparison of that household to 100 neighbors, provides analysis of usage, and offers household-specific energy-saving tips. This combination yields average savings of 2 percent.

5.1 Reliable, Persistent, and Significant Energy Savings

Demographically, Connexus's service territory does not appear to be tailor-made for a behavioral energy efficiency and customer engagement program: Republican Congresswoman Michele Bachmann's district overlaps with much of their service territory. But within just 6 months of program launch, report recipients were saving more than 2 percent — with savings increases every month.

Power System Engineering (PSE) independently evaluated this program's impact over its first 11 months in a report released July 28, 2010.[37] In this

---

36. See, e.g.: (i) Allcott H. Social norms and energy conservation. J Public Eco 2011 October;95 (9–10):1082–1095, (ii) Ivanov C. Measurement and verification report of OPOWER energy efficiency pilot program. Pow Syst Eng 2010 July and (iii) Allcott H. Social norms and energy conservation. Working paper, Massachusetts institute of technology's center for energy and environmental policy research; February 2010.
37. Ivanov C. Measurement and verification report of opower energy efficiency pilot program, connexus energy, <http://opower.com/uploads/library/file/14/power_systems_engineering.pdf>; July 2010.

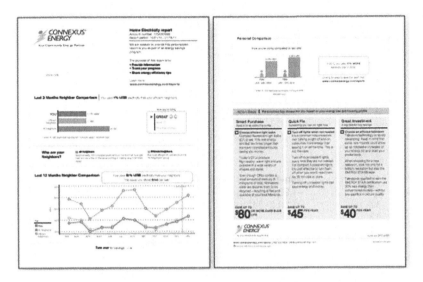

FIGURE 23.5   Sample home energy report. *Source: Opower (2011).*

evaluation, PSE measures an average impact of 2.07 percent per household, or the equivalent of 229 kWh per household in annual savings. This is on par with the average achieved. Hunt Allcott evaluated 17 deployments, and determined savings averaged 2 percent.[38] Figure 23.6 provides a table included in this report. This table shows the breakdown of average treatment effects (ATEs) – the average savings by each of these 17 deployments, organized by frequency of mailed report. For example, the second experiment in the table below shows an average savings of 1.83 percent. This table demonstrates the consistency of savings across deployments and report frequencies.

In the 3 years since the program began, Connexus households have collectively saved approximately 30,000 MWh at an average savings rate of 2.5 percent per household. These savings have not only persisted, but increased steadily over time. The dollar value of these savings is enough to feed over 600 families of four for a year, and the carbon abated is equivalent to avoiding over 350 cross-country flights.

Further evidence of the value of these savings, the Environmental Defense Fund ("EDF") evaluated eleven Opower deployments in its report, "Behavior and Energy Savings."[39] In this report, EDF extrapolates Opower

---

38. Allcott H. Social norms and energy conservation. J Public Econ, <http://web.mit.edu/allcott/ www/Allcott%202011%20JPubEc%20-%20Social%20Norms%20and%20Energy%20Conservation. pdf>; October 2011.
39. Davis M. Behavior and energy savings, environmental defense fund. <http://www.edf.org/ news/study-concludes-information-based-energy-efficiency-can-save-americans-billions>; 2011.

Experiment	ATEs (%)		
Number	Monthly	BiMonthly	Quarterly
1	Non–Exper	–	–
2	–1.83 (0.20)	–	–
3	–	–1.40 (0.19)	–1.37 (0.19)
4	–2.72 (0.18)	–	–2.26 (0.21)
5	–	–2.70 (0.44)	–
6	–	–	–1.64 (0.33)
7	–	–2.48 (0.25)	–
8	–	–3.32 (0.54)	–
9	–	–1.63 (0.15)	–
10	Non–Exper	–	–
11	–1.96 (0.14)	–	–1.49 (0.20)
12	–1.39 (0.34)	–	–
13	–	–	–1.44 (0.51)
14	–	Non–Exper	–
15	–	–1.89 (0.21)	–
16	–3.14 (0.37)	–	–
17	–	–	–1.84 (0.43)
Mean ATE	–2.03		

FIGURE 23.6   Average treatment effects for seventeen Opower programs evaluated in Hunt Allcott's article, "Social Norms and Energy Conservation" (2011). *Source: Allcott (2011).*

savings and calculates that if this program were deployed to the entire United States, households would save $3 billion annually.

5.2 Cost Effectiveness

The Opower platform has provided Connexus with a cost-effective way to meet their energy efficiency targets. In 2010, the Opower program accounted for just 25 percent of Connexus' efficiency budget, but more than 50 percent of the savings committed to in the utility's Conservative Improvement Plan. Figure 23.7 provides a comparison of Opower's proportion of the total Connexus efficiency portfolio savings and budget.

This program has achieved average cost effectiveness of $0.039/kWh of energy saved, which is comparable to the most cost-effective efficiency programs and only a fraction the cost of generating the same amount of energy.

5.3 Customer Engagement

The Opower customer web portal complements and reinforces Home Energy Reports. The online platform allows Connexus members to dive deeper into their energy usage, compare variations in bill prices, and explore a database of efficiency tips tailored to their household profile. Users can commit to taking particular steps to improve their efficiency, and compare their

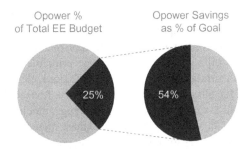

FIGURE 23.7   Opower as a proportion of Connexus' total efficiency portfolio budget and savings.

FIGURE 23.8   Sample Opower web portal design.

commitments against those of their neighbors. Figure 23.8 is a sample Opower web portal.

Commitments and profile updates made online are also reflected on the reports customers receive in the mail, creating a coherent and accurate customer experience.

6 CONCLUSIONS

Customer engagement has been a chronic problem for utilities. As utilities begin to engage their customers more deeply, the quality of service will improve, household satisfaction will increase, and utilities will develop better relationships with their customers.

This engagement holds enormous energy saving potential. With minimal engagement through mobile applications and email, savings of 2 percent per household can be achieved and sustained. If applied nationwide, the savings could amount to $3 billion annually. The same principles can be applied in other countries with similar outcomes. In the UK alone, the projected savings could amount to £1.7 billion over 3 years for British households.

Using reasonable assumptions for increased engagement through web and mobile applications, and through the use of smart metering data to develop better information and tools for engagement, the resulting savings per household could conceivably be doubled.

When combined, the potential for this type of customer engagement is substantial. By providing better and more targetedd energy information to households, consumers not only become more engaged but can make significant efforts that deliver robust, verifiable and sustainable savings.

How Do We Get There From Here?

As the editor of the volume I had the privilege to read – several times – and learn from the vast and diverse experiences and insights of the contributing authors. And the most important thing I learned is that, despite the disagreements among the authors on many points, the underlying optimism is that we have learned a lot and can do more on energy efficiency going forward. The opportunities are boundless; it is mostly a question of having the will and the means.

On a personal note, I am convinced that humanity has reached or will soon arrive at satiation levels on many things – certainly in mature economies. The same will eventually happen to developing economies as they evolve. Since so much of the projected growth in global energy demand is expected to come from the developing economies, the challenge is to find a shortcut to where they will eventually go but in a more efficient and sustainable way.

As described in an article in *The Economist*[1] on green growth, today's mature economies grew first and focused on cleaning up later. This strategy, however, is unlikely to work for masses of humanity who have rightful aspirations to achieve the same high standards of life that people in rich countries take for granted. Future development in these countries must be green from the start – with efficient utilization of energy as its cornerstone.

Fereidoon P. Sioshansi
Menlo Energy Economics

1. Shoots, greens, and leaves, *The Economist*, June 16, 2012.

Index

Printed in the United States
By Bookmasters